H. Abé · K. Hayashi · M. Sato (Eds.)

Data Book on Mechanical Properties of Living Cells, Tissues, and Organs

With 659 Figures and 180 Tables

Springer

Hiroyuki Abé, Ph.D.
Department of Machine Intelligence and Systems Engineering, Graduate School of Engineering, Tohoku University, Aoba-ku, Sendai, 980-77 Japan

Kozaburo Hayashi, Ph.D.
Department of Mechanical Engineering, Faculty of Engineering Science, Osaka University, 1-3 Machikaneyama, Toyonaka, Osaka, 560 Japan

Masaaki Sato, Ph.D.
Department of Mechatronics and Precision Engineering, Graduate School of Engineering, Tohoku University, Aoba-ku, Sendai, 980-77 Japan

ISBN 978-4-431-65864-1   ISBN 978-4-431-65862-7 (eBook)
DOI 10.1007/978-4-431-65862-7

Typesetting: Best-set Typesetter Ltd., Hong Kong

Preface

Biomechanics has been defined by Y. C. Fung as "mechanics applied to biology." It seeks to understand the mechanics of living systems, aiming to apply that knowledge to a wide variety of fields including biomedical engineering, medical science, clinical medicine, applied mechanics, and engineering. Biomechanics is a modern discipline with ancient roots, and its scope has expanded tremendously in the past decade, with progress being made in all areas. Now the field has become widely recognized as one of the most important and interesting areas in basic science. For the past quarter-century, many scientists have conducted research in biomechanics. A vast amount of data on the mechanical properties of biological tissues and organs has been collected, and much of it has been published in journals. Several books on biomechanics have also been published, but few of them have been devoted exclusively to data on the mechanical properties of biological materials. Such a book should provide accurate and useful data on the mechanical properties of living cells, tissues, and organs primarily for biomechanical scientists, biomedical engineers, medical scientists, and clinicians. Biomedical engineers, for example, can effectively utilize the data for the analysis of living systems; clinicians need the data for diagnosis and therapy in medical practice. In addition, with recent advances in computational biomechanics, fundamental data on the mechanical properties of living tissues such as bone, blood vessels, muscle, and tendons are indispensable for computer analysis.

This volume contains basic data collected from published articles and also presents original data obtained from research financially supported by a 3-year Grant-in-Aid for Scientific Research on a Priority Area (Biomechanics, Nos. 04237101~04237106) from the Ministry of Education, Science and Culture, Japan, with Kozaburo Hayashi as the principal investigator. The data are shown in graphs and tables for the reader's convenience, and the origin of materials, material structures, experimental conditions, and other relevant information are documented in detail. A set of data on a single topic is printed on one or two pages, with useful keywords. We hope that this book can be utilized effectively by the many students and researchers who, whatever their field, are interested in biomechanics.

To the many associate editors and contributors listed on the following pages we express our special appreciation for their hard work in selecting and collecting the data. We are also grateful to Dr. Takeo Matsumoto, Tohoku University, and Dr. Noritaka Yamamoto, Osaka University, for their efforts in preparing the format and editing the manuscripts by computer. We thank our students E. Yamamoto, T. Goto, M. Inaoka, K. Takahama, W. Iwanaga, and T. Tadami of Osaka University, and H. Abe, E. Okumura, H. Ogawa, S. Ujita, and T. Iyoda of Tohoku University for typing and preparing camera-ready copy. This publication was financially supported in part by a Grant-in-Aid for Publication of Scientific Research Result (No. 77013). Finally, we wish to thank the editorial and production staff of Springer-Verlag for their care and cooperation in producing this book.

Hiroyuki Abé
Kozaburo Hayashi
Masaaki Sato

V

Contents

Associate Editors

K. Akazawa	Department of Computer and Systems Engineering, Faculty of Engineering, Kobe University
T. Matsumoto	Department of Mechatronics and Precision Engineering, Graduate School of Engineering, Tohoku University
K. Okamoto	Department of Biochemical Engineering and Science, Faculty of Computer Science and Systems Engineering, Kyushu Institute of Technology
M. Okazaki	Department of Dental Technology, Faculty of Dentistry, Osaka University
M. Sugawara	Department of Surgical Science, The Heart Institute of Japan, Tokyo Women's Medical College
M. Tanaka	Department of Mechanical Engineering, Faculty of Engineering Science, Osaka University
K. Tanishita	Department of Mechanical Engineering, Faculty of Science and Technology, Keio University
N. Yamamoto	Department of Mechanical Engineering, Faculty of Engineering Science, Osaka University

List of Contributors

T. Adachi	Faculty of Engineering, Kobe University
K. Akazawa	Faculty of Engineering, Kobe University
N. Hara	Hokkaido University School of Medicine
T. Hara	Faculty of Engineering, Niigata University
T. Hara	Tokyo Dental College
K. Hayashi	Faculty of Engineering Science, Osaka University
H. Higaki	Faculty of Engineering, Kyushu University
Y. Ide	Tokyo Dental College
A. Ishida	Institute for Medical and Dental Engineering, Tokyo Medical and Dental University
H. Ishida	Hokkaido University School of Medicine
S. Ishiwata	School of Science and Engineering, Waseda University
K. Ito	School of Information Engineering, Toyohashi Institute of Technology
K. Kaibara	Faculty of Science, Kyushu University
N. Kataoka	Graduate School of Engineering, Tohoku University
M. Keira	Hokkaido University School of Medicine
S. Kimura	Hokkaido University School of Medicine
T. Kimura	Institute of Development, Aging, and Cancer, Tohoku University
Y. Kurobe	Hokkaido University School of Medicine
T. Majima	Hokkaido University School of Medicine
T. Matsumoto	Graduate School of Engineering, Tohoku University
N. Miyagi	Hokkaido University School of Medicine
K. Miyata	Hokkaido University School of Medicine
T. Murakami	Faculty of Engineering, Kyushu University
E. Nakamachi	Faculty of Engineering, Osaka Institute of Technology
T. Ohashi	Graduate School of Engineering, Tohoku University
K. Okamoto	Faculty of Computer Science and Systems Engineering, Kyushu Institute of Technology
M. Okazaki	Faculty of Dentistry, Osaka University
H. Okuyama	Kawasaki Medical School
M. Sato	Graduate School of Engineering, Tohoku University
H. Suga	Okayama University Medical School
H. Sugi	School of Medicine, Teikyo University
S. Tadano	Faculty of Engineering, Hokkaido University
S. Takai	Kyoto Prefectural University of Medicine
Y. Tamatsu	Tokyo Dental College
T. Tameyasu	St. Marianna University School of Medicine
E. Tanaka	Faculty of Engineering, Nagoya University

M. Tanaka   Faculty of Engineering Science, Osaka University
K. Tanishita   Faculty of Science and Technology, Keio University
H. Tohyama   Hokkaido University School of Medicine
M. Tokuda   Faculty of Engineering, Mie University
H. Toyota   Kawasaki Medical School
K. Tsujioka   Kawasaki Medical School
Y. Uemura   Faculty of Computer Science and Systems Engineering, Kyushu Institute of
                 Technology
S. Wada   Research Institute for Electronic Science, Hokkaido University
N. Yamamoto   Faculty of Engineering Science, Osaka University
K. Yasuda   Hokkaido University School of Medicine

1. Introduction

This book contains data on the mechanical properties of cells, tissues, and organs from humans and animals, useful for biomechanical scientists, biomedical engineers, and clinicians. Numerous data on the mechanical properties of living materials from various parts of the body have been published in the past decades. Because we cannot include all of it here, the editors have selected important and useful data from reports published mostly after 1970. In particular, we have focused on reliable data whose mechanical parameters are well defined.

The contents of this book are divided into three parts. The first, entitled "Soft Tissues," deals with the mechanical properties of the heart and heart muscle, blood vessels, skeletal muscle, smooth muscle, tendon and ligament, articular cartilage, and other soft tissues such as valve leaflet, skin, and nerve. In the section on heart and heart muscle, morphological parameters are included because they are indispensable to stress and strain analyses. The section on blood vessels contains the mechanical properties not only of normal tissues but also of diseased vascular walls such as those associated with hypertension and atherosclerosis. The second major part, entitled "Hard Tissues," is devoted to the mechanical properties of bone and teeth. Although the data on bones were obtained from humans and animals, the data on teeth are mainly from humans. The third part is entitled "Cells, Proteins, and Polymers" and includes the mechanical properties of isolated living cells, biological proteins and polymers, and biological fluids such as blood and synovial fluid.

In principle, data selected from each report are summarized on one or two pages. In some cases, data selected from several reports are summarized on one page. As shown in the example, keywords related to mechanical properties head each page. These are in alphabetical order (e.g., **Anisotropy, Compressibility, Creep**, etc.). Following the keyword heading, there are three boxes in which other keywords appear. In the first box, on the left, the reader can find more specific keywords on mechanical properties, such as elastic modulus and tensile strength. These, too, are in alphabetical order (e.g., "**Anisotropy** •Pressure–diameter relation •Longitudinal force" is followed on the next page by "**Anisotropy** •Stress–strain curve •Axial mechanical property," then by "**Anisotropy** •Stress–strain curve •Distensibility," etc.). The keywords in the middle box are related to materials and specimens, and show species, organs, anatomical sites of tissues, and so on. In the third box, on the right, supplemental information is given. The data sheets for the same mechanical property are grouped by animal and then by organ and anatomical site, and are arranged in order of year of publication in each group. Following the three boxes for keywords, brief summaries of Materials and Methods, Data, Comments, and Reference(s) appear. All figures and tables were redrawn and rewritten from originals; mechanical parameters were expressed mostly in SI units. All data are used by permission of the authors and publishers of the originals as indicated at the end of the references.

- *Layout of a typical page*

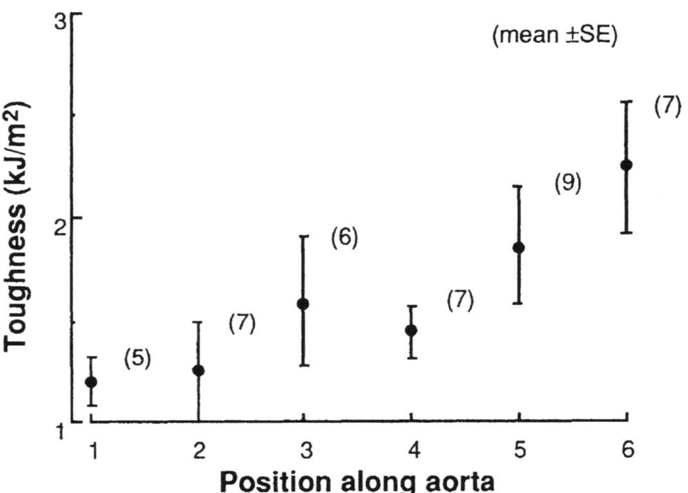

2. Soft Tissues

Contractility

• Total mechanical energy •	• Rabbit, canine, ferret • Papillary muscles	• •

Materials
- Canine and rabbit left ventricles
- Canine right ventricles
- Ferret papillary muscles

Testing Methods and Experimental Conditions
- Pressure, volume, and oxygen consumption of the ventricles of the excised blood-cross-circulated canine and rabbit hearts were measured
- From the instantaneous pressure and volume data of every contraction, a contractility index (E_{max}, maximum elastance) and a total mechanical energy measure (PVA, pressure–volume area) were determined
- Ventricular oxygen consumption (VO_2) was correlated with PVA at a constant E_{max}, and the VO_2–PVA relation was obtained for different E_{max}
- From all the VO_2–PVA relations with E_{max} as a parameter, the oxygen cost of PVA ($\Delta VO_2/\Delta PVA$) and that of E_{max} (ΔPVA-independent $VO_2/\Delta E_{max}$) were obtained
- The reciprocal of the oxygen cost of PVA is the contractile efficiency of the ventricle
- The reciprocal of the oxygen cost of E_{max} is the economy of ventricular contractility

Data

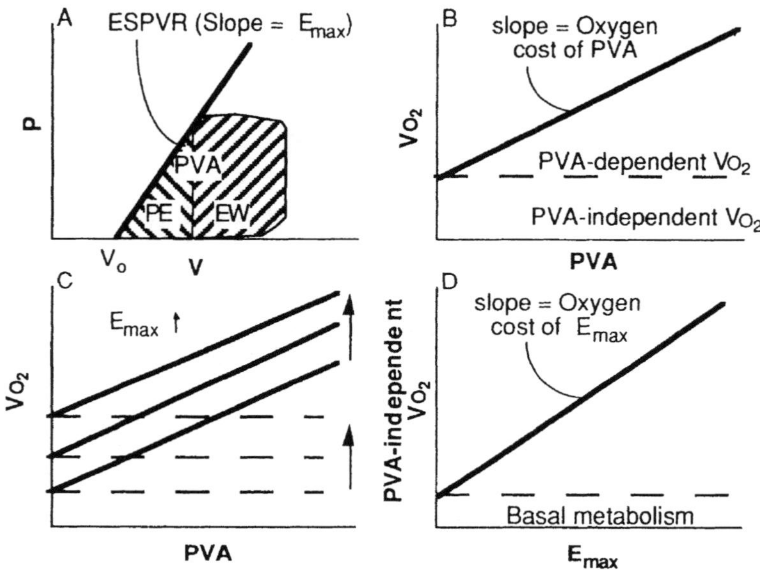

Comments
- E_{max} and PVA are the mechanical and energetic terms that are based on the time-varying elastic model of the ventricle. This model is considered to be a reasonable approximation of not only the ventricle and atrium, whether right or left, but also a cardiac wall region and even a papillary muscle.
- Left ventricle (dog and rabbit). End-diastolic pressure, – 5–15 mmHg; end-systolic pressure, 50–250 mmHg. End-diastolic volume, 30–60 ml/100 g left ventricle; end-systolic volume, 10–30 ml/100 g. E_{max} (= end-systolic pressure/end-systolic volume above V_o), 5–15 mmHg/(ml/100 g). PVA, 1000–3000 mmHg ml/100 g per beat. Oxygen consumption, 7–15 ml O_2/min per 100 g. Oxygen cost of PVA, 1.5-2.2×10^{-5} ml O_2/ (mmHg ml), 2.3–3.5 (dimensionless because of 1 ml O_2 = 20 J and 1 mmHg ml =

1.33×10^{-4} J). Its reciprocal, contractile efficiency, 30–50%. Oxygen cost of E_{max}, 0.001–0.003 ml O_2 ml/mmHg per beat per 100 g. Its reciprocal, economy of contractility, 300–1000 (ml/mmHg)/ml O_2 per beat per 100 g.

• Right ventricle (dog). End-systolic pressure is 1/4-1/3 of that of the left ventricle. The other
  weight-normalized data are comparable to those of the left ventricle.
• Papillary muscle (ferret). Peak developed force, 500–1200 g/cm². The VO_2–FLA (linear version of PVA) relation is comparable in its slope (ie, oxygen cost of FLA or contractile efficiency) to the ventricular Vo_2–PVA relation. Essentially the same holds for a ventricular wall region.

Reference(s)

1. Suga H (1990) Ventricular energetics. Physiol Rev 70:247–277 (with permission)
2. Suga H, Goto Y (1991) Cardiac oxygen costs of contractility (E_{max}) and mechanical energy (PVA): New key concepts in cardiac energetics. In: Sasayama S, Suga H (eds) Recent Progress in failing heart syndrome. Springer-Verlag, Berlin Heidelberg Tokyo, pp 61–115

Dynamic Property (1)

• Circumferential stress • Longitudinal stress	• Human • Left ventricle	• Angiographic study •

Materials

- Human
- Left ventricle (data were obtained from the angiographic study of Gould et al. Am J Cardiol 34, 1974)

Testing Methods and Experimental Conditions

- Many mathematical models used to calculate left ventricular wall stress were compared

Data

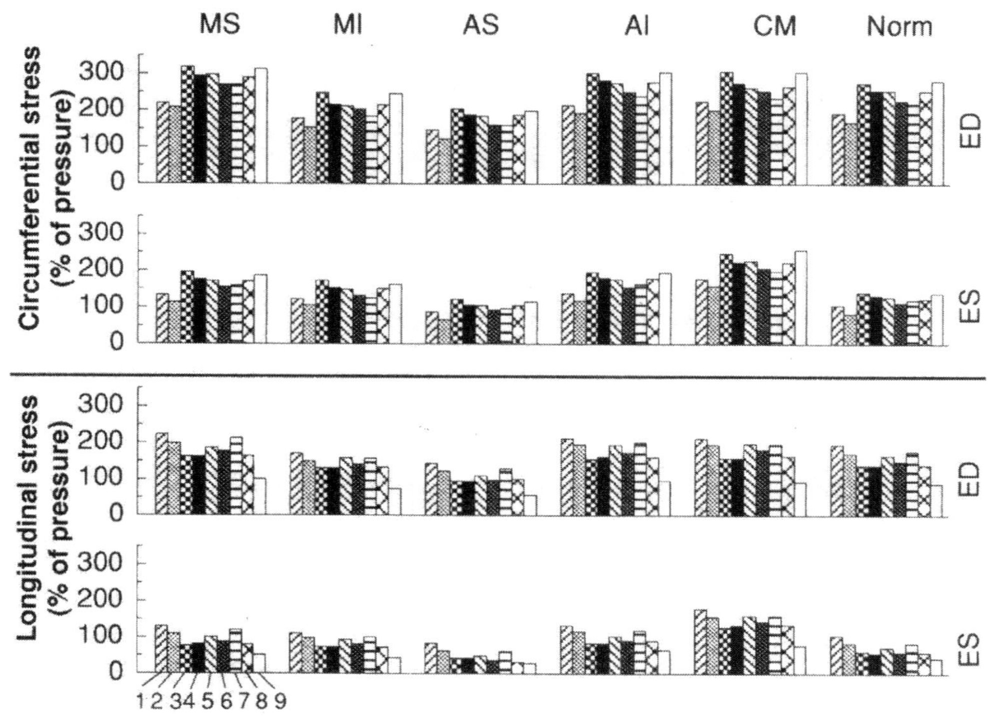

Comments

- Stress is expressed as a percentage of pressure.
- Falsetti's ellipsoidal model is recommended in this paper.
- MS, mitral stenosis; MI, mitral insufficiency; AS, aortic stenosis; AI, aortic insufficiency; CM, cardiomyopathy; Norm, normal human heart; ED, end-diastolic; ES, end-systolic.
- The stress values are represented by columns: Column 1, thin-walled sphere; 2, thick-walled sphere; 3, Sandler and Dodge model; 4, Falsetti model; 5, Wong model; 6, WONCO model; 7, Ghista model; 8, Mirsky model; 9, Streeter model.

Reference(s)

Huisman RM, Sipkema P, Westerhof N, Elzinga G (1980) Comparison of models used to calculate left ventricular wall force. Med Biol Eng Comput 18:133–144 (with permission)

Dynamic Property (2)

• Free shortening • Isovelocity shortening	• Cat • Ventricle	• Trabecula • Myocyte

Materials

• Rat and cat ventricular trabeculae and rat ventricular myocyte

Testing Methods and Experimental Conditions

• Slack-test, after contraction and free shortening and isovelocity shortening

Data

Velocity (length/s)	Temperature (°C)	Material and methods [reference]
0.35 ± 0.03	22	Myocyte Free and afterloaded shortening Mechanical dissection [1]
1.65 ± 0.42	15	Myocyte (2.3 μm sarcomere length) Slack-test Enzymatical digestion [2]
2.8	14	Myocyte Slack-test Enzymatical digestion [3]
6.65 ± 0.16	20	Rat trabeculae Isovelocity [4]
4.9 ± 0.1	20	Cat trabeculae Isovelocity [4]

Comments

• Shortening velocity depends on the length, temperature, and measuring instrument (especially for cells).
• If it is normalized by length, velocity is no longer length-dependent.

Reference(s)

1. De Clerck NM, Claes VA, Brutsaert DL (1977) Force velocity relations of single cardiac muscle cells. J Gen Physiol 69:221–241
2. Strang KT, Moss RL (1995) a_1-adrenergic receptor stimulation decreases maximum shortening velocity of skinned single ventricular myocytes from rats. Circ Res 77:114–120
3. Herland JS, Julian FJ, Stephenson DG (1990) Unloaded shortening velocity of skinned rat myocardium: Effects of volatile anesthetics. Am J Physiol 259:H1118–H1125 (with permission)
4. de Tombe PP, ter Keurs HEDJ (1991) Sarcomere dynamics in cat cardiac trabeculae. Circ Res 68:588–596

Dynamic Property (3)

• Tension •	• Cat, rat, and guinea pig • Ventricle	• Trabecula • Myocyte

Materials
- Cat and rat ventricular trabeculae
- Guinea pig and rat ventricular myocyte

Testing Methods and Experimental Conditions
- Thin trabeculae or enzymatically isolated myocytes

Data

Maximum force (mN/mm^2)	Sarcomere length (μm)	Temperature $(^\circ C)$	Materials [reference]
108 ± 13.8	1.85 ± 20.04	25	Cat trabeculae [1]
79 ± 9	2.1–2.2	25	Rat trabeculae [1]
0.35	1.77 ± 0.05	37	Guinea pig myocyte [2]
5.3 ± 2.6	120^b	35	Guinea pig myocyte [3]
$70.6 \pm 6.7\,^a$	230 ± 0.02	15	Rat myocyte [4]

[a] mN
[b] Cell length

Comments
- It is quite difficult to measure the cross–sectional area of the cell, the maximal forces have some degree of error. Moreover, sarcomere length has also wide variety.

Reference(s)

1. de Tombe PP, ter Keurs HEDJ (1991) Sarcomere dynamics in cat cardiac trabeculae. Circ Res 68:588–596
2. Le Guennec JY, Peineau N, Argibay JA, Mongo KG, Garnier D (1990) A new method of attachment of isolated mammalian ventricular myocytes for tension recording: Length dependence of passive and active tension. J Mol Cell Cardiol 22:1083–1093
3. Shepherd N, Vornanen M, Isenberg G (1990) Force measurements from voltage clamped guinea pig ventricular myocytes. Am J Physiol 258:H452–H459
4. Sweitzer NK, Moss RL (1993) Determinants of loaded shortening velocity in single cardiac myocytes permeabilized with a hemolysin. Circ Res 73:1150–1162

Dynamic Property (4)

• Wall stress •	• Human • Left ventricle	• Chronic heart disease •

Materials
- Human left ventricle (187 patients with chronic heart disease)

Testing Methods and Experimental Conditions
- Data were obtained from cardiac catheterization
- Left ventricular volume (V) was calculated as follows:

$$V = \pi LM^2/6$$

L, longest chamber axis; M, internal transverse diameter at the midpoint of L
- Left ventricular mass (LV-mass) was calculated as follows:

$$LV\text{–}mass = 4/3\pi(L/2 + d)(M/2 + d)^2 - EDV$$

d, wall thickness; EDV, end-diastolic volume
- Peak systolic stress was calculated using the Laplace equation

Data

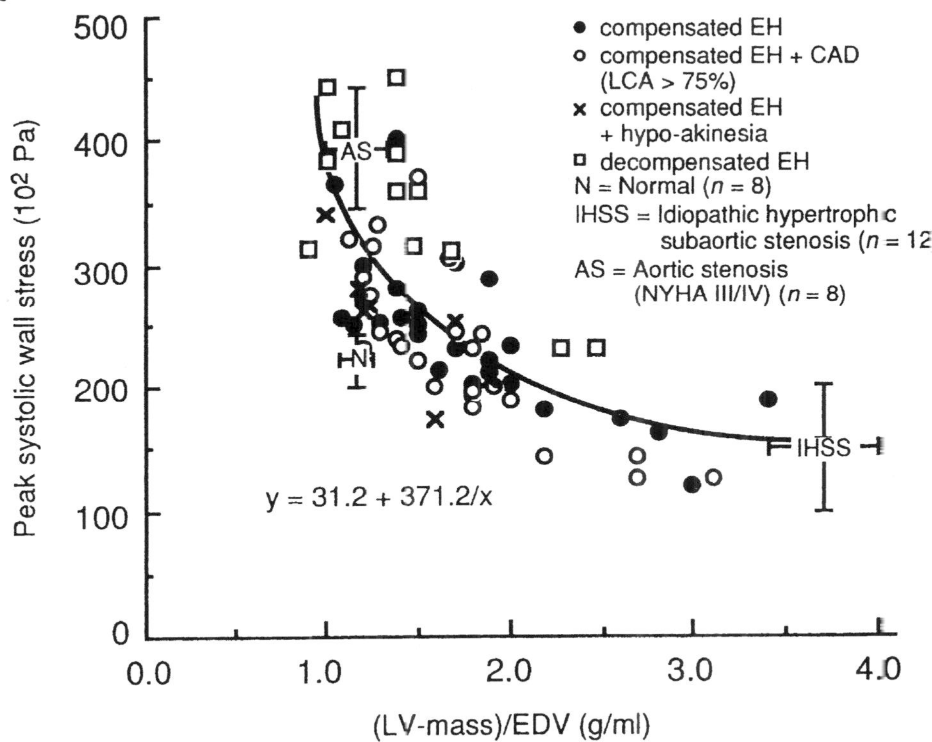

Comments
- EH, essential hypertension; CAD, coronary artery disease; LCA, left coronary artery.
- An inverse relation exists between mass to volume ratio and peak systolic wall stress.

Reference(s)
Strauer BE (1979) Myocardial oxygen consumption in chronic heart disease: Role of wall stress, hypertrophy and coronary reserve. Am J Cardiol 44:730–740 (with permission)

Dynamic Property (5)

• Wall stress • Pressure	• Dog • Left ventricle	• Model dependency •

Materials

- Dog
- Left ventricle

Testing Methods and Experimental Conditions

- Wall stress was directly measured with an auxotonic strain gauge in the left ventricle of open-chest dogs, and the results were compared with values calculated from spherical and ellipsoidal models

Data

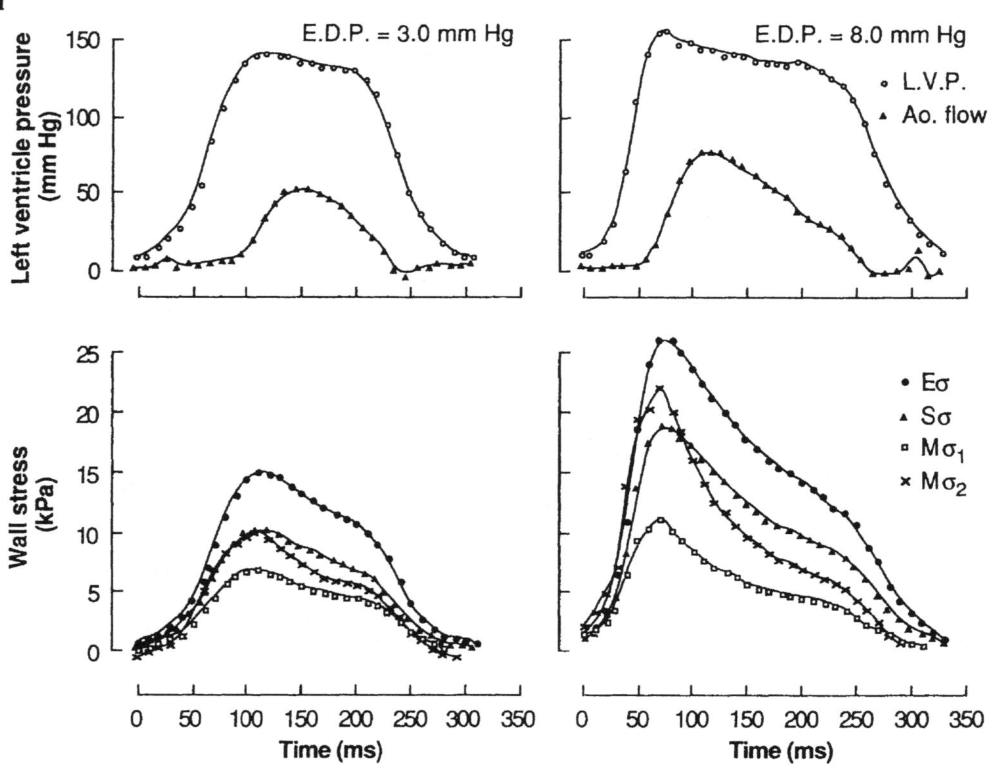

Comments

- Eσ, stress calculated using a thick-walled ellipsoidal model; Sσ, stress calculated using a thick-walled spherical model; Mσ₁, measured stress assuming a constant wall thickness; Mσ₂, measured stress corrected for dynamic wall thickness change.

Reference(s)

Burns JW, Covell JW, Myers R, Ross J Jr (1971) Comparison of directly measured left ventricular wall stress and stress calculated from geometric reference figures. Circ Res 28:611–621 (with permission)

Dynamic Property (6)

| • Wall stress
• Pressure | • Human
• Left ventricle | • Pressure overload
• Volume overload |

Materials
- Animal, organ, tissue etc.
- Human left ventricles in vivo (6 with pressure overload, 18 with volume overload, and 6 with no heart disease)

Testing Methods and Experimental Conditions
- Data were obtained throughout the cardiac cycle by cardiac catheterization and M-mode echo cardiography

Data

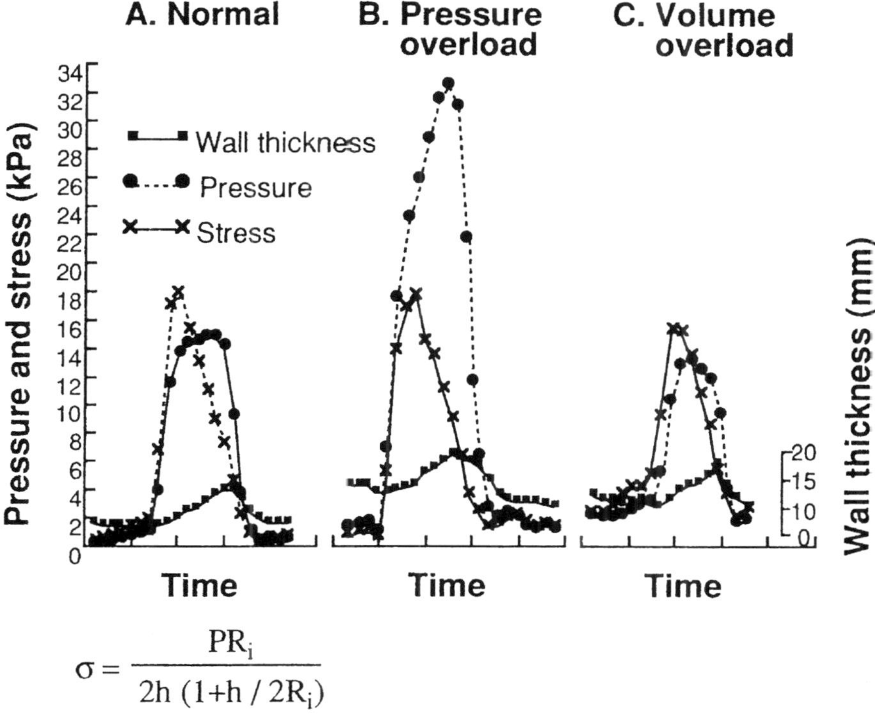

$$\sigma = \frac{PR_i}{2h \, (1 + h / 2R_i)}$$

h : wall thickness, P : LV pressure, R_i : inner radius

Comments
- LV mass index and wall thickness were significantly increased for both overload groups compared with control.
- In the pressure overload group, both peak-systolic and end-diastolic stresses were normal, but in the volume overload group, end-diastolic stresses were higher than control.

Reference(s)
Grossman W, Jones D, McLaurin LP (1975) Wall stress and patterns of hypertrophy in the human left ventricle. J Clin Invest 56:56–64 (with permission)

Ejection Fraction

• Circumferential stress • Velocity of fiber shortening	• Human • Left ventricle	• Normal • Aortic stress

Materials
- Human, left ventricles
- 14 patients with aortic stenosis, 6 with cardiomyopathy, and 6 normal

Testing Methods and Experimental Conditions
- Data were obtained from cardiac catheterization
- Left ventricular volume was computed according to the method of Kennedy et al. and left ventricular mass was calculated using the methods of Kennedy and Rackley et al.
- Midwall circumferential stress, σ_c, was computed by the formula of Mirsky:

$$\sigma_c = Pb \, / \, h \; (1 - h \, / \, 2b - b^2 \, / \, 2a^2)$$

P, left ventricular pressure; h, wall thickness; a and b, the midwall semiaxes

Data

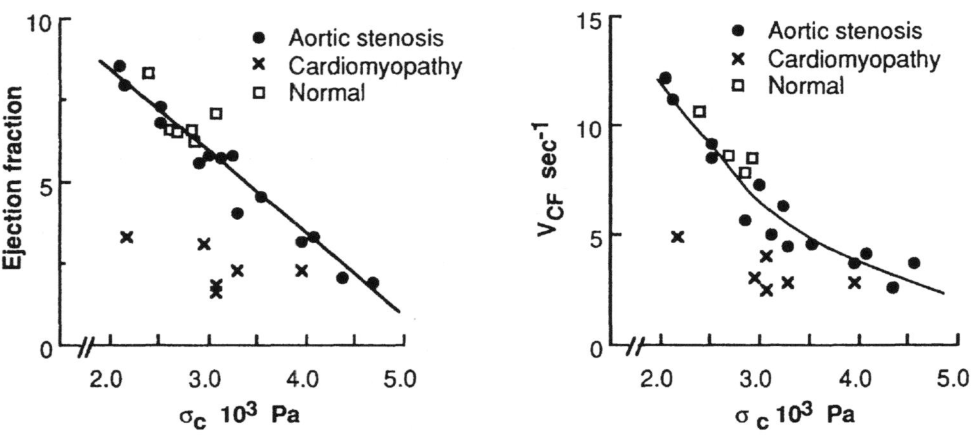

Comments
- There were close correlations between circumferential wall stress and both ejection fraction and velocity of fiber shortening in patients with aortic stenosis and normal control subjects, while both relationships clearly differed in patients with cardiomyopathy.

Reference(s)
Gunther S, Grossman W (1979) Determinants of ventricular function in pressure-overload hypertrophy in man. Circulation 59:679–688 (with permission)

Length–Tension Relation (1)

• Active tension • Resting tension	• Dogs • Ventricular myocytes	• Skinned myocytes •

Materials
- Dogs
- Isolated ventricular myocytes

Testing Methods and Experimental Conditions
- Active, resting, and rigor tensions *vs* sarcomere lengths in chemically skinned myocytes

Data

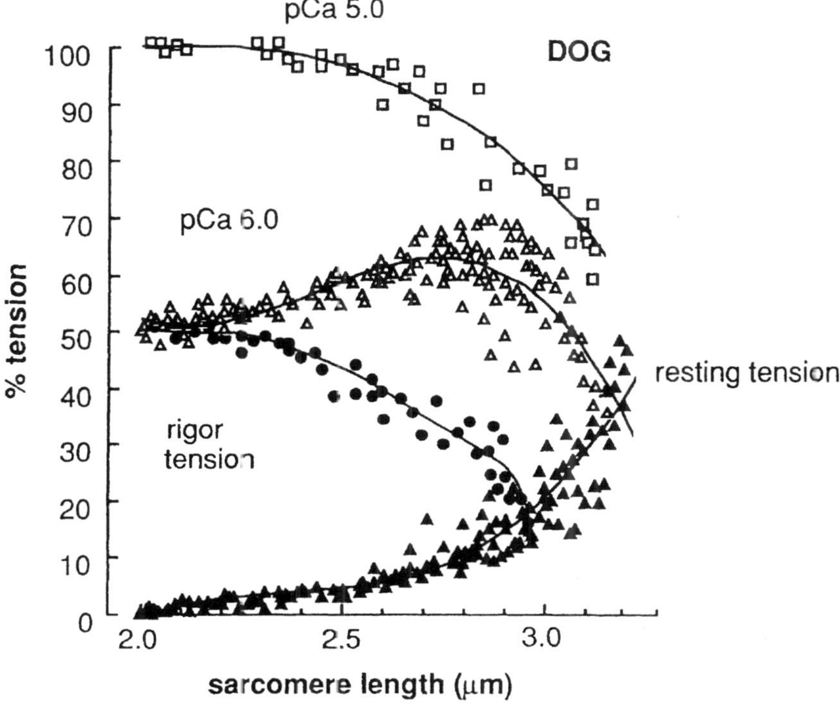

Comments
- Tension is normalized to the maximum active tension at pCa 5.0.

Reference(s)
Fabiato A, Fabiato F (1978) Myofilament-generated tension oscillations during partial calcium activation and activation dependence of the sarcomere length–tension relation of skinned cardiac cells. J Gen Physiol 72:667–699 (with permission)

Length–Tension Relation (2)

• Active tension • Resting tension	• Rats • Ventricular myocyte	• Skinned myocyte •

Materials

- Sprague–Dawley rats (200 ± 20 g)
- Isolated ventricular myocyte

Testing Methods and Experimental Conditions

- Active and resting tensions *vs* sarcomere lengths in chemically skinned myocyte

Data

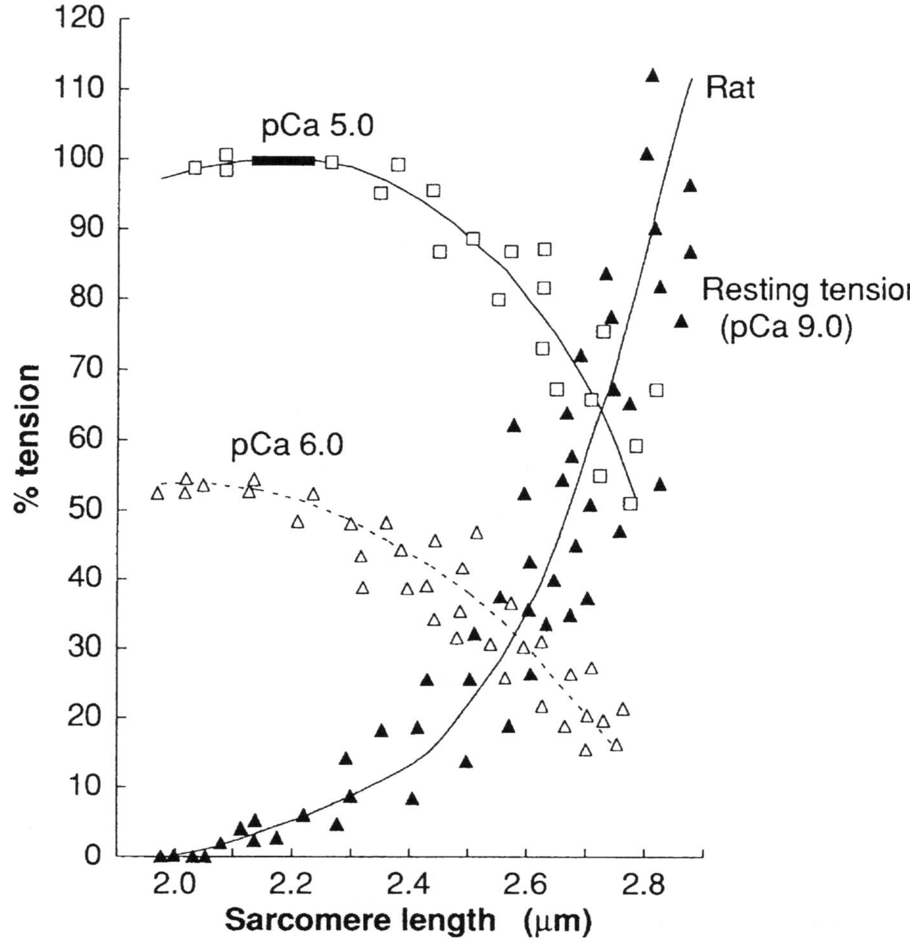

Comments

- Tension is normalized to the maximum active tension at pCa 5.0.

Reference(s)

Fabiato A, Fabiato F (1978) Myofilament-generated tension oscillations during partial calcium activation and activation dependence of the sarcomere length–tension relation of skinned cardiac cells. J Gen Physiol 72:667–699 (with permission)

Length–Tension Relation (3)

• Resting force •	• Rats • Ventricular myocyte	• Relaxed state •

Materials
• Male Wistar rats (age, 3–5 months), enzymatically isolated single myocyte of ventricle

Testing Methods and Experimental Conditions
• Length–static force relation measured in relaxing solution
• Before and after treatment with a detergent, Brij 58 (0.5%) for 30 min

Data

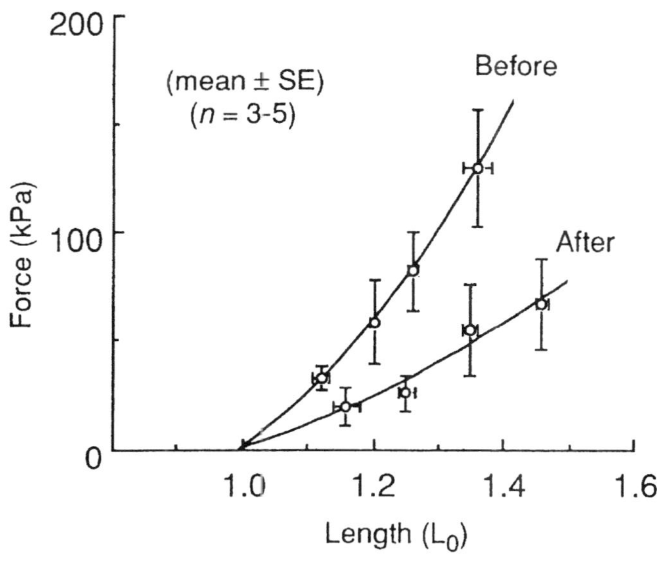

Comments
• Length is normalized to L_o, at which the resting force is just barely detectable.

Reference(s)
Tameyasu T (unpublished data)

Length–Tension Relation (4)

• Resting tension •	• Hamster • Ventricle	• •

Materials
- Male Syrian golden hamster (2–5 months old, 100–300 g), isolated left ventricular myocytes

Testing Methods and Experimental Conditions
- Resting tension–sarcomere length relation of cardiac myocytes together with that of skeletal muscle for comparison

Data

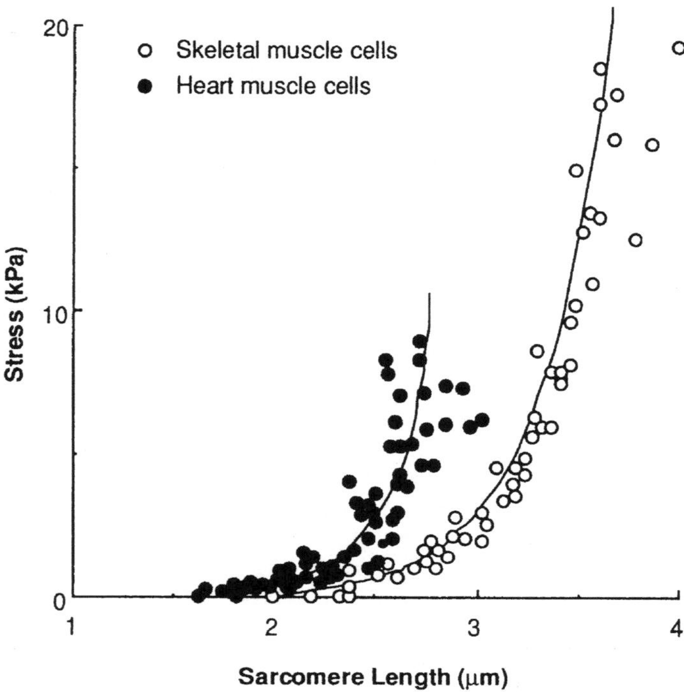

Comments
- Composite length–tension curve for eight cardiac myocytes and five skeletal myocyte fragments.

Reference(s)
Fish D, Orenstein J, Bloom S (1984) Passive stiffness of isolated cardiac and skeletal myocytes in the hamster. Circ Res 54:267–276 (with permission)

Morphological Property (1)

| • Fiber orientation | • Dogs | • Normal |
| • | • Left ventricle | • Systole, diastole |

Materials
- Dogs
- Left ventricle

Testing Methods and Experimental Conditions
- Fiber orientation across the left ventricular wall was measured with light microscopy
- Specimens were obtained from ventricles fixed in situ in systole, diastole, and dilated diastole

Data

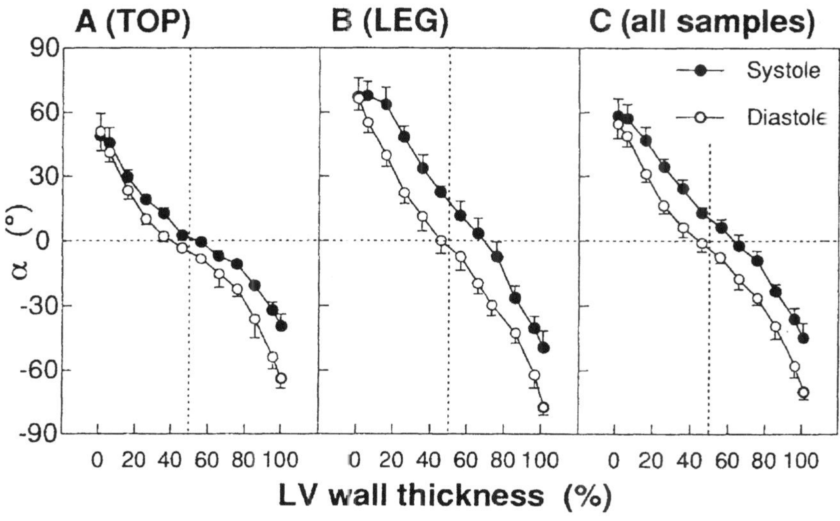

Comments
- TOP, near the base; LEG, near the apex.
- Fiber angle (α) was measured in the plane parallel to the epicardium relative to a line of latitude.
- Fiber angles, viewed from the epicardial side, were positive in the upper right quadrant and negative in the lower right.
- Percent represents the distance from the endocardium as a percent of wall thickness.
- Zero percent implies the endocardium and 100 % implies the epicardium.

Reference(s)
Streeter DD Jr, Spotnitz HM, Patel DP, Ross J Jr, Sonnenblick EH (1969) Fiber orientation in the canine left ventricle during diastole and systole. Circ Res 24:339–347 (with permission)

Morphological Property (2)

• Residual strain • Stretch ratio	• Rat • Left ventricle	• Potassium-arrested •

Materials
- Sprague–Dawley rat (male, BW 200–300 g)
- Left ventricle

Testing Methods and Experimental Conditions
- Potassium-arrested left ventricles
- Deformation of equatorial cross-sectional rings (2–3 mm thickness) after a radial cut through the left ventricular free wall was measured

Data

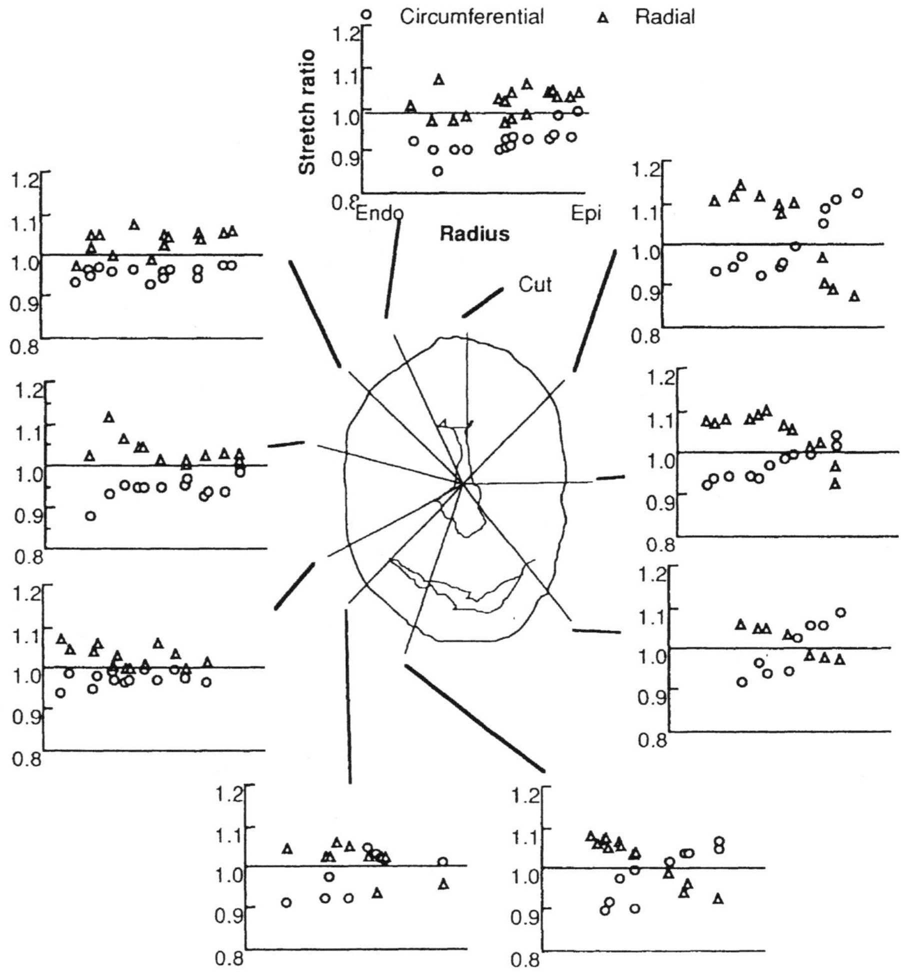

Comments
- There is compressive circumferential residual strain at the inner layers of the ventricle.

Reference(s)
Omens JH, Fung YC (1989) Residual strain in rat left ventricle. Circ Res 59:37–45 (with permission)

Morphological Property (3)

• Sarcomere length • Diastole pressure	• Dogs • Left ventricle	• •

Materials
- Dogs
- Left ventricle

Testing Methods and Experimental Conditions
- Hearts were arrested with KCl (40 mEq), and fixed with glutaraldehyde, 2 at 2 mmHg, 2 at 6 mmHg, 2 at 12 mmHg, and 3 at 20 mmHg of left ventricular diastolic pressure

Data

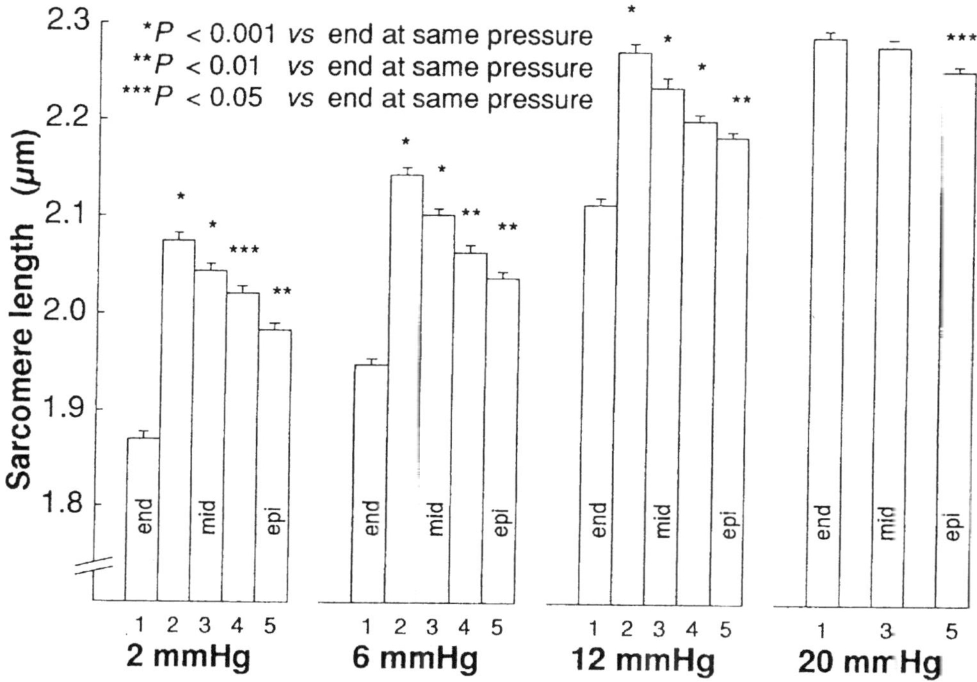

Comments
- Sarcomere lengths from five sites across the left ventricular free wall were measured with an electron microscope.
- At the lower filling pressures (2, 6, and 12 mmHg), sarcomere lengths were shortest at the subendocardium, longest at a site between the subendocardium and the midwall layer, and progressively shorter towards the epicardium.
- At the higher pressure (20 mmHg), the difference was small.

Reference(s)
Yoran C, Covell JW, Ross J Jr (1973) Structural basis for the ascending limb of left ventricular function. Circ Res 32:297–303 (with permission)

Morphological Property (4)

• Ventricular weight •	• Sheep, swine, dog, ... • Ventricle	• Normal •

Materials

- Animal, organ, tissue etc.
- Normal mammals of seven species (sheep, swine, dogs, cats, rabbits, guinea pigs, rats) heart

Testing Methods and Experimental Conditions

- The weights of anatomic segments of the hearts were studied and compared between newborn, young, and adult age groups

Data

Species	Age	N	BW	AT (combined)	RVF	S	LVF	LV volume, (mm^3)
Sheep	2 h–1 D	10	3.61	3.97	7.74	4.17	12.54	5909
(Dorset)			± 0.13	± 0.23	± 0.27	± 0.15	± 5.0	± 334
	2-8 D	25	4.43	4.82	7.87	5.22	15.54	7621
			± 0.20	± 0.4	± 0.37	± 0.31	± 1.04	± 402
	Adult	8	59.76	20.98	34.39	32.06	97.94	36030
			± 2.60	± 0.87	± 1.40	± 2.17	± 3.80	± 2121
Swine	2–14 D	14	3.09	2.38	4.10	3.94	6.91	4434
			± 0.29	± 0.24	± 0.38	± 0.44	± 2.63	± 441
	Adult	6	139.75	61.38	85.40	74.24	198.04	72845
			± 5.70	± 5.19	± 6.80	± 6.40	± 16.23	± 3406
Dogs	< 1 D †	80			0.5840		0.5844	
					± 0.1313		± 0.1389	
	6–17 D	8	910.00 *	0.4500	1.000	0.9000	1.9176	735
			± 1.01	± 0.0236	± 0.0934	± 0.1546	± 0.2520	± 44
	Adult	9	25.40	17.15	36.58	38.85	67.30	31963
			± 2.62	± 1.63	± 1.96	± 3.96	± 4.85	± 2411
Cats	1 D	8	125.80 *	0.0571	0.2626	0.1185	0.3410	115
			± 0.50	± 0.0032	± 0.0137	± 0.0091	± 0.0162	± 1
	65 D	4	932.50 *	0.7250	1.1750	1.2875	2.7375	1449
			± 0.15	± 0.0960	± 0.0893	± 0.0974	± 0.1690	± 52
	Adult	19	3.40	1.3474	2.2000	2.4658	5.2605	2495
			± 0.13	± 0.0988	± 0.1484	± 0.1304	± 0.3103	± 109
Rabbits	1–3 D	12	71.00 *	0.0528	0.1068	0.0559	0.1297	57
(NZW)			± 7.53	± 0.0050	± 0.0115	± 0.0066	± 0.0135	± 8
	Adult	8	3.07	0.8313	1.3688	1.4125	2.5688	1569
			± 0.30	± 0.1436	± 0.1620	± 0.1141	± 0.3416	± 191
Guinea	4 D	12	91.57 *	0.0487	0.0907	0.0504	0.1624	77
pigs			± 5.09	± 0.0027	± 0.0035	± 0.0037	± 0.0081	± 5
(Hartley)	Adult	5	841.86 *	0.2081	0.3222	0.3765	0.7251	412
			± 56.66	± 0.0474	± 0.0587	± 0.0591	± 0.1235	± 56
Rats	< 12 h	15	6.13 *	0.0053	0.0075	0.0042	0.0077	
(Wistar)			± 0.20	± 0.0002	± 0.0006	± 0.0003	± 0.0004	
	11–12 D	21	21.13 *	0.0111	0.0185	0.0250	0.0358	21
			± 0.88	± 0.0006	± 0.0013	± 0.0016	± 0.0022	± 1
	Adult	30	462.70 *	0.0852	0.2500	0.2840	0.5720	200
			± 6.60	± 0.0070	± 0.0080	± 0.0090	± 0.0130	± 6

Comments

- Values are mean ± SE. *N*, number of hearts in respective groups; BW, body weight (kg except as noted); AT, atria (g); RVF, right ventricular free wall (g); S, interventricular septum (g); LVF, left ventricular free wall (g); h, hour; D, day; NZW, New Zealand White. * Body weighting. † From Latimer.

Reference(s)

Lee JC, Taylor JFN, Downing SE (1975) A comparison of ventricular weights and geometry in newborn, young, and adult mammals. J Appl Physiol 38:147–150 (with permission)

Stiffness (1)

• Passive and active stiffness •	• Hamsters, rats, rabbits, .. • Ventricular	• Trabecula • Myocyte

Materials
- Hamsters, rats, guinea pigs, and rabbits
- Ventricular trabeculae, ventricular myocyte

Testing Methods and Experimental Conditions
- Passive and active stiffness

Data

Material	Sarcomere length (µm)	Passive stiffness (kPa)
Trabeculae		
Hamster	2.20	3-8
Rat	2.20	8.0
Rat	2.20	4.7
Rat	2.20	1.6
Rat	2.20	15.7
Myocyte		
Guinea pig	2.20	2.33
Hamster	2.20	0.88
Rabbit	2.20	2.84
Rat	2.20	4.68
Rat	2.20	2.97

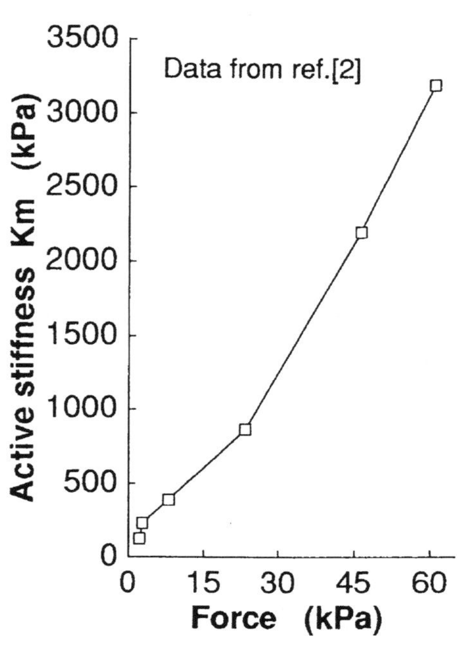

Comments
- Passive stiffness indicates the series elastic component (SEC) of the three-element mechanical model.
- Owing to the difficulty of measuring the cross–sectional area of cells, stiffness values have a wide range.
- Active stiffness (Km in figure) increases with the increase of active force development.

Reference(s)

1. Brady AJ (1991) Length dependence of passive stiffness in single cardiac myocytes. Am J Physiol 260:H1062–H1071 (with permission)
2. Leijendekker WJ, Gao WD, ter Keurs HEDJ (1990) Unstimulated force during hypoxia of rat cardiac muscle: Stiffness and calcium dependence. Am J Physiol 258:H861–H869 (with permission)

Stiffness (2)

• Stress–strain relation • Pressure–volume relation	• Dog • Ventricle	• Normal • Hypertrophy

Materials

• Dog (control and hypertrophied hearts)

Testing Methods and Experimental Conditions

• Hypertrophy was produced by coarctation of the aorta
• Pressure–volume relationship was measured in potassium-arrested left ventricle
• Stress–strain relation was computed based on the finite deformation theory

Data

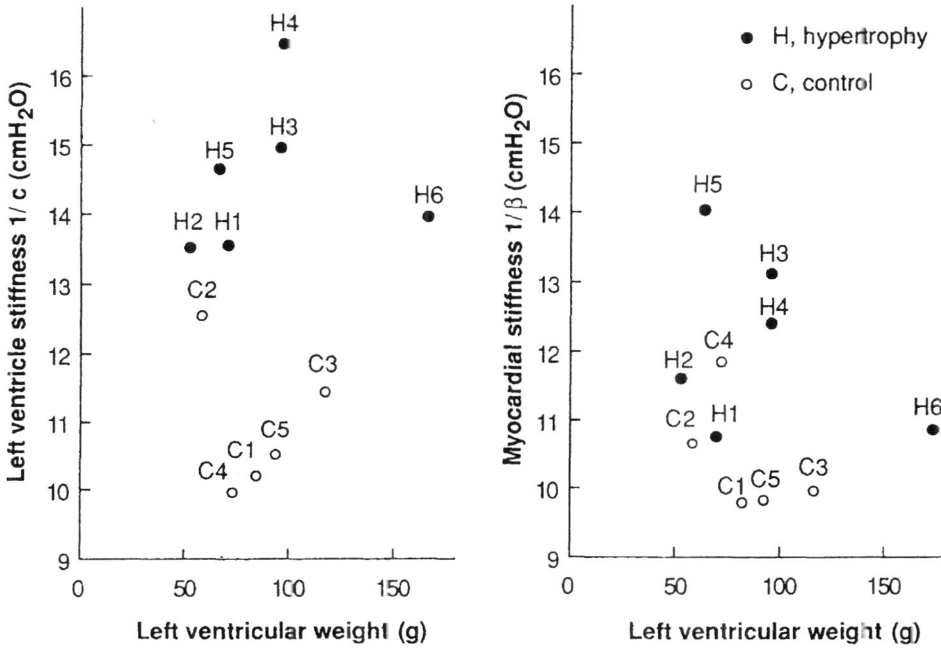

Comments

• Overall left ventricular stiffness was higher in pressure-overloaded hypertrophied heart.
• There is no significant difference in myocardial stiffness.
• Pressure–volume relationship was expressed by
$$V = a - b \cdot \exp(-cp).$$
• Stress–strain relation was expressed by
$$\varepsilon = \delta \cdot [1 - \exp(-\beta\sigma)].$$

Reference(s)

Abe H, Nakamura T, Motomiya M, Konno K, Arai S (1978) Stresses in left ventricular wall and biaxial stress–strain relation of the cardiac muscle fiber for the potassium-arrested heart. Trans ASME J Biomech Eng 100:116–121

Stiffness (3)

• Transverse stiffness • Stress	• Dogs • Interventricular septum	• •

Materials
- Dogs, interventricular septum

Testing Methods and Experimental Conditions
- The center of the arterially perfused septum, during cardioplegia or active contraction, was indented by a probe of 7 mm diameter while transverse and in-plane stress and strain were measured
- The transverse stiffness was defined as the slope of the relation between the indentation stress and the strain

Data

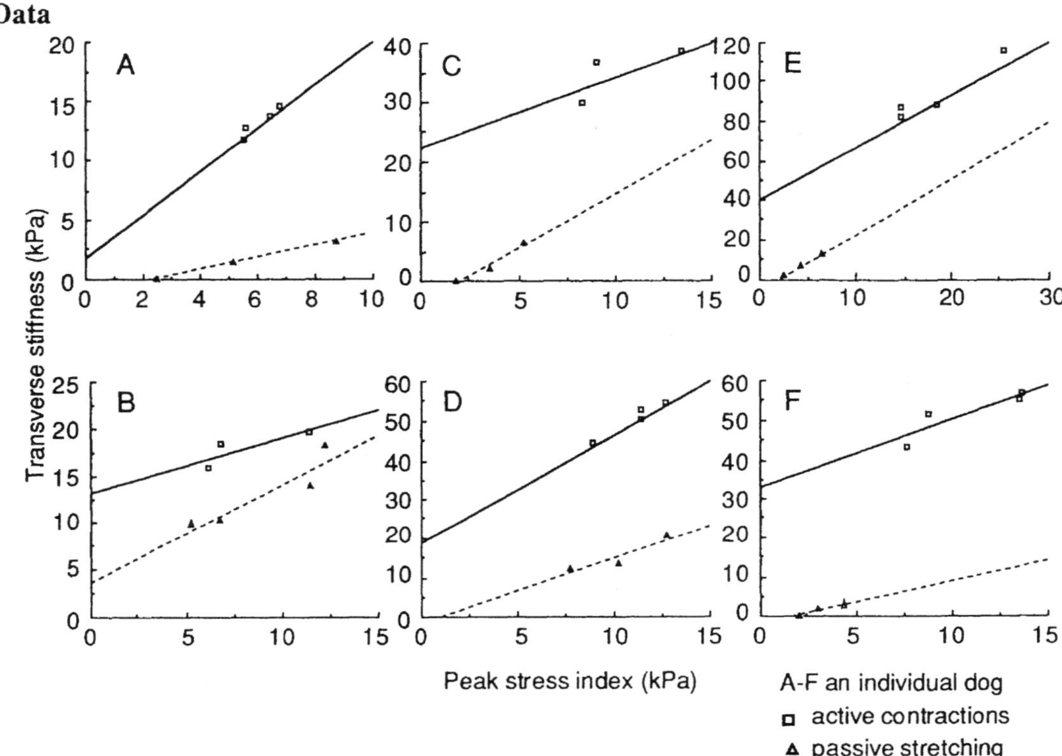

Comments
- Stress index is the square root of the sum of the squares of the magnitudes of the two in-plane stresses.
- Transverse stiffness is linearly related to the in-plane stress and can distinguish between actively generated and passively applied stress.

Reference(s)
Halperin HR, Chew PH, Weisfeidt ML, Sagawa K, Humphrey JD, Yin FCP (1987) Transverse stiffness: A method for estimation of myocardial wall stress. Circ Res 61:695–703 (with permission)

Anisotropy (1)

• Axial property • Stress–strain curve	• Dogs • Aorta, carotid, femoral	• Regional variation •

Materials
- Adult mongrel dogs (weight, 7–14 kg)
- Abdominal aortas, and common carotid and femoral arteries

Testing Methods and Experimental Conditions
- Cylindrical specimen (in situ length, ca. 50 mm)
- Medium: physiological saline at 37°C was filled in the lumen
- Uniaxial elongation with a tensile tester in air (crosshead speed, 0.1–10.0 mm/s)
- Intraluminal pressure held constant at 50, 100, 150, and 200 mmHg

Data

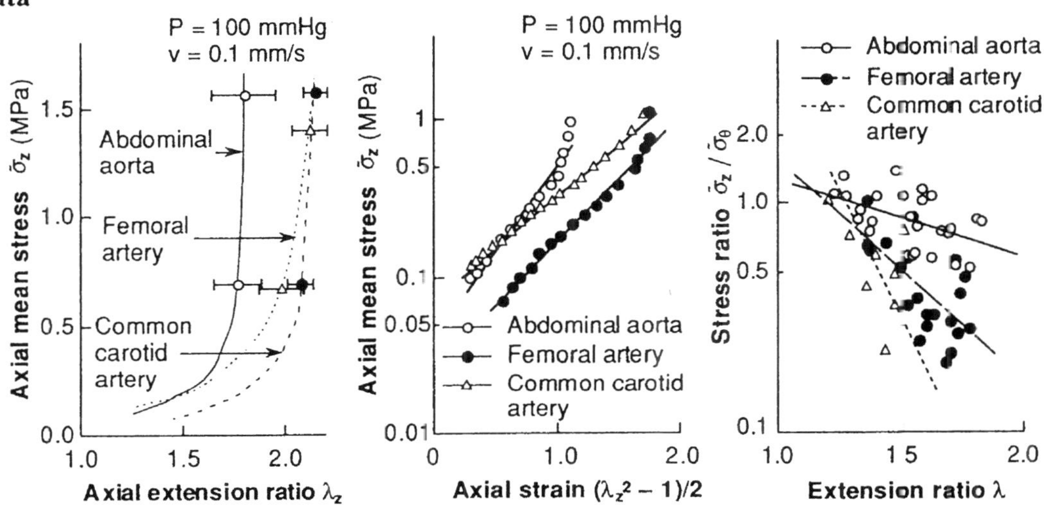

Comments
- Right panel shows stress ratio at the same extension ratio in axial and circumferential directions.
- Effect of crosshead speed on stress–strain relation was insignificant.

Reference(s)
Sato M, Hayashi K, Niimi H, Moritake K, Okumura A, Handa H (1979) Axial mechanical properties of arterial walls and their anisotropy. Med Biol Eng Comput 17:170–176 (with permission)

Anisotropy (2)

• Compliance • Poisson's ratio	• Rats • Aorta, carotid artery	• In vivo measurement • Effect of smooth muscle

Materials
- Sprague–Dawley rats weighing an average of 300 g
- Carotid artery and abdominal aorta

Testing Methods and Experimental Conditions
- Anesthetized with 5 mg/kg xylazine and 50 mg/kg ketamine (i.m.)
- Exposition of carotid artery and abdominal aorta
- Diametrical change and longitudinal displacement monitored with a video-tracked motion analyzer
- Blood pressure measurement via cannulation into the femoral artery
- Photosensitive drug (CASPc) to eliminate vascular cell function
- Biaxial compliances were measured at 100 mmHg

Data

Pressure (mmHg)	C_{circ} ($\%$/mmHg $\times 10^{-2}$)	C_{long}	Poisson's ratio
Carotid artery ($n = 22$)			
50	44.3 ± 14.0	4.8 ± 2.5	0.11 ± 0.06
100	16.5 ± 3.7	3.2 ± 1.4	0.20 ± 0.07
150	6.7 ± 2.5	2.5 ± 1.7	0.46 ± 0.28
Abdominal aorta ($n = 5$)			
50	36.4 ± 6.5	4.5 ± 0.8	0.12 ± 0.01
100	12.1 ± 3.9	3.2 ± 1.5	0.28 ± 0.14
150	4.6 ± 2.9	2.4 ± 2.1	0.80 ± 0.70

Values are means $\pm$ SD.
C_{circ}, circumferential compliance.
C_{long}, longitudinal compliance.

	Laser only	CASPc + laser
Circumferential compliance		
Pre	14.0 ± 4.1	15.6 ± 3.3
Post	12.8 ± 5.2	8.6 ± 2.3*
Longitudinal compliance		
Pre	2.8 ± 0.7	2.3 ± 0.6
Post	3.0 ± 1.1	3.1 ± 0.7
Poisson's ratio		
Pre	0.21 ± 0.05	0.19 ± 0.04
Post	0.25 ± 0.06	0.39 ± 0.23**

Values are means $\pm$ SD; $n = 6$.
Pre, before operation of laser.
Post, after operation of lasesr.
* $P < 0.004$ *vs* Pre.
** $P < 0.03$ *vs* Pre.

Comments
- $C_{circ} = (D_{sys} - D_{dia})/\{(P_{sys} - P_{dia}) \times D_{dia}\}$.
- $C_{long} = (L_{sys} - L_{dia})/\{(P_{sys} - P_{dia}) \times L_{dia}\}$.
- Pressure–diameter and elastic modulus–stress relationships are also shown .

Reference(s)
L'Italien GJ, Chandrasekar NR, Lamuraglia GM, Pevec WC, Dhara S, Warnock DF, Abbott WM (1994) Biaxial elastic properties of rat arteries in vivo: Influence of vascular wall cells on anisotropy. Am J Physiol 267:H574–H579 (with permission)

Anisotropy (3)

<table>
<tr><td>• Creep
• Stress relaxation</td><td>• Dogs
• Aorta, femoral, iliac</td><td>• Relaxation function
• Regional variation</td></tr>
</table>

Materials
- Dogs of about 20 kg weight
- Thoracic and abdominal aortas, external iliac and femoral arteries

Testing Methods and Experimental Conditions
- Strips of 5 mm in width taken at 8 different regions between the aortic arch and femoral artery in either longitudinal or circumferential directions
- Medium: aerated Krebs–Ringer solution at 37°C
- Tensile test at constant strain rate
- Stress relaxation test
- Creep test

Data

Region	α	T^* (kPa) at $\lambda^* = 1.6$	E_0	$G(300)$	$G(t)$-slope
Circumferential segments					
Arch	0.88 ± 0.93	52.4 ± 5.4	89.2 ± 47.8	75.52 ± 5.53	-0.0326
Prox. thoracic	0.92 ± 0.68	57.3 ± 12.7	86.7 ± 31.3	79.17 ± 5.55	-0.0316
Mid-thoracic	1.24 ± 0.50	61.7 ± 8.3	88.4 ± 17.2	75.13 ± 6.75	-0.0345
Dist. thoracic	1.48 ± 0.04	65.2 ± 19.6	84.9 ± 8.4		-0.0369
Prox. abdom	3.24 ± 0.45	75.0 ± 14.7	34.2 ± 19.4	65.93 ± 10.06	-0.0432
Dist. abdom	3.88 ± 0.70	55.9 ± 11.3	15.7 ± 10.2	65.79 ± 5.93	-0.0452
Ext. iliac	4.70 ± 1.52	75.5 ± 19.6	6.6 ± 35.7	58.24 ± 9.39	-0.0423
Femoral	4.15 ± 0.83	70.6 ± 9.8	-1.4 ± 22.1	65.91 ± 12.15	-0.0426
Longitudinal segments					
Arch	1.50 ± 0.91	47.5 ± 6.9	55.4 ± 26.2	82.2 ± 2.02	-0.0256
Prox. thoracic	2.93 ± 0.83	61.2 ± 17.6	37.7 ± 27.5	86.3 ± 3.7	-0.0256
Mid-thoracic	2.41 ± 0.46	43.3 ± 2.9	49.8 ± 55.0	84.8 ± 5.73	-0.0195
Dist. thoracic	2.12 ± 0.38	45.1 ± 3.9	57.0 ± 44.2	85.4 ± 8.39	-0.0152
Prox. abdom	4.36 ± 0.39	69.8 ± 20.6	22.2 ± 7.9	88.1 ± 4.46	-0.0135
Dist. abdom	3.81 ± 0.51	56.6 ± 2.5	62.3 ± 19.7	75.0 ± 12.6	-0.0280
Ext. iliac	1.30 ± 0.28	67.8 ± 18.6	124.5 ± 6.1	48.3 ± 6.2	-0.0402
Femoral	1.06 ± 0.18	52.3 ± 12.7	83.5 ± 4.6	62.7 ± 8.4	-0.0400

All data are given as mean $\pm$ SD, $n = 7$.

Comments
- Stress–strain relationship:
  $$dT/d\lambda = \alpha T + E_o \quad (T, \text{ Lagrangian stress}; \ \lambda, \text{ extension ratio}; \ \alpha \text{ and } E_o, \text{ constants});$$
  α and E_o are given in the range 19 6 kPa $< T < T^*$, and average of values in strain rates between 4 and 40 mm/min.
- Normalized relaxation function:
  $$G(t) = T(t, \lambda)/T(0, \lambda) \ .$$
- The $G(t)$-slope, $dG/d \log_e t$ is evaluated at $t = 1$ s.

Reference(s)
Tanaka T, Fung YC (1974) Elastic and inelastic properties of the canine aorta and their variation along the aortic tree. J Biomech 7:357–370 (with permission)

Anisotropy (4)

• Distensibility • Stress–strain curve	• Dogs • Various veins	• Regional variation • Bilinear analysis

Materials
• Dog
• Superior vena cava, intrathoracic and abdominal portions of inferior vena cava, and jugular, axillary, common iliac, and femoral veins

Testing Methods and Experimental Conditions
• Longitudinal and circumferential strips, 5 mm in width
• Tensile test at crosshead speed of 5 mm/min
• Medium: aerated physiological salt solution at 30°C

Data

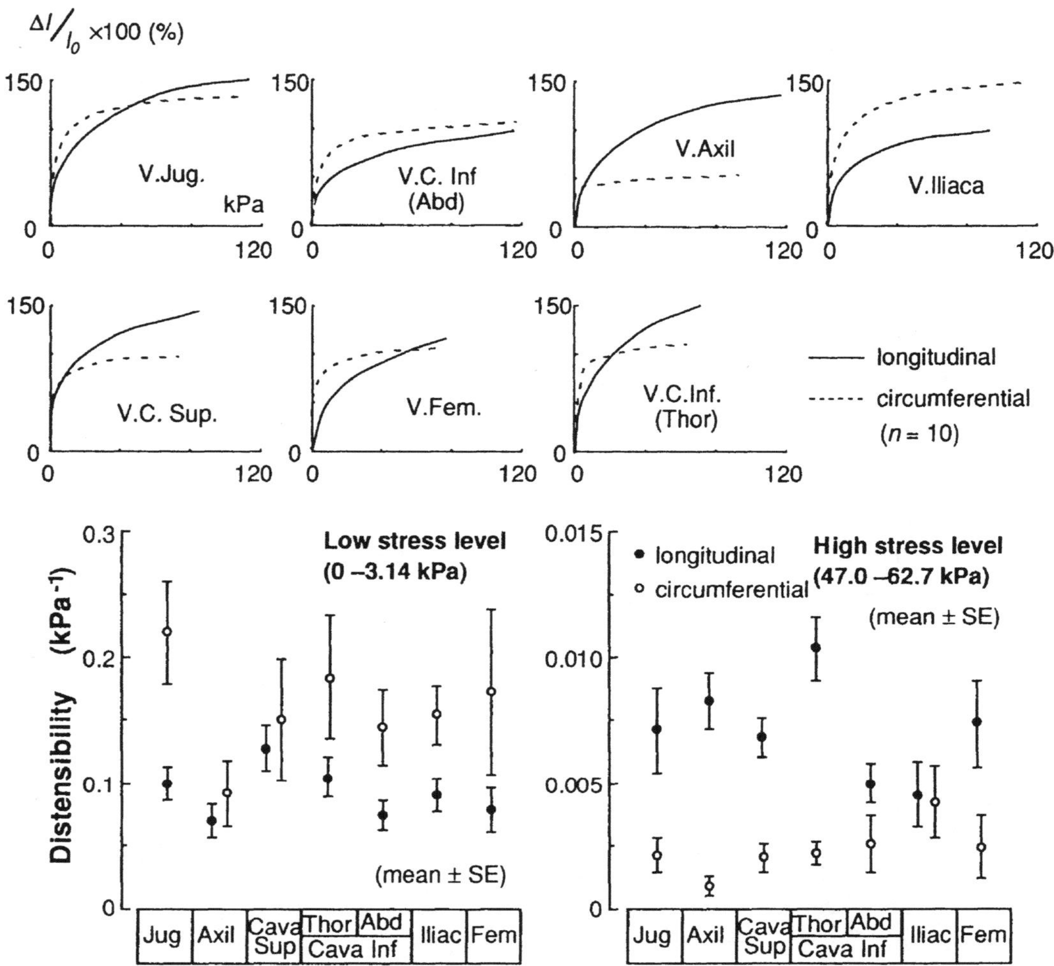

Comments
None.

Reference(s)
Azuma T, Hasegawa M (1973) Distensibility of the vein: From the architectural point of view. Biorheology 10:469–479 (with permission)

Anisotropy (5)

• Elastic modulus • Dynamic P–D–Fz test	• Dogs • Aorta	• In vivo measurement •

Materials
- Mongrel dogs weighing an average of 28.6 kg
- Thoracic aorta

Testing Methods and Experimental Conditions
- Static relations among pressure–diameter–axial force were measured in vivo
- A segment 7.5 cm long isolated and bypassed from the main systemic circulation
- Incremental elastic moduli in circumferential, axial, and radial directions were calculated based on constitutive relations formulated in this paper
- Temperature, 26°–27°C

Data

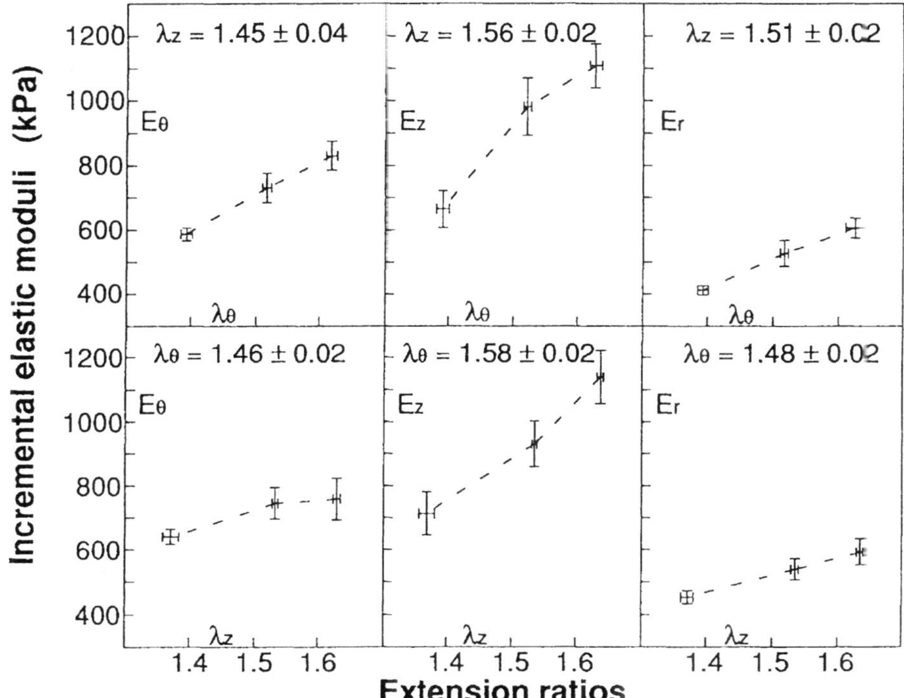

Comments
- Vasa vasorum intact.
- E_θ, circumferential; E_z, longitudinal; E_r, radial elastic moduli.
- λ_θ, circumferential; λ_z, longitudinal; λ_r, radial extension ratios.
- All the elastic moduli decreased when the specimen was studied in vitro; E_z decreased by 32% ($P < 0.01$), E_r by 20% ($0.2 < P < 0.05$, *sic*), E_θ by 9% ($0.05 < P < 0.1$). The decrease in E_z can be explained by removal of longitudinal tethering.

Reference(s)
Patel DJ, Janicki JS, Carew TE (1969) Static anisotropic elastic properties of the aorta in living dogs. Circ Res 25:765–779 (with permission)

Anisotropy (6)

• Elastic modulus • Stress–strain curve	• Puppies • Carotid artery	• Developmental change •

Materials

- A litter of puppies and their mother (to ensure homogeneity of animal group)
- Carotid artery

Testing Methods and Experimental Conditions

- Tubular specimen in aerated physiological salt solution at 37°C
- Pressure–diameter–axial force relation obtained at inflation rate of 1–2 mmHg/s
- Anisotropic elastic modulus calculated using a method of Vaishnav et al.

Data

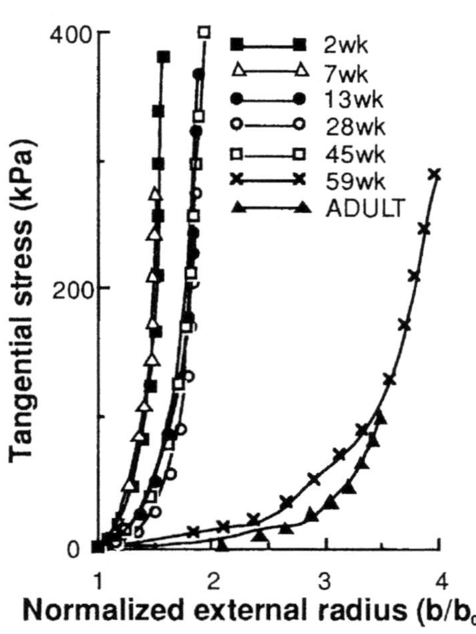

Age	λ_θ	λ_x	E_θ (kPa)	E_r (kPa)	E_x (kPa)
2	1.011	1.822	245	262	3742
7	1.067	1.953	630	656	2912
13	1.282	1.817	971	1416	669
28	1.318	1.660	1047	773	956
45	1.247	1.959	1228	348	424
59	1.541	1.432	1501	1244	1686
Mother	1.268	1.643	1552	1443	2242
Adult	1.373 ± 0.043	1.577 ± 0.042	1358 ± 75	1001 ± 130	1359 ± 225

Adult values are means ± SD, $n = 7$. Age is the age of dogs in weeks. λ_θ and λ_x are the tangential and axial extension ratios. E_θ, E_r, and E_x are the tangential, radial, and axial elastic moduli.

Comments

- Change in connective tissue and electrolyte contents during growth are also presented.

Reference(s)

1. Cox RH, Jones AW, Fischer GM (1974) Carotid artery mechanics, connective tissue, and electrolyte changes in puppies. Am J Physiol 227:563–568 (with permission)
2. Vaishnav RN, Young JT, Janicki JS, Patel DJ (1972) Non-linear anisotropic elastic properties of the canine aorta. Biophys J 12:1008–1027 (with permission)

Anisotropy (7)

• Extensibilities • Volume elasticity	• Dogs • Thoracic aorta	• In vivo measurement • Regional variation

Materials
- Six mongrel dogs weighing 18.8 ± 3.2 kg (mean $\pm$ SE)
- Ascending, upper descending, and middle descending thoracic aortas

Testing Methods and Experimental Conditions
- Right lateral thoracotomy under anesthesia with a combination of Innovar and pentobarbital sodium under mechanical ventilation
- Pressure measurement with a catheter-tip manometer
- Diameter and axial length measurement with an ultrasonic dimension gauge system
- Segments dissected after measurement for wall volume calculation
- Inner radius calculation from outer diameter and wall volume

Data

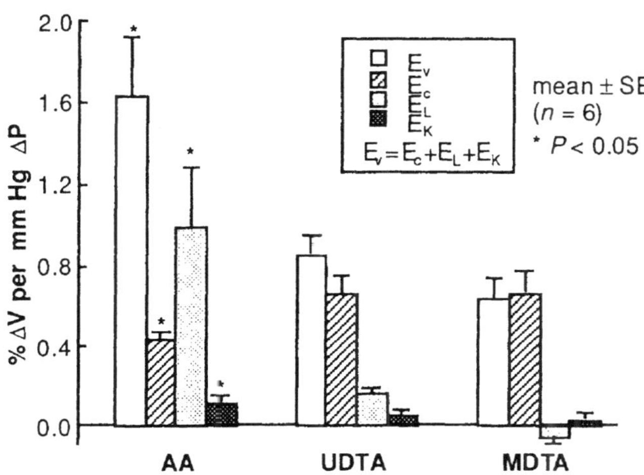

Location		AA	UDTA	MDTA
Mean pressure	(mmHg)	71 ± 3	70 ± 6	70 ± 6
Pulse pressure	(mmHg)	23 ± 2	24 ± 3	25 ± 2
Heart rate	(bpm)	134 ± 6	130 ± 3	124 ± 3

All data are given as mean $\pm$ SE.

Comments
- Volume distensibility (E_v), and circumferential (E_c), longitudinal (E_L), and high-order extensibilities (E_K) of various segments of thoracic aortas.
- AA, ascending thoracic aorta; UDTA, upper descending thoracic aorta; MDTA, middle descending thoracic aorta.
- Volume distensibility, E_v:

$$E_v = (\Delta V/V)(100/\Delta P)$$
$$= \underbrace{2\Delta R_i/R_i \cdot 100/\Delta P}_{E_C} + \underbrace{\Delta L/L \cdot 100/\Delta P}_{E_L} + \underbrace{[(\Delta R_i/R_i)2 + 2\Delta R_i/R_i \cdot \Delta L/L] \cdot 100/\Delta P + \dots}_{E_K}$$

where R_i and L are inner radius and axial length, respectively, at diastolic pressure, and $V = \pi R_i L$. Δ denotes increments from diastole to systole.

Reference(s)

Gentle BJ, Gross DR, Chuong CJC, Hwang NHC (1988) Segmental volume distensibility of the canine thoracic aorta in vivo. Atherosclerosis 22:385–389 (with permission)

Anisotropy (8)

• Incremental elastic modulus • Incremental Poisson's ratio	• Rats • Abdominal vena cava	• •

Materials
- Cylindrical segment of abdominal vena cavae of adult Wistar rats

Testing Methods and Experimental Conditions
- Medium: calcium-free Tyrode solution with EGTA at 37°C
- Quasi-static inflation test between 0 and 200 mmHg under constant axial length
- Pressure–diameter–axial force relation obtained at several successive levels of axial strain

Data

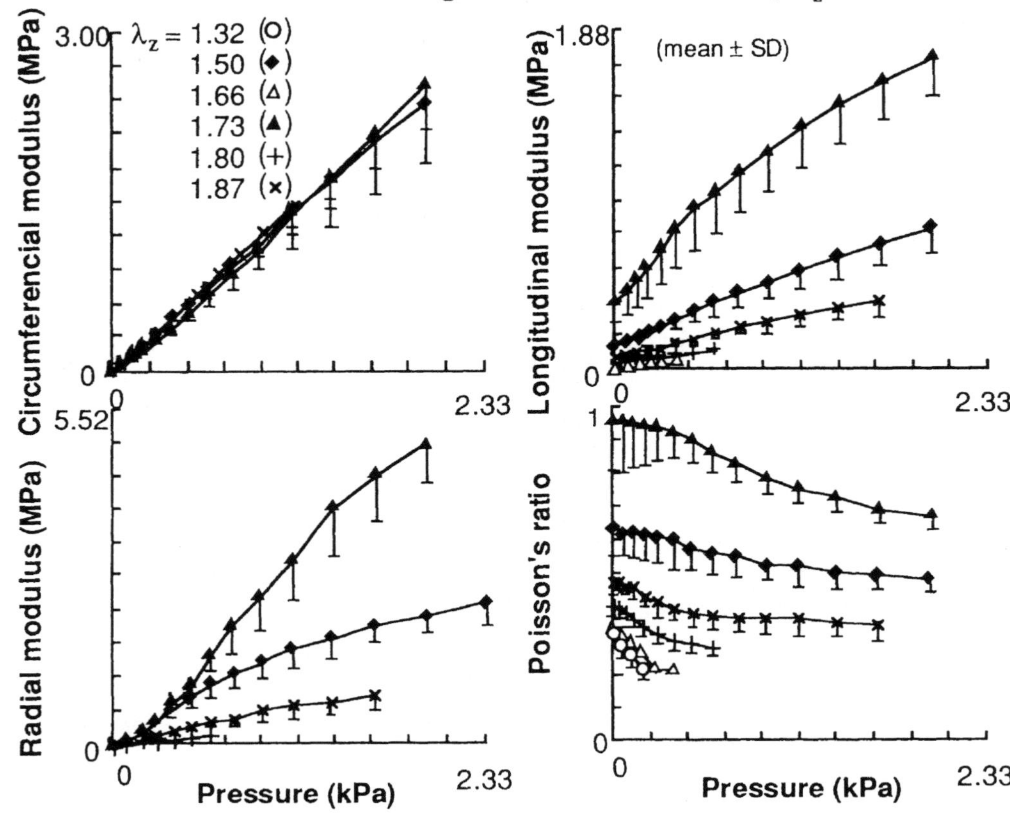

Comments
None.

Reference(s)
Weizsacker HW (1988) Passive elastic properties of the rat abdominal vena cava. Pflügers Arch 412:147–154

Anisotropy (9)

<table>
<tr><td>• Incremental elastic modulus
• Inflation–extension test</td><td>• Dogs, human
• Jugular, saphenous vein</td><td>• Transplantation
•</td></tr>
</table>

Materials

- Dog external jugular vein
- Human saphenous vein (average age, 33 years)
- Dog jugular vein transplanted into autologous arterial circuits for at least 10 months

Testing Methods and Experimental Conditions

- Quasi-static pressure–diameter–axial length relation obtained with constant axial force
- Tubular specimen held vertically in air with weight at the lower end to restore in vivo length
- Specimen inflated with dogs' heparinized blood or 5% dextrose in normal saline

Data

n	Pressure (mmHg)	Incremental modulus		R^2/h (cm)	E_{inc} (MPa)
		Circumferential, E_θ (MPa)	Longitudinal, E_z (MPa)		
		Canine jugular vein			
7	7.4	1.5 ± 0.3†‡	0.12 ± 0.018	11.41 ± 0.37	3.2 ± 1.4
	18.4	4.7 ± 0.6†‡	0.44 ± 0.035*	12.24 ± 0.42	4.4 ± 1.1‡
	36.8	8.8 ± 0.7†‡	1.18 ± 0.21‡	13.51 ± 0.7	11.5 ± 2.8‡
	55.2	9.8 ± 0.7†‡	4.6 ± 1.3*	13.95 ± 1.28	15.8 ± 4.8‡
	73.6	11.7 ± 1.0‡	6.7 ± 2.5*	14.45 ± 1.48	19.6 ± 2.4‡
	91.9	13.4 ± 2.4‡	8.9 ± 3.2*	14.73 ± 1.65	25.2 ± 4.0‡
	110.3	17.1 ± 0.9†‡	11.3 ± 1.3‡	15.93 ± 1.98	28.1 ± 3.2‡
		Human saphenous vein			
4	7.4	0.027 ± 0.012§	0.161 ± 0.032	0.800 ± 0.27	0.118 ± 0.018
	18.4	0.065 ± 0.013§	0.203 ± 0.039	0.909 ± 0.27	0.152 ± 0.022
	36.8	0.189 ± 0.041	0.275 ± 0.078	1.090 ± 0.29	0.231 ± 0.031
	55.2	0.985 ± 0.16§	0.318 ± 0.076	1.616 ± 0.06	0.354 ± 0.059
	73.6	1.50 ± 0.26§	0.356 ± 0.058	1.871 ± 0.06	0.569 ± 0.048
	91.9	2.04 ± 0.16†	0.398 ± 0.096	2.100 ± 0.19	1.050 ± 0.140
	110.3	2.51 ± 0.75	0.475 ± 0.12	2.508 ± 0.40	2.160 ± 0.157

All values are given as mean ± SE.
* $P < 0.05$, ‡ $P < 0.01$ for the comparison between jugular and saphenous segments
† $P < 0.01$, § $P < 0.01$ for the comparison between circumferential and longitudinal moduli for the same type segment.

Comments

- Incremental modulus: $E_i = (1 - \sigma^2)\lambda\,(dS_i/d\lambda_i)$.

  σ, Poisson's ratio (= 0.5); S_i and λ_i, stress and extension ratio, respectively, in direction i ($i = \theta, z$).

  $S_\theta = P(R/h - 1/2)$, $S_z = P(R/h - 1)/2 + F/(2\pi Rh)$.

  P, pressure; R, midwall radius; h, wall thickness; F, axial force.

- Incremental elastic modulus: $E_{inc} = 0.75(\Delta P/\Delta R)(R^2/h)$.
- All of the jugular segments transplanted into arterial circuits were virtually nondistensible and remained at extension ratios of 1.0–1.3 throughout the pressure range from 0 to 150 mmHg.

Reference(s)

Wesly RLR, Vaishnav RN, Fuchs JCA, Patel DJ, Greenfield JC Jr (1975) Static linear and nonlinear elastic properties of normal and arterialized venous tissue in dog and man. Circ Res 37:509–520 (with permission)

Anisotropy (10)

• Static P–D–Fz relation •	• Rats • Carotid artery	• Latex rubber tube • Incremental elastic modulus

Materials
- Cylindrical segment of carotid arteries of adult Wistar rats

Testing Methods and Experimental Conditions
- Medium: calcium-free Tyrode solution with EGTA at 37°C
- Quasi-static inflation test between 1 and 200 mmHg under constant axial length
- Pressure–diameter–axial force relation obtained at several successive levels of axial strain

Data

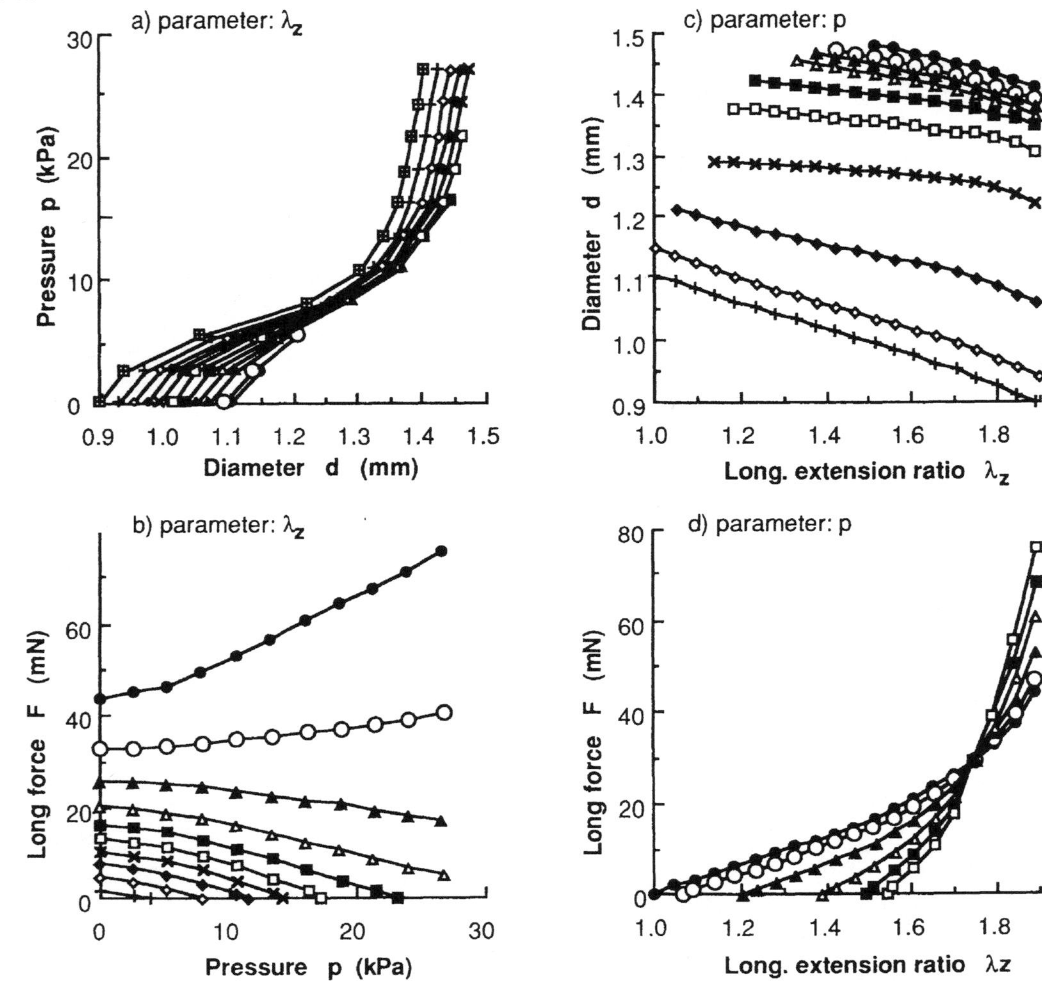

Comments
- A part of the data is shown in the figure from the original.
- Also presented:
  Quasi-static pressure–diameter–axial force relation of a thin-walled latex rubber tube;
  Incremental elastic modulus.

Reference(s)
Weizsacker HW, Pinto JG (1988) Isotropy and anisotropy of the arterial wall. J Biomech
21:477–487 (with permission)

Anisotropy (11)

• Stress–strain curve • Inflation–extension test	• Dogs • Thoracic aorta	• Hypertension • Hydrostatic pressure

Materials

- Male mongrel dogs, ranging in weight from 20.0 to 29.2 kg
- Hypertension induced either by constriction of one renal artery and removal of the contralateral kidney (1K-1C), or by constriction of both renal arteries (2K-2C)
- Animal killed with T-61 after 6 to 8 weeks after the development of hypertension
- Thoracic aorta obtained through thoracotomy

Testing Methods and Experimental Conditions

- Cylindrical specimen held vertically in air with a weight on the distal end
- Specimen pressurized with physiological saline at room temperature
- Diameter and length change during pressurization and axial loading were measured
- Several steps of intraluminal pressure (25 mmHg) and longitudinal force (40 g) to a maximum of 200 mmHg pressure and 200 g of axial force were used
- Eight steps of constant pressure levels used, with five force steps at each pressure level
- Circumferential stress, S_θ, and longitudinal stress, S_z, calculated assuming incompressibility of wall

Data

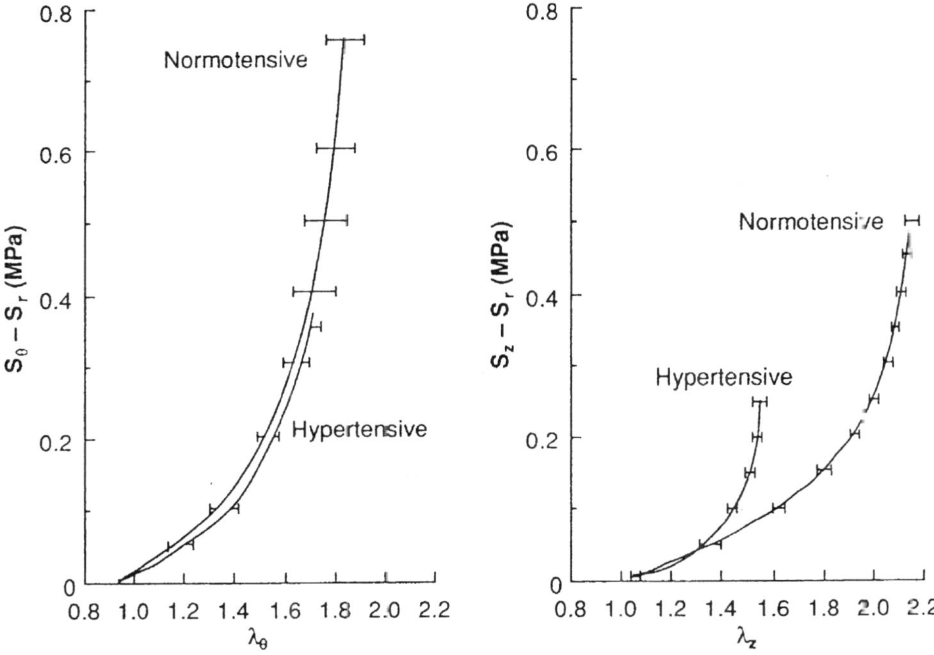

Comments

- To eliminate hydrostatic pressure, stress differences $(S_\theta - S_r)$ and $(S_z - S_r)$, where $S_r = P/2$ is the radial stress, were used (P, intraluminal pressure).

Reference(s)

Vaishnav RN, Vossoughi J, Patel DJ, Cothran LN, Coleman BR, Ison-Franklin EL (1990) Effect of hypertension on elasticity and geometry of aortic tissue from dogs. J Biomech Eng 112:70–74

Anisotropy (12)

• Stress–strain curve • Isotropic condition	• Dogs • Carotid artery	• Mean of 120 specimens •

Materials
- Adult mongrel dogs of either sex weighing 18–26 kg
- One hundred and twenty segments of carotid artery

Testing Methods and Experimental Conditions
- Segments kept under refrigeration for 24–28 h in physiological salt solution
- Quasi-static pressure–diameter test in physiological salt solution at 37°C
- Specimens exhibited contractile activity to supramaximal dose of norepinephrine were excluded
- Pressure–diameter relation obtained at longitudinal length of 1.2, 1.4, 1.6, 1.8 times unloaded length in systematically randomized order
- Pressure increased up to 300 mmHg or until the longitudinal force became negative

Data

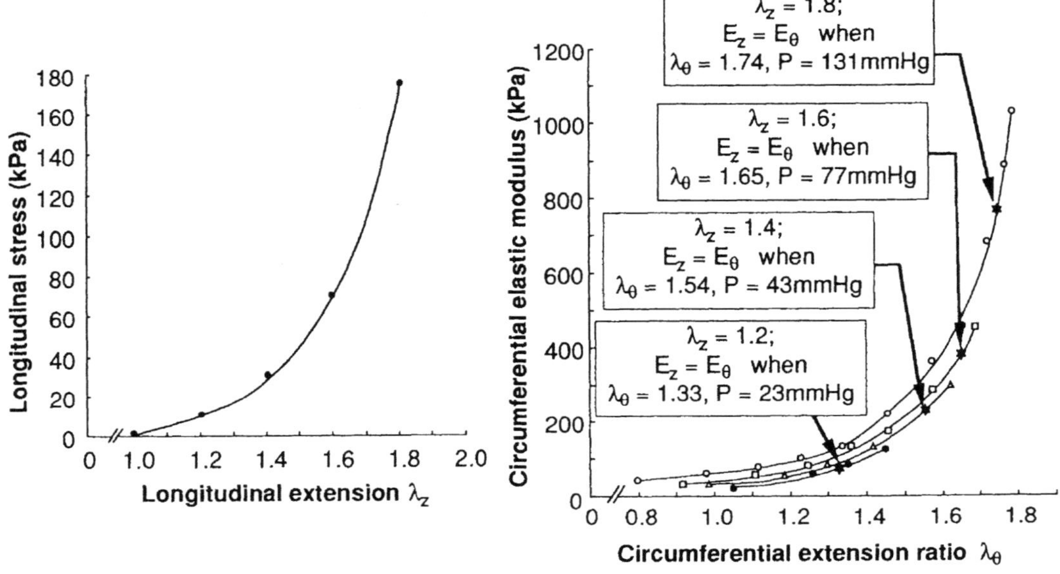

Comments
- Circumferential elastic modulus, E_θ:

  $$E_\theta = \Delta\sigma_\theta/(\Delta\lambda_\theta + \nu_{\theta Z}\Delta\sigma_Z/E_Z).$$

  E_Z, longitudinal elastic modulus measured on unpressurized vessel;

  $\nu_{\theta Z} = \Delta\sigma_Z/\Delta\sigma_\theta$, Poisson's ratio at constant longitudinal length.
     See Dobrin and Doyle (Circ Res 27: 105–119, 1970) for details.
- Each data point is the mean for 120 vessels. Standard errors are smaller than the symbols used to designate the means.
- In the right-hand diagram, stars identify isotropy with conditions specified in the rectangles.

Reference(s)
Dobrin PB (1986) Biaxial anisotropy of dog carotid artery: Estimation of circumferential elastic modulus. J Biomech 19:351–358 (with permission)

Anisotropy (13)

• Stress–strain curve • Tensile test	• Dogs • Aorta and its branch	• Local property • Branching site

Materials
- Dogs weighing 12–35 kg
- Abdominal aorta with its major branches

Testing Methods and Experimental Conditions
- Longitudinal strip including the aorta, the bifurcation region, and the branch
- Circumferential strips that were cut circumferentially on the junction and obliquely on the aorta
- Grid applied to each region with aldehyde fuchsin stain
- Uniaxial loading/unloading in a load range from 0 to 200 kPa while taking photographs of the specimen at crosshead speed of 10 mm/min (= strain rate of 0.3 per minute)
- Thickness measurement in each of the three regions

Data

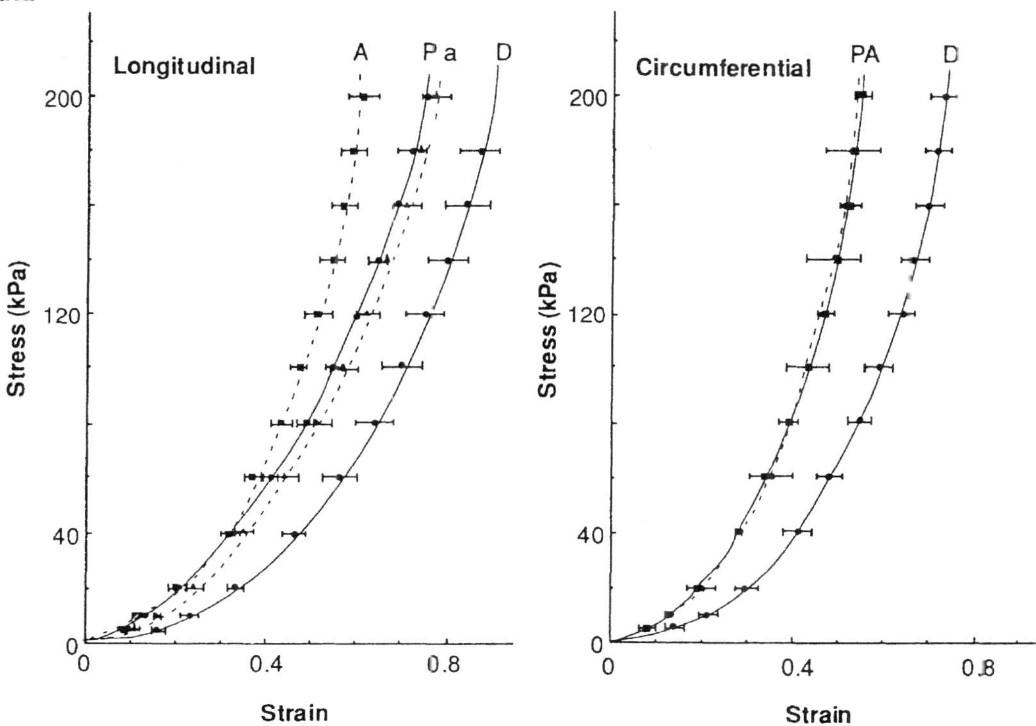

Comments
- A, aorta; P, proximal junction; a, branch artery; D, distal junction.

Reference(s)
Touw DM, Sherebrin MH, Roach MR (1985) The elastic properties of canine abdominal aorta at its branches. Can J Physiol Pharmacol 63:1378–1383 (with permission)

Anisotropy (14)

• Ultimate stress/strain • Strain rate effect	• Human • Aorta	• Failure theory •

Materials

- Human (mean age, 54.6 years, range from 4 to 89)
- Descending mid-thoracic aorta

Testing Methods and Experimental Conditions

- Dumbbell specimens 6.35 or 4.57 mm in width and 19.05 or 7.87 mm in length
- Tensile test in air at room temperature, Ringer solution sprayed to keep specimen wet
- Quasi-static test with crosshead speed of 12.7 mm/min
- Dynamic test with crosshead speed of 92 m/min

Data

	Quasi-static		Dynamic (~100/s)	
	Longitudinal	Transverse	Longitudinal	Transverse
σ (MPa)	1.47 ± 0.91 ($n = 18$)	1.72 ± 0.89 ($n = 18$)	3.59 ± 2.04 ($n = 18$)	5.07 ± 3.29 ($n = 16$)
λ_1	1.47 ± 0.23 ($n = 18$)	1.53 ± 0.28 ($n = 18$)	1.64 ± 0.28 ($n = 21$)	1.60 ± 0.28 ($n = 17$)

All data are given as mean ± SD.

σ, ultimate stress; λ_1, ultimate strain.

Comments

- The most reasonable failure theory for aortic tissue is maximum tensile strain theory.

Reference(s)

Mohan D, Melvin JW (1982) Failure properties of passive human aortic tissue. I—Uniaxial tension test. J Biomech 15:887–902 (with permission)

Anisotropy (15)

• Viscoelasticity • Dynamic elastic modulus	• Dogs • Thoracic aorta	• In situ measurement • Frequency dependence

Materials
- Dogs (weight, 24.2–39.1 kg)
- Middle descending thoracic aorta

Testing Methods and Experimental Conditions
- A segment 8 cm long isolated and bypassed from the main systemic circulation after thoracotomy
- Dynamic pressure–diameter–axial force relation measured in situ
- Variable length, constant volume test and constant length, variable volume test
- Frequencies: 0.5, 1, 2, 3, 4, and 5 Hz
- Temperature: 26°–27°C

Data

	In vivo data				In vitro data
	Group 1 (low λ_θ, low λ_z)	Group 2 (high λ_θ, low λ_z)	Group 3 (low λ_θ, high λ_z)	Group 4 (high λ_θ, high λ_z)	
λ_θ	1.39 ± 0.05	1.59 ± 0.02	1.41 ± 0.03	1.59 ± 0.03	1.53 ± 0.03
λ_z	1.40 ± 0.05	1.50 ± 0.02	1.63 ± 0.02	1.63 ± 0.02	1.50 ± 0.06

For each group, average values $\pm$ SE of various parameters are shown. The values represent averages over all frequencies for the number of arterial segments included in each group. λ_θ and λ_z are the extension ratios (stretches) in the circumferential and longitudinal directions, respectively.

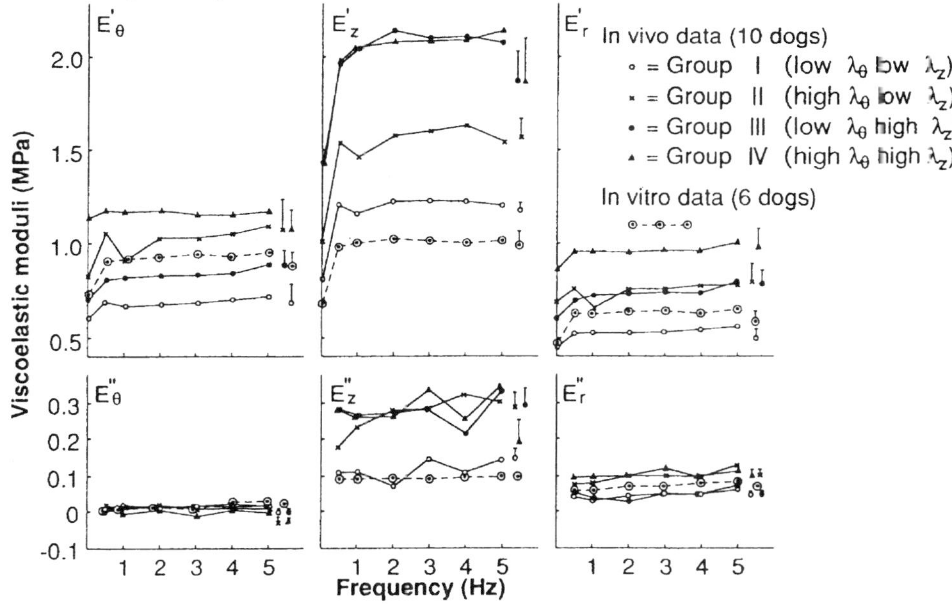

Comments
- The vertical lines to the right represent the average SE for each curve; the symbols at the lower ends of the line segments merely identify the appropriate curves.
- $E_i = E_i' + jE_i''$, where E_i is complex elastic moduli in direction i (= θ, z, r).
- Vasa vasorum intact.

Reference(s)
Patel DJ, Janicki JS, Vaishnav RN, Young JT (1973) Dynamic anisotropic viscoelastic properties of the aorta in living dogs. Circ Res 32:93–107 (with permission)

Anisotropy (16)

• Viscoelasticity • Dynamic Young's modulus	• Rabbits • Thoracic aorta	• Atherosclerosis • Pathological change

Materials
- New Zealand White male rabbits weighing 2.5–3.0 kg
- Thoracic aorta

Testing Methods and Experimental Conditions
- Normal diet or normal diet +1.5% (w/w) cholesterol for 0, 2, 4, 6 weeks
- Aortic wall segments about 9 mm in length and 4 mm wide cut out in both longitudinal and circumferential directions
- Dynamic Young's modulus measurement with a Rheovibron dynamic viscoelastometer at 110 Hz
- Medium: an isotonic saline solution at 37°C

Data

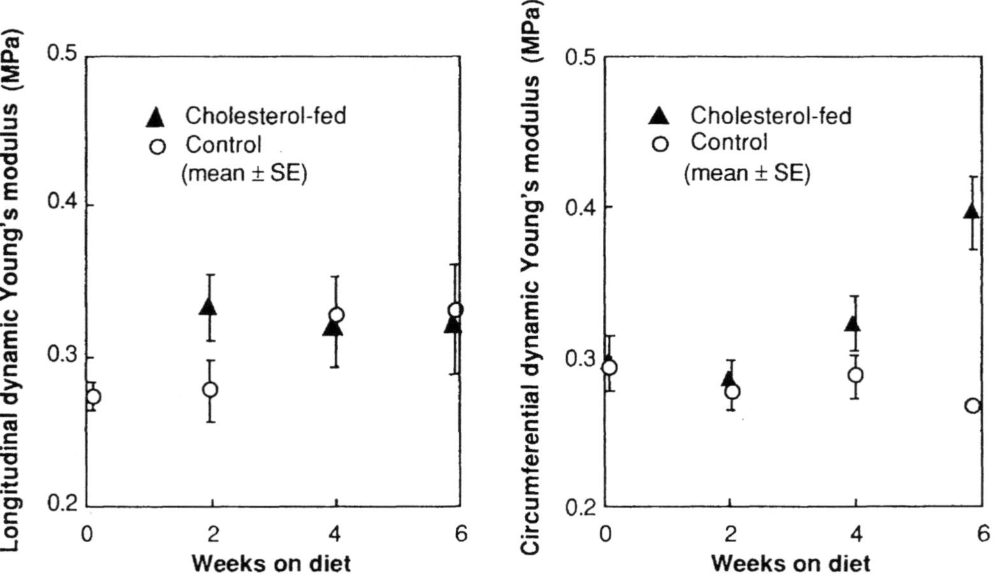

Comments
- Cholesterol feeding had no effect of the longitudinal dynamic Young's modulus.
- Circumferential modulus was very much influenced by the cholesterol diet.

Reference(s)
Pynadath TI, Mukherjee DP (1977) Dynamic mechanical properties of atherosclerotic aorta. Atherosclerosis 26:311–318 (with permission)

Compressibility (1)

• Bulk modulus • Volumetric change	• Dogs • Aorta	• Incompressibility •

Materials
- Mongrel dogs; weight, 27.7–34.3 kg
- Descending thoracic aorta

Testing Methods and Experimental Conditions
- Tubular segment was stretched to a length of 3%–10% greater than the in vivo length and inflated with pressures greater than those in vivo (range 160–200 mmHg) for the measurement of pressure, outer diameter, and the longitudinal force
- Similar deformation in a closed chamber to measure volumetric change of the specimen
- Calculation of the bulk modulus

Data

Parameter	Value
Longitudinal strain $\Delta L/L_0$ (%)	40.0 ± 1.5
Circumferential strain $\Delta R/R_0$ (%)	70.0 ± 2.4
Intraluminal pressure P (kPa)	24.10 ± 0.36
Radial stress T_r (kPa)	-12.10 ± 0.19
Circumferential stress T_θ (kPa)	272.3 ± 9.3
Longitudinal stress T_z (kPa)	269.6 ± 33.5
Hydrostatic stress T_h (kPa)	176.3 ± 10.7
Volumetric strain $\Delta V/V_0$ (%)	0.0602 ± 0.0134
Bulk modulus k (MPa)	435 ± 90

All data are given as mean $\pm$ SE.

Comments
- Definition of parameters:

  Bulk modulus, $k = T_h/(\Delta V/V_o)$; hydrostatic stress, $T_h = (T_r+T_\theta+T_z)/3$; radial stress,

  $T_r = -P/2$; circumferential stress, $T_\theta = PR/H$; longitudinal stress, $T_z = (F+P\pi R^2)/2\pi RH$;
  F, longitudinal force; H, wall thickness; P, pressure; R, middle radius; ΔV, specimen volume change; V_o, original specimen volume.

Reference(s)
Carew TE, Vaishnav RN, Patel DJ (1968) Compressibility of the arterial wall. Circ Res 23:61–68 (with permission)

　　　　　　　　　　　　　2.2 Blood Vessels

Compressibility (2)

• Radial compression • Static stiffness	• Human • Fibrous cap	• Atherosclerosis • Pathological change

Materials
- Human
- Fibrous cap from atherosclerotic plaque of abdominal aorta (at least 9 mm in diameter)

Testing Methods and Experimental Conditions
- Medium: normal saline at room temperature
- Radial compressive stress of 30 mmHg applied to specimen and kept until strain equilibrium was achieved (strain increase in 5 min became less than 2% of total strain)
- The stress was increased to 90 mmHg, and maintained until another strain equilibrium
- Specimens classified by ultrasound imaging as nonfibrous, fibrous, or calcified

Data

Specimen	Creep strain[a] (%)	Radial static stiffness (kPa)
Nonfibrous	23.5 ± 10.5	41.2 ± 18.8
Fibrous	11.4 ± 4.5	81.7 ± 33.2
Calcified	3.1 ± 1.7	345.5 ± 245.4

Values are mean ± SD.

[a]Strain at second equilibrium in reference to the first equilibrium.

Comments
- Static stiffness was calculated as the ratio of increase in equilibrium stress to strain.

Reference(s)
Lee RT, Richardson G, Loree HM, Grodzinsky AJ, Gharib SA, Schoen FJ, Pandian N (1992) Prediction of mechanical properties of human atherosclerotic tissue by high-frequency intravascular ultrasound imaging: An in vitro study. Arterioscler Thromb 12:1–5 (with permission)

Compressibility (3)

• Radial compression • Stress–strain curve	• Rabbits • Thoracic aorta	• Incompressibility • Fluid extrusion

Materials
- Rabbits weighing approximately 2 kg
- Thoracic aorta, longitudinally cut

Testing Methods and Experimental Conditions
- Slab-like specimen sandwiched between transparent glass plates in air
- Radial compressive force-deformation relationship obtained after preconditioning
- Circumferential and longitudinal strains calculated from dimensional change of a mark made at the center of the adventitia surface of the specimen
- Amount of extruded fluid (Vf) under compression was measured to evaluate incompressibility under radial compression

Data

F (g)	T_r (kPa)	h (mm)	V_f/V_o
0	0	0.275 ± 0.008	0
100	5.71 ± 0.31	0.259 ± 0.007	0.0033 ± 0.0002
200	11.56 ± 0.74	0.252 ± 0.008	0.0083 ± 0.0006
300	17.13 ± 0.93	0.248 ± 0.008	0.0140 ± 0.0012
400	22.84 ± 1.24	0.245 ± 0.009	0.0185 ± 0.0012
500	28.55 ± 1.55	0.240 ± 0.009	0.0228 ± 0.0016
600	33.51 ± 2.16	0.236 ± 0.009	0.0255 ± 0.0015

All data are given as mean ± SE, $n = 4$.
V_f/V_o, relative amount of extruded fluid to specimen volume.
Specimen dimension: $l_{\theta o} = 12.141 \pm 0.595$ mm; $l_{zo} = 14.290 \pm 0.641$ mm.

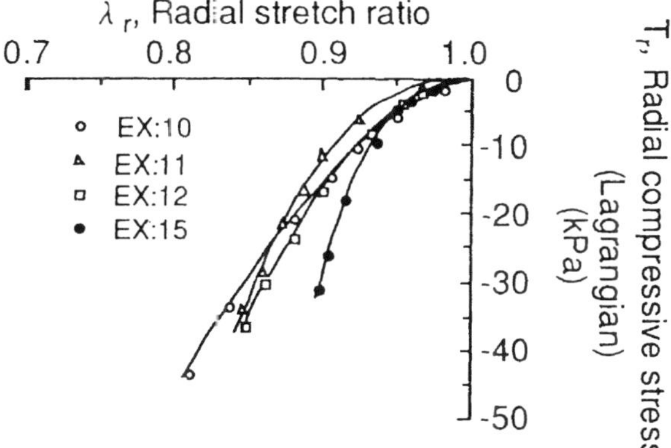

Comments
- The fluid extrusion was in the range of 0.50%–1.26% of the undeformed tissue volume per 10 kPa compressive stress, suggesting the tissue is only slightly compressible (or nearly incompressible).
- Figure shows loading paths from four radial compression experiments on rabbit thoracic arteries.

Reference(s)
Chuong CJ, Fung YC (1984) Compressibility and constitutive equation of arterial wall in radial compression experiments. J Biomech 17:35–40 (with permission)

Creep (1)

• Creep function • Static compliance	• Human • Aorta	• Five component model • Inflation vs deflation

Materials

- Human
- Thoracic and abdominal aortas obtained at autopsy

Testing Methods and Experimental Conditions

- Medium: aerated Tyrode's solution at $37 \pm 1°C$
- Pressure–diameter relation of tubular specimens at in situ length obtained by changing pressure stepwise every 20 mmHg from 20 to 200 mmHg
- A five-component model employed to fit creep responses of cross-sectional area following step change in pressure

Data

Parameter[a]	Thoracic aorta ($n = 35$)	Abdominal aorta ($n = 16$)
α^1	0.076 ± 0.017	0.078 ± 0.017
α^2	0.102 ± 0.028	0.101 ± 0.025
τ^1 (s)	0.73 ± 0.29	0.61 ± 0.12
τ^2 (s)	14.0 ± 4.1	12.1 ± 3.4
$C \left(10^{-3} \dfrac{cm^2}{mmHg} \right)$	5.1 ± 2.5	1.5 ± 0.78
Age	63 ± 14	58 ± 15

All data are given as mean ± SD.

C, aortic compliance per cm length.

[a]Values are at 110 mmHg inflation pressure.

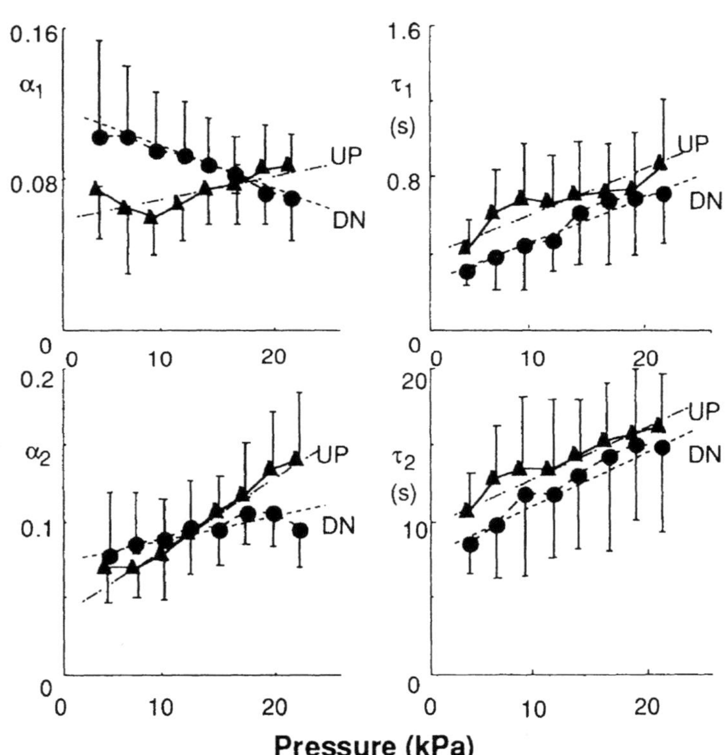

Comments

- Five-component model (C-model):

  $A^c(t) = \Delta A[1 - \alpha_1 exp(-t/\tau_1) - \alpha_2 exp(-t/\tau_2)]$, where ΔA, α_1, α_2, τ_1, τ_2 are constants obtained for each creep curve.
- Comparison of C-model parameters estimated during inflation (UP) and deflation (DN) experiments is shown in the above figure for mean $\pm$ SD of 32 thoracic aortas.

Reference(s)

Langewouters GJ, Wesseling KH, Goedhard WJA (1985) The pressure-dependent dynamic elasticity of 35 thoracic and 16 abdominal human aortas in vitro described by a five-component model. J Biomech 18: 613–620 (with permission)

Creep (2)

• Indentation property • Compliance	• Rabbits • Thoracic aorta	• Local property • Atherosclerosis

Materials
- Four to five-month-old New Zealand White rabbits weighing 2.6–3.5 kg
- Fed 1% cholesterol diet for 2–8 weeks
- Animals killed with an overdose of pentobarbitone
- Approx. 5.5 cm length of descending thoracic aorta

Testing Methods and Experimental Conditions
- Cylindrical segment was slit longitudinally in a line diametrically opposed to the intercostal ostia, opened out flat and stretched in a rack with its intimal surface uppermost to approximate to in vivo dimensions corresponding to a transmural pressure of 100 mmHg
- Local viscoelastic properties characterized with a microindentor consisting of a small spherical probe (25 μm in diameter) which is positioned gently on the intimal surface by preloading it with a force of 2 μN and then loading it with a force equivalent to 1 mg weight (~ 10 μN)
- Local compliance was calculated as the ratio of the resulting displacement of the probe measured 10 s after loading to the force applied to the probe

Data

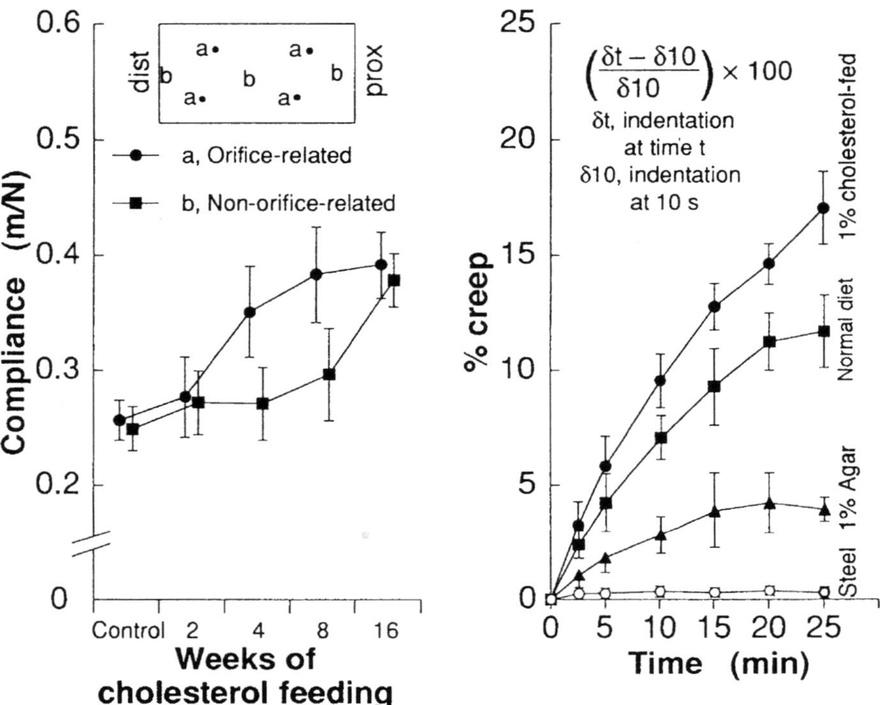

Comments
None.

Reference(s)
Castle WD, Gow BS (1983) Changes in the microindentation properties of aortic intimal surface during cholesterol feeding of rabbits. Atherosclerosis 47:251–261 (with permission)

Elastic Modulus (1)

• Bending test • Young's modulus	• Pigs • Aorta	• Local property • Neutral axis location

Materials

- Pigs
- Thoracic aorta 3 cm below the aortic arch

Testing Methods and Experimental Conditions

- Tubular segments of 4–6 mm in length cut radially to obtain their zero-stress state
- Medium: physiological saline solution at room temperature
- Specimens bent in a plane perpendicular to the vessel axis
- Strain calculated from displacements of markers sprayed on the specimen surface

Data

Parameter	Value
Thickness (mm)	2.44 ± 0.32
c	0.3031 ± 0.0169
E_i (kPa)	43.254 ± 15.826
E_o (kPa)	4.697 ± 1.717

Values are means $\pm$ SD, $n = 4$.
c, normalized position of neutral axis from inner wall;
E_i and E_o, Young's moduli for inner and outer layers,
respectively.

Comments

None.

Reference(s)

Yu Q, Zhou J, Fung YC (1993) Neutral axis location in bending and Young's modulus of
different layers of arterial wall. Am J Physiol 265:H52–H60 (with permission)

Elastic Modulus (2)

• Compliance distribution • Pressure–volume relation	• Human • Coronary circulation	• Compartment analysis •

Materials
- Human
- Coronary circulation

Testing Methods and Experimental Conditions
- A compliance distribution of the coronary circulation was estimated on the basis of a compartment analysis assuming pressure, volume, and distensibility distributions

Data

Compartment	Pressure (mmHg)	Volume (ml/100 gLV)	Distensibility (/kPa)	Compliance (ml/kPa/100 gLV)
Arteries > 200 μm	100	1.6	0.011	0.017
Small arteries	35	2.7	0.034	0.098
Capillaries	20	6.3	0.041	0.263
Small veins	12	5.4[a]	0.060	0.323[a]
Large veins	8	3.2[a]	0.105	0.338[a]
Total		19.2		1.038
Intramyocardial		14.4		0.684

LV, left ventricular.

[a]These data are based on the assumption that coronary venous volume equals twice the coronary arterial volume.

Comments
- For the distensibility of coronary vessels, the data from studies on mesenteric vessels were used.
- For all but the venous division of the circulation, experimental data on volume were available.
- Intramyocardial compliance, i.e., 0.091 ml/mmHg/100 gLV is very close to the value of 0.08 ml/mmHg/100 gLV which was predicted by Kajiya et al.

Reference(s)
1. Spaan JAE (1985) Coronary diastolic pressure-flow relation and zero flow pressure explained on the basis of intramyocardial compliance. Circ Res 56:293–309 (with permission)
2. Kajiya F, Tsujioka K, Goto M, Wada Y, Chen X-L, Nakai M, Tadaoka S, Hiramatsu O, Ogasawara Y, Mito K, Tomonaga G (1986) Functional characteristics of intramyocardial capacitance vessels during diastole in the dog. Circ Res 58:476–485

Elastic Modulus (3)

• Compliance • Static pressure–volume test	• Human • Aorta	• Atherosclerosis • Pathological change

Materials
- Human
- Aorta (between the bulbus aortae and the bifurcation)

Testing Methods and Experimental Conditions
- Quasi-static pressure–volume test at room temperature
- No longitudinal restraint
- Volume increment by 10 ml until pressure reaches 250 mmHg
- Severity of atherosclerosis graded according to WHO recommendation
- Pressure dependence of compliance, C, of the investigated aortae with and without arteriosclerosis of different degree

Data

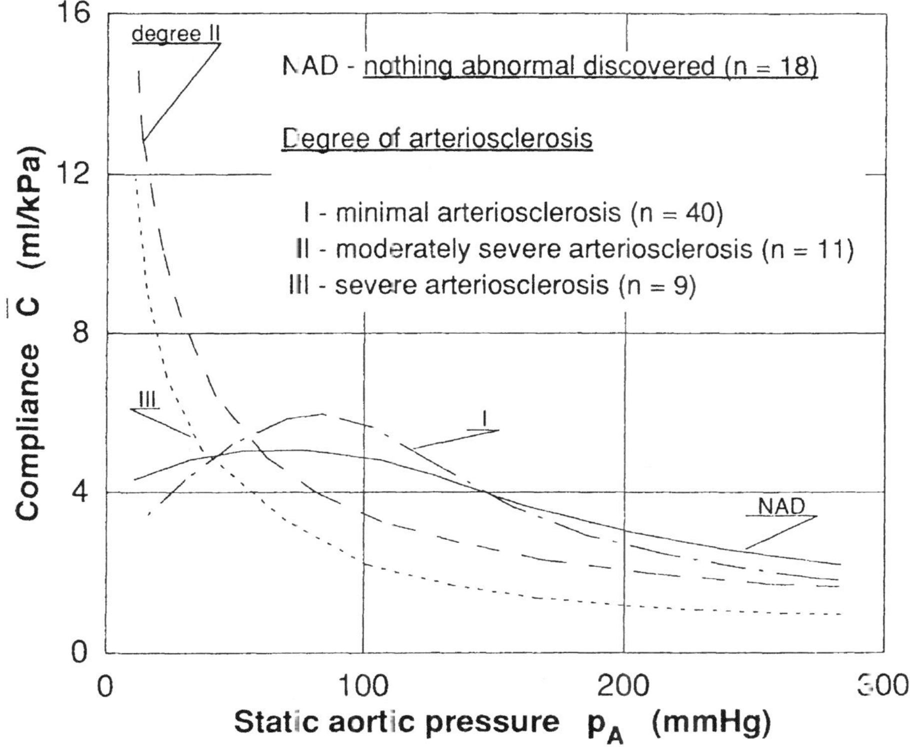

Comments
- Best-fit pressure–volume relationships: $p_A = -107.6 + 150.5V_r - 43.2V_r^2 + 8.0V_r^3$ for NAD; $p_A = -343.3 + 616.9V_r - 331.8V_r^2 + 71.9V_r^3$ for Degree I; $p_A = 13.3V_r^{3.30}$ for Degree II; $p_A = 12.9V_r^{5.86}$ for Degree III. p_A, inner pressure (mmHg); V_r, $(V_o+\Delta V)/V_o$; V_o, aortic volume at $p_A = 5$ mmHg; ΔV, volume increment).
- $C = V_o(dV_r/dp_A)$.

Reference(s)
Richter HA, Mittermayer CH (1984) Volume elasticity, modulus of elasticity and compliance of normal and arteriosclerotic human aorta. Biorheology 21:723–734 (with permission)

Elastic Modulus (4)

• Incremental elastic modulus • Static P–D test	• Cockerels • Aorta	• Atherosclerosis • In situ measurement

Materials

- Cockerels
- Abdominal aorta

Testing Methods and Experimental Conditions

- Diet: an atherogenic crushed-egg diet or control diet
- Feeding period, 14–60 weeks
- Static pressure–diameter relation between 0 and 180 mmHg in 20 mmHg steps
- In situ, post mortem measurement with a radiographic method
- Correlation of distensibility with the extent of atheroma on the intimal surface

Data

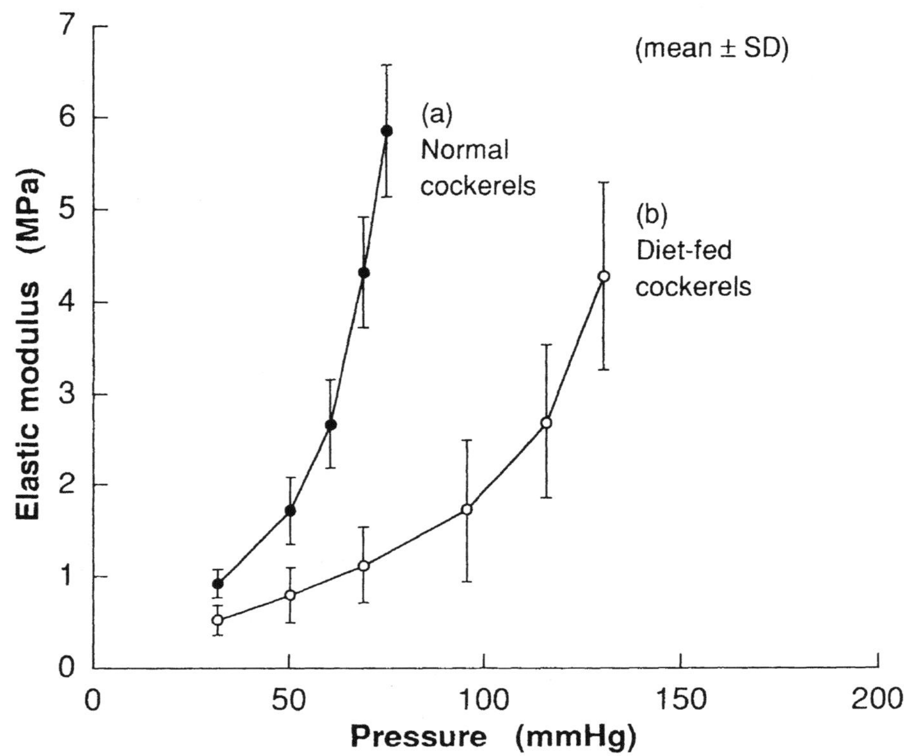

Comments

- Aortic distensibility increased once and then decreased with the progression of atherosclerosis.

Reference(s)

Newman DL, Gosling RG, Bowden NLR (1971) Changes in aortic distensibility and area ratio with the development of atherosclerosis. Atherosclerosis 14:231–240 (with permission)

Elastic Modulus (5)

• Incremental elastic modulus • Static P–D test	• Dogs • Aorta, femoral, carotid	• Classics • Regional variation

Materials
- Dogs
- Thoracic and abdominal aortas, femoral and carotid arteries

Testing Methods and Experimental Conditions
- Static pressure–diameter relation between 0 and 240 mmHg
- Pressure increased at each step of 20 mmHg in 2 min interval
- Specimens stretched to their in vivo length
- Medium, isotonic saline at room temperature

Data

Pressure (mmHg)	Static incremental modulus of elasticity (MPa)			
	Thoracic aorta	Abdominal aorta	Femoral artery	Carotid artery
40	0.12 ± 0.01 (6)	0.16 ± 0.04 (4)	0.12 ± 0.02 (6)	0.10 ± 0.02 (7)
100	0.43 ± 0.04 (12)	0.89 ± 0.35 (8)	0.69 ± 0.10 (9)	0.64 ± 0.10 (12)
160	0.99 ± 0.05 (6)	1.24 ± 0.22 (4)	1.21 ± 0.24 (6)	1.22 ± 0.27 (7)
220	1.81 ± 0.28 (5)	1.30 ± 0.55 (3)	2.04 ± 0.44 (6)	1.22 ± 0.15 (7)

All data are mean ± SE (n).

Comments
- Some additional specimens were studied at 100 mmHg before making dynamic measurements, and these have been included.

Reference(s)
Bergel DH (1961) The static elastic properties of the arterial wall. J Physiol 156:445–457 (with permission)

Elastic Modulus (6)

• Incremental elastic modulus • Static P–D test	• Human • Aorta, carotid, iliac, femoral	• Age-related variation • Regional difference

Materials
- Human
- Thoracic and abdominal aortas, and carotid, iliac, and femoral arteries

Testing Methods and Experimental Conditions
- Stored frozen in physiological saline after excision at postmortem until use
- Grossly atheromatous vessels were discarded
- Static pressure–diameter measurement at in vivo length
- Pressure increased at each step of 10–20 mmHg in 2 min interval
- Medium: physiological saline at room temperature

Data

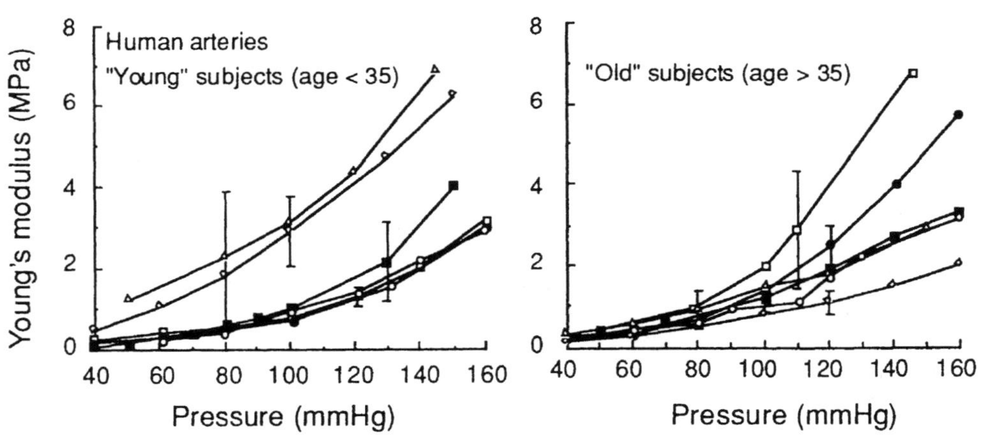

Comments
- Incremental elastic modulus, E_{inc}:

$$E_{inc} = 1.5R_i^2 R_o/(R_o^2 - R_i^2) \cdot \Delta P_i / \Delta R_o$$

    R_i, inner radius; R_o, outer radius; ΔR_o, radius increment over inner pressure increment of ΔP_i.

Reference(s)
Learoyd BM, Taylor MG (1966) Alterations with age in the viscoelastic properties of human arterial walls. Circ Res 18:278–292 (with permission)

Elastic Modulus (7)

• Incremental elastic modulus • Static P–D test	• Human • Cerebral arteries	• Atherosclerosis • Pathological change

Materials
- Human
- Twenty anterior cerebral arteries (ACA) and 19 internal carotid arteries (ICA)

Testing Methods and Experimental Conditions
- Specimen excised from cadavers not later than 2 days after death
- Medium: aerated Krebs–Ringer solution at 37°C
- ICA stretched to their in vivo length, ACA not stretched axially (assuming no axial stretch in vivo)
- Quasi-static pressure–diameter test between 5 and 250 mmHg ($\approx$ 30 s in a cycle)
- Specimen classified into fibrosclerotic and normal groups by visual inspection of histological sections

Data

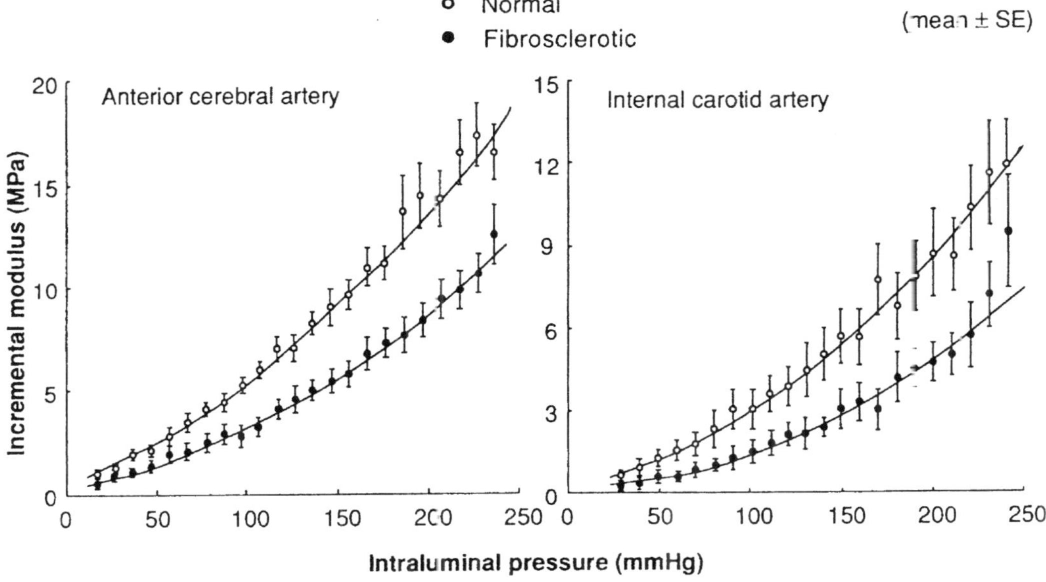

Comments
- Average age in years of the normal and fibrosclerotic groups were 73 ± 15 (mean $\pm$ SD, $n = 7$) and 72 ± 7 ($n = 13$) for ACA, and 51 ± 13 ($n = 8$) and 73 ± 7 ($n = 11$) for ICA.
- Incremental elastic modulus, E:

$$E = 2R_i^2 R_o/(R_o^2 - R_i^2)\cdot\Delta P/\Delta R_o \quad (R_i, \text{ inner radius; } R_o, \text{ outer radius; } P, \text{ inner pressure}).$$

Reference(s)
Hudetz AG, Mark G, Kovach AGB, Kerenyi T, Fody L, Monos E (1981) Biomechanical properties of normal and fibrosclerotic human cerebral arteries. Atherosclerosis 39:353–365 (with permission)

Elastic Modulus (8)

• Incremental elastic modulus • Static P–D test	• Human • Cerebral arteries	• Age-related variation • Smooth muscle activation

Materials
- Autopsy subjects who had neither systemic nor cerebral vascular lesions
- Intracranial (basilar and vertebral) and extracranial (common carotid, internal carotid, and vertebral) arteries

Testing Methods and Experimental Conditions
- Quasi-static pressure–diameter test
- In aerated Krebs–Ringer solution at 37°C
- Response to potassium or norepinephrine
- Biochemical analysis for collagen and elastin contents

Data

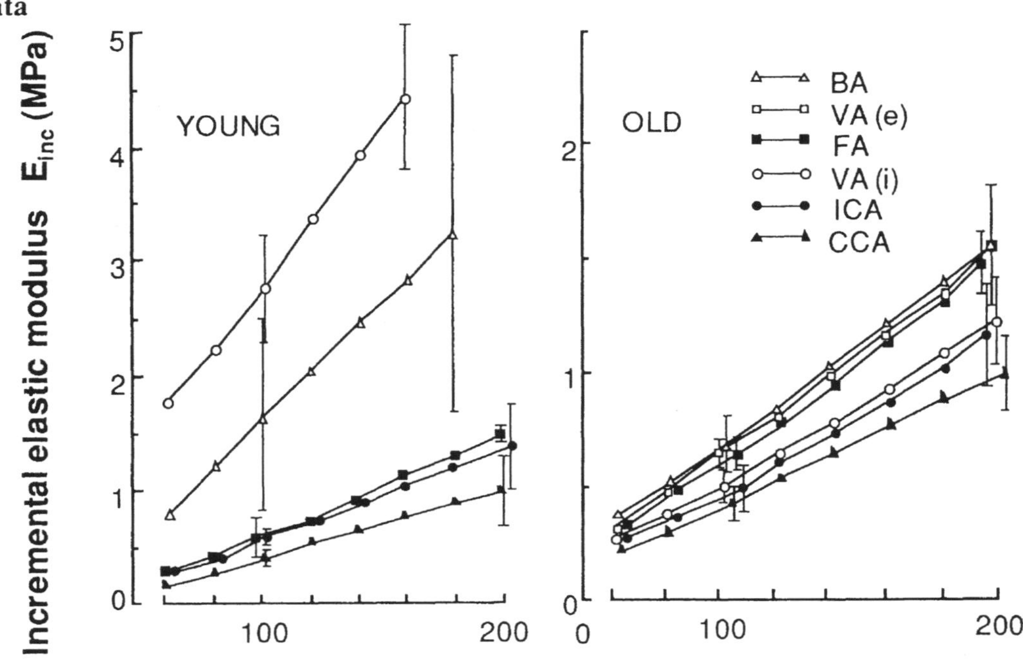

Comments
- Incremental elastic modulus, E_{inc}:

$$E_{inc} = 1.5R_i^2R_o/(R_o^2-R_i^2) \cdot \Delta P_i/\Delta R_o$$

  R_i, inner radius; R_o, outer radius; ΔR_o, radius increment over inner pressure increment of ΔP_i.
- Legends: BA, basilar; CCA, common carotid; FA, femoral; ICA, internal carotid; VA(e), extracranial vertebral; VA(i), intracranial vertebral.

Reference(s)
Nagasawa S, Handa H, Okumura A, Naruo Y, Moritake K, Hayashi K (1979) Mechanical properties of human cerebral arteries. Part I: Effects of age and vascular smooth muscle activation. Surg Neurol 12:297–304 (with permission)

Elastic Modulus (9)

• Incremental elastic modulus • Stress–strain curve	• Rats, rabbits, dogs • Carotid artery	• Species variation • Smooth muscle activation

Materials
- Wistar rats (age, 12 weeks), New Zealand White rabbits (age, ≈ 6 months), and healthy adult mongrel dogs (age, unknown)
- Carotid artery

Testing Methods and Experimental Conditions
- Tubular specimen (in vivo length) in an aerated physiological salt solution (PSS) at 37°C + norepinephrine (5 µg/ml) to activate smooth muscle
- Inflation at a rate of 0.2 mmHg/s (between 0 and 250 mmHg)
- Medium changed to a calcium-free physiologic salt solution with 2 mM EGTA at 37°C
- Quasi-static pressure–diameter relation at inflation rate of 1 mmHg/s between 0 and 250 mmHg

Data

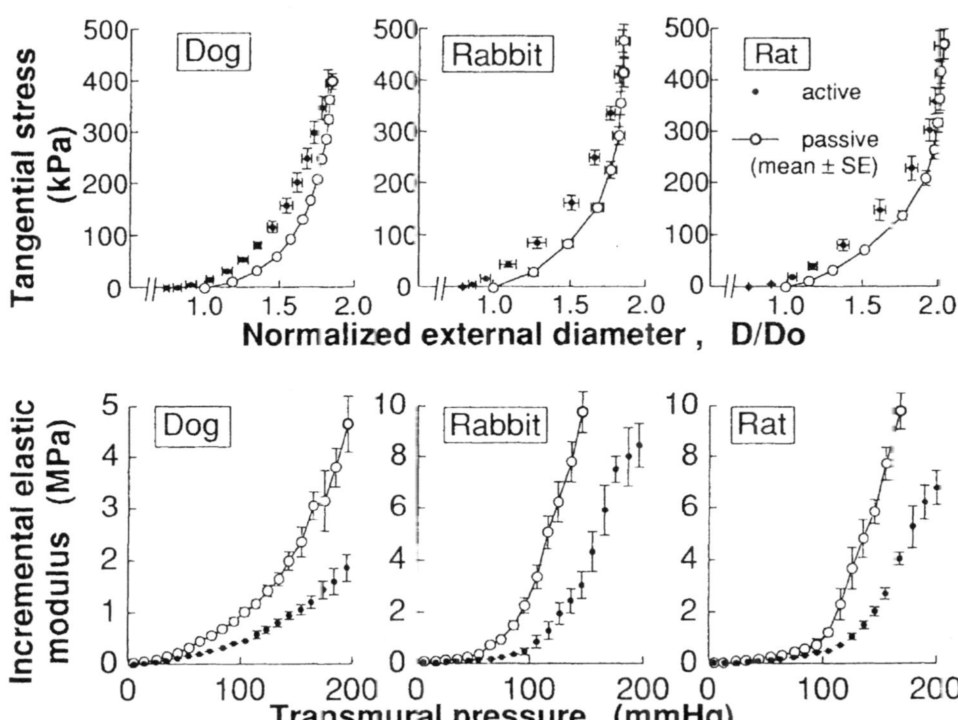

Comments
- Compliance: rat > rabbit > dog.
- Collagen/elastin ratio: rat < rabbit < dog.
- Incremental elastic modulus, E:

$$E = \{2a^2b/(b^2 - a^2)\}\{\Delta P/\Delta b\}.$$

P, transmural pressure; a and b, internal and external radii, respectively.

Reference(s)
Cox RH (1978) Comparison of carotid artery mechanics in the rat, rabbit, and dog. Am J Physiol 234:H280–H288 (with permission)

Elastic Modulus (10)

• Incremental elastic modulus • Tensile test	• Human • Aorta	• Atherosclerosis • Pathological change

Materials
- Human autopsies (21 patients)
- Age, 68 ± 12 years
- Atherosclerotic intimal plaques of thoracic and abdominal aortas
- Kept at 4°C in physiological saline solution until use, within 36 h of death

Testing Methods and Experimental Conditions
- Tensile test of dumbbell-shape specimen (circumferentially oriented)
- Room temperature
- Plaques were classified by histological examination as
  (a) cellular, (b) hypocellular, (c) calcified

Data

	Average stiffness for histologic class (mean ± SD)		
Strain (%)	Cellular ($n = 12$) (kPa)	Hypocellular ($n = 9$) (kPa)	Calcified ($n = 5$) (kPa)
1	183 ± 248	1740 ± 2630	871 ± 1720
4	594 ± 656	4110 ± 5140	2860 ± 5500
8	1680 ± 1870	10,700 ± 13,800	7220 ± 13,300
12	3380 ± 3800	21,100 ± 28,600	13,500 ± 24,300
16	5690 ± 6450	35,400 ± 49,600	21,800 ± 38,500
20	8620 ± 9820	53,600 ± 76,800	32,000 ± 55,900

Comments
- Static circumferential modulus was calculated as the slope of the stress–strain curve.
- Stress–strain relationship was expressed by:

$$\sigma = A\varepsilon + B\varepsilon^2 + C\varepsilon^3 \quad (A, B, C; \text{constants}).$$

Reference(s)
Loree HM, Grodzinsky AJ, Park SY, Gibson LJ, Lee RT (1994) Static circumferential tangential modulus of human atherosclerotic tissue. J Biomech 27:195–204 (with permission)

Elastic Modulus (11)

• Incremental elastic modulus • Vessel dimension	• Cats • Capillary and venule	• In vivo measurement •

Materials
- Cats of either sex; weight, 1.2–2.5 kg
- Capillary and venule in mesentry

Testing Methods and Experimental Conditions
- Laparotomy under anesthesia with pentobarbitone (up to 35 mg/kg) + heparinization
- Mesentry placed on microscope stage, irrigated with an aerated Krebs solution heated to 37°C
- Pressure–diameter relation measured proximally to a position where the vessel was occluded temporarily
- Pressure measured through a glass micropipette with a resistance null-balance servo system
- Diameter measured on photographs of vessels taken through microscope

Data

	Capillaries ($n = 9$)		Venules ($n = 7$)	
	Whole wall	Whole mesentery	Whole wall	Whole mesentery
r (μm)	3.97 ± 0.09		8.1 ± 0.59	
ΔP (Pa)	428 ± 75		304 ± 24.7	
Δr (μm)	0.03 ± 0.007		0.028 ± 0.003	
$E_{inc}x$ (Pa·m)	0.296 ± 0.067		0.747 ± 0.102	
x (μm)	0.5	31	1.5	27
E_{inc} (kPa)	370	10.4	388	32.2

Data are given as mean ± SE.
r, vessel radius; ΔP, pulse pressure; Δr, change in vessel radius calculated from equation $\Delta r = r(\Delta l - \Delta l')/2l$; Δl, observed red cell ocilation; $\Delta l'$, distance moved by red cell during single pulse as a consequence of filtration; E_{inc}, Young's modulus; x, wall thickness.

Comments
- $E_{inc}x$ is calculated from the equation $\Delta Pr^2/\Delta r$.
- Present elastic modulus was compared with those reported by others:
  Cat omental capillaries (Intaglietta et al., 1971)   150 kPa (low strain).
  Frog mesenteric arterioles (Wiederhielm, 1965)   120 kPa (low strain),
  4500 kPa (high strain).
  Frog whole mesentery (Fung et al., 1966)   90 kPa (low strain),
  1800 kPa (high strain).

Reference(s)
1. Smaje LH, Fraser PA, Clough G (1980) The distensibility of single capillaries and venules in the cat mesentery. Microvasc Res 20:358–370 (with permission)
2. Intaglietta M, Richardson DR, Tompkins WR (1971) Blood pressure, flow, and elastic properties in microvessels of cat omentum. Am J Physiol 221:922–928
3. Wiederhielm CA (1965) Distensibility characteristics of small blood vessels. Fed Proc 24:1075–1084
4. Fung YC, Zweifach BW, Intaglietta M (1966) Elastic environment of the capillary bed. Circ Res 19:441–461

Elastic Modulus (12)

• Incremental elastic modulus • Vessel dimension	• Human • Radial artery	• Non-invasive measurement • Hypertension

Materials

- Normotensive subjects and untreated patients with mild or moderate essential hypertension
- Radial artery at the wrist

Testing Methods and Experimental Conditions

- Subject in the supine position after at least 20 min rest
- Inner diameter and posterior wall intima-media thickness measured at 2 cm from the wrist with a high-resolution, pulse–echo tracking device
- Blood pressure measurement at brachial artery with a sphygmomanometer and at finger artery with a plethysmograph

Data

Parameters	Control normotensive (n = 22)	Hypertensive (n = 25)
Age (years)	44 ± 11	48 ± 12
Systolic blood pressure[a] (mmHg)	128 ± 21	165 ± 25*
Diastolic blood pressure[a] (mmHg)	71 ± 13	96 ± 24*
Mean blood pressure[a] (mmHg)	90 ± 15	121 ± 24*
Internal diastolic diameter at MAP (mm)	2.53 ± 0.32	2.50 ± 0.56
Intima-media thickness at MAP (mm)	0.28 ± 0.05	0.40 ± 0.063**
Elastic modulus at MAP (MPa)	2.68 ± 1.81	2.25 ± 2.14**

Values are mean ± SD.

$*P < 0.0001; **P < 0.001.$

[a]Brachial artery blood pressure was measured with a sphygmomanometer.

MAP, mean arterial pressure.

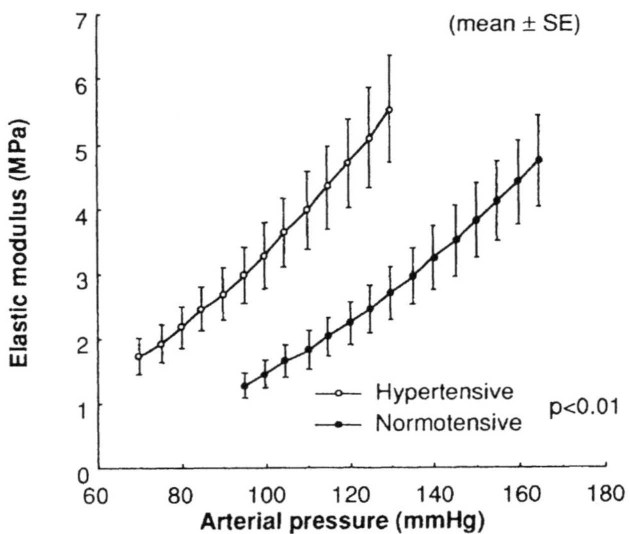

Comments

None.

Reference(s)

Laurent S, Girerd X, Mourad JJ, Lacolley P, Beck L, Boutouyrie P, Mignot JP, Safar M (1994) Elastic modulus of the radial artery wall material is not increased in patients with essential hypertension. Arterioscler Thromb 14:1223–1231 (with permission)

Elastic Modulus (13)

• Incremental elastic modulus • Vessel dimension	• Rabbits • Aorta	• Hypertension •

Materials

- Normal and hypertensive rabbits of both sexes weighing an average of 3.2 kg
- Circumferential strips of ascending aorta and aortic arch (width, 1.3 mm)

Testing Methods and Experimental Conditions

- Hypertension induced by wrapping one kidney with silk + contralateral nephrectomy
- Uniaxial tensile test in aerated physiological salt solution at 38°C

Data

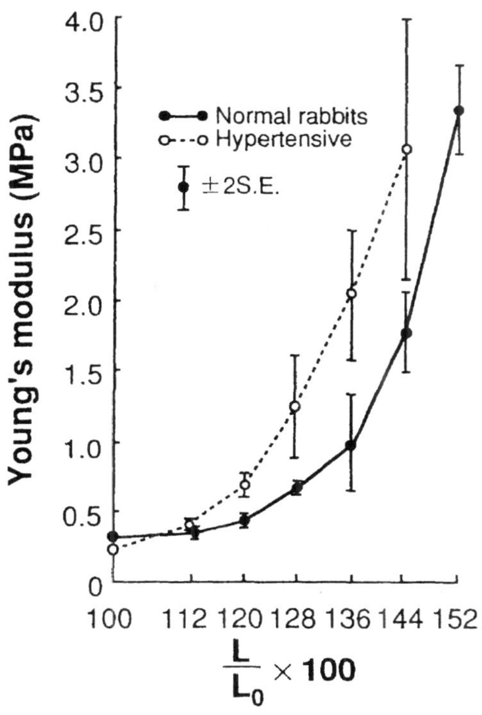

$$\frac{L}{L_0} \times 100$$

Comments

- Mean arterial pressure AP (mmHg) and strip dimensions (mm):

	AP	Position	n	Thickness	Length at load of 3 g
Normal	200/159	ASC	6	0.36	18.6
		ARC	5	0.26	17.3
Hypertensive	132/100	ASC	7	0.50	20.1
		ARC	6	0.40	18.1

ASC, ascending thoracic aorta; ARC, aortic arch.

Reference(s)

Aars H (1968) Static load–length characteristics of aortic strips from hypertensive rabbits. Acta Physiol Scand 73:101–110 (with permission)

Elastic Modulus (14)

• Incremental elastic modulus • Vessel dimension	• Sheep • Thoracic aorta	• Unanesthetized • Age-related change

Materials
- Near-term fetal lamb, newborn (< 1 week) lamb, and adult sheep
- Proximal third of the descending thoracic aorta

Testing Methods and Experimental Conditions
- Animals previously instrumented with pressure gauges and ultrasonic dimension crystals for measurements of internal pressure and external diameter
- Experiment performed when the animals had fully recovered from surgery (1–3 weeks for adults, 2–5 days for lambs)
- Experiments on the fetus performed 24–48 h after surgery, when the ewes were fully awake

Data

Parameter	Adult ($n = 9$)	Newborn ($n = 7$)	Fetus ($n = 5$)
Mean aortic pressure (mmHg)	83 ± 2 *,**,#	70 ± 5 *	50 ± 2
Mean external aortic diameter (mm)	21.07 ± 0.69 *,**,#	9.55 ± 0.56	9.58 ± 0.80
Wall thickness (mm)	1.56 ± 0.07 *,#	1.25 ± 0.16	0.88 ± 0.06
Wall thickness/external radius	0.14 ± 0.11 **,#	0.26 ± 0.02	0.19 ± 0.03
Heart rate (beats/min)	88 ± 3 *,**,#	123 ± 7 *	171 ± 5
Midwall stress (kPa)	69.0 ± 5.9 *,**,#	33.6 ± 2.7	36.2 ± 5.6
Elastic modulus (kPa)	367 ± 50 #	242 ± 38	258 ± 58

All data are given as mean ± SE.

*$P < 0.05$ vs fetus; **$P < 0.05$ vs newborn; #$P < 0.05$ vs newborn and fetus.

Comments
- Elastic modulus:

  $E_{inc} = 0.75Rd\sigma/dR$;

  $\sigma = 2P(ab/R^2)/(b^2 - a^2)$;

  P, the aortic pressure; a and b, the inner and outer aortic radii; R=(a+b)/2.

Reference(s)
Pagani M, Mirsky I, Baig H, Manders WT, Kerkhof P, Vatner SF (1979) Effects of age on aortic pressure–diameter and elastic stiffness–stress relationships in unanesthetized sheep. Circ Res 44:420–429 (with permission)

Elastic Modulus (15)

• Pressure–strain elastic modulus • Diametrical change	• Human • Abdominal aorta	• Non-invasive measurement • Ultrasonic imaging

Materials
- Volunteers clinically free from hypertension, arteriosclerosis obliterans, cardiac disease, diabetes mellitus, and hyperlipidemia
- Abdominal aorta

Testing Methods and Experimental Conditions
- Each subject rested supine for 10 min before the measurement
- Pulsatile wall motion of the abdominal aorta observed at 3–5 cm distal to the branching site of the superior mesenteric artery
- Pulsatile diameter change measured with a new echo tracking device linked to real time ultrasonic B mode equipment
- Brachial artery pressure measured by the auscultatory method

Data

Group	1 (young adults)	2 (middle-aged subjects)	3 (elderly subjects)
n	20	21	20
Age (years)	24.6 ± 3.3	46.0 ± 7.0	75.5 ± 8.6
Med P (mmHg)	82.8 ± 10.4	95.6 ± 10.8	103.3 ± 13.6
ΔP (mmHg)	53.0 ± 13.9	49.5 ± 16.9	73.0 ± 19.8
D (mm)	16.7 ± 3.6*,#	17.0 ± 1.9 #	15.3 ± 3.1
ΔD (mm)	1.22 ± 0.31**,##	0.80 ± 0.31 ##	0.45 ± 0.15
Strain (ΔD/D)	0.076 ± 0.024**,##	0.048 ± 0.024 ##	0.030 ± 0.010
E_p (kPa)	99 ± 34*,##	155 ± 68 ##	380 ± 205

All data are given as mean $\pm$ SD.
Med P, mediated blood pressure {(systolic + diastolic blood pressure)/2}; ΔP, pulse pressure; D, mean diameter of the abdominal aorta; ΔD, pulsatile diameter change in the abdominal aorta; E_p, pressure–strain elastic modulus.
*NS *vs* group 2; **$P < 0.001$ *vs* group 2; #NS *vs* group 3; ##$P < 0.001$ *vs* group 3.

Comments
None.

Reference(s)
Imura T, Yamamoto K, Kanamori K, Mikami T, Yasuda H (1986) Non-invasive ultrasonic measurement of the elastic properties of the human abdominal aorta. Cardiovasc Res 20:208–214 (with permission)

Elastic Modulus (16)

• Pressure–strain elastic modulus • Diametrical change	• Human • Pulmonary artery	• In vivo measurement • Pulmonary hypertension

Materials
- Human
- Main pulmonary artery

Testing Methods and Experimental Conditions
- Pressure–diameter relation measured on the main pulmonary artery of patients undergoing open-heart surgery
- Diameter measurement with a recording caliper sutured to the vessel wall
- Intravascular pressure measured with a 20 G needle connected directly to pressure transducer

Data

Parameter	Normal	Pulmonary hypertension
n	8	3
Age (years)	41.4 ± 41.4	39.3 ± 9.0
Heart rate (beat/min)	81.3 ± 3.0	82.7 ± 3.9
Systolic blood pressure (mmHg)	20.9 ± 1.2	64.3 ± 8.1
Diastolic blood pressure (mmHg)	8.1 ± 1.1	29.2 ± 2.6
ΔR (mm)	1.59 ± 0.26	0.53 ± 0.07
E_p (kPa)	15.7 ± 1.0	199.3 ± 59.2

All data are given as mean $\pm$ SE.
ΔR, the value of the greatest change in radius; E_p, pressure–strain elastic modulus.

Comments
- $E_p = \Delta P R_0 / \Delta R$.

ΔP, the value of the greatest change in pulse pressure; R_0, the diastolic radius.

Reference(s)
Greenfield JC, Griggs DM (1963) Relation between pressure and diameter in main pulmonary artery of man. J Appl Physiol 18:557–559 (with permission)

Elastic Modulus (17)

• Pressure–strain elastic modulus • Diametrical change	• Monkeys • Carotid artery	• Atherosclerosis • In vivo measurement

Materials
- Adult male *Macaca fascicularis* monkeys
- Common carotid artery

Testing Methods and Experimental Conditions
- Diet: a high-cholesterol diet (1 mg cholesterol/kcal) with 45% of the calories from lard, or a nonatherogenic diet composed of commercially available monkey chow
- Anesthesia with ketamine hydrochloride (10 mg/kg) maintained by sodium pentobarbital
- Aortic pressure measurement at aortic arch with a Millar catheter tip transducer
- Diameter measurement with a 5 MHz linear array ultrasonic imaging system

Data

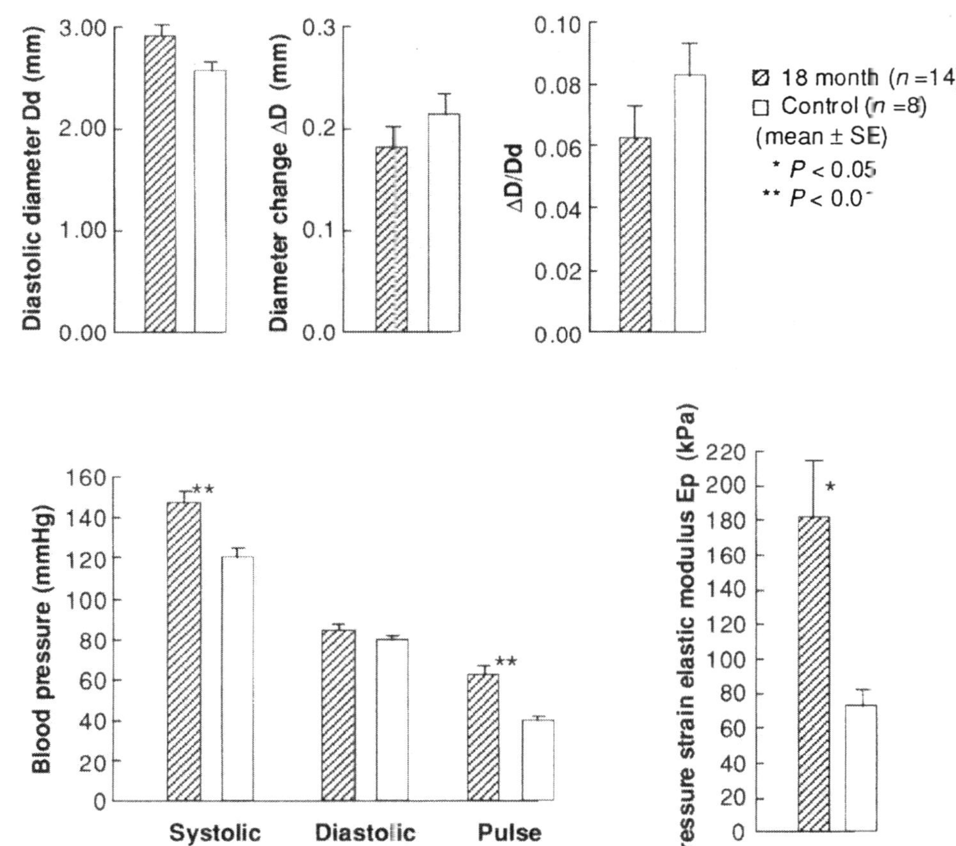

Comments
None.

Reference(s)
Farrar DJ, Riley WA, Bond MG, Barnes RN, Love LA (1982) Detection of early atherosclerosis in *M. fascicularis* with transcutaneous ultrasonic measurement of the elastic properties of the common carotid artery. Texas Heart Inst J 9:335–343 (with permission)

Elastic Modulus (18)

• Pressure–strain elastic modulus • Incremental elastic modulus	• Rats • Aorta	• Hypertension • Temporal change

Materials
- Male Wistar rats (age, 10, 12, 16, 24 weeks)
- Thoracic aorta

Testing Methods and Experimental Conditions
- Static pressure–diameter test in physiological saline solution at 37°C
- Hypertension (systolic pressure $\geq$ 160 mmHg) induced by constricting left renal arteries of animals aged 8–9 weeks (Goldblatt hypertension)
- Comparison with age-matched control
- Mechanical parameters were calculated from mean diameters of loading and unloading curves

Data

Weeks after operation	2	4	8	16
Control				
No. of animals	5	5	7	7
P_{sys} (mmHg)	134 ± 3	134 ± 4	129 ± 3	137 ± 3
E_p (kPa)				
at 100 mmHg	0.51 ± 0.03	0.45 ± 0.04	0.44 ± 0.01	0.43 ± 0.01
at P_{sys}	1.86 ± 0.31	0.45 ± 0.04	0.44 ± 0.01	0.43 ± 0.01
at 200 mmHg	7.10 ± 0.77	6.97 ± 0.69	7.37 ± 0.81	6.76 ± 0.90
$H_{\theta\theta}$ (kPa)				
at 100 mmHg	8.9 ± 0.5	8.2 ± 0.8	9.2 ± 0.5	7.8 ± 0.8
at P_{sys}	31.5 ± 5.1	28.9 ± 5.2	26.3 ± 3.3	25.3 ± 4.8
at 200 mmHg	117.8 ± 12.0	121.0 ± 12.5	147.5 ± 16.2	123.8 ± 20.0
Hypertensive ($P_{sys} \geq$ 160 mmHg)				
No. of animals	8	8	8	5
P_{sys} (mmHg)	199 ± 7	190 ± 9	206 ± 8	192 ± 9
E_p (kPa)				
at 100 mmHg	0.64 ± 0.04	0.52 ± 0.03	0.57 ± 0.05	0.53 ± 0.04
at P_{sys}	5.47 ± 0.84	3.55 ± 0.70	4.69 ± 0.64	2.98 ± 0.37
at 200 mmHg	5.32 ± 0.57	4.12 ± 0.35	4.91 ± 0.88	4.16 ± 1.06
$H_{\theta\theta}$ (kPa)				
at 100 mmHg	7.7 ± 0.8	6.8 ± 0.4	5.7 ± 0.4	5.5 ± 0.8
at P_{sys}	68.2 ± 9.2	52.8 ± 10.0	54.0 ± 6.3	36.3 ± 5.2
at 200 mmHg	67.3 ± 6.9	63.8 ± 6.3	57.7 ± 11.7	51.7 ± 15.7

Values are expressed as mean $\pm$ SE.

E_p, pressure–strain elastic modulus; P_{sys}, systolic blood pressure before death; $H_{\theta\theta}$, incremental elastic modulus.

Comments
- Incremental elastic modulus, $H_{\theta\theta} = 2D_o(\Delta P_i D_i^2/\Delta D_o + P_i D_o)/(D_o^2 - D_i^2)$; D_i and D_o, inner and outer diameter, respectively; P_i, inner pressure; ΔD_o, diameter increment for pressure increment ΔP_i.

Reference(s)
Matsumoto T, Hayashi K (1994) Mechanical and dimensional adaptation of rat aorta to hypertension. J Biomech Eng 116:278–283

Elastic Modulus (19)

• Pressure–strain elastic modulus • Vessel Dimension	• Dogs • Femoral artery	• Hypertension • In vivo measurement

Materials

- Mongrel dogs over 12 kg
- Femoral artery

Testing Methods and Experimental Conditions

- Control measurement of pressure and diameter on one femoral artery under local anesthesia of the skin and subcutaneous tissue (made possible by training of dogs)
- Hypertension induced with wrapping both kidneys
- Measurement on the contralateral vessel after an elevation of mean blood pressure of at least 30 mmHg for at least 4 weeks

Data

Parameter	Normotensive	Hypertensive
n	17–20	7–8
Mean blood pressure (mmHg)	114.8	172.0
E_p (kPa)	177.7 ± 13.6	357.7 ± 33.3
r/d	6.4 ± 0.31	7.1 ± 0.47

All data are given as mean $\pm$ SE.

E_p, pressure–strain elastic modulus; r, vessel radius; d, vessel wall thickness.

Comments

- $E_p = \Delta P / \Delta \varepsilon$.

  ΔP, the pressure change between diastole and systole; $\Delta \varepsilon$, the change in strain during the same period estimated in the form $\varepsilon = (r - r_0)/r_0$; r_0, radius at zero strain.

Reference(s)

Feigl EO, Peterson LH, Jones AW (1963) Mechanical and chemical properties of arteries in experimental hypertension. J Clin Invest 42:1640–1647 (with permission)

Elastic Modulus (20)

• Pressure–strain elastic modulus • Vessel dimension	• Dogs • Aorta	• In vivo measurement • Regional variation

Materials

- Dogs with an average weight of 23.3 kg (range 17.3–35.5 kg)
- Aorta, external iliac artery, brachiocephalic artery

Testing Methods and Experimental Conditions

- Thoracotomy under chloralose–urethane and morphine anesthesia
- Diameter measurement with an electrical recording caliper sewn on a diameter of the vessel
- Intravascular pressure was measured with a polyethylene catheter connected to a Statham B23Db transducer

Data

Position		$\Delta R/\Delta P \times 10^4$ (cm/mmHg)	Mean radius R (cm)	E_p (kPa)
Ascending aorta		20.01 ± 2.07	1.129 ± 0.040	78.4 (6)
Descending thoracic aorta	High	12.82 ± 1.37	0.916 ± 0.046	79.7 (5)
	Middle	7.60 ± 0.41	0.740 ± 0.033	120.2 (6)
	Low	4.09 ± 0.90	0.687 ± 0.034	156.6 (5)
Abdominal aorta	High	2.87 ± 0.63	0.504 ± 0.055	200.0 (6)
	Low	1.71 ± 0.26	0.488 ± 0.033	353.0 (8)
External iliac artery		0.37 ± 0.14	0.265 ± 0.040	961.9 (3)
Brachiocephalic artery		2.84 ± 0.38	0.608 ± 0.042	283.0 (6)

Average value of pulse pressure, ΔP, encountered in 11 dogs was 28 mmHg ($\pm$ 0.7 SE).
$\Delta R/\Delta P$ and mean radius are given as mean $\pm$ SE, $n = 11$.
Numbers in parentheses indicate number of observations for a particular site.

Comments

- $E_p = \Delta P \times R / \Delta R$.

Reference(s)

Patel DJ, DeFreitas F, Greenfield JC, Fry DL (1963) Relationship of radius to pressure along the aorta in living dogs. J Appl Physiol 18:1111–1117 (with permission)

Elastic Modulus (21)

• Stiffness parameter • Incremental elastic modulus	• Human • Cerebral arteries	• Regional variation • Connective tissue content

Materials
- Eighteen autopsy subjects who had neither systemic nor cerebral vascular lesions
- Intracranial artery (basilar and vertebral artery)
- Extracranial artery (common carotid, internal carotid, and vertebral artery)
- Young (< 45), and old (> 45) groups

Testing Methods and Experimental Conditions
- Quasi–static pressure–diameter relation
- Medium: Krebs–Ringer solution at 37°C
- Biochemical analysis for connective tissue contents

Data

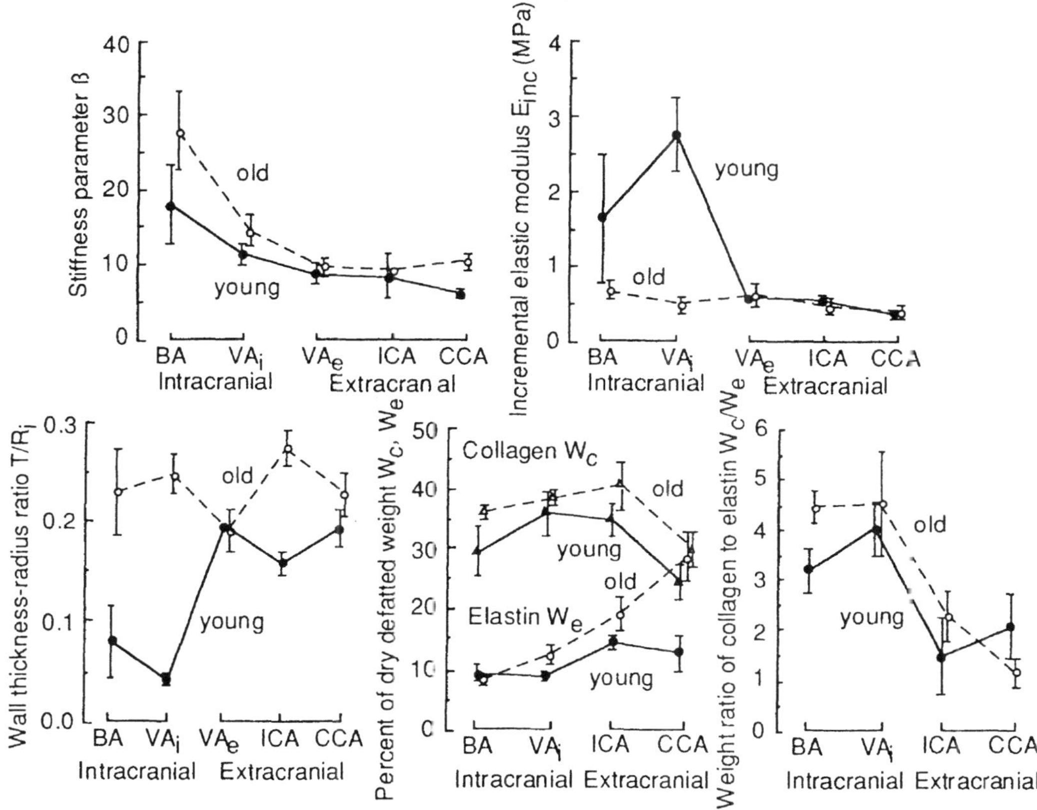

Comments

- Stiffness parameter, β: $\ln (P/P_s) = \beta\, (D/D_s - 1)$.

  P_s, standard pressure (=100 mmHg for this study); D_s, wall diameter at P_s.

- Incremental elastic modulus, E_{inc}: $E_{inc} = 2(1-v)^2(\Delta P/\Delta R_o)R_i^2 R_o/(R_o^2 - R_i^2)$.

  R_o and R_i, external and internal radius, respectively; v, Poisson's ratio.
- Data under smooth muscle activation were also presented.

Reference(s)
Hayashi K, Nagasawa S, Naruo Y, Okumura A, Moritake K, Handa H (1980) Mechanical properties of human cerebral arteries. Biorheology 17:211–218 (with permission)

Elastic Modulus (22)

• Stiffness parameter • Incremental elastic modulus	• Rabbits • Aorta, carotid, femoral	• Atherosclerosis • Elastase

Materials
- Japanese White female rabbits
- Common carotid arteries, femoral arteries, thoracic aortas

Testing Methods and Experimental Conditions
- Experimental groups: RA, Chol x 14 w; RB, Chol x 7 w and (Chol + Ela) x 7 w; RC, Chol x 7 w and (Con + Ela) x 7 w; RD, Chol x 7 w and Con x 7 w; RE, Con x 14 w (Chol, 1% cholesterol diet; Ela, 450 elastase units/kg/day; Con, normal diet)
- Quasi-static pressure–diameter test in aerated Krebs–Ringer solution at 37°C

Data

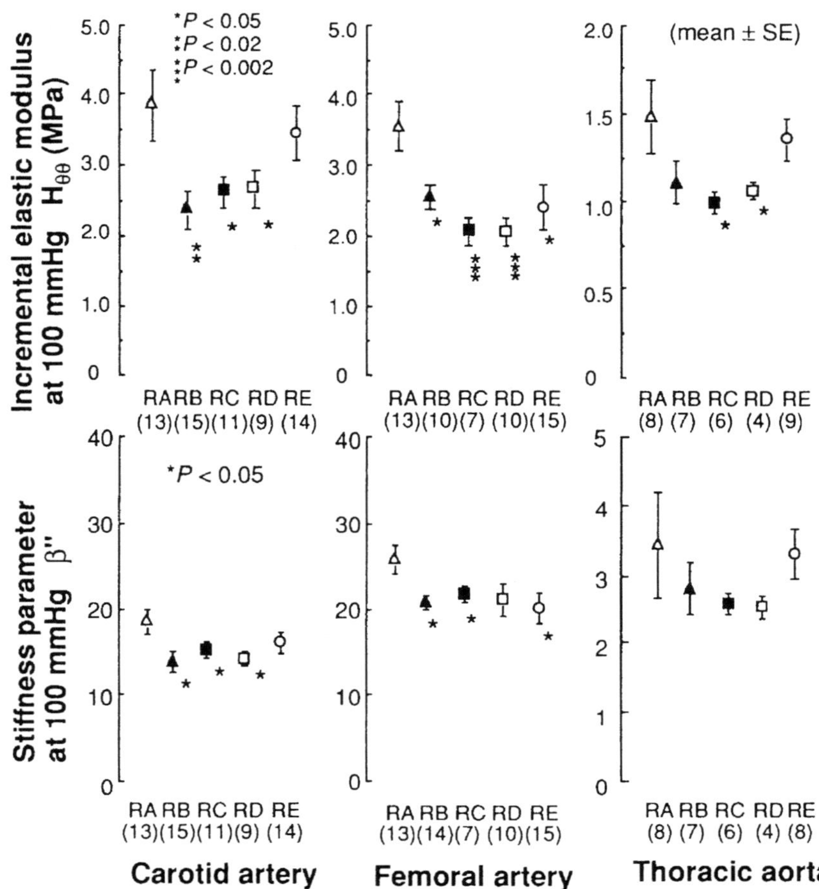

Comments
- Stiffness parameter, β'', at intraluminal pressure P:

  $\beta'' = (\Delta P/P) / (\Delta D_o/D_s)$ where ΔD_o is the diameter increment for pressure increment ΔP and D_s is the diameter at 100 mmHg.

Reference(s)
Hayashi K, Takamizawa K, Nakamura T, Kato T, Tsushima N (1987) Effects of elastase on the stiffness and elastic properties of arterial walls in cholesterol-fed rabbits. Atherosclerosis 66: 259–267 (with permission)

Elastic Modulus (23)

• Stiffness parameter • Incremental elastic modulus	• Rabbits • Thoracic aorta	• Atherosclerosis • Temporal change

Materials
- Mature male Japanese White rabbits weighing 3.1–3.5 kg
- Thoracic aorta

Testing Methods and Experimental Conditions
- Atherosclerosis induced by regular chow + 1% cholesterol and/or aortic endothelium denudation by ballooning (4, 8, 16, 32 weeks)
- Quasi-static pressure–diameter test in aerated Krebs–Ringer solution at 37°C
- Mechanical parameters were calculated from mean diameters of loading and unloading curves
- Correlation with intraluminal area fraction of atherosclerotic lesion, A_s
- A, control; B, 1% cholesterol; C, endothelial cell (EC) denuded; D, 1% cholesterol + EC denudation B0 and D0, $A_s < 80\%$ and $\beta'' < 10$; B2 and D2, $A_s \geq 80\%$ and $\beta'' \geq 10$

Data

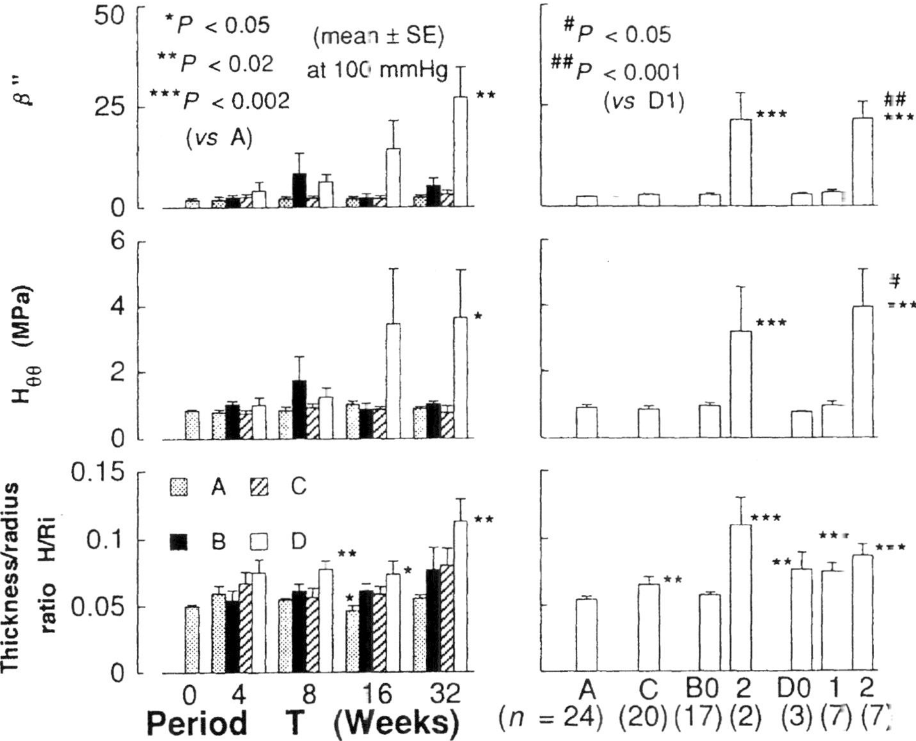

Comments

- Stiffness parameter β'' at intraluminal pressure P_i: $\beta'' = (\Delta P_i/P_i)/(\Delta D_o/D_s)$; ΔD_o, outer diameter (D_o) increment for pressure increment ΔP_i; D_s, diameter at 100 mmHg.
- Incremental elastic modulus, $H_{\theta\theta} = 2D_o(\Delta P_i D_i^2/\Delta D_o + P_i D_o)/(D_o^2 - D_i^2)$; D_i, inner diameter.
- Elastic parameters were not affected by atherosclerosis until $A_s \geq 80\%$.

Reference(s)
Hayashi K, Ide K, Matsumoto T (1994) Aortic walls in atherosclerotic rabbits —mechanical study. J Biomech Eng 116:284–293

Elastic Modulus (24)

• Stiffness parameter • Pressure–strain elastic modulus	• Human, dogs, calves, horses • Coronary artery	• Review •

Materials

- Human, dog, calf, and horse
- Right, left, left anterior descending, and left circumflex coronary arteries

Testing Methods and Experimental Conditions

- Brief review on coronary artery mechanics
- Data obtained by the authors and other researchers were compared by converting them to stiffness parameter β: $\ln (P/P_s) = \beta (D/D_s - 1)$

  D, outer diameter at pressure P; D_s, outer diameter at standard pressure $P_s = 100$ mmHg

Data

Study	Species (age)	Site (n)	Measurement	Data (pressure)		β
Gow and Hadfield (1)	Human (14–40)	RGA (1)	In vitro	E_P (stat)=0.663 MPa	(100 mmHg)	49.7
				E_P (dyn)=0.72 MPa	(100 mmHg)	54.0
		LAD (3)	In vitro	E_P (stat)=0.624 MPa	(100 mmHg)	46.8
				E_P (dyn)=1.38 MPa	(100 mmHg)	103.5
		LGGA (2)	In vitro	E_P (stat)=0.568 MPa	(100 mmHg)	42.6
				E_P (dyn)=0.795 MPa	(100 mmHg)	59.6
Hayashi et al.(2)	Human (50–77)	RGA (7)	In vitro			36.8
		LAD (9)	In vitro			39.8
		LGGA (5)	In vitro			29.8
Gow and Hadfield (1)	Dog	LGGA (9)	In vivo	E_P (dyn)=0.225 MPa	(108 mmHg)	15.6
Gow et al. (3)	Dog	LGGA (12)	In vitro	E_P (stat)=0.341 MPa	(100 mmHg)	25.6
		LGGA (11)	In vitro	E_P (dyn)=0.701 MPa	(100 mmHg), 2 Hz	53.1
Cox (4)	Dog	LGGA (12)	In vitro	E_{inc} (stat)=1.741 MPa	(125 mmHg)	
Vatner et al. (5)	Dog	LGGA (8)	In vivo	E_{inc} (dyn)=0.17 MPa	(90 mmHg)	
Patel and Janicki (6)	Dog	LGGA (7)	In vitro	C_v(stat)=34.1/MPa	(123 mmHg)	3.6
Douglas and Green-field (7)	Dog	All (11)	In vitro	C_v(stat)=5.6/MPa	(100 mmHg)	26.8
Gross et al. (8)	Calf	LGA (5)	In vivo	E_P (dyn)=0.156 MPa	(91 mmHg)	12.9
Rumberger et al. (9)	Horse	LGA	In vivo	c=6.24 m/s	(100 mmHg)	6.2

Values are expressed as mean. RGA, right coronary artery; LAD, left anterior descending coronary artery; LGGA, left circumflex coronary artery; LGA, left coronary artery; E_P, pressure–strain elastic modulus; E_{inc}, incremental elastic modulus; C_v, compliance; c, pressure wave speed; stat, static data; β, stiffness parameter; n, number of specimens.

Comments

None.

Reference(s)

1. Gow BS, Hadfield CD (1979) Circ Res 45:588
2. Hayashi K, Igarashi Y, Takamizawa K (1986) Kitamura K, Abe H, Sagawa K (eds) New Approaches in Cardiac Mechanics. Gordon and Breach, Tokyo, pp 285–294
3. Gow BS, Schonfeld D, Patel DJ (1974) J Biomech 7:389
4. Cox RH (1978) Am J Physiol 234:H533
5. Vatner SF, Pagani M, Manders WT, Pasipoularides AD (1980) J Clin Invest, 65:5
6. Patel DJ, Janicki JS (1970) Circ Res 27:149
7. Douglas JE, Greenfield JC Jr. (1970) Circ Res 27:921
8. Gross DR, Hunter JF, Allert JA (1981) J Biomech 14:613
9. Rumberger IA, Nerem RM, Muir WW (1979) Cardiovasc Res, 13:413

Elastic Modulus (25)

• Stiffness parameter • Static P–D test	• Human • Cerebral arteries	• Age-related changes •

Materials
- Eighteen autopsy subjects who had neither systemic nor cerebral vascular lesions
- Intracranial artery (basilar and vertebral artery)
- Extracranial artery (common carotid, vertebral, and femoral artery)

Testing Methods and Experimental Conditions
- Quasi-static pressure–diameter relation
- Medium: Krebs–Ringer solution at 37°C

Data

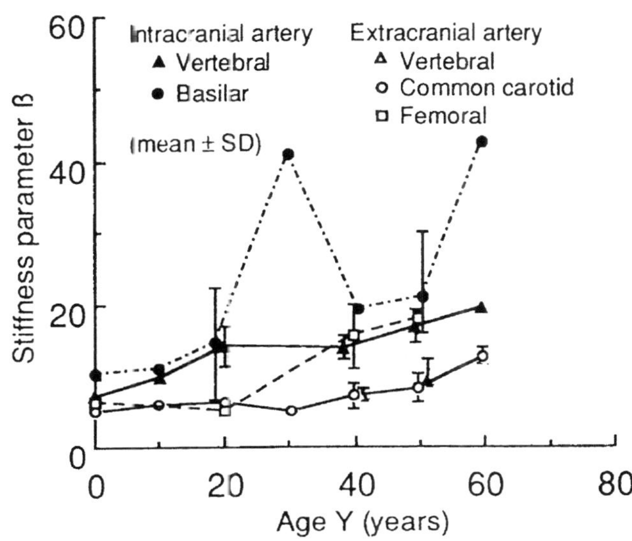

Comments

- Stiffness parameter, β:

  $\ln (P/P_s) = \beta (D/D_s - 1)$.

  P_s, standard pressure ($=100$ mmHg for this study); D_s, wall diameter at P_s.

Reference(s)

Hayashi K, Handa H, Nagasawa S, Naruo Y, Okumura A, Moritake K (1980) Stiffness and elastic behaviour of human intracranial and extracranial arteries. J Biomech 13:175–184 (with permission)

Elastic Modulus (26)

• Strain energy density func. • Exponential vs logarithmic	• Dogs • Carotid artery	• Uniform strain hypothesis • Zero initial stress hypothesis

Materials
- Male and female mongrel dogs
- Common carotid artery

Testing Methods and Experimental Conditions
- Quasi–static pressure–diameter–axial force test
- In aerated Krebs–Ringer solution at 37°C
- Ascending limbs of pressure–diameter and axial force–diameter curves were used for the analysis
- Strain energy density functions:

  Exponential function: $W = C(\exp\psi - 1)$

  Logarithmic function: $W = -C\ln(1 - \psi)$, $\psi = (1/2)(a_{\theta\theta}\varepsilon_\theta^2 + a_{zz}\varepsilon_z^2 + 2a_{\theta z}\varepsilon_\theta\varepsilon_z)$

- Stress/strain distributions obtained assuming (1) zero stress under no load condition (zero initial stress hypothesis), or (2) uniform strain under in vivo condition (uniform strain hypothesis)

Data

Strain energy density function	Condition	C (kPa)	$a_{\theta\theta}$	$a_{\theta z}$	a_{zz}	$E(P_i)$ (kPa)	$E(F_z)$ (mN)
Exponential	Zero	17.478	1.3000	0.1061	1.6604	1.90	47.5
		12.965	0.5211	0.1205	0.7345	0.65	18.7
Logarithmic	Zero	65.44	0.4532	0.0710	0.5982	1.85	52.2
		13.20	0.1471	0.0575	0.2509	0.65	27.5
Exponential	Uniform	16.687	1.8703	0.0754	1.7578	2.01	47.6
		12.709	0.7754	0.1528	0.8954	0.70	18.5
Logarithmic	Uniform	44.52	0.8053	0.0450	0.7694	1.92	40.9
		16.12	0.2538	0.0898	0.3277	0.77	18.8

All data are given as mean (upper row) and SD (lower row; $n = 7$).
$E(x)$, root mean square error of x between calculation and experiment; P_i, intraluminal pressure; F_z, axial force.
Zero, zero initial stress hypothesis; Uniform, uniform strain hypothesis.

Comments
- Vessel dimensions at physiological condition (100 mmHg, in vivo axial length):

External diameter (mm):	4.557 ± 0.573
Internal diameter (mm):	4.093 ± 0.539
Axial extension ratio:	1.61 ± 0.17
Circumferential extension ratio:	1.571 ± 0.097 (mean $\pm$ SE, $n = 8$).

Reference(s)
Takamizawa K, Hayashi K (1987) Strain energy density function and uniform strain hypothesis for arterial mechanics. J Biomech 20:7–17 (with permission)

Elastic Modulus (27)

• Strain energy density func. • Exponential vs polynomial	• Rabbits • Aorta, carotid, iliac	• •

Materials
• Rabbits
• Carotid and iliac arteries, and lower and upper aortas

Testing Methods and Experimental Conditions
• Quasi–static pressure–diameter–axial force test
• Medium: aerated calcium-free Krebs solution at 37°C

Data

	Carotid	Left iliac	Lower aorta	Upper aorta
Exponential function				
C_1 (kPa)	29.307	21.575	21.744	33.856
a_1	2.5084	8.1674	9.5660	2.8173
a_2	0.4615	1.2173	3.0913	0.5239
a_4	0.1764	1.0546	0.8805	0.5790
$E_{\theta\theta}{}^*$	0.5191	0.2418	0.2743	0.4061
$E_{zz}{}^*$	0.9939	0.8834	0.6495	0.9566
$S_{\theta\theta}{}^*$ (kPa)	34 000	45.000	49.000	52.000
$S_{zz}{}^*$ (kPa)	13.000	19.000	30.000	19.000
Polynomial function				
A (kPa)	-71.889	-163.871	-147.22	-120.062
B (kPa)	31.255	-3.854	-41.606	51.405
C (kPa)	1.911	29.122	44.821	-15.936
D (kPa)	13.711	160.463	165.753	-21.292
E (kPa)	102.775	296.790	299.390	235.706
F (kPa)	-33.677	11.872	62.093	-73.431
G (kPa)	0.787	-22.552	-18.999	22.069

All data are given as means.

Comments
• Strain energy functions:
  Exponential function:

$$\rho_o W = (C_1/2)\, \exp[a_1(E_{\theta\theta}^2 - E_{\theta\theta}^{*2}) + a_2(E_{zz}^2 - E_{zz}^{*2}) + 2a_4(E_{\theta\theta}E_{zz} - E_{\theta\theta}^* E_{zz}^*)]$$

  Polynomial function:

$$\rho_o W = AE_{\theta\theta}^2 + BE_{\theta\theta}E_{zz} + CE_{zz}^2 + DE_{\theta\theta}^3 + EE_{\theta\theta}^2 E_{zz} + FE_{\theta\theta}E_{zz}^2 + GE_{zz}^3$$

  A, B, C, D, E, F, G, C_1, a_1, a_2, a_4 – constants; $E_{\theta\theta}$, E_{zz} – Green's strain.

Reference(s)
Fung YC, Fronek K, Patitucci P (1979) Pseudoelasticity of arteries and the choice of its mathematical expression. Am J Physiol 237:H620–H631 (with permission)

Elastic Modulus (28)

| • Strain energy density func. | • Dogs | • Taylor series expression |
| • Stress distribution | • Aorta | • |

Materials
- Healthy dogs of varying sex, age, and weights, killed by exsanguination
- Aortas stored in a normal saline at 5°C, and used within 24 h of removal
- Cylindrical segments in the range of 2.26–2.98 cm obtained from various aortic locations

Testing Methods and Experimental Conditions
- Medium: normal saline at room temperature (22 ± 2°C)
- Continuous axial loading and unloading at a fixed stretch rate (0.25 mm/s) to produce length changes in the range of 5–10 mm
- Luminal pressure simultaneously varied so as to maintain a fixed diameter
- Elastic response analyzed with a Taylor series expression of strain energy function:

$$W(I_1, I_2) = b_1(I_1 - 3) + b_2(I_1 - 3)^2 + ...$$

b_i ($i = 1,2, ..., n$), constants; I_1 and I_2, strain invariants (considered to be $I_1 \sim I_2$)

$$I_1 = (x/\Lambda)^2 + 1/x^2 + \Lambda^2, \quad I_2 = (\Lambda/x)^2 + x^2 + 1/\Lambda^2, \quad x = R/r, \quad \Lambda = L/L_0$$

R and r, undeformed and deformed radius, respectively;
L and L_0, deformed and undeformed length, respectively

Data

Specimen no.	ΔL (cm)	i	b_i-load (kPa)	b_i-unload (kPa)	RMS errors-load P (mmHg)	RMS errors-load F (mN)	RMS errors-unload P (mmHg)	RMS errors-unload F (mN)
1	1.0	1	-13.03	-23.27	4.46	29.0	4.98	30.8
		2	82.01	109.29				
		3	-96.78	-120.05				
2	1.0	1	3.19	2.07	2.01	17.2	1.57	13.9
		2	30.49	31.60				
		3	-35.37	-33.76				
3	1.0	1	13.24	4.99	2.04	16.7	1.74	12.0
		2	51.52	25.48				
		3	146.20	26.92				
4	1.0	1	15.73	7.26	2.82	33.9	4.89	52.8
		2	-142.19	32.90				
		3	921.88	743.80				
5	1.0	1	8.96	6.05	0.53	3.7	0.26	1.9
		2	16.74	17.65				
		3	-26.82	-33.50				
		4	43.29	62.90				
6	1.0	1	13.99	8.70	0.42	4.6	0.54	6.0
		2	-8.47	16.56				
		3	25.20	-10.62				
7	1.0	1	8.73	7.98	0.93	5.2	0.57	4.6
		2	42.30	40.40				
		3	-19.60	25.50				
		4	-53.00	-134.00				

ΔL, length change; RMS, root mean square.

Comments
- Strain energy function containing up to 4 terms resulted in good agreement with the data.

Reference(s)
Vito RP, Hickey J (1980) The mechanical properties of soft tissues. II. The elastic response of arterial segments. J Biomech 13:951–957 (with permission)

Elastic Modulus (29)

• Stress–strain curve • Vessel dimension	• Rats • Cerebral arteriole	• Age-related variation • In vivo measurement

Materials
- Fischer 344 rats, adult (age, 9–12 months; $n = 10$) and aged (age, 24–27 months; $n = 11$)
- First-order pial arteriole

Testing Methods and Experimental Conditions
- Anesthesia with sodium pentobarbital (5 mg/100 g BW, i.p.) under mechanical ventilation
- Craniotomy over the left parietal cortex
- Smooth muscle deactivation with artificial cerebrospinal fluid with EDTA (25 mg/ml) at 37–38°C
- Pressure–diameter relation obtained while lowering blood pressure by hemorrhage
- Pressure-fixation at 60–80 mmHg for the measurement of blood vessel dimension
- Wall composition measurement by histometry

Data

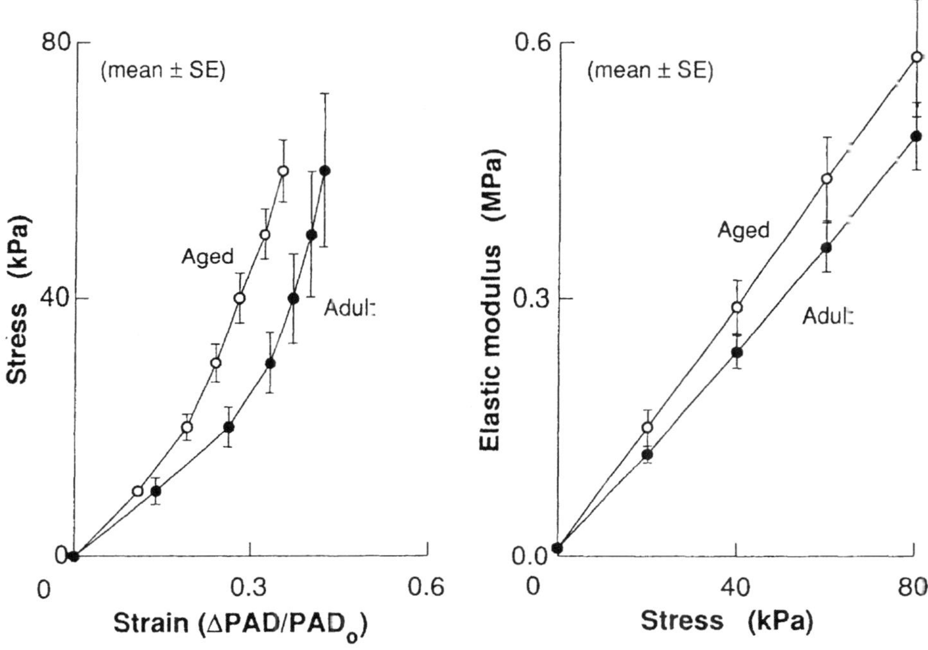

Comments
	Adult	Aged
• Baseline values:		
Pial arteriolar pressure (mmHg)	73 ± 2	71 ± 3
Inner diameter of pial arterioles at 70 mmHg (μm)	83 ± 7	$66 \pm 5^*$
Cross-sectional area (μm^2)	1718 ± 142	$1260 \pm 117^*$

$^*P < 0.05$ vs adult.

- Tangential elastic modulus E_T was calculated by differentiating stress–strain data fitted to an exponential curve ($\sigma = \sigma_0 e^{\beta\varepsilon}$), i.e., $E_T = \beta\sigma_0 e^{\beta\varepsilon}$; β, constant obtained by the fitting.

Reference(s)
Hajdu MA, Heistad DD, Siems JE, Baumbach GL (1990) Effects of aging on mechanics and composition of cerebral arterioles in rats. Circ Res 66:1747–1754 (with permission)

Elastic Modulus (30)

• Yield stress / strain • Secant modulus	• Human • Cerebral aneurysm	• Local property •

Materials

- Human (age, 39–64 years)
- Cerebral aneurysm at anterior communicating, middle cerebral, and internal carotid arteries
- Intracranial arteries (anterior cerebral, vertebral, basilar, and middle cerebral arteries)

Testing Methods and Experimental Conditions

- Specimen obtained at surgery or at autopsy (within 10 h after death)
- One-dimensional incremental elongation (0.1 mm/step, 30 s interval) on isolated strips
- Elongation calculated from crosshead displacement
- Specimen kept moist by irrigation with normal saline solution at room temperature

Data

	Aneurysm fundi	Aneurysm necks	Arteries
Yield strain	0.37 ± 0.15	0.57 ± 0.39	0.34 ± 0.10
Yield stress	0.50 ± 0.26	1.21 ± 0.49	1.06 ± 0.13
Elastic moduli	1.7 ± 0.8	3.1 ± 0.9	2.5 ± 1.1
Systolic in vivo stress	0.33 ± 0.17	0.37 ± 0.14	0.12 ± 0.04
Mean in vivo stress	0.26 ± 0.11	0.30 ± 0.13	0.10 ± 0.04

All values are mean ± SD.
Stresses and elastic moduli in MPa.

Comments

- Secant modulus at yield point is used as the elastic modulus.
- In vivo stresses calculated from Laplace's formula on the basis of blood pressures of each patient and the radius of the aneurysm and artery as measured at surgery or autopsy.

Reference(s)

Steiger HJ, Aaslid R, Keller S, Reulen HJ (1989) Strength, elasticity and viscoelastic properties of cerebral aneurysms. Heart Vessels 5:41–46

Elastic modulus (31)

• Stiffness parameter • Diametrical change	• Human • Aorta, large arteries	• Non-invasive measurement • Regional variation

Materials

- Human in four different age groups
  (6–81 years, normotensive and clinically free of vascular symptoms)

Testing Methods and Experimental Conditions

- Blood vessel wall motion measured with an ultrasonic phase locked echo tracking system
- Stiffness parameter calculated from the wall motion

Data

	Abdominal aorta	Common carotid artery	Brachial artery	Femoral artery
		Group 1 (0–19 years)		
P_s	109 (12)	107 (13)	106 (12)	106 (10)
P_d	58 (6)	61 (8)	61 (6)	61 (8)
D_d	10.6 (2.7)	6.1 (0.8)	3.1 (0.7)	7.1 (1.4)
D_s-D_d	1.49 (0.40)	0.84 (0.24)	0.22 (0.10)	0.46 (0.18)
$(D_s-D_d)/D_d$	0.15 (0.06)	0.14 (0.04)	0.07 (0.03)	0.07 (0.03)
β	4.29 (1.40)	4.32 (1.00)	8.56 (2.32)	9.41 (2.71)
(n)	(9)	(18)	(17)	(18)
		Group 2 (20–39 years)		
P_s	120 (18)	117 (13)*	116 (12)*	116 (12)**
P_d	69 (11)**	70 (8)**	69 (8)**	68 (9)*
D_d	11.7 (2.5)	7.1 (0.9)**	3.6 (0.9)	8.1 (1.1)
D_s-D_d	1.08 (0.25)**	0.64 (0.17)***	0.24 (0.16)	0.45 (0.18)
$(D_s-D_d)/D_d$	0.09 (0.02)**	0.09 (0.02)***	0.07 (0.04)	0.06 (0.02)
β	6.05 (1.01)*	5.90 (1.29)**	8.68 (3.79)	10.60 (3.32)
(n)	(11)	(22)	(19)	(20)
		Group 3 (40–59 years)		
P_s	117 (15)	119 (16)*	120 (18)**	118 (12)**
P_d	72 (10)**	72 (13)***	75 (12)***†	70 (10)**
D_d	15.3 (3.0)***††	7.3 (1.5)**	4.1 (1.1)**	8.8 (2.1)**
D_s-D_d	0.93 (0.15)***	0.49 (0.12)***††	0.16 (0.06)	0.39 (0.12)
$(D_s-D_d)/D_d$	0.07 (0.01)***†	0.07 (0.02)***††	0.04 (0.02)**††	0.05 (0.01)**
β	7.45 (1.23)***†	7.75 (1.66)***†††	13.15 (5.39)**††	12.63 (3.77)**
(n)	(10)	(20)	(17)	
		Group 4 (> 60 years)		
P_s	131 (7)**‡	126 (15)***†	126 (13)***†	124 (12)***†
P_d	74 (7)***	73 (6)**	74 (8)***	73 (6)***
D_d	14.4 (2.3)**†	8.1 (1.0)***††‡	4.9 (1.2)***††††‡	10.2 (1.6)***†††‡
D_s-D_d	0.86 (0.17)***†	0.40 (0.14)***†††	0.20 (0.06)	0.38 (0.10)
$(D_s-D_d)/D_d$	0.06 (0.01)***†	0.05 (0.01)***†††‡	0.05 (0.02)*†	0.04 (0.01)***†
β	9.83 (2.07)***††††‡	11.31 (2.01)***††††‡‡	13.73 (5.36)***†††	15.31 (4.52)***†††‡
(n)	(9)	(19)	(18)	(17)

Data are given as mean (SD).

β, stiffness index; D_d, end-diastolic diameter (mm); $D_s - D_d$, stroke change of the arterial diameter during systole (mm); P_d, diastolic arterial pressure (mmHg); P_s, systolic arterial pressure (mmHg).

* $P < 0.05$, ** $P < 0.01$, *** $P < 0.001$ vs group 1.
† $P < 0.05$, †† $P < 0.01$, ††† $P < 0.001$ vs group 2.
‡ $P < 0.05$, ‡‡ $P < 0.01$, ‡‡‡ $P < 0.001$ vs group 3.

Comments

- Stiffness parameter was calculated as stiffnes index β: $\ln (P_s/P_d) /[(D_s - D_d)/D_d]$.

Reference(s)

Kawasaki T, Sasayama S, Yagi S, Asakawa T, Hirai T (1987) Non-invasive assessment of the age-related changes in stiffness of major branches of the human arteries. Cardiovasc Res 21:678–687 (with permission)

Pressure–Diameter Relation (1)

• Incremental distensibility • Vessel dimension	• Rats • Saphenous vein	• Long-term tilt •

Materials
- Male Sprague–Dawley rats weighing 300–400 g
- Great saphenous vein

Testing Methods and Experimental Conditions
- Animals kept in tube-like cages to increase femoral vein pressure for 2 weeks
- Pressure–diameter relation between 0 and 20 mmHg (pressure increased at each step of 5 mmHg)
- Medium: aerated physiological salt solution at 37°C

Data

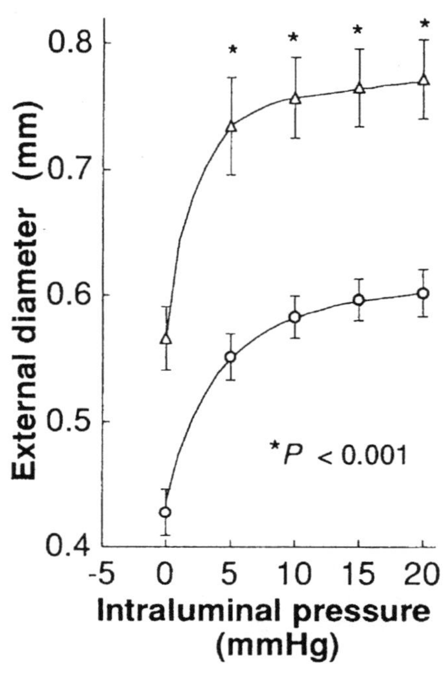

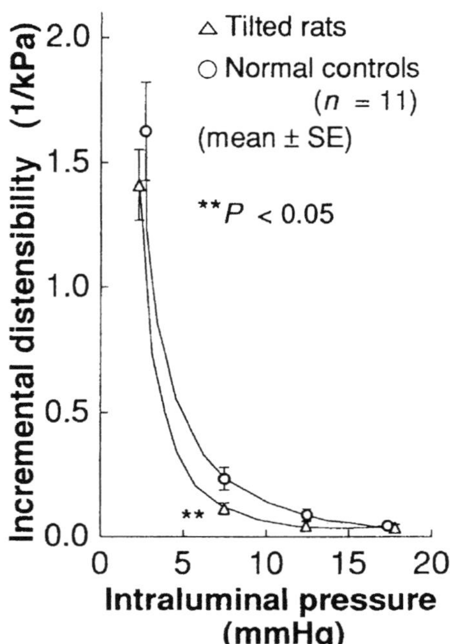

Comments
- Incremental distensibility, D:

$$D = [\{(R_i(j + 1)/R_i(j))\}^2 - 1]/\Delta P.$$

  $R_i(j)$, inner radius at pressure $P(j)$; $\Delta P = P(j + 1) - P(j)$.
- Change in femoral vein pressure due to head-up tilt: $2.9 \pm 0.2 \to 5.9 \pm 0.2$ mmHg (mean ± SE).

Reference(s)
Monos E, Contney SJ, Cowley AW, Stekiel WJ (1989) Effect of long-term tilt on mechanical and electrical properties of rat saphenous vein. Am J Physiol 256:H1185–H1191 (with permission)

Pressure–Diameter Relation (2)

• Incremental elastic modulus • Static P–D test	• Dogs • Carotid, iliac artery	• Atherosclerosis • Smooth muscle activation

Materials
- Racing greyhound (age, approx. 2–4 years)
- Carotid and iliac arteries

Testing Methods and Experimental Conditions
- A normal diet or a diet containing 16% coconut oil and 5% cholesterol for 12 months
- Quasi-static pressure–diameter relation between 0 and 250 mmHg
- Norepinephrine + an aerated physiological salt solution (PSS) at 37°C for active smooth muscle
- A calcium-free, glucose-free PSS with 2 mM EGTA for passive smooth muscle

Data

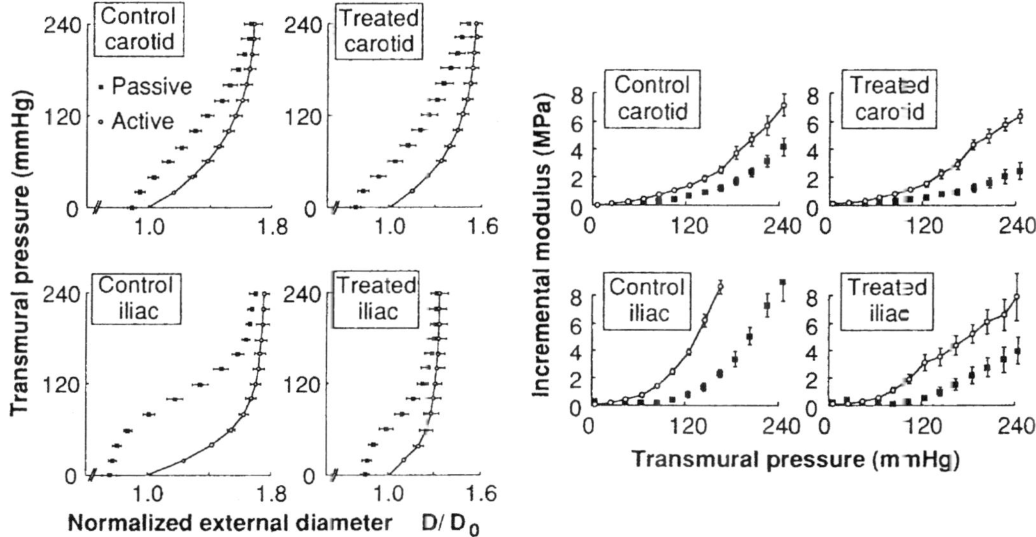

Comments
- Incremental elastic modulus, E:

$$E = 2a^2b/(b^2 - a^2) \cdot \Delta P/\Delta b.$$

  P, transmural pressure; a, internal radius; b, external radius.
- Series elasticity was also evaluated with Voigt and Maxwell models.

Reference(s)
Cox RH, Detweiler DK (1979) Arterial wall properties and dietary atherosclerosis in the racing greyhound. Am J Physiol 236:H790–797 (with permission)

Pressure–Diameter Relation (3)

• Incremental elastic modulus • Static P–D test	• Rats • Carotid artery	• Age-related variation • Smooth muscle activation

Materials
- Male rats (CD strain; age, 8 weeks)
- Carotid artery

Testing Methods and Experimental Conditions
- Medium: aerated physiological salt solution at 37°C
- Smooth muscle contraction by 5 µg/ml norepinephrine
- Quasi-static pressure–diameter relation at a rate of 0.5 mmHg/s from 0 to 200 mmHg (Norepinephrine)
- Deactivation of smooth muscle by cyanide, iodoacetate, and dinitrophenol
- Quasi-static pressure–diameter relation at a rate of 1 mmHg/s (Passive)

Data

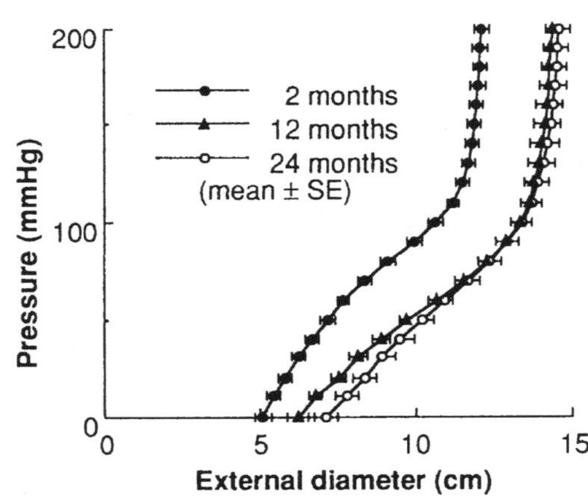

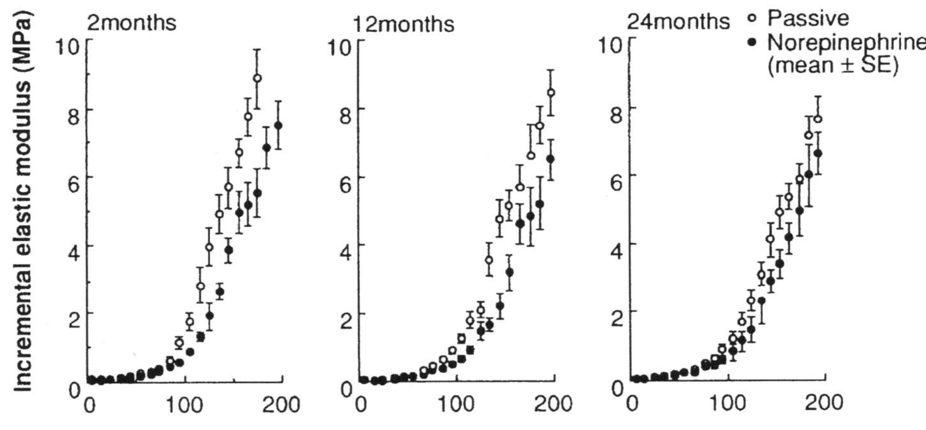

Comments
- Incremental elastic modulus:

$$E = 2a^2b/(b^2-a^2) \cdot \Delta P/\Delta b \quad (a, \text{ inner radius; } b, \text{ outer radius}).$$

Reference(s)

Cox RH (1977) Effects of age on the mechanical properties of rat carotid artery. Am J Physiol 233:H256–H263 (with permission)

Pressure–Diameter Relation (4)

• Incremental elastic modulus • Static P–D test	• Rats • Thoracic aorta	• Hypertension • Chemical composition

Materials
- Male albino Wistar rats
- Thoracic aorta

Testing Methods and Experimental Conditions
- Hypertension induced by unilateral nephrectomy + DOCA salt at age 4 weeks
- Age-matched control
- In situ static pressure–diameter test (filling the aorta with radio-opaque material)

Data

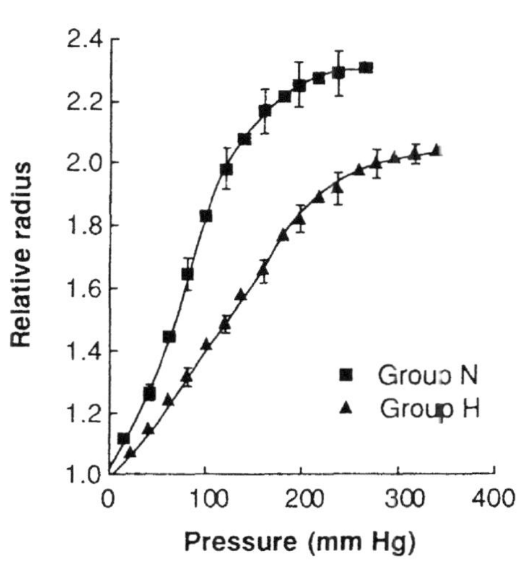

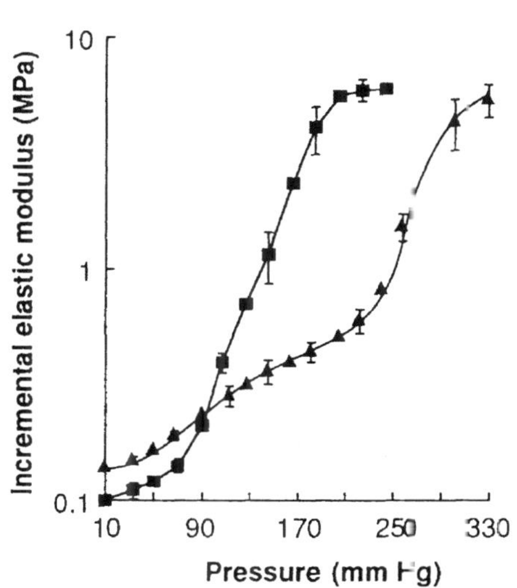

Comments
- Incremental elastic modulus, E_{inc}:

  $E_{inc} = 0.75 \times \Delta P/(\Delta R/R) \times (R/h)$.
  R, average radius; h, average wall thickness; ΔR, radius increment over pressure increment of ΔP.
- Relative wall thickness of the thoracic aorta at a pressure of 110 mmHg was 0 076 ± 5.2

  (mean ± SE × 10³) for control and 0.136 ± 6.8 for hypertensive rats aged 20 weeks.
- Systolic blood pressure at the caudal artery: 110 mmHg for control and 180 mmHg for treated animals at the age of 20 weeks.

Reference(s)
Berry CL, Greenwald SE (1976) Effects of hypertension on the elastic mechanical properties and chemical composition of the rat aorta. Cardiovasc Res 10:437–451 (with permission)

Pressure–Diameter Relation (5)

• Incremental elastic modulus • Vessel dimension	• Rats • Carotid artery	• Hypertension • Hemodynamics

Materials
- Normal Wistar (NW), Kyoto Wistar (WKY), and spontaneous hypertensive (SHR) rats (age, 15 weeks)
- Common carotid artery

Testing Methods and Experimental Conditions
- Quasi-static pressure–diameter test at a pressure increase of 1 mmHg/s
- Medium: a calcium- and glucose-free physiologic salt solution with 2 mM EGTA at 37°C

Data

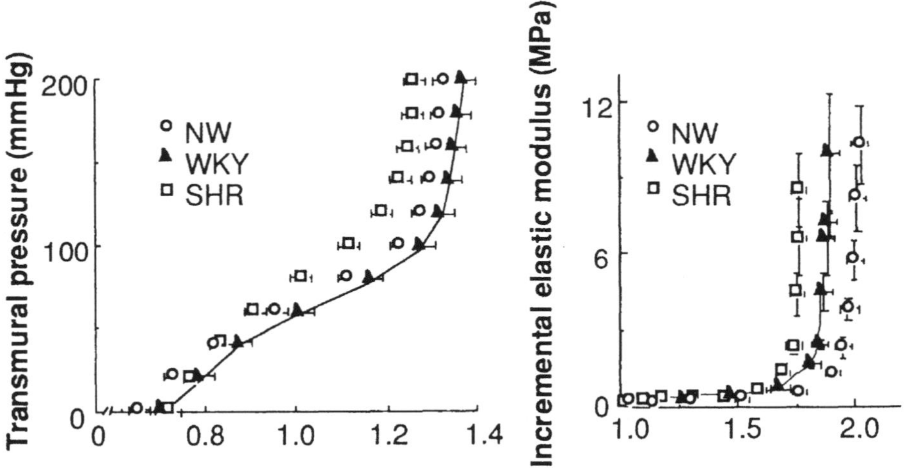

Group	n	Bw (g)	SP (mmHg)	HR (bpm)	a (mm)	a/h
Normal Wistar	15	305 ± 9	121 ± 3	368 ± 13	0.574 ± 0.019	16.9 ± 1.4
Kyoto Wistar	15	269 ± 13	124 ± 4	361 ± 15	0.603 ± 0.018	17.3 ± 1.3
SHR	15	$245 \pm 5^*$	$187 \pm 5^*$	$401 \pm 12^*$	$0.508 \pm 0.012^*$	10.3 ± 0.7

BW, body weight; SP, systolic pressure; HR, heart rate; a, internal radius at 100 mmHg; a/h, radius–wall thickness ratio at 100 mmHg; n, number of animals per group. $^*P < 0.05$ *vs* Kyoto Wistar.

Comments
- Also presented: mechanical properties under smooth muscle activation, water and connective tissue content.

Reference(s)
Cox RH (1979) Comparison of arterial wall mechanics in normotensive and spontaneously hypertensive rats. Am J Physiol 237:H159–H167 (with permission)

Pressure–Diameter Relation (6)

• Static P–D test •	• Human, rats • Finger, tail artery	• Smooth musc.e activation • Inflation vs deflation

Materials
- Human finger arteries obtained at autopsy (age, 57–85 years)
- Ventral tail artery of Wistar rats; weight, 250–350 g

Testing Methods and Experimental Conditions
- Medium: aerated Tyrode's solution at 37°C
- Pressure–diameter relation of tubular specimen at in situ length obtained by changing pressure stepwise every 5 or 10 mmHg from –40 to 100 mmHg

Data

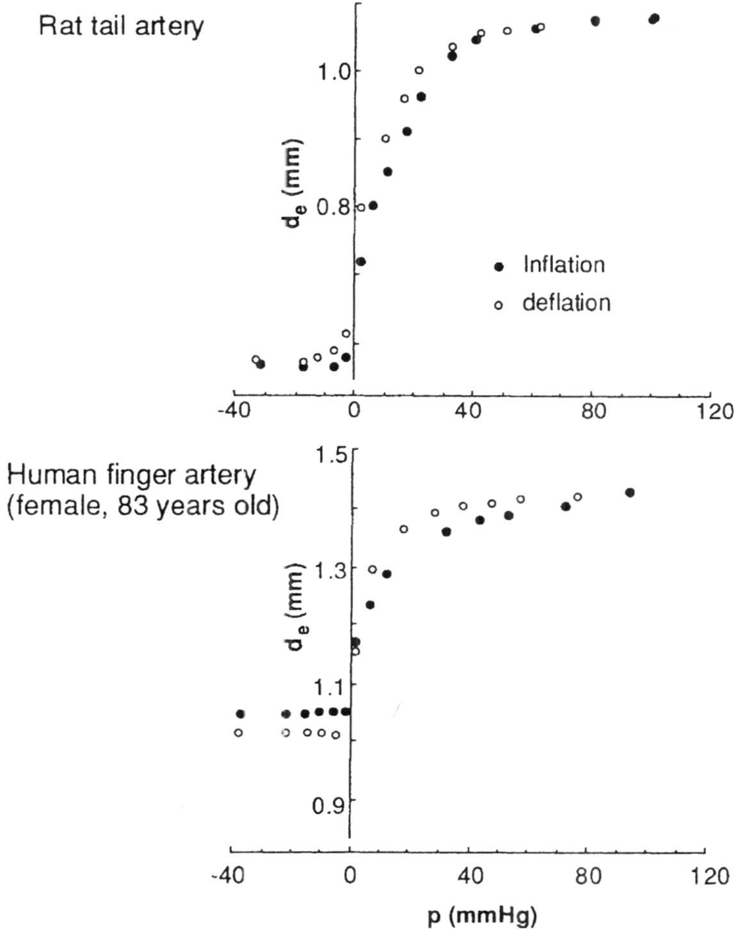

Comments
- Spontaneous rhythmic contractions observed in both specimens.
- Human finger arteries could contract to complete closure both spontaneously and after addition of noradrenaline, while rat tail arteries did not.

Reference(s)
Langewouters GJ, Zwart A, Busse R, Wesseling KH (1986) Pressure-diameter relationships of segments of human finger arteries. Clin Phys Physiol Meas 7:43–55 (with permission)

Pressure–Diameter Relation (7)

• Static P–D test • In vivo pressure	• Rats • Thoracic aorta	• Hypertension • Temporal change

Materials
- Male Wistar rats (age, 10, 12, 16, 24 weeks)
- Thoracic aorta

Testing Methods and Experimental Conditions
- Quasi-static pressure–diameter test in physiological saline solution at 37°C
- Hypertension induced by constricting left renal arteries of animals aged 8–9 weeks (Goldblatt hypertension)
- Comparison with age-matched control
- Outer diameter was normalized by the diameter at 0 mmHg
- Mean diameter during inflation and deflation

Data

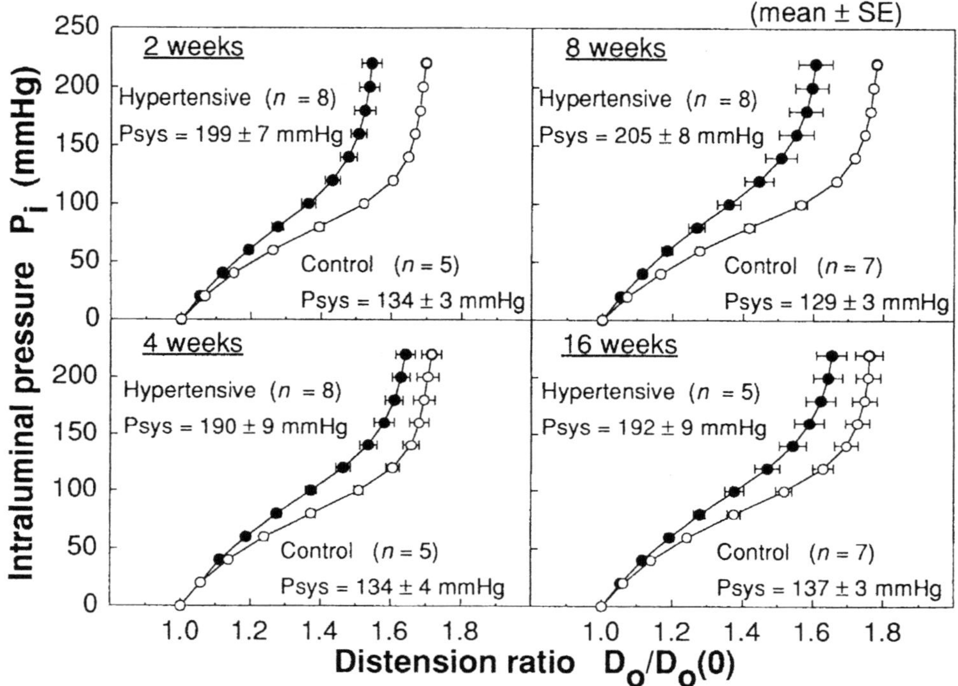

Comments
- Change in pressure–diameter relation following the induction of hypertension.
- Psys, in vivo systolic pressure.

Reference(s)
Matsumoto T, Hayashi K (1994) Mechanical and dimensional adaptation of rat aorta to hypertension. J Biomech Eng 116:278–283

Pressure–Diameter Relation (8)

• Static pressure–volume test •	• Human • Cerebral aneurysm	• Pathological change •

Materials
- Human
- Intracranial arteries and aneurysms obtained at autopsy

Testing Methods and Experimental Conditions
- Pressure–volume test operated manually

Data

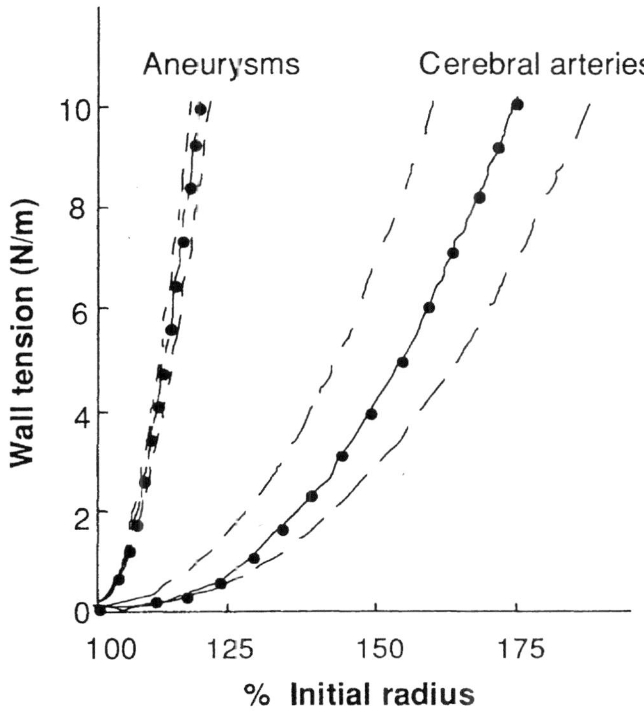

Comments
- Wall tension, T, was calculated from the law of Laplace.
  $T = P / (1/R_1 + 1/R_2)$.
  P, intraluminal pressure; R_1 and R_2, principal radii of curvature ($1/R_2=0$ in the case of an artery).
- Elastic diagrams summarizing the results from all the cases. The mean elasticity of seven aneurysms is compared to that of 16 major intracranial arteries. The dashed lines represent one SEM.

Reference(s)
Scott S, Ferguson GG, Roach MR (1972) Comparison of the elastic properties of human intracranial arteries and aneurysms. Can J Physiol Pharmacol 50: 328–332 (with permission)

Pressure–Diameter Relation (9)

• Strain rate effect • Static P–D test	• Dogs • Iliac artery	• Smooth muscle activation •

Materials
- Healthy adult mongrel dogs
- Iliac artery

Testing Methods and Experimental Conditions
- Tubular specimen at its in vivo length
- Quasi-static pressure–diameter test between 0 and 250 mmHg at various inflation rates
- Medium: an aerated physiological salt solution (PSS) at 37°C
- Norepinephrine (5 µg/ml) for active smooth muscle

Data

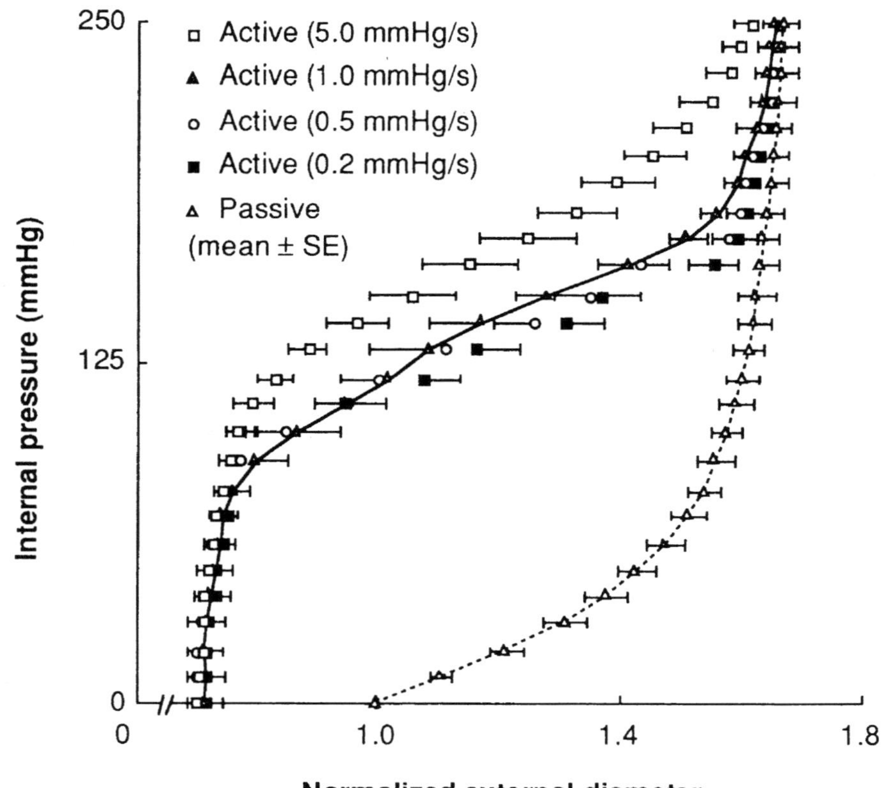

Comments
- Pressure–diameter curves of unactivated iliac arteries were essentially independent of inflation rate in the range 0.2–5 mmHg/s.

Reference(s)
Cox RH (1976) Mechanics of canine iliac artery smooth muscle in vitro. Am J Physiol 230:462–470 (with permission)

Pressure–Diameter Relation (10)

• Stress–strain curve • Incremental elastic modulus	• Dogs • Various arteries	• Regional variation • Connective tissue

Materials
- Sixteen healthy adult mongrel dogs of unknown age
- Carotid, renal, mesenteric, iliac, and coronary arteries

Testing Methods and Experimental Conditions
- Tubular specimen in aerated physiological salt solution at 37°C
- Pressure–diameter relation obtained at inflation rate of 1 mmHg/s (between 0 and 250 mmHg)

Data

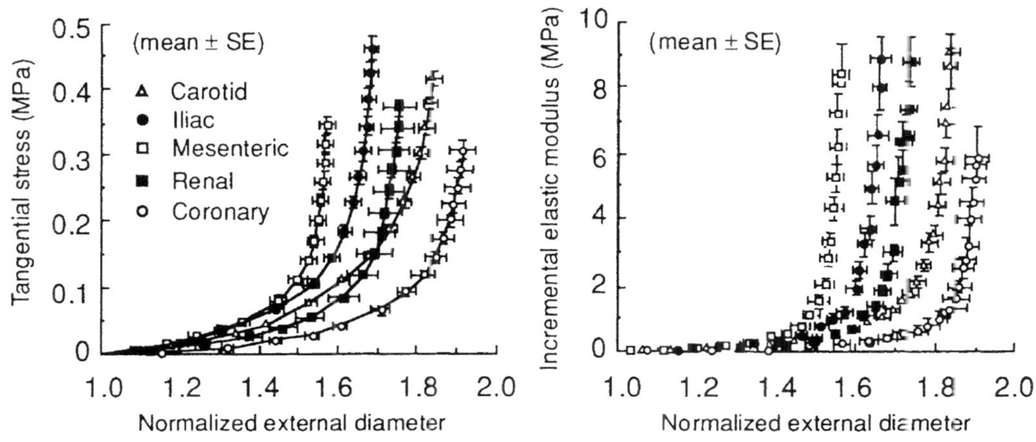

Comments
- Also presented:
  Typical examples of pressure–diameter relationships;
  Strain energy density–strain/pressure relationships;
  Connective tissue composition at various arterial sites.

Reference(s)
Cox RH (1978) Passive mechanics and connective tissue composition of canine arteries. Am J Physiol 234:H533–H541 (with permission)

Pressure–Diameter Relation (11)

• Stress–strain curve • Incremental elastic modulus	• Rats • Cerebral arteriole	• Stress unloading • In vivo measurement

Materials
- Wistar–Kyoto (WKY) and stroke-prone spontaneously hypertensive (SHRSP) rats (age, 10–12 months)
- First-order pial arterioles

Testing Methods and Experimental Conditions
- Partial occlusion of left common carotid artery with a clip (gap size, 0.30 mm) at 1 month of age
- Craniotomy over the left parietal cortex
- Smooth muscle deactivation with artificial cerebrospinal fluid with EDTA (25 mg/ml) at 37–38°C
- Pressure–diameter relation obtained while lowering blood pressure by hemorrhage

Data

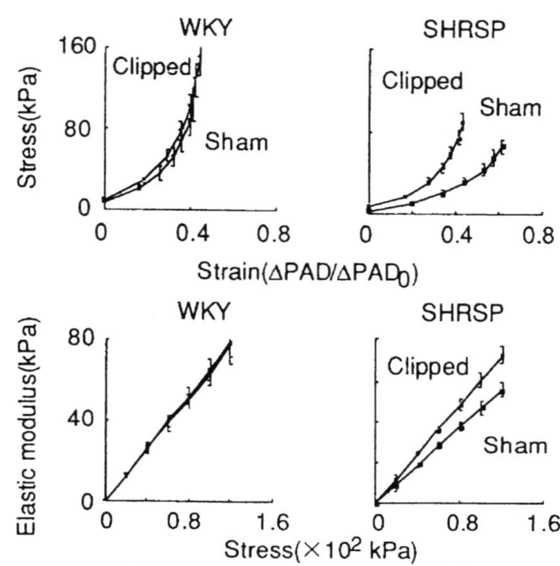

	WKY		SHRSP	
	Sham	Clipped	Sham	Clipped
Mean pial arteriolar pressure before SM deactivation (mmHg)	73 ± 2	63 ± 2*	110 ± 4*	94 ± 4†‡
Inner diameter at 70 mmHg (μm)	102 ± 7	113 ± 6	86 ± 6*	89 ± 4
Wall thickness (μm)	5.2 ± 0.5	3.9 ± 0.3*	6.8 ± 0.5*	5.0 ± 0.3‡

*$P < 0.05$ *vs* sham WKY; †$P < 0.05$ *vs* sham SHRSP; ‡$P < 0.05$ *vs* clipped WKY.

Comments
- Elastic modulus E was calculated by differentiating stress–strain data fitted to an exponential curve ($\sigma = \sigma_o e^{\beta\varepsilon}$), i.e., $E = \beta\sigma_o e^{\beta\varepsilon}$ (β, constant obtained by the fitting).

Reference(s)
Baumbach GL, Siems JE, Heistad DD (1991) Effects of local reduction in pressure on distensibility and composition of cerebral arterioles. Circ Res 68:338–351 (with permission)

Pressure–Diameter Relation (12)

• Stress–strain curve • Static P–D test	• Dogs • Iliac and carotid arteries	• Smooth muscle activation • Characteristic impedance

Materials
- Fifteen young, healthy, adult mongrel dogs
- Carotid and iliac arteries

Testing Methods and Experimental Conditions
- Tubular specimen (in vivo length) in an aerated physiological salt solution at 37°C
- Quasi-static pressure–diameter relation at inflation rate of 1 mmHg/s between 0 and 250 mmHg (Passive)
- Norepinephrine (5 µg/ml) added to medium to activate smooth muscle
- One-time inflation at a rate of 0.2 mmHg/s from 0 to 250 mmHg (Active)

Data

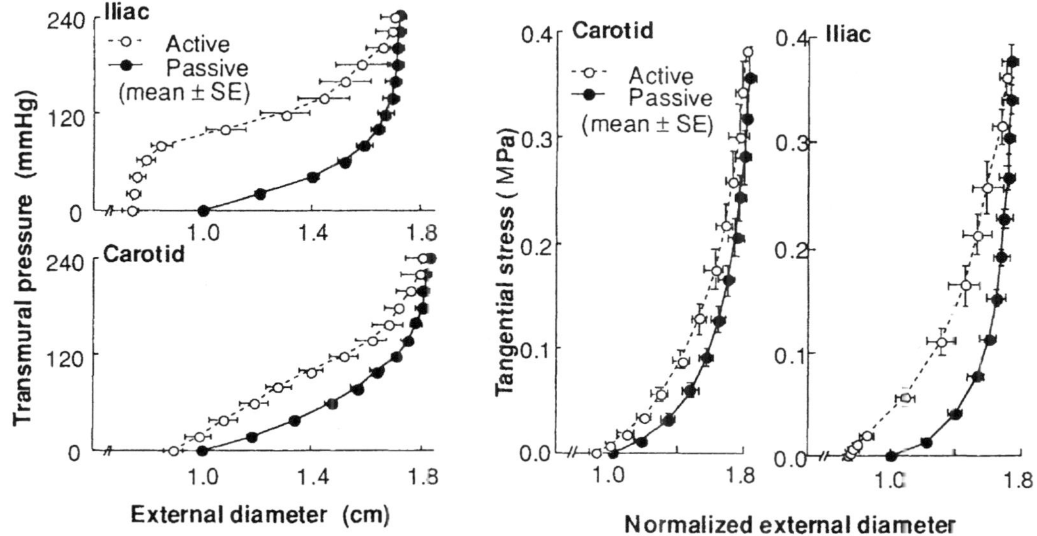

Comments
- Also presented:
  Incremental elastic modulus *vs* strain;
  Characteristic impedance *vs* transmural pressure.
- Values of external diameter were normalized for each segment by dividing by the value at zero pressure for passive conditions before averaging.

Reference(s)
Cox RH (1976) Effects of norepinephrine on mechanics of arteries in vitro. Am J Physiol 231:420–425 (with permission)

Pressure–Diameter Relation (13)

• Stress–strain curve • Static P–D test	• Puppies • Mesenteric artery	• Developmental change • Biochemical analysis

Materials
- A litter of puppies and their mother (to ensure homogeneity of animal group)
- Mesenteric artery

Testing Methods and Experimental Conditions
- Tubular specimen in aerated physiological salt solution at 37°C
- Pressure–diameter relation obtained at inflation rate of 1 mmHg/s
- Diameter measured with a cantilever transducer

Data

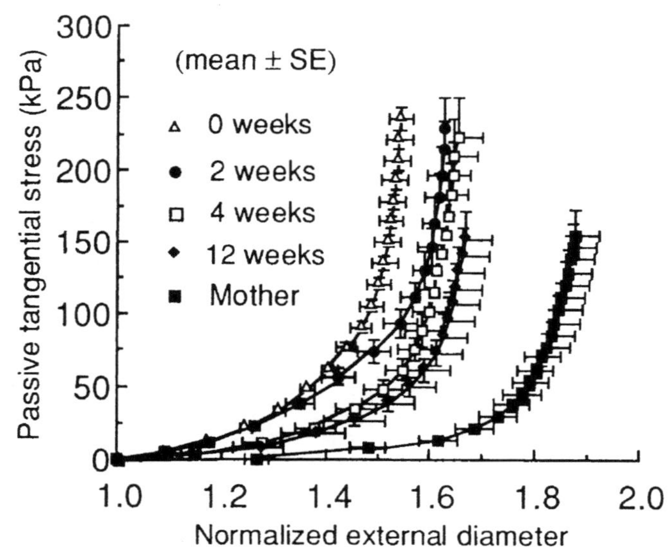

Comments
- Also presented:
  Response to smooth muscle contraction;
  Changes in connective tissue and electrolyte contents during growth.
- Relation between passive tangential wall stress and external diameter for superior mesenteric artery at various ages.
- Diameter is normalized by dividing by value at zero pressure and in vivo length.
- Values of wall stress and normalized diameter were averaged at specific values of transmural pressure for all animals in each age group.

Reference(s)
Cox RH, Jones AW, Swain ML (1976) Mechanics and electrolyte composition of arterial smooth muscle in developing dogs. Am J Physiol 231:77–83 (with permission)

Pressure–Diameter Relation (14)

• Stress–strain curve • Vessel dimension	• Rats • Cerebral arteries	• Hypertension • In vivo measurement

Materials

- Age-matched 6 to 8 months old stroke-prone spontaneously hypertensive rats (SHRSP) and Wistar–Kyoto rats (WKY)
- Second- and third-order branches of the posterior cerebral artery

Testing Methods and Experimental Conditions

- Anesthesia with sodium pentobarbital (5 mg/100 g BW, i.p.) under mechanical ventilation
- Craniotomy over the left parietal cortex
- Smooth muscle deactivated with 67 mM EDTA
- Pressure–diameter relation obtained while lowering blood pressure by hemorrhage
- Fixed at 70 mmHg after experiment for the measurement of vessel dimension

Data

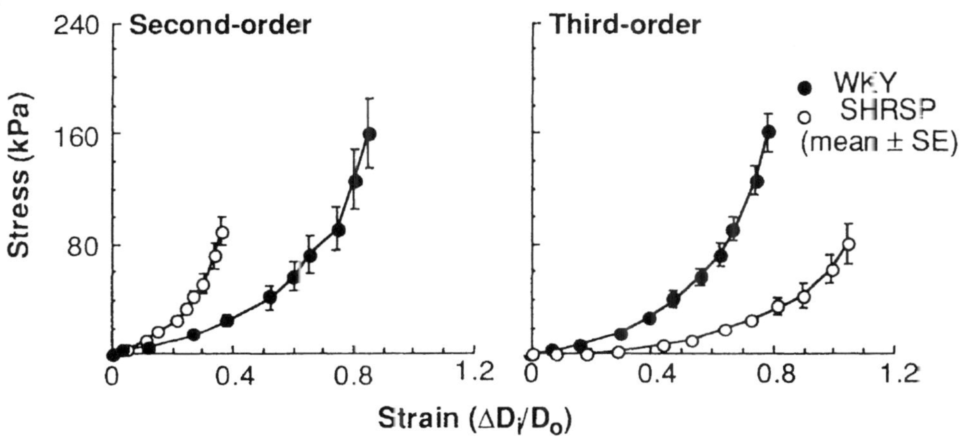

	Large		Small	
	WKY	SHRSP	WKY	SHRSP
Internal diameter (μm)	201 ± 23	148 ± 7*	127 ± 12	84 ± 10**
External diameter (μm)	222 ± 22	176 ± 6*	140 ± 12	104 ± 10**
Wall thickness (μm)	10.8 ± 0.9	13.9 ± 1.2*	6.7 ± 0.6	9.7 ± 1.3**
Wall-to-lumen ratio	0.06 ± 0.01	0.10 ± 0.01*	0.06 ± 0.01	0.14 ± 0.04**
Slope of tangential elastic modulus *vs* stress, ß	5.4 ± 0.5	11.1 ± 1.8*	6.2 ± 0.2	4.8 ± 0.5**
Cross-sectional area (μm^2)	7070 ± 787	7009 ± 582	2809 ± 339	2881 ± 537

Values are means $\pm$ SE in 7 WKY and 7 SHRSP. Internal diameter, wall thickness and wall-to-lumen ratio were obtained at intravascular pressure of 70 mmHg after deactivation of smooth muscle. * $P < 0.05$ *vs* large arteries in WKY. **$P < 0.05$ *vs* small arteries in WKY.

Comments

- Tangential elastic modulus *vs* strain and pressure–diameter relation are also presented.

Reference(s)

Hajdu MA, Baumbach GL (1994) Mechanics of large and small cerebral arteries in chronic hypertension. Am J Physiol 266:H1027–H1033 (with permission)

Pressure–Diameter Relation (15)

• Tension–length curve • Static pressure–volume test	• Human • Iliac artery	• Effect of digestion • Effect of preservation

Materials

- Human
- External iliac artery obtained at autopsy
- Stored in the refrigerator in 1/10000 saline solution of methiolate

Testing Methods and Experimental Conditions

- Static pressure–volume test on fresh specimen at room temperature
- No restraint on the longitudinal direction
- Fresh specimen digested by formic acid to remove collagen or by trypsin to remove elastin
- Study effects of collagen and elastin digestions

Data

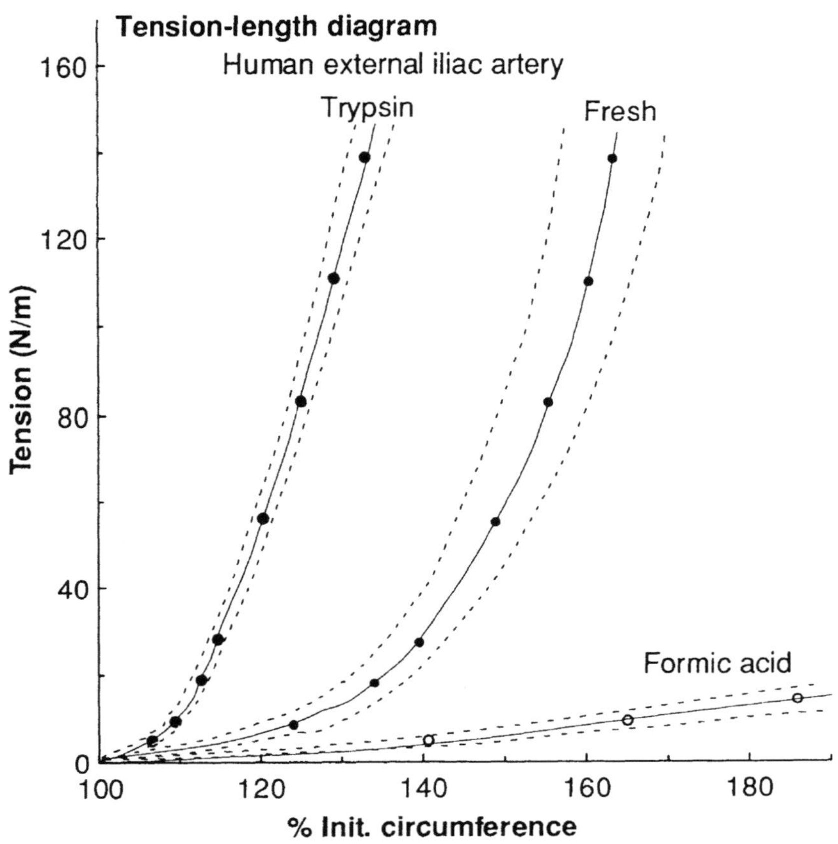

Comments

- The figure shows the elastic diagram of an iliac artery representing the mean of the results on nine vessels of the young to middle age group (20–50 years). The broken lines indicate the standard error.
- No change in the distensibility curves was detectable during periods up to 10 days.

Reference(s)

Roach MR, Burton AL (1957) The reason for the shape of the distensibility curves of arteries. Can J Biochem Physiol 35:681–690

Pressure–Diameter Relation (16)

• Vessel dimension • Static P–D test	• Rabbits • Thoracic aorta	• Atherosclerosis • Pathological change

Materials

- Matured male Japanese White rabbits weighing 3.1–3.5 kg
- Descending thoracic aorta

Testing Methods and Experimental Conditions

- Atherosclerosis induced by regular chow + 1% cholesterol and/or aortic endothelium denudation by ballooning (4, 8, 16, 32 weeks)
- Quasi-static pressure–diameter test in aerated Krebs–Ringer solution at 37°C
- Mechanical parameters were calculated from mean diameters of loading and unloading curves
- Correlation with intraluminal area fraction of atherosclerotic lesion (A_s)

Data

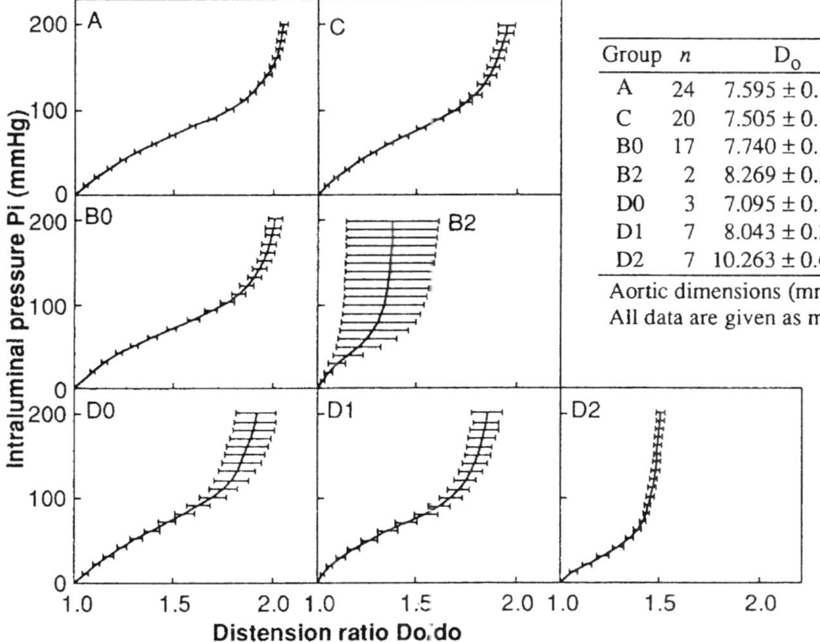

Group	n	D_0	H
A	24	7.595 ± 0.157	0.1918 ± 0.0067
C	20	7.505 ± 0.124	0.2284 ± 0.0169
B0	17	7.740 ± 0.198	0.2064 ± 0.0073
B2	2	8.269 ± 0.582	0.3990 ± 0.1320
D0	3	7.095 ± 0.193	0.2533 ± 0.0535
D1	7	8.043 ± 0.316	0.2641 ± 0.0230
D2	7	10.263 ± 0.677	0.4100 ± 0.0381

Aortic dimensions (mm) at 100 mmHg.
All data are given as mean ± SE.

Comments

- Experimental groups: A, control; B, 1% cholesterol; C, EC denudation; D, 1% cholesterol + EC denudation; B0 and D0, A_s < 80%; D2, A_s ≥ 80% and β″<10; B2 and D2, A_s ≥ 80% and β″ ≥ 10.

Reference(s)

Hayashi K, Ide K, Matsumoto T (1994) Aortic walls in atherosclerotic rabbits–mechanical study. J Biomech Eng 116:284–293

Shear Property (1)

• Multi-axial test • Torsion test	• Dogs • Aorta, carotid artery	• Test system •

Materials

• A common carotid artery and a descending thoracic aorta

Testing Methods and Experimental Conditions

• Cylindrical specimen
• Medium: oxygenated calcium-free Krebs solution with 2 mM EGTA at 37°C
• Quasi-static inflation test under constant axial length
• Quasi-static axial extension test under constant diameter
• Quasi-static torsion test under constant axial length
• Quasi-static torsion test under constant diameter

Data

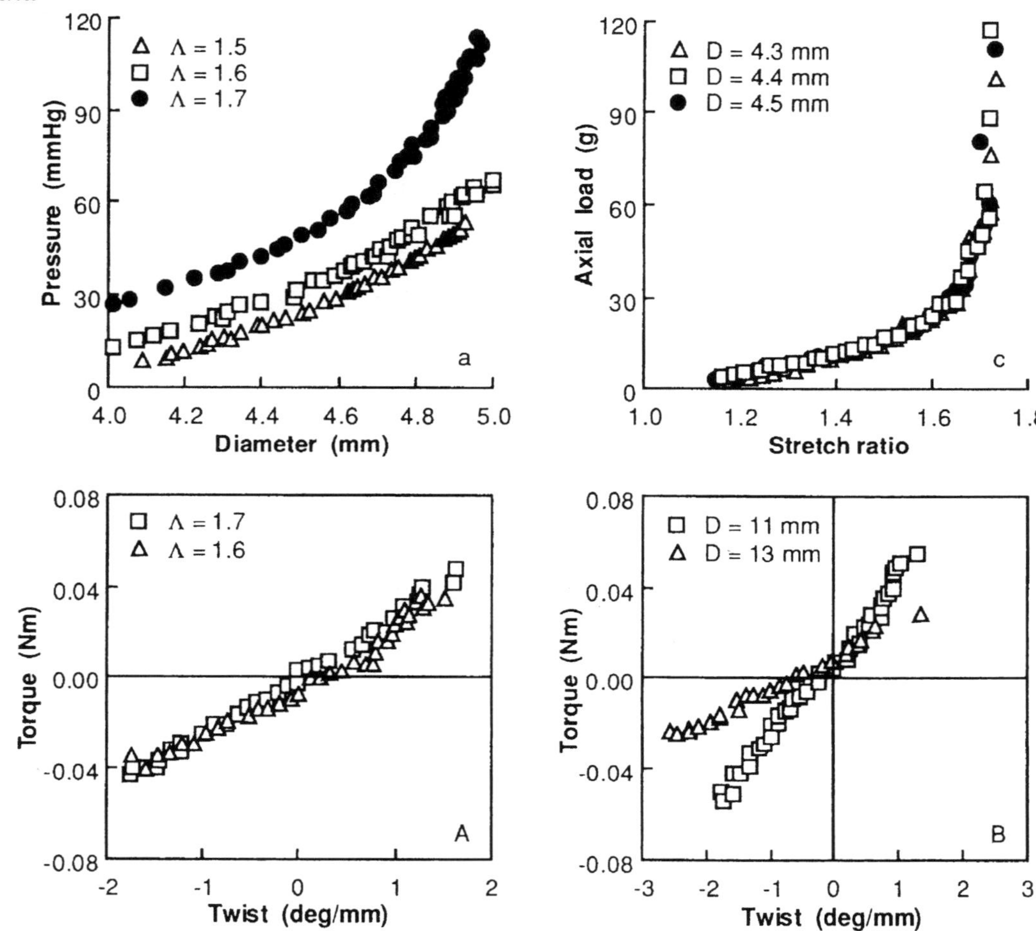

Comments

None.

Reference(s)

Humphrey JD, Kang T, Sakarda P, Anjanappa M (1993) Computer-aided vascular experimentation: A new electromechanical test system. Ann Biomed Eng 21:33–43 (with permission)

Shear Property (2)

• Shear modulus • Torsion test	• Rats • Thoracic aorta	• Hypertension • Effect of temperature

Materials
- Sprague–Dawley rats
- Thoracic aorta

Testing Methods and Experimental Conditions
- Tubular specimen subjected to a transmural pressure, a longitudinal force, and a torque
- Relation between twisted angle and torque was measured under various transmural pressure and longitudinal force conditions
- Experiments were done in aerated Krebs solution at 24°C

Data

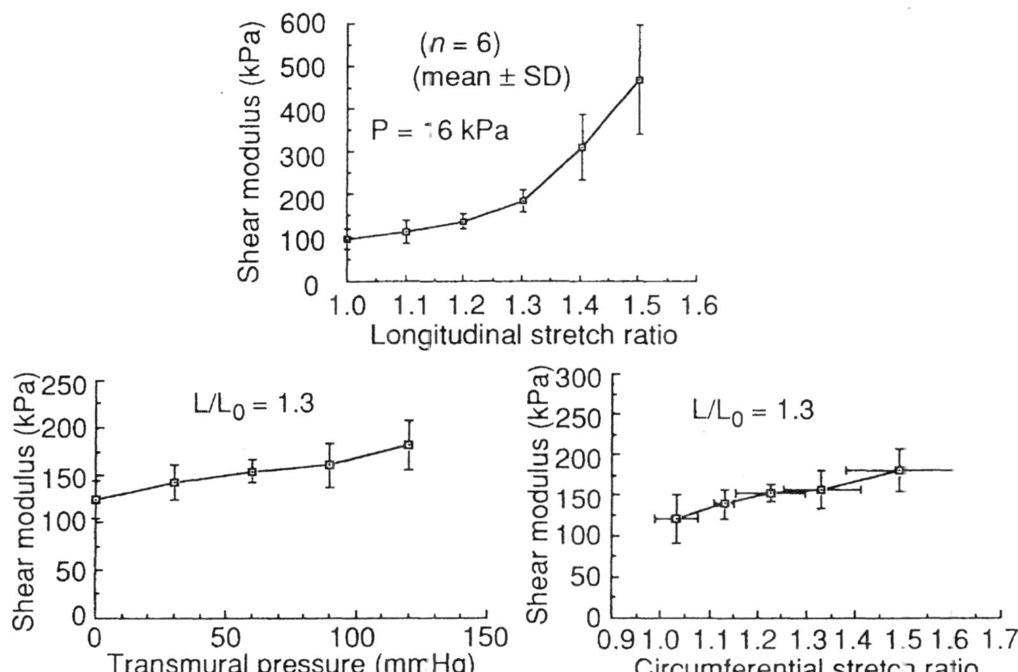

Comments
- Change in shear modulus in the course of developing hypertension was not significant.
- Change in shear modulus due to temperature increase from 24° to 37°C was within 10%.

Reference(s)
Deng SX, Tomioka J, Debes JC, Fung YC (1994) New experiments on shear modulus of elasticity of arteries. Am J Physiol 266:H1–H10 (with permission)

Tensile Property (1)

• Axial property • Force–length relation	• Dogs • Carotid and femoral	• Axial tethering •

Materials
- Mongrel dogs
- Carotid and femoral arteries

Testing Methods and Experimental Conditions
- Medium: Ringer–Tyrode solution at 37°C with 100 mg/l of KCN to deactivate smooth muscle
- Axial force–length relation measured on tubular segments at different constant pressures (0, 20, 40, 80, 120, 160, and 200 mmHg)

Data

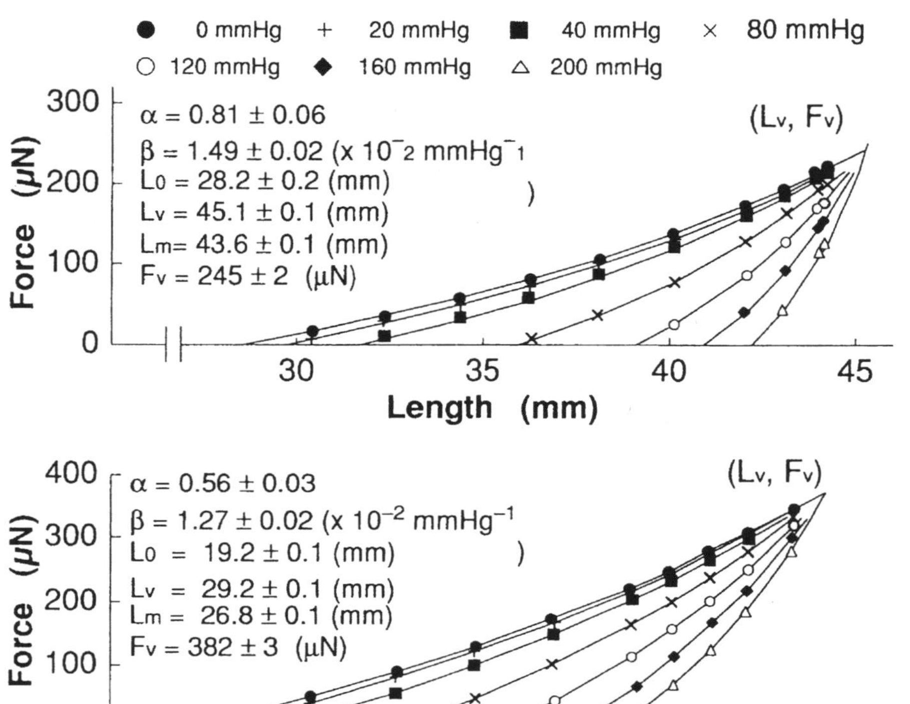

Comments
- Formula used to fit length-force data:

$$F = F_v(L - L_s)/\{L_v - L_s + \alpha(L_v - L)\}$$

$$L_s = L_o + (L_m - L_o)(1 - e^{-\beta P})^2; \quad (L_o, L_v, L_m, F_v, \alpha, \text{ and } \beta \text{ are constants}).$$

- The length–force curves for the different pressures tend to have a common point of intersection, which corresponds to the length of the artery in vitro.
- Similar results were obtained for canine aorta, human carotid and iliac arteries.

Reference(s)
van Loon P, Klip W, Bradley EL (1977) Length–force and volume–pressure relationships of arteries. Biorheology 14:181–201 (with permission)

Tensile Property (2)

• Axial property • Stress–strain curve	• Human • Cerebral blood vessels	• •

Materials
- Human
- Cerebral blood vessels (middle, anterior, and posterior cerebral arteries and veins)

Testing Methods and Experimental Conditions
- Uniaxial tensile test of tubular specimen in longitudinal direction
- Inner pressure kept at 80 mmHg (approximate in vivo value)
- Medium: lactate Ringer solution at 37°C

Data

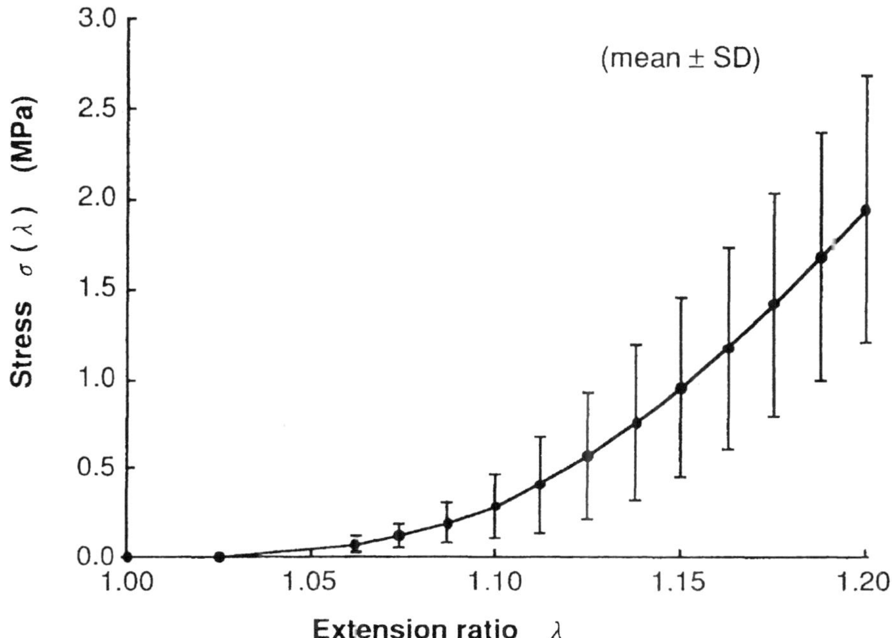

Comments
None.

Reference(s)
Chalupnik JD, Daily CH, Merchant HC (1971) Material properties of cerebral blood vessels.
Final Report of Contract NoNIH-69-2232, Report No. ME 71–11

Tensile Property (3)

• Axial property • Stress–strain curve	• Sheep • Aorta and its branch	• Local property • Aorta branch junction

Materials
- Ewes weighing 40–50 kg
- Abdominal aorta with its major branch

Testing Methods and Experimental Conditions
- Longitudinal strip including the aorta, the bifurcation region, and the branch
- Grid applied to each region with India ink
- Uniaxial loading/unloading in a load range corresponding to 40–200 mmHg while taking photographs of the specimen at crosshead speed of 20 mm/min

Data

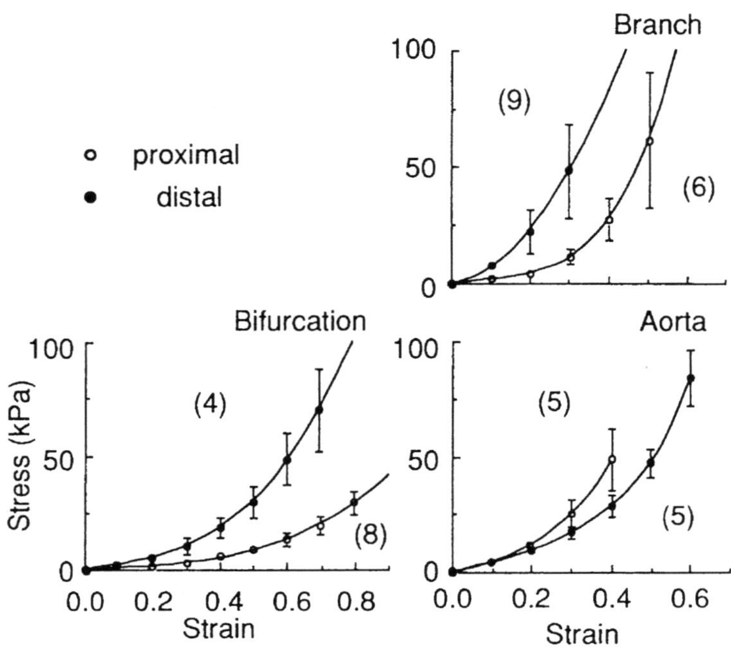

Comments
- Figures show plots of the averaged data for each region.
- The bifurcation and branch regions show statistically significant differences using the Student t-test.

Reference(s)
Cleave J, Roach MR (1982) Comparison of longitudinal elastic properties of proximal and distal strips of aorta-branch junctions from the abdominal aorta of sheep. Can J Physiol Pharmacol 61:614–618 (with permission)

Tensile Property (4)

• Fracture toughness • Tensile strength	• Pigs • Aorta	• Positional variation •

Materials
- Landrace x Large White pigs
- Descending thoracic aorta

Testing Methods and Experimental Conditions
- Aortic specimens kept frozen at -20°C until use, when they were thawed in normal saline
- Segments obtained from six different positions along the aorta (proximal, 1; distal, 5)
- Tear test on notched specimens in circumferential or longitudinal direction at an extension rate of 50 mm/min, unloaded before complete fracture
- Tensile test on unnotched specimens in circumferential and longitudinal directions
- Specimen kept wet, in air, at room temperature during mechanical tests

Data

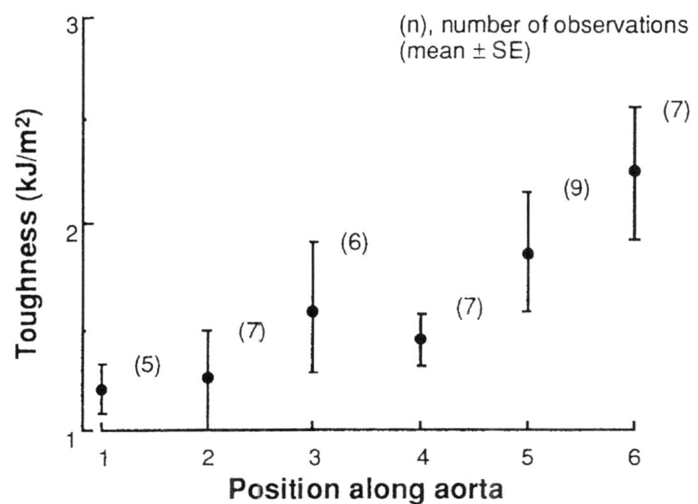

Longitudinal			Circumferential		
Position	Breaking stress (MPa)		Position	Breaking stress (MPa)	
1	0.18 ± 0.44	(7)	2 and 3	2.19 ± 0.57	(11)
4 and 5	0.87 ± 0.34	(9)	6	3.64 ± 0.53	(5)

All data are given as mean $\pm$ SE (number of observations).

Comments
- Fracture toughness was calculated by dividing the total work given to a specimen by the nominal area through which crack propagation has occurred.
- Longitudinal toughness was so much greater than circumferential toughness that it was difficult to achieve crack propagation right in the longitudinal direction.
- Stress–strain curves in circumferential and axial directions were also presented

Reference(s)
Purslow PP (1983) Positional variations in fracture toughness, stiffness and strength of descending thoracic pig aorta. J Biomech 16:947–953 (with permission)

Tensile Property (5)

• Stiffness • Failure stress/strain	• Dogs • Thoracic aorta	• Hypercholesterolemia • Bilinear analysis

Materials
- Mongrel dogs of both sexes (weight, 15–20 kg)
- Hyperlipidemic group: bilateral subtotal thyroidectomy (95%) + cholesterol diet (a normal ad libitum diet complemented by 20 g cholesterol, 400 mg propylthiouracil, and 5 g suet) for approx. 11 months
- Control group: a normal ad libitum diet for approx. 11 months
- Thoracic aorta obtained after death and kept in an icy normal saline until test, which was conducted within 24 h postmortem

Testing Methods and Experimental Conditions
- Aortic segment between the ligamentum arteriosum and the first intercostal artery cut into 4-mm-wide serial ring-like sections
- Medium: normal saline at room temperature
- Tensile test at a rate of 600 mm/s until failure
- (Stress) = (recorded load) / (cross-sectional area)
      where (cross-sectional area) = (post-test width) x (wall thickness) / 2
- (Strain) = (input stretch) / (test gage length)
- Stress–strain curve considered to be bilinear to calculate initial and secondary stiffnesses

Data

	Normolipemic	Hyperlipidemic
Number of animals[a]	7	6
Wall thickness (mm)	1.98 ± 0.22	1.88 ± 0.25
Mean blood pressure at death (mmHg)	127 ± 15	116 ± 14
Initial stiffness (kPa)	150 ± 17	132 ± 36
Secondary stiffness (kPa)	4390 ± 644	3860 ± 666*
Secondary slope intercept (mm/mm)	1.023 ± 0.102	0.899 ± 0.055*
Failure stress (kPa)	1891 ± 256	1568 ± 285*
Failure strain (mm/mm)	1.43 ± 0.10	1.29 ± 0.06*
Energy to rupture (kJ/m^3)	482 ± 88	362 ± 71*

All data are given as mean $\pm$ SD.
*$P < 0.05$ *vs* normolipemic (2-sided *t*-test).
[a]Three specimens were obtained from each animal.

Comments
- Animals had been subjected to an aortocoronary bypass for another experiment.

Reference(s)
Haut RC, Garg BD, Metke M, Josa M, Kaye MP (1980) Mechanical properties of the canine aorta following hypercholesterolemia. J Biomech Eng 102:98–102

Tensile Property (6)

• Stress–strain curve • Ultimate strain	• Human • Coronary plaque	• Atherosclerosis • Pathological change

Materials
- Human coronary artery obtained at autopsy
- Fibrous caps from atherosclerotic plaques
- Nearby intima not involved in a lesion

Testing Methods and Experimental Conditions
- Medium: physiological saline at room temperature
- Specimen dimension, ≈ 0.5 mm in cross-section (sic) and 3–4 mm long with its major axis in the longitudinal direction
- Uniaxial tensile test

Data

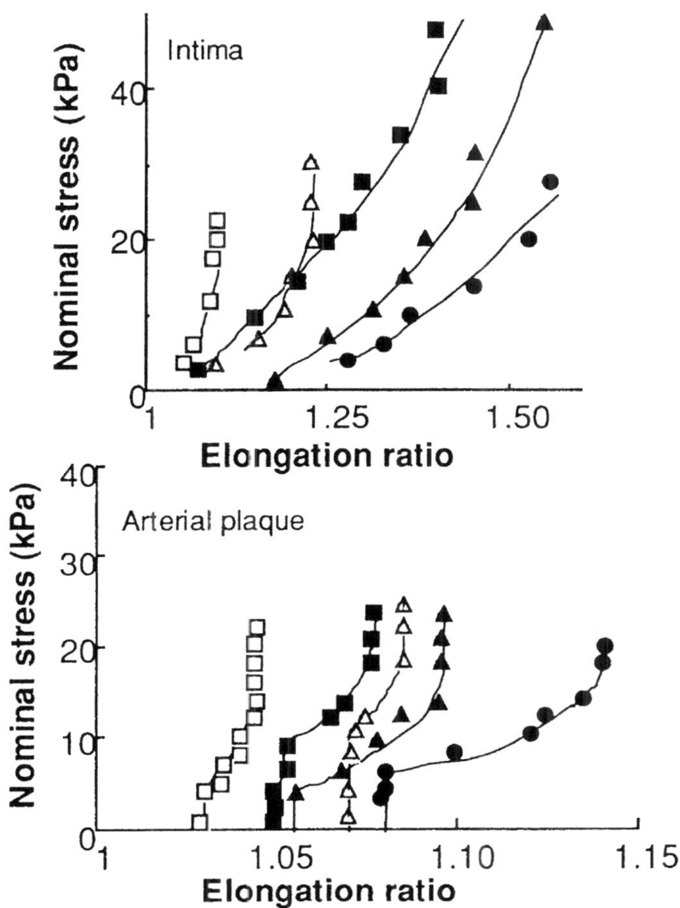

Comments
- Stress–elongation ratio curve of human coronary artery intima (not locally involved in atherosclerotic plaque) and plaque.

Reference(s)
Born GVR, Richardson PD (1990) Mechanical properties of human atherosclerotic lesions. In: Glagov S et al. (eds) Pathology of the human atherosclerotic plaque. Springer, Berlin, Heidelberg, New York, pp 413–423

Tensile Property (7)

• Tensile strength • Incremental elastic modulus	• Rats • Aorta	• Effect of exercise • Biochemical properties

Materials

- Twelve male Kyoto–Wistar rats (age, 9 weeks; mean weight, 243 g)
- 6 sedentary, 6 trained
- Descending thoracic aorta

Testing Methods and Experimental Conditions

- Rats in training group forced to swim 1 h/day, 6 days/week for 16 weeks in 35°C water
- Tensile test on ring specimens in a saline solution at 24 ± 1°C (tensile speed, 2 mm/min)
- Stress calculated based on the initial load-free cross-sectional area

Data

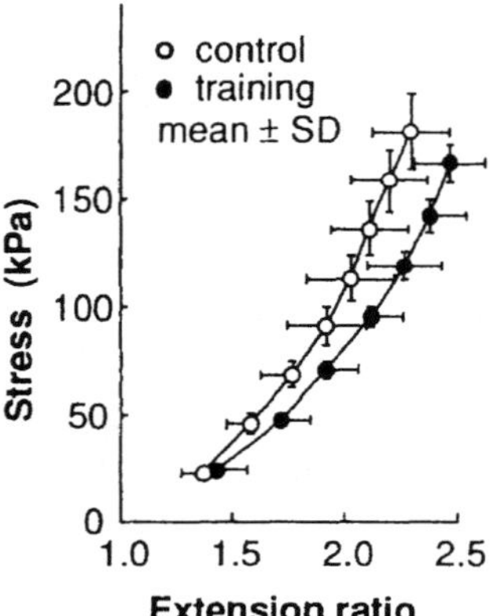

	Control Rats ($n = 6$)	Training Rats ($n = 6$)
Tensile strength (kPa)	1448.4 ± 254.8	1884.5 ± 469.4*
Ultimate tensile extension ratio	31.16 ± 2.65	17.54 ± 1.47*
Incremental elastic modulus (kPa)		
at extension ratio =1.5	124.5 ± 39.2	88.2 ± 22.5*
2.0	245.0 ± 91.1	158.8 ± 45.1*
2.5	980.0 ± 655.6	494.9 ± 205.8

Values are means $\pm$ SD. *$P < 0.05$; **$P < 0.01$.

Comments

None.

Reference(s)

Matsuda M, Nosaka T, Sato M, Oshima N, Fukushima H (1988) Effect of prolonged voluntary running on biochemical and biomechanical properties of rat aorta. J Jpn Coll Angiol 28:477–480

Tensile Property (8)

• Tensile strength • Maximum extension	• Human • Aortic plaque cap	• Atherosclerosis • Macrophage

Materials
- Human adults over the age of 40 excluding subjects with hypertension and diabetes
- Plaque tissue with or without rupture (ulceration)

Testing Methods and Experimental Conditions
- Uniform strips (7 x 1.5 mm) taken from each plaque in a circumferential direction
- Medium: physiological saline at 37°C
- Constant rate uniaxial elongation to fracture
- Determination of macrophage density

Data

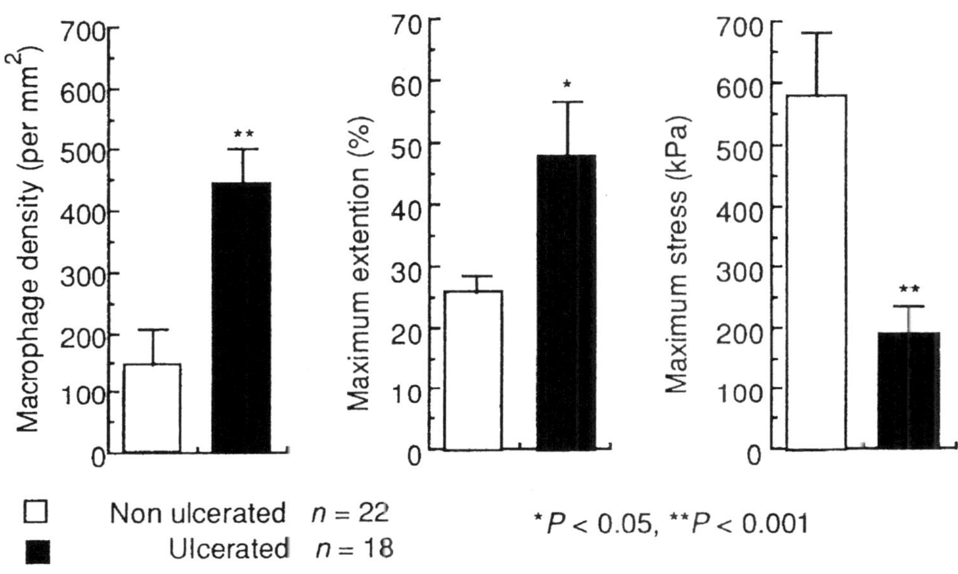

Comments
- Macrophage density had close correlation with mechanical conditions (stress and elongation) at cap tissue fracture.

Reference(s)
Lendon CL, Davies MJ, Born GVR, Richardson PD (1991) Atherosclerotic plaque caps are locally weakened when macrophage density is increased. Atherosclerosis 87:87–90 (with permission)

Tensile Property (9)

• Tensile strength • Ultimate strain	• Rabbits • Aortas	• Atherosclerosis • Pathological change

Materials
- New Zealand White rabbits (male; body weight, 2.8 ± 0.3 kg [mean ± SD])
- Thoracic and abdominal aortas

Testing Methods and Experimental Conditions
- Diet, normal chow + 0.5% cholesterol + 0.5% olive oil
- Feeding period, 8–25 weeks
- Tensile test of ring specimens 2–3 mm in width
- Cross-head speed, 2 mm/min
- Room temperature

Data

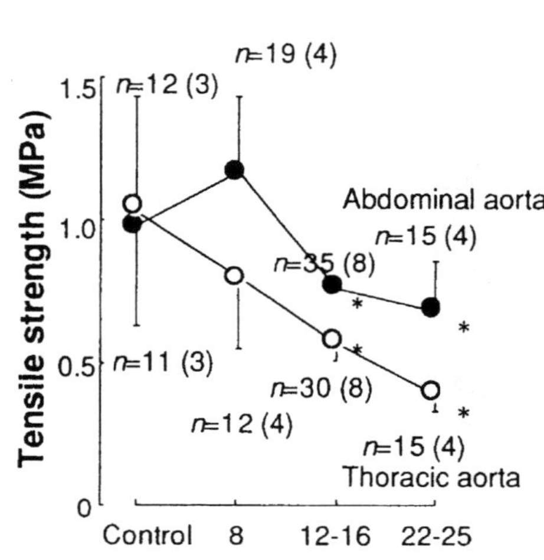

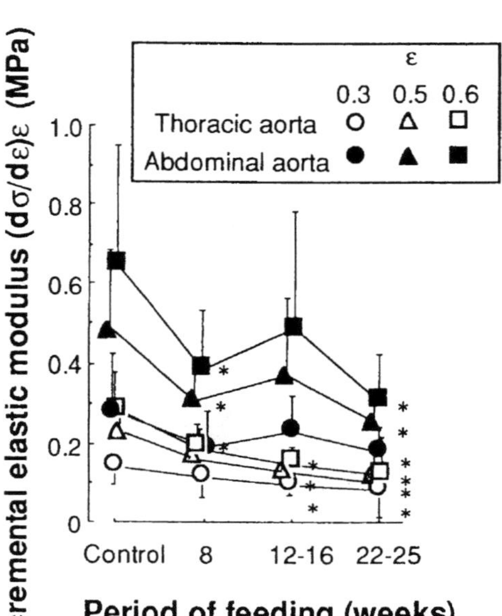

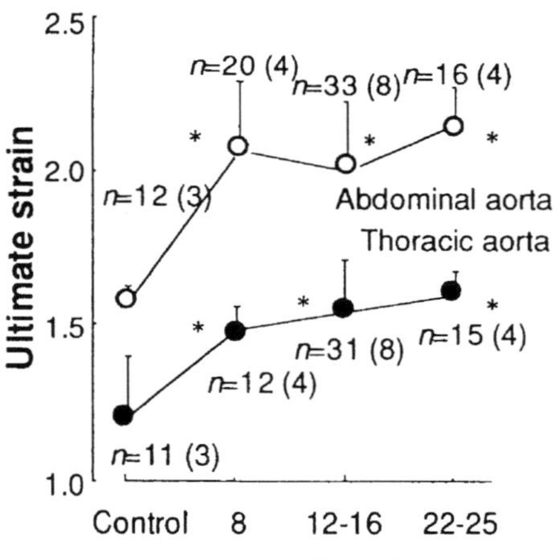

Comments

- Incremental elastic modulus was defined at $\varepsilon = 0.3$, 0.5, and 0.6.
- Stress–strain relationship was expressed by:

$$\sigma = -k_0 + k_0\exp(k_1\varepsilon), \quad (k_0, k_1; \text{constants}).$$

- *P* < 0.05 *vs* Control.
- Numbers in parentheses indicate numbers of animals.

Reference(s)

Abd El-Haleem MA, Sato M, Oshima N (1994) Effect of cholesterol feeding periods on aortic mechanical properties of rabbits. JSME Int J, Ser A 37:79–86

Viscoelasticity (1)

• Dynamic elastic modulus • Dynamic P–D test	• Dogs • Aorta, femoral, carotid	• Classics • Frequency dependency

Materials

- Dogs
- Thoracic and abdominal aortas, femoral and carotid arteries

Testing Methods and Experimental Conditions

- Dynamic pressure–diameter relation
  (mean pressure, 100 mmHg; amplitude, ± 5–10 mmHg; frequency, 2–18 Hz)
- Specimens stretched to their in vivo length
- Medium: isotonic saline at room temperature

Data

Vessel	E_{dyn} at each frequency (MPa)			
	0 Hz	2 Hz	5 Hz	18 Hz
Thoracic aorta	0.44 ± 0.040 (10)	0.47 ± 0.042 (10)	0.49 ± 0.045 (10)	0.53 ± 0.080 (4)
Abdominal aorta	0.92 ± 0.094 (7)	1.09 ± 0.088 (7)	1.10 ± 0.082 (7)	1.22 ± 0.046 (4)
Femoral artery	0.90 ± 0.115 (5)	1.20 ± 0.081 (5)	1.20 ± 0.082 (5)	1.06 ± 0.139 (5)
Carotid artery	0.60 ± 0.048 (6)	1.10 ± 0.100 (6)	1.13 ± 0.099 (6)	1.15 ± 0.103 (6)

All data are given as mean ± SE (n).

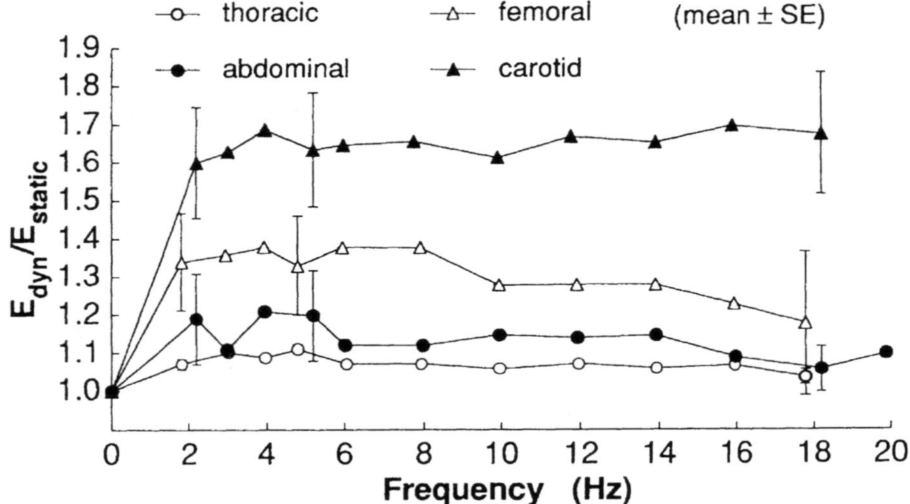

Comments

- $E_{dyn} = E' \cos \phi$; $E' = \Delta P\, 2(1 - \sigma^2)R_o R_i^2 / \{\Delta R\, (R_o^2 - R_i^2)\}$.

  ϕ, phase angle between pressure changes and radius changes; P, pressure; R_o, outer radius; R_i, inner radius; σ, Poisson's ratio.

Reference(s)

Bergel DH (1961) The dynamic elastic properties of the arterial wall. J Physiol 156:458–469
(with permission)

Viscoelasticity (2)

• Dynamic elastic modulus • Pulse wave velocity	• Dogs • Aorta, iliac, femoral	• In vivo measurement • Hemodynamics

Materials

- Mongrel dogs and greyhounds (age, approximately 2–4 years), weight 12.5–31 kg (mean 22 kg)
- Upper descending, midthoracic, lower thoracic, and midabdominal aortas, and iliac, femoral, and saphenous arteries

Testing Methods and Experimental Conditions

- Anesthetized with pentobarbital sodium, 20–35 mg/kg i.v.
- Mechanical ventilation
- Simultaneous measurement of pressure and diameter
- Fourier analysis up to 10 harmonics

Data

Weight (kg)	D_o (mm)	γ	BP (mmHg)	Frequency (Hz)	E_{dyn} (kPa)	$\eta\omega$ (kPa)	c (m/s)
Upper descending thoracic aorta							
25 ± 2	19.0 ± 0.8	0.12 ± 0.02	110 ± 7	1.86 ± 0.37	290 ± 50	33 ± 15	4.5 ± 0.2
Midthoracic aorta							
23 ± 1	14.3 ± 0.9	0.14 ± 0.01	105 ± 4	2.09 ± 0.16	300 ± 30	26 ± 7	4.9 ± 0.2
Lower thoracic aorta							
26 ± 1	13.8 ± 0.6	0.11 ± 0.02	112 ± 9	1.77 ± 0.04	570 ± 100	58 ± 17	5.9 ± 0.3
Midabdominal aorta							
23 ± 1	10.6 ± 0.7	0.12 ± 0.01	115 ± 6	2.03 ± 0.20	980 ± 130	123 ± 23	8.5 ± 0.4
Iliac artery							
21 ± 2	4.6 ± 0.3	0.16 ± 0.01	105 ± 7	1.93 ± 0.26	1100 ± 240	226 ± 69	10.2 ± 0.7
Femoral artery							
22 ± 1	4.4 ± 0.1	0.13 ± 0.01	118 ± 4	1.83 ± 0.16	1460 ± 260	224 ± 70	10.4 ± 0.5
Saphenous artery							
26	1.5	0.12	100	1.08	6060	300	22.0

Values are mean $\pm$ SE.

Comments

- Pressure–strain elastic modulus: $E_p = (\Delta P)/(\Delta D/D_o)$, where ΔP and ΔD are pressure and diameter changes, respectively, and D_o mean outer diameter.
- Dynamic incremental elastic modulus: $E_{inc} = 1.5E_p(1 - \gamma)^2/\{1 - (1 - \gamma)^2\}$, where $\gamma = h/R_o$; h, wall thickness; R_o, outer radius.
- Complex elastic modulus: $E_{dyn} = E_{inc} \cdot \cos\phi$, where ϕ is the phase of the complex elastic modulus.
- Viscous modulus: $\eta\omega = E_{inc} \cdot \sin\phi$, where $\eta\omega$ is the viscous modulus, and ω is the angular frequency.
- Pulse wave velocity: $c = (1.5E_{dyn} \cdot \gamma/\rho)^{1/2}$, where r is the density of blood.

Reference(s)

Gow BS, Taylor MG (1968) Measurement of viscoelastic properties of arteries in the living dog. Circ Res 23:111–122 (with permission)

Viscoelasticity (3)

• Dynamic elastic modulus • Viscous retarding force	• Human • Aorta, carotid, iliac, femoral	• Age-related variation • Frequency dependency

Materials
- Human
- Thoracic and abdominal aortas, and carotid, iliac, and femoral arteries

Testing Methods and Experimental Conditions
- Stored frozen in physiological saline after excision at post mortem until use
- Grossly atheromatous vessels were discarded
- Medium, physiological saline at room temperature
- Dynamic pressure–diameter measurement
  (mean pressure, 100 mmHg; amplitude, ±5–15 mmHg; frequency, 1–10 Hz)

Data

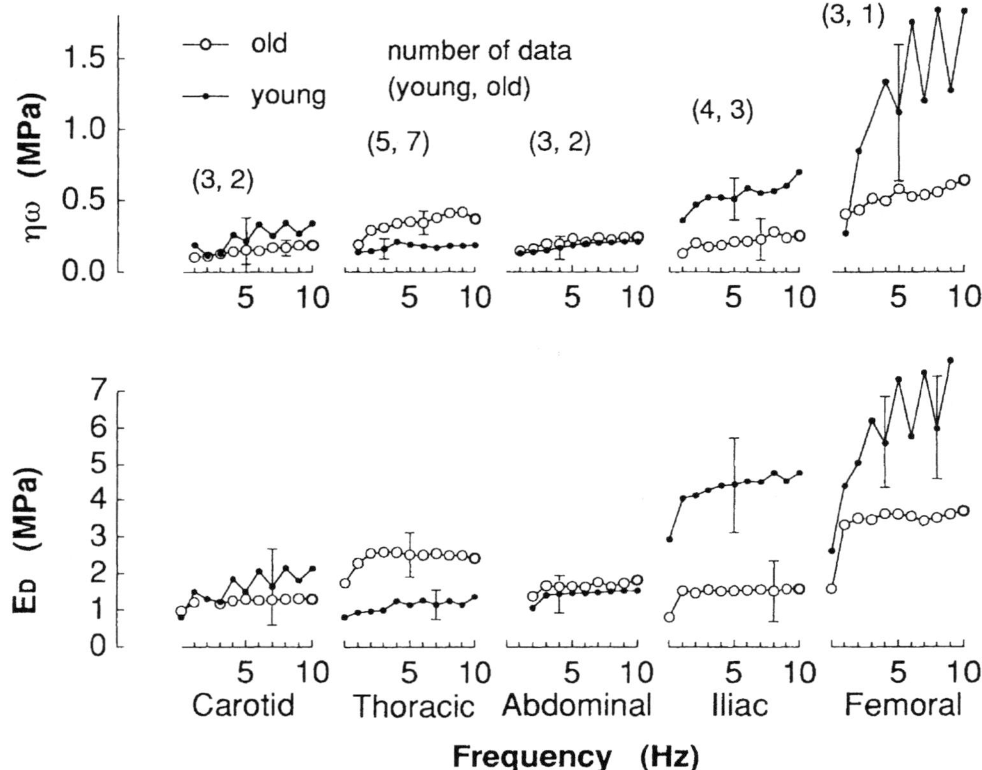

Comments
- Dynamic modulus:

  $E_D = E_C \cos \phi$, dynamic elastic component;

  $\eta\omega = E_C \sin \phi$, viscous retarding force.

  E_C, the amplitude of complex elastic modulus; ϕ, phase difference.

Reference(s)
Learoyd BM, Taylor MG (1966) Alterations with age in the viscoelastic properties of human arterial walls. Circ Res 18:278–292 (with permission)

Viscoelasticity (4)

• Incremental elastic modulus • Dynamic P–D test	• Rabbits • Carotid artery	• Smooth muscle activation • Inflation vs deflation

Materials
- Rabbit weighing approximately 2 kg
- Left common carotid artery

Testing Methods and Experimental Conditions
- Tubular specimen perfused with aerated Tyrode's solution for 45 min then filled with blood
- 10^{-5} M noradrenaline to activate smooth muscle
- Static pressure–diameter relation obtained every 20 mmHg, dynamic test ($\pm$ 10 mmHg, 4 Hz) at each pressure
- Identical test in replenished Tyrode's solution in the absence of noradrenaline

Data

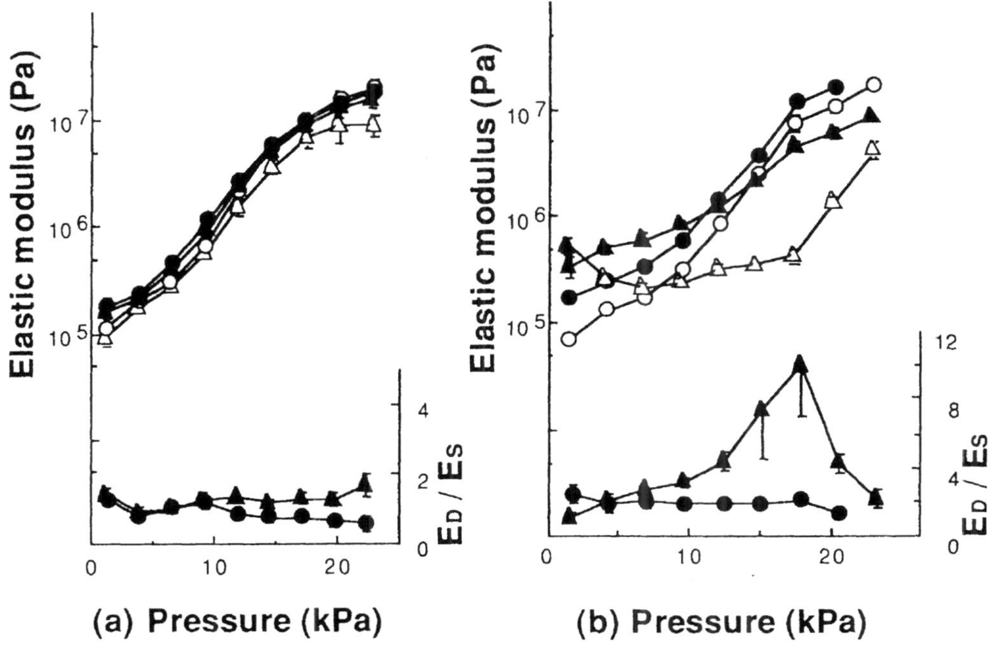

Comments
- Incremental elastic modulus, E:
$$E = 2R_i^2 R_o/(R_o^2 - R_i^2) \cdot \Delta P/\Delta R_o \quad (R_i, \text{ inner radius}; R_o, \text{ outer radius}; P, \text{ inner pressure})$$
- Variation of incremental elastic modulus with pressure during inflation (triangular symbols) and deflation (circles). Solid symbols and lines, dynamic; open symbols and broken lines, static. Points represent mean values for 10 specimens. Vertical bar $\pm$ 1 SE. Panel (a), passive; panel (b), active.

Reference(s)
Greenwald SE, Newman DL, Denyer HT (1982) Effect of smooth muscle activity on the static and dynamic elastic properties of the rabbit carotid artery. Cardiovasc Res 16: 86–94 (with permission)

Viscoelasticity (5)

• Pressure–strain elastic modulus • Vessel dimension	• Dogs • Femoral artery	• In vivo measurement • Postmortem change

Materials
- Six greyhounds (weight, 20–27 kg)
- Femoral artery

Testing Methods and Experimental Conditions
- Femoral artery exposed under pentobarbital anesthesia (30 mg/kg)
- An electrical caliper sewn on to the adventitia for diameter measurement
- Pressure lowered with constriction of a snare around the proximal artery
- In vivo static pressure–diameter measurement at 110 and 70 mmHg
- Dynamic pressure–diameter measurement with mean pressure of 90–95 mmHg and peak-to-peak pulse of 10–15 mmHg at 2 Hz
- Similar static and dynamic measurements in Dulbecco's buffered saline solution at room temperature just after excision and after storing the specimen in a refrigerator at 4°C overnight

Data
See next page.

Comments
- Elastic parameters:

    Pressure–strain elastic modulus, $E_p = (\Delta P/\Delta D_o)D_o$;

    Elastic modulus, $E = 1.5E_p[D_i^2/(D_o^2 - D_i^2)]$

    D_o and D_i, outer and inner diameters; ΔD_o, diametrical change caused by pressure change ΔP

Reference(s)
Gow BS, Hadfield CD (1979) The elasticity of canine and human coronary arteries with reference to postmortem changes. Circ Res 45:588–594 (with permission)

State	BP (mmHg)	ΔP or ΔP_d (mmHg)	D_o (mm)	γ	E_p(stat) or E_p(dyn) (MPa)	E_{stat} or E_{dyn} (MPa)	Frequency (Hz)
Static							
In vivo ($n = 6$)	93 ± 3.9	39 ± 0.7	5.15 ± 0.21	0.11 ± 0.01	0.133 ± 0.006	0.79 ± 0.10	
In vitro ($n = 6$)	92 ± 1.1	40 ± 0.3	5.61 ± 0.20	0.092 ± 0.008	0.187 ± 0.027	1.34 ± 0.18	
In vitro after cold storage ($n = 5$)	92 ± 1.2	40 ± 0.2	5.77 ± 0.19	0.09 ± 0.01	0.266 ± 0.033	1.87 ± 0.08	
Dynamic							
In vivo ($n = 6$)	93 ± 1.2	50 ± 9.5	5.20 ± 0.22	0.11 ± 0.01	0.200 ± 0.018	1.18 ± 0.13	2.1 ± 0.3
In vitro ($n = 6$)	93 ± 1.6	50 ± 7.2	5.67 ± 0.20	0.09 ± 0.01	0.226 ± 0.014	1.16 ± 0.14	2.2 ± 0.1
In vitro after cold storage ($n = 5$)	92 ± 1.6	53 ± 10.1	5.82 ± 0.21	0.09 ± 0.01	0.316 ± 0.040	2.32 ± 0.26	2.1 ± 0.07

Values are expressed as mean $\pm$ SE. BP, mean blood pressure; ΔP, quasistatic change in pressure; ΔP_d, dynamic pressure change; D_o, mean outside diameter; γ, ratio of wall thickness to outside radius; E_p (stat), static pressure/strain elastic modulus; E_p (dyn), dynamic pressure/strain elastic modulus; E_{stat}, static elastic modulus; E_{dyn}, dynamic elastic modulus.

Viscoelasticity (6)

• Pressure–strain elastic modulus • Vessel dimension	• Human, dogs • Coronary artery	• In vivo measurement •

Materials
- Coronary arteries without gross atheromatous deposits from 5 subjects; age, 14–40 years
- Left circumflex coronary arteries (LCCA) of greyhounds (weight, 20–27 kg)

Testing Methods and Experimental Conditions
- Human coronary artery within 24–96 h postmortem
- In vivo length not known (stretched until no buckling occurred during pressurization)
- Static pressure–diameter test in normal saline at room temperature (20–150 mmHg)
- Dynamic measurement at amplitudes of 15 mmHg at frequencies of 0.5–10 Hz
- Dog LCCA exposed through thoracotomy under pentobarbital anesthesia and ventilation
- Pressure–diameter relation on the beating heart measured with an electrical caliper and a pressure transducer connected to an anterior ventricular branch via a catheter

Data

Artery	Age (years)	D_o (cm)	γ	E_p (stat) (MPa)	E_{stat} (MPa)	E_p (dyn) (MPa) (2Hz)	E_{dyn} (MPa) (2Hz)	E_{dyn}/E_{stat} (2Hz)
LAD 1	30	0.57	0.13	0.701	3.20	1.78	8.38	2.62
LAD 2	40	0.52	0.25	0.619	1.22	1.15	2.64	2.16
LAD 3	20	0.39	0.10	0.551	3.50	1.21	7.72	2.20
LCCA 1	40	0.56	0.23	0.472	1.01	0.66	1.40	1.39
LCCA 2	36	0.44	0.12	0.663	1.45	0.93	2.11	1.45
RCA 1	14	0.44	0.10	0.663	3.50	0.72	4.50	1.15
Mean	30	0.49	0.17	0.602	2.31	1.07	4.46	1.83
SE	4	0.03	0.03	0.032	0.49	0.17	1.21	0.24

D_o, mean outside diameter; γ, wall thickness ÷ outside radius; E_p (stat), static pressure/strain elastic modulus; E_{stat}, static elastic modulus; E_p (dyn), dynamic pressure/strain elastic modulus; E_{dyn}, dynamic elastic modulus.
LAD, left anterior descending artery; LCCA, left circumflex coronory artery; RCA, right coronary artery. LAD 2 and LCCA 1 from same subject.

Dog	Wt. (kg)	Frequency (Hz)	BP (mmHg)	Δp (mmHg)	D_o (cm)	E_p (MPa)	γ	E_{dyn} (MPa)
1	25.4	2.78	101	32.0	0.32	0.201		
2	28.1	2.50	105	18.0	0.37	0.262	0.174	0.82
3	21.8	2.40	102	18.7	0.36	0.164	0.093	1.48
4	24.5	1.56	126	25.1	0.32	0.274	0.126	1.33
5	22.7	1.66	98	10.9	0.39	0.195	0.083	1.53
6	26.3	2.66	124	18.1	0.37	0.249	0.128	1.18
7	25.0	2.60	102	23.1	0.33	0.149	0.081	1.21
8	25.4	2.24	102	19.0	0.35	0.168	0.142	0.70
9	28.2	2.70	119	25.8	0.36	0.364	0.095	2.47

Δp, pulse pressure; E_p, pressure–strain elastic modulus.

Comments
- Elastic parameters:
  Pressure–strain elastic modulus, $E_p = (\Delta P/\Delta D_o)D_o$
  Elastic modulus, $E = 1.5E_p[D_i^2/(D_o^2 - D_i^2)]$
  D_o and D_i, outer and inner diameters; ΔD_o, diametrical change caused by pressure change ΔP.

Reference(s)
Gow BS, Hadfield CD (1979) The elasticity of canine and human coronary arteries with reference to postmortem changes. Circ Res 45:588–594 (with permission)

Viscoelasticity (7)

• Radial compression • Dynamic elastic modulus	• Dogs • Various aortas	• Positional variation • Frequency dependency

Materials
- Dogs weighing 26.0–32.6 kg (average weight 28.7 kg)
- Upper (UDTA), middle (MDTA), lower (LDTA) descending thoracic aorta, and abdominal aorta (AA)

Testing Methods and Experimental Conditions
- Dynamic (sinusoidal) compression test on longitudinally cut specimens
- Specimen was stretched longitudinally and circumferentially and/or compressed radially to restore its in vivo thickness
- Specimen kept wet with Ringer solution
- Temperature, 26°–27°C

Data

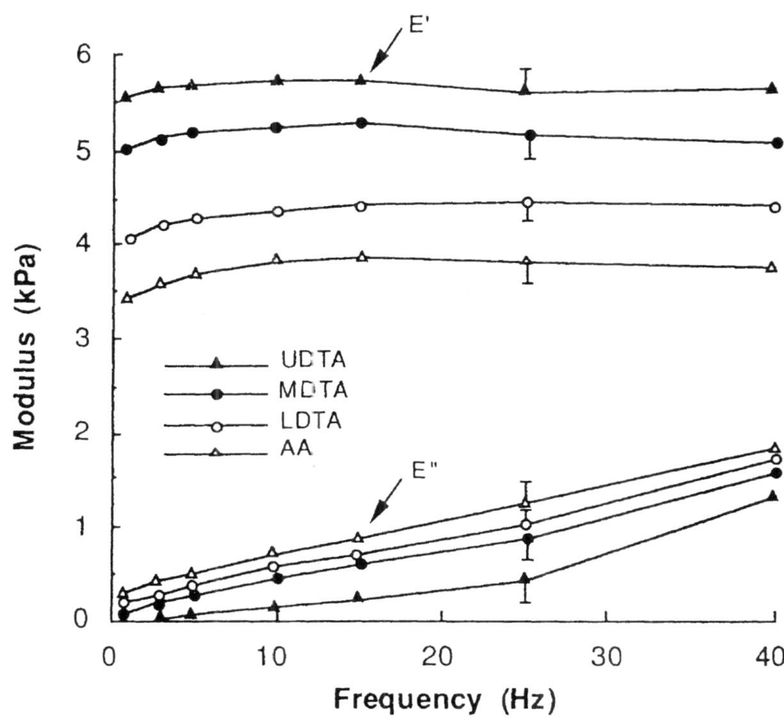

Comments
- $E = E' + jE''$, where E is the complex elastic moduli in the radial direction.
- Data were selected from several dogs in such a way that the initial strain for each site averaged aproximately –0.50 (range –0.49 to –0.51).

Reference(s)
Patel DJ, Tucker WK, Janicki JS (1970) Dynamic elastic properties of the aorta in radial direction. J Appl Physiol 28:578–582 (with permission)

Viscoelasticity (8)

• Radial compression • Dynamic stiffness	• Human • Fibrous cap	• Atherosclerosis • Plaque rupture

Materials
- Human (age, 71 ± 9 years)
- Fibrous cap from atherosclerotic plaque of abdominal aorta

Testing Methods and Experimental Conditions
- Plaques obtained at autopsy performed within 12 h of death
- Medium: normal saline at room temperature
- Radial compressive stress of 9.3 kPa (= 70 mmHg) applied to specimen for 30 min
- Radial compressive stiffness obtained by applying sinusoidal load of 9.3 ± 0.5 kPa (0.05–10 Hz)
- Specimens classified by histological examination as cellular, hypocellular, or calcified

Data

Frequency (Hz)	Composition/histologic structure		
	Cellular (MPa)	Hypocellular (MPa)	Calcified (MPa)
0.5	0.48 ± 0.20	0.86 ± 0.22	2.03 ± 0.90
1.0	0.51 ± 0.22	0.90 ± 0.22	2.19 ± .099
2.0	0.55 ± 0.24	0.96 ± 0.25	2.34 ± 1.09

All data are given as mean ± SD.
By analysis of variance, the effects of histological class ($P = 0.001$) and frequency ($P = 0.0001$) are both highly significant.

Comments
- Hypocellular fibrous caps were 1–2 times stiffer than cellular caps, and calcified caps were 4–5 times stiffer than cellular caps.
- Stiffness increases with frequency in the range of physiological heart rates.

Reference(s)
Lee RT, Grodzinsky AJ, Frank EH, Kamm RD, Schoen FJ (1991) Structure-dependent dynamic mechanical behavior of fibrous caps from human atherosclerotic plaques. Circulation 83:1764–1770 (with permission)

Force–Length Relation (1)

• Force–extension relation •	• Frog • Semitendinosus muscle	• Isolated muscle fiber • Series elastic component

Materials

- Frogs, *Rana nigromaculata*
- Semitendinosus muscle
- Small bundle of fast fibers prepared from ventral caput

Testing Methods and Experimental Conditions

- Temperature, 10°C
- Tension–extension curve of the series elastic component (SEC) was directly determined by a controlled quick-release method
- Two kinds of preparations, short tendons and normal long tendons, were used

Data

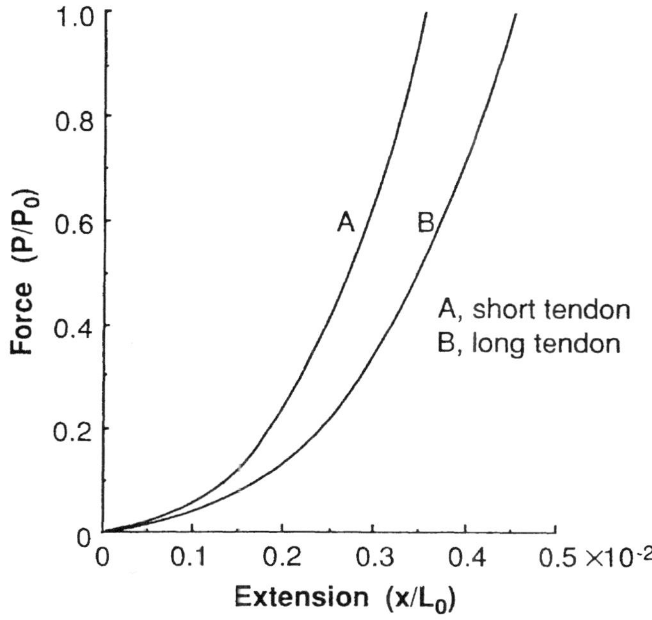

Comments

- P, tension; P_0, maximum isometric tension; x, extension; L_0, standard length.
- In the short tendon preparation, the extension of SEC was about 3.5% L_0 at the tension of P_0 but it was 4.5% L_0 in the long tendon preparation.
- The tension–extension curves are expressed by
  $$P/P_0 = 2.9X + 51.8\,X^2 + 1.9 \times 10^6\,X^3 \text{ for short tendon preparation, and}$$
  $$P/P_0 = 2.3X + 31.4\,X^2 + 0.9 \times 10^6\,X^3 \text{ for long tendon preparation,}$$
  where $X = x/L_0$.

Reference(s)

Mashima H, Akazawa K, Kushima H, Fujii K (1973) Graphical analysis and experimental determination of the active statein frog skeletal muscle. Jpn J Physiol 23:217–240

Force–Length Relation (2)

• Force–extension relation •	• Rat • Gracilis anticus muscle	• Isolated muscle fiber • Series elastic component

Materials
- White Wistar rats (male; age, 50 days; weight, 140–165 g)
- Gracilis anticus muscle (rest length, 2.7 cm; weight, 60 mg)

Testing Methods and Experimental Conditions
- Krebs–Ringer solution, pH 7.3; temperature, 17.5°C
- Modified Wilkie's quick-release method, releasing from isometric condition to some fixed isometric load
- Muscle stimulation: current density = 0.08 amp/cm^2, train of 322 ms pulses with a pulse separation of 10.5 ms

Data

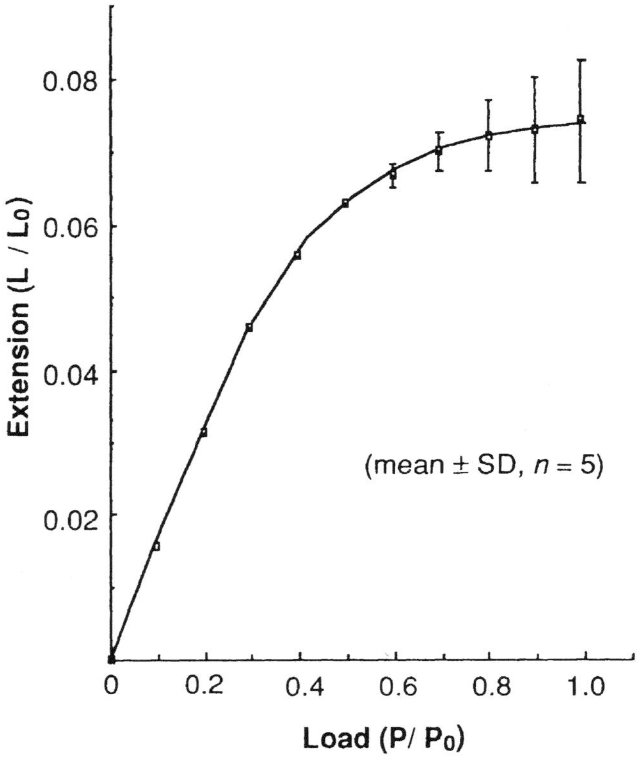

Comments
- Experimental results typically showed the uncorrected Wilkie method to underestimate the extension of the series elastic component (SEC) at P_0 by 1% of L_0.
- SEC is highly nonlinear (compliance varying from 1.1×10^{-6} cm/dyne at 1.0 P_0 to 17.6×10^{-6} cm/dyne at zero tension).
- The extension of SEC at the maximum isometric tension is 0.07 rest length.

Reference(s)
Bahler AS (1967) Series elastic component of mammalian skeletal muscle. Am J Physiol 213:1560–1564 (with permission)

Force–Length Relation (3)

• Force–length relation •	• Cat • Gastrocnemius muscle	• Intact whole muscle •

Materials
- Cats (adult; weight, 3.26 kg [mean])
- Medial gastrocnemius (MG) muscle and tibialis anterior (TA) muscle

Testing Methods and Experimental Conditions
- Cats were anesthetized with chloralose (60 mg/kg) and maintained at 37°C
- A bipolar nerve cuff electrode was placed on the nerve branch of the muscle under investigation
- Pulse duration was 0.1 ms, and pulse rate was approximately 56 pps for the MG and 68 pps for the TA

Data

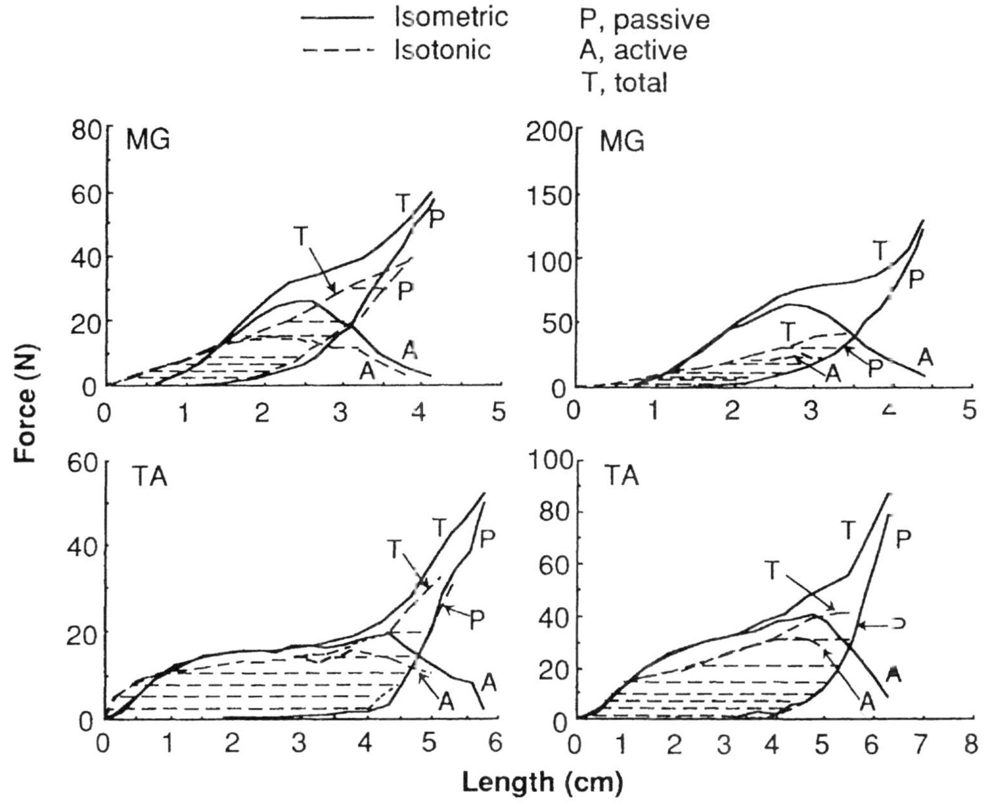

Comments
- The horizontal lines depict the values of force in isotonic contraction originating from the points on the passive length–tension curve and terminating at the points on the total length–tension curve, as obtained in the constant mass load trials.
- The above figures apparently show that the isotonic tension–length curve does not coincide with an ascending limb of the isometric tension–length curve.

Reference(s)
Vance TL, Solomonow M, Baratta R, Zembo M, D'Ambrosia D (1994) Comparison of isometric and load moving length–tension models of two bicompartmental muscles. IEEE Trans Biomed Eng 41:771–781 (with permission)

Force–Length Relation (4)

• Force–length relation	• Cat	• Intact whole muscle
•	• Gastrocnemius muscle	•

Materials
- Cats (weight, 3.26 kg [mean])
- Medial gastrocnemius muscle

Testing Methods and Experimental Conditions
- Isometric and isotonic length–tension relations were measured
- Injection of isotonic saline solution at 37°C

Data

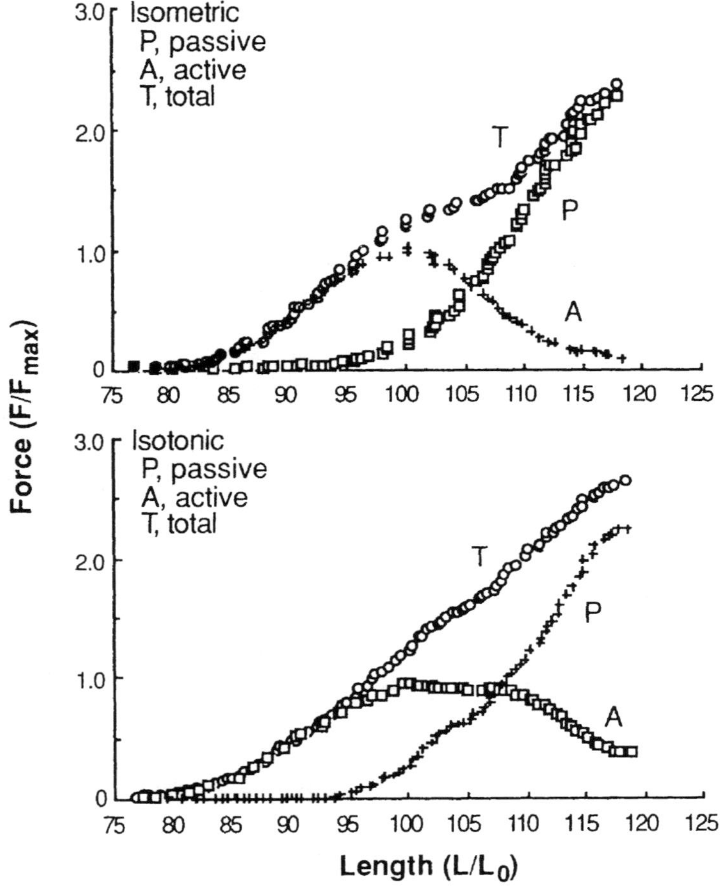

Comments
None.

Reference(s)

Vance TL, Solomonow M, Baratta R, Zembo M, D'Ambrosia RD (1994) Comparison of
isometric and load moving length–tension models of two biocompartmental muscles. IEEE
Trans Biomed Eng 41:771–781 (with permission)

Force–Length Relation (5)

• Load–extension curve • Maximum isometric tension	• Frog • Semitendinosus muscle	• Isolated muscle fiber • Resting state

Materials

- Frogs, *Rana nigromaculata*
- Semitendinosus muscle
- Small bundle of fast muscle fibers prepared from ventral caput

Testing Methods and Experimental Conditions

- Relation between force and extension of the resting muscle was measured in Ringer's solution at 10°C
- Standard length of 14 mm in all preparations

Data

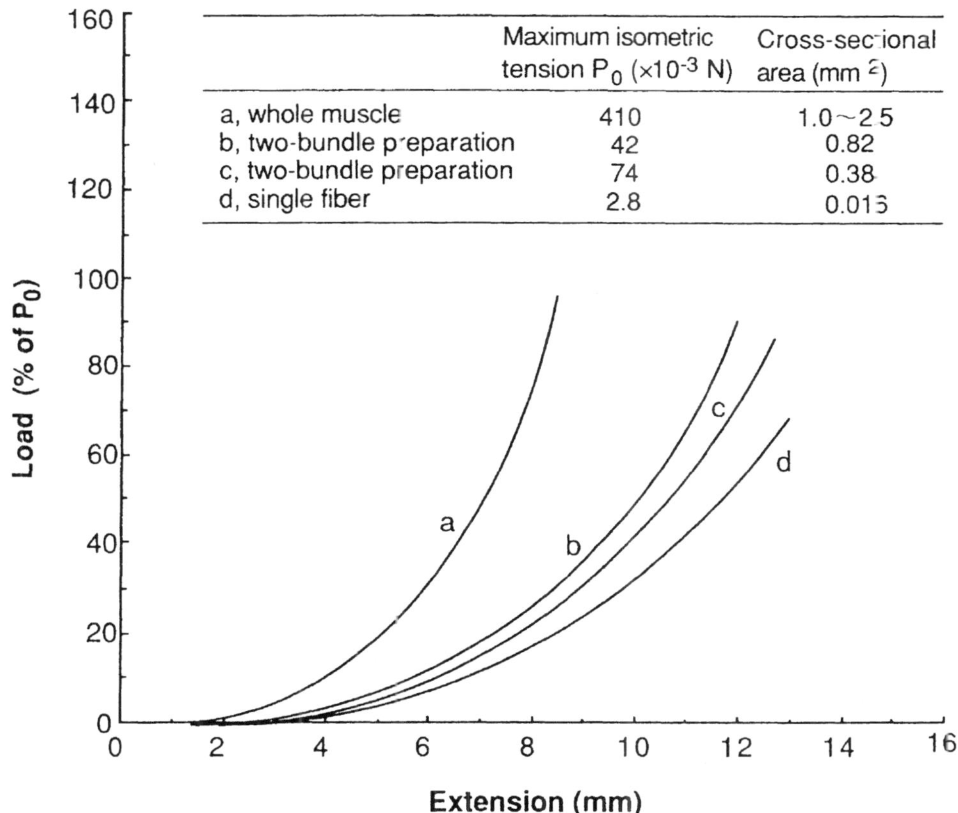

	Maximum isometric tension P_0 ($\times 10^{-3}$ N)	Cross-sectional area (mm^2)
a, whole muscle	410	1.0~2.5
b, two-bundle preparation	42	0.82
c, two-bundle preparation	74	0.38
d, single fiber	2.8	0.013

Comments

None.

Reference(s)

Mashima H, Akazawa K, Kushima H, Fujii K (1972) The force–velocity relation and the viscous-like force in the frog skeletal muscle. Jpn J Physiol 22:103–120

Force–Length Relation (6)

• Stress–sarcomere length •	• Frog • Semitendinosus muscle	• Isolated muscle fiber • Resting state

Materials
- Frogs, *Rana pipiens*
- Intact whole semitendinosus fiber

Testing Methods and Experimental Conditions
- Resting muscle in phosphate-buffered Ringer's solution
- Temperature, 11.2°, 16.3°, and 26.2°C

Data

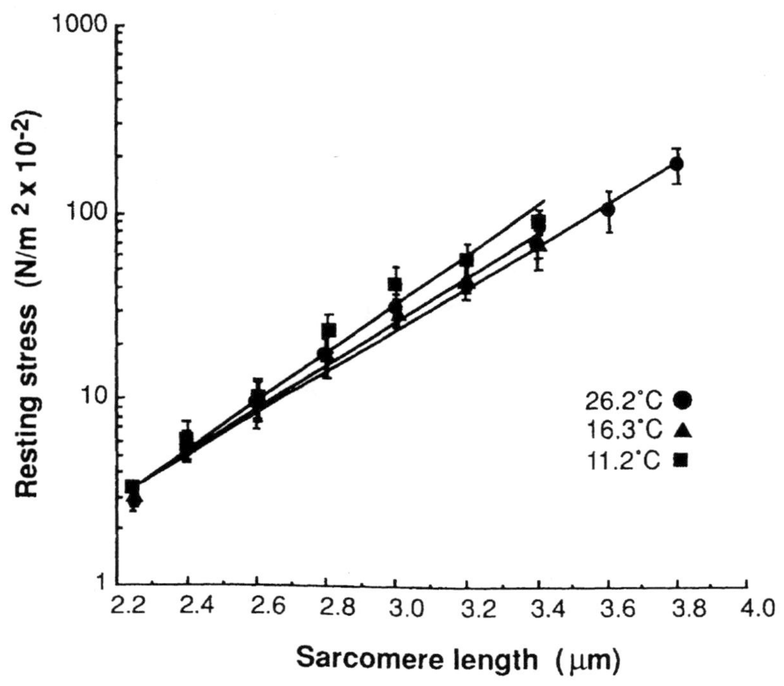

Comments
- The data fit to $\sigma = a \cdot \exp(b\varepsilon)$, where σ = resting stress, ε = sarcomere strain, a and b = constants.
- The mean static resting stress at resting length for the three temperatures was 290 ± 18 N/m^2 (mean $\pm$ SE).

Reference(s)

Moss RL, Halpern W (1977) Elastic and viscous properties of resting frog skeletal muscle. Biophys J 17:213–228 (with permission)

Force–Length Relation (7)

• Tensile strength •	• Rabbit • White muscle	• Myo-filament •

Materials
- Rabbits
- Skeletal white muscle
- Reconstituted single actin filament with and without tropomyosin

Testing Methods and Experimental Conditions
- Force measurements by micromanipulation with a pair of glass microneedles urder an optical microscope
- Temperature, $25 \pm 2°C$

Data

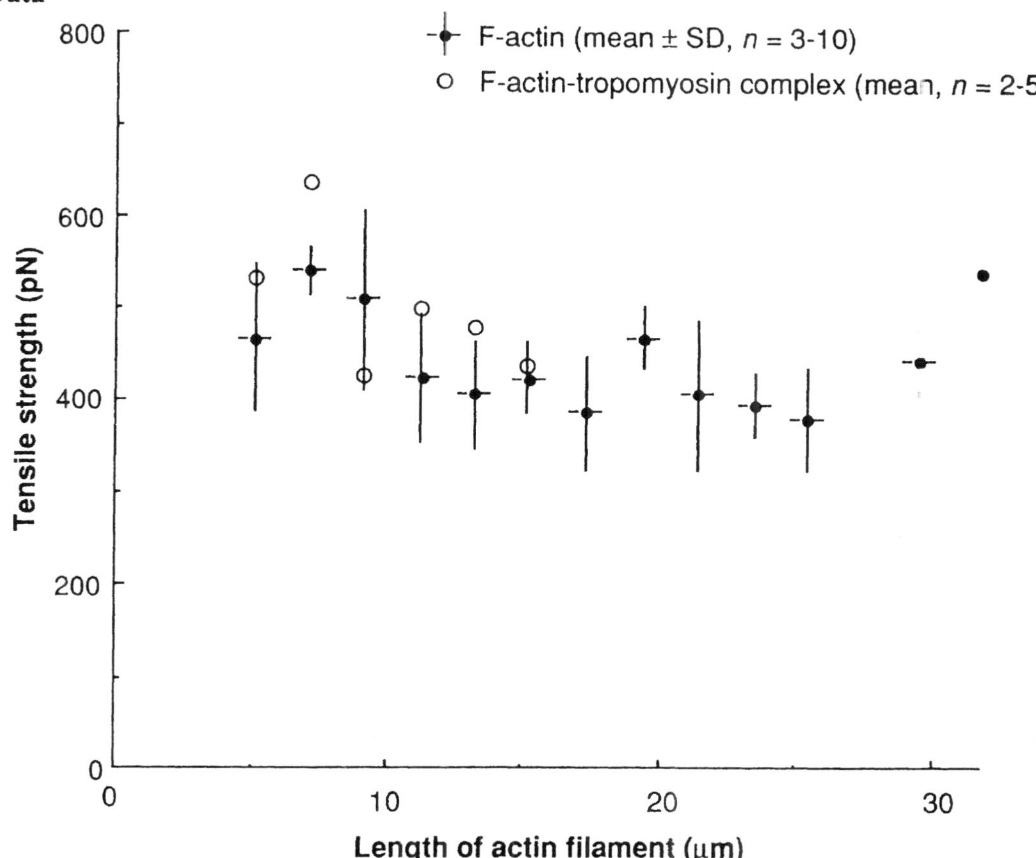

Comments
- The tensile strength of actin filament was 430 ± 20 pN and binding of tropomyosin increased it to 480 ± 30 pN.
- The numbers in the graph ordinate and the values of the tensile strength were ccrrected by authors' request.

Reference(s)
Kishino A, Yanagida T (1988) Force measurements by micromanipulation of a single actin filament by glass needles. Nature 334:74–76 (with permission)

Force–Length Relation (8)

• Tension–length relation •	• Cat • Soleus muscle	• Intact whole muscle •

Materials
- Cats (1.7–3.8 kg)
- Soleus muscle

Testing Methods and Experimental Conditions
- Relation between isometric force and extension of the tetanized muscle was measured
- The muscle was stimulated through its nerve supply using distributed and synchronous stimulation pulses
- Body temperature

Data

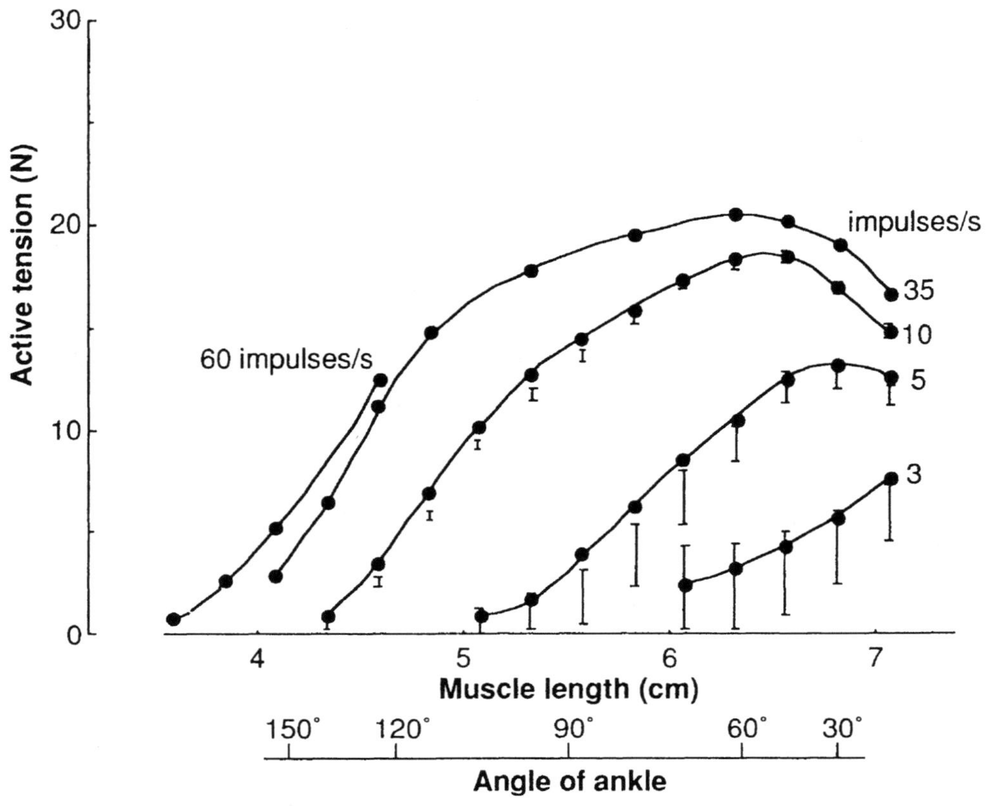

Comments
- Continuous lines show the tension during distributed stimulation at the rates indicated (five stimulating channels were used).
- The vertical lines show the limits of tension fluctuations during synchronous stimulation.
- Isometric tension increased with increasing muscle length at the fixed stimulus rate.
- There was a reciprocal relationship between stimulus rate and muscle length: when the muscle was long, low rates of stimulation gave near maximal tension, whereas at short lengths the maximum tension was reached only when the stimulus rate was very high.

Reference(s)
Rack PMH, Westbury DR (1969) The effects of length and stimulus rate on tension in the isometric cat soleus muscle. J Physiol 204:443–460 (with permission)

Force–Length Relation (9)

• Tension–sarcomere length •	• Frog • Semitendinosus muscle	• Isolated muscle fiber • Contracting state

Materials

- Frogs, *Rana temporaria*
- Semitendinosus muscle and tibialis anterior muscle

Testing Methods and Experimental Conditions

- Applying a constant-velocity (ramp) shortening to an activated fiber and measuring tension at the time it first became nearly constant
- The initial sarcomere lengths for decreasing average length by 0.1 μm with ramp were 2.05, 2.10, 2.15, 2.20, 2.30, 2.40, 2.60, 2.80, 3.00, and 3.20 μm
- Temperature, $2.5 \pm 0.2°C$
- Length–tension relationships at four different shortening velocities (10, 20, 30, and 50% V_u; V_u, unloaded shortening velocity) and under isometric conditions were determined
- P_0 is defined as the maximum isometric tension attained during 500 ms contraction at a sarcomere length of 2.15 μm for the semitendinosus fiber and 2.05 μm for the tibialis anterior fiber

Data

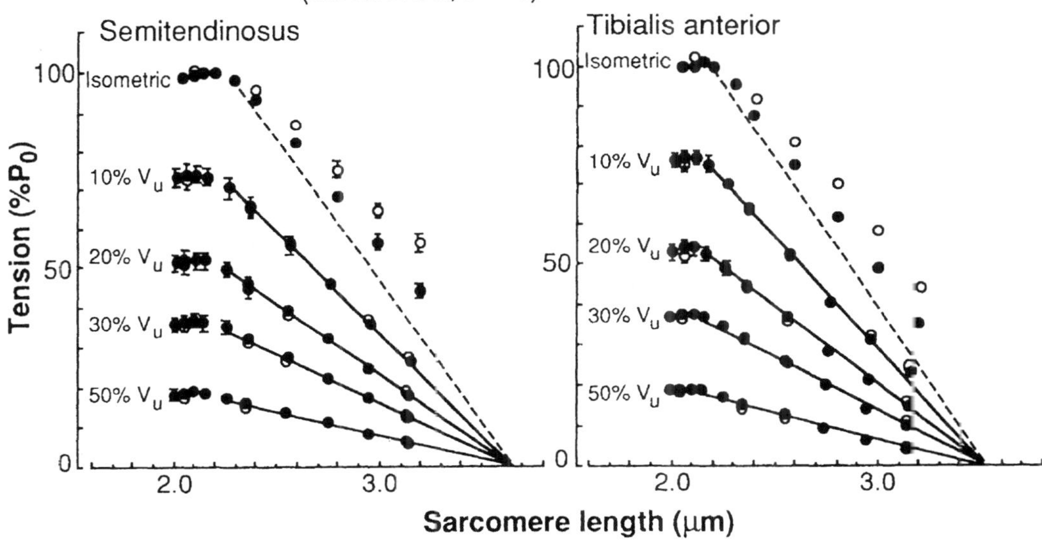

Comments

- For zero velocity, tension at sarcomere length greater than 2.2 μm significantly increased with the increase in the duration of tetanus.
- For non-zero velocities, tension was relatively independent of the duration of tetanus.

Reference(s)

Morgan DL, Clafin DR, Julian FJ (1991) Tension as a function of sarcomere length and velocity of shortening in single skeletal muscle fibres of the frog. J Physiol 441:719–732 (with permission)

Force–Length Relation (10)

• Tension–sarcomere length •	• Human • Extensor muscle	• •

Materials
- Human (age, 35–50 years)
- Extensor carpi radialis brevis muscle

Testing Methods and Experimental Conditions
- Sarcomere length was measured intraoperatively in five subjects using laser diffraction
- Hypothetical length–tension curve was obtained using measured filament lengths and assuming the sliding filament mechanism proposed by Gordon et al. (1966)

Data

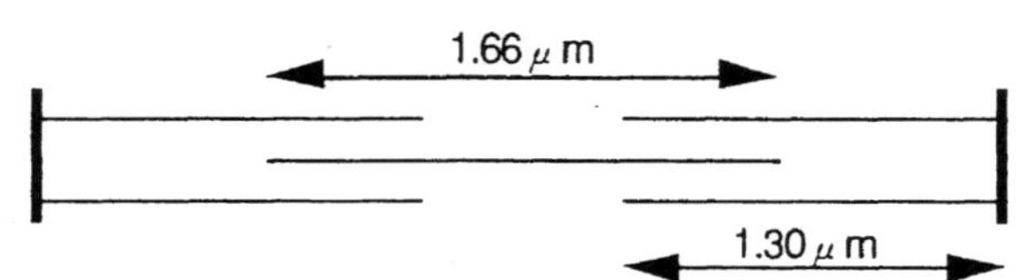

Schematic of measured filament lengths

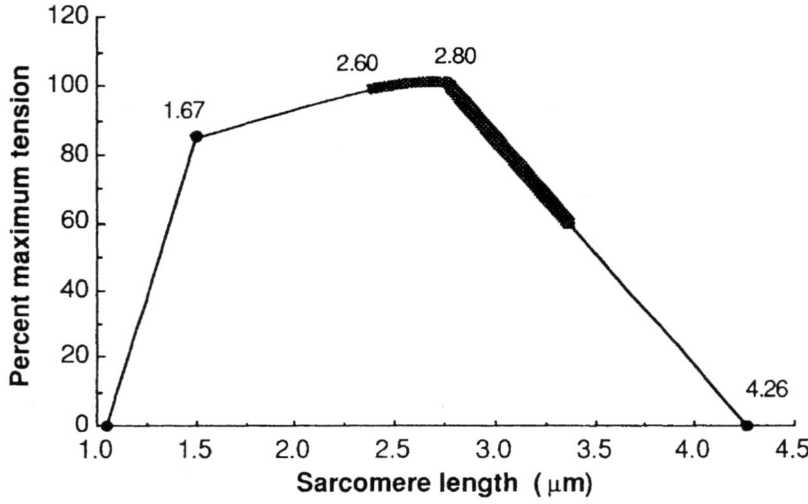

Comments
- Shaded area represents sarcomere length change during wrist flexion (causing sarcomere length increase) and wrist extension (causing sarcomere length decrease).
- With the wrist full extension, sarcomere length was ~2.6 μm, which was significantly shorter than the 3.4 μm sarcomere length measured in the flexed position.
- Sarcomere length with the joint in the neutral position was 3.0 μm.
- The relation between sarcomere length (SL) and wrist joint angle was approximated by the linear function SL (μm) = –0.0076 μm/° × Joint Angle (°) + 3.01 μm ($r^2 = 0.68$, $P < 0.001$).

Reference(s)
Lieber RL, Loren GJ, Friden J (1994) In vivo measurement of human wrist extensor muscle sarcomere length changes. J Neurophysiol 71:874–991 (with permission)

Force–Length Relation (11)

• Tension–sarcomere length •	• Rabbit • Adductor magnus muscle	• Isolated musc.e fiber • Resting state

Materials
- Rabbits (adult)
- Mechanically skinned single fiber prepared from six different muscle tissues (adductor magnus [AM], psoas [PS], longissimus dorsi [LD], sartorius [SA], soleus [SO], semitendinosus [ST])

Testing Methods and Experimental Conditions
- Stress–strain relation under relaxing conditions
- Temperature, 24°C

Data

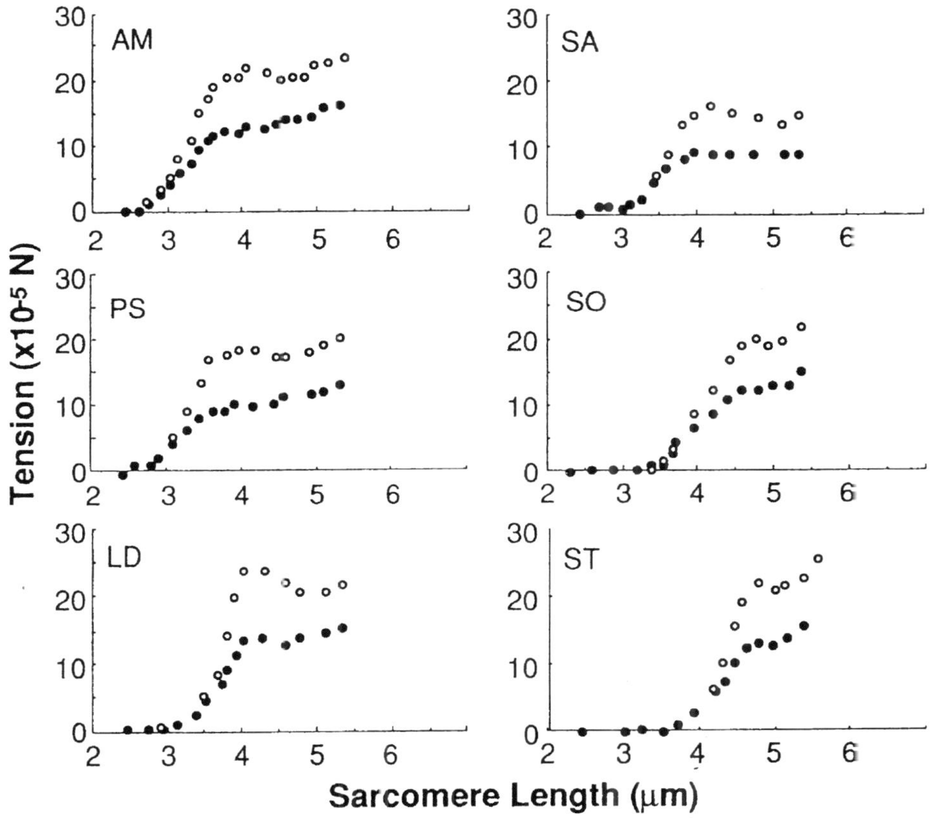

Comments
- The stress–strain relation under relaxing conditions is due to the reversible extension of a segment of titin (connectin) between the Z line and the ends of thick filaments.
- The data suggest that skeletal muscle cells may control and modulate elasticity and compliance and the elastic limit of the sarcomere by selective expression of spec.fic titin size isoforms.

Reference(s)
Wang K, McCarter R, Wright J, Beverly J, Ramirez-Mitchell R (1991) Regulation of skeletal muscle stiffness and elasticity by titin isomers: A test of the segmental extension model of resting tension. Proc Natl Acad Sci USA 88:7101–7105 (with permission)

Force–Length Relation (12)

• Tension–sarcomere length •	• Rabbit • Psoas muscle	• Isolated muscle fiber • Resting state

Materials
- Rabbits
- Psoas glycerinated single myofibril

Testing Methods and Experimental Conditions
- Both ends of a single myofibril were held with a pair of flexible and stiff glass microneedles under a phase-contrast microscope; a stiff needle was displaced with a piezo element and the deflection of the flexible needle was measured with an image processor
- Room temperature, about 25°C

Data

✗	Rigor tension of intact myofibril
■	Passive tension under rigor condition after selective removal of thin filaments at the I band
□	Passive tension under rigor condition after complete removal of thin filaments
○	Resting tension in the absence of butanedione-monoxime (BDM)
●	Resting tension in the presence of BDM

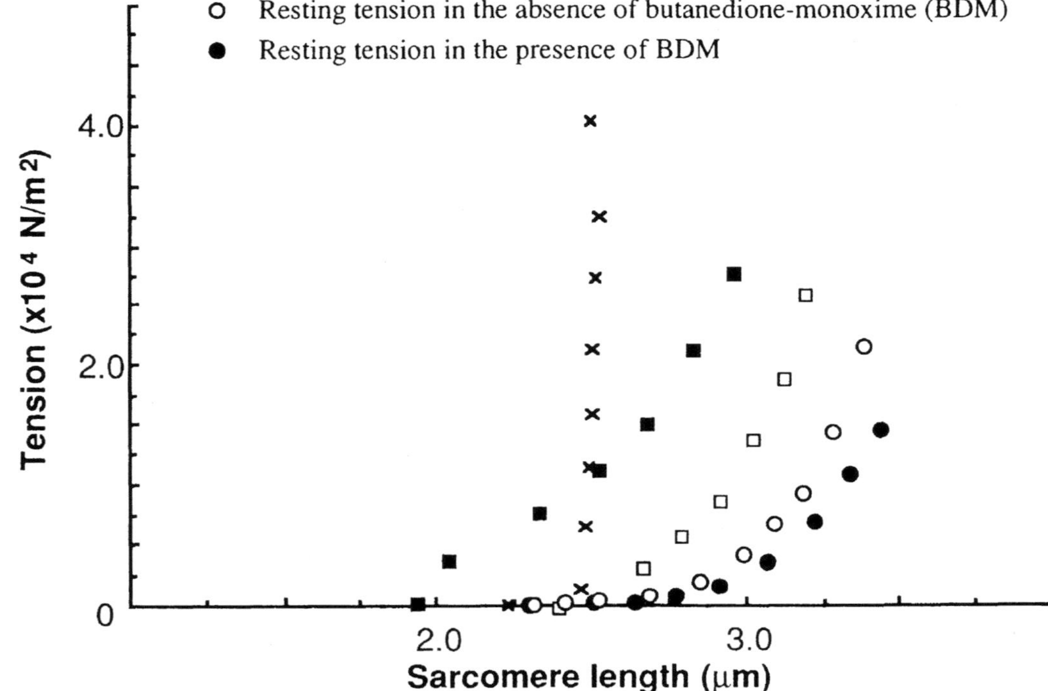

Comments
- Elastic modulus of connectin (titin) 1 μm long at a sarcomere length of 3.4 μm was estimated to be 0.1 pN/nm.
- Elastic modulus of nebulin 1 μm long was estimated to be 0.01 pN/nm.

Reference(s)
Yasuda K, Anazawa T, Ishiwata S (1995) Microscopic analysis of the elastic properties of nebulin in skeletal myofibrils. Biophys J 68:598–608 (with permission)

Force–Length Relation (13)

• Tension–sarcomere length •	• Waterbug • Dorso-longitudinal	• Isolated muscle fiber • Resting state

Materials

- Waterbugs (*Lethocerus griseus* and *Lethocerus uhleri* from Florida; *Lethocerus indicus* from Thailand)
- Fresh fiber from dorso-longitudinal muscle

Testing Methods and Experimental Conditions

- Resting muscle
- Temperature, 4°C

Data

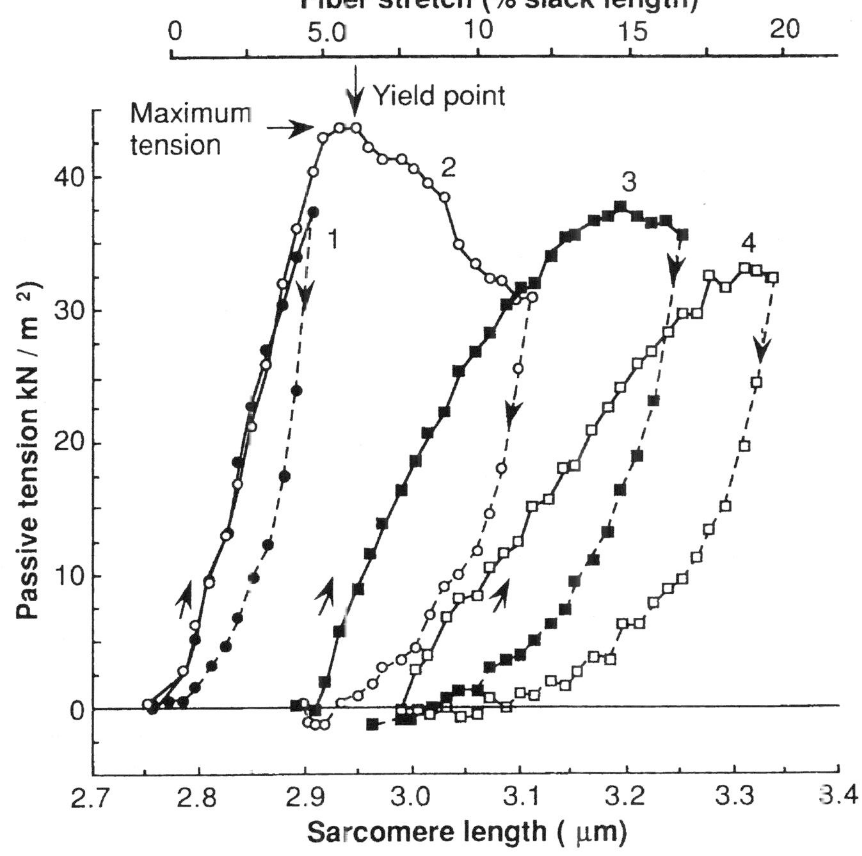

Comments

- A functional dissection of the relative contribution of cross-bridges, actin filament and connecting (connectin [titin]) filaments to tension and stiffness of passive, activated, and rigor fibers was carried out by comparing mechanical properties at different ionic strengths of sarcomeres with and without thin filaments.

Reference(s)

Granzier HLM, Wang K (1993) Interplay between passive tension and strong and weak binding cross-bridges in insect indirect flight muscle. A functional dissection by gelsolin-mediated thin filament removal. J Gen Physiol 101:235–270 (with permission)

Force–Length Relation (14)

• Stiffness • Yield tension	• Rabbit, waterbug • Psoas muscle	• Isolated muscle fiber • Contracting state

Materials
- Rabbits and waterbugs
- Glycerinated single fiber from rabbit psoas and semitendinosus muscles
- Glycerinated fiber from waterbug indirect flight muscle (IFM)

Testing Methods and Experimental Conditions
- Tension and stiffness at resting, active, and rigor conditions
- Tension and stiffness were measured before (control) and after (treated) the treatment with noncalcium-requiring gelsolin fragment which selectively removed thin filaments
- Temperature, 22°C

Data

	IFM		Psoas		Semitendinosus
	Control	Treated	Control	Treated	Control
Slack length (μm)	2.68 ± 0.02	2.67 ± 0.04	2.00 ± 0.01	1.91 ± 0.03	2.12 ± 0.01
Active	(4% prestretch)[a]		(10% prestretch)[a]		
Tension (kN/m^2)	28.0 ± 2.0	0.2 ± 0.2	110.2	0.5	
Stiffness (MN/m^2)	1.7 ± 0.3	0.01 ± 0.02	5.5	0.1	
Rigor	(2% prestretch)[a]		(10% prestretch)[a]		(10% prestretch)[a]
Tension (kN/m^2)	28.1 ± 5.0	0.3 ± 0.2	8.7 ± 3.1	0.29 ± 0.1	7.3 ± 2.2
Stiffness (MN/m^2)	11.3 ± 1.3	0.3 ± 0.1	5.5 ± 0.9	0.1 ± 0.1	6.2 ± 0.8
Passive					
E_0 (kN/m^2)[b]	169 ± 43	215 ± 37	7.4 ± 1.4	11.7 ± 0.8	3.5 ± 2.0
α[b]	56 ± 8	47 ± 4	6.9 ± 0.1	6.8 ± 0.02	6.35 ± 0.21
Tension yield point (μm)	3.07 ± 0.01	3.06 ± 0.02	3.63 ± 0.03	3.60 ± 0.05	4.59 ± 0.09
Yield tension (kN/m^2)	111 ± 10	90 ± 12.6	110 ± 7	148 ± 42	126 ± 10.0
Stiffness yield point (μm)	2.96 ± 0.04	2.93 ± 0.05	3.53 ± 0.07	3.53 ± 0.07	4.44 ± 0.09
Yield stiffness (MN/m^2)	7.5 ± 0.8	4.5 ± 0.7	3.4 ± 0.5	1.9 ± 0.7	5.9 ± 1.4

[a] Stretch prior to activation or rigor induction.

[b] Passive tension (σ) and sarcomere strain (ε) relation fitted to: $\sigma = E_0\alpha^{-1}(e^{\alpha\varepsilon} - 1)$.

Comments
None.

Reference(s)
Granzier HLM, Wang K (1993) Passive tension and stiffness of vertebrate skeletal and insect flight muscles: The contribution of weak crossbridges and elastic filaments. Biophys J 65:2141–2159 (with permission)

Force–Length Relation (15)

• Stress–sarcomere length • Initial elastic modulus	• Frog • Semitendinosus muscle	• Isolated muscle fiber • Resting state

Materials
- Frogs, *Rana pipiens* (6.3–8.5 cm from nose to vent)
- Intact and mechanically skinned single fiber from semitendinosus muscle

Testing Methods and Experimental Conditions
- Resting muscle in normal or calcium-free Ringer's solution
- Room temperature, 20°–22°C

Data

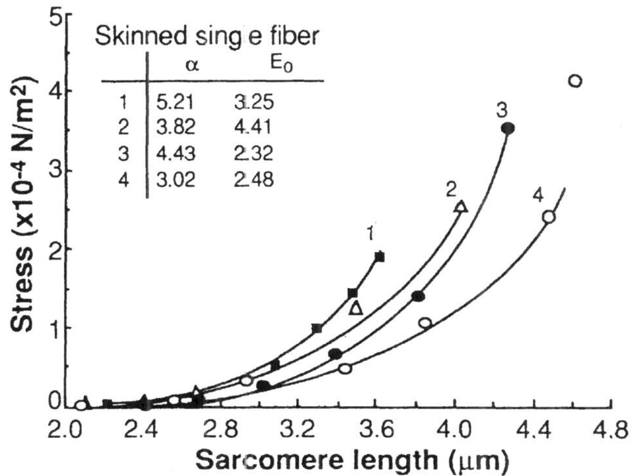

Preparation	Number of samples	Exponent α	Initial elastic modulus E_0 ($\times 10^3$ N/m^2)
Whole muscle	11	4.28 ± 0.19	2.6 ± 0.25
Intact single fiber	10	4.04[a]	9.8[a]
Skinned single fiber	12	4.13 ± 0.10	5.4 ± 0.75

Date are given as mean $\pm$ SE.

[a] No SEs are reported because these values were obtained from a single analysis of pooled data.

Comments
- Stress–strain relation was expressed by
  $$\sigma = (E_0/\alpha)(\exp(\alpha\varepsilon) - 1)$$
  σ, stress (force per cross-sectional area); E_0, initial elastic modulus; α, empirical constant; ε, sarcomere strain.

Reference(s)
Magid A, Law DJ (1985) Myofibrils bear most of the resting tension in frog skeletal muscle. Science 230:1280–1282 (with permission)

Force–Length Relation (16)

• Tension–length relation •	• Frog • Semitendinosus muscle	• Isolated muscle fiber • Contracting state

Materials
- Frogs, *Rana temporaria*
- Semitendinosus muscle

Testing Methods and Experimental Conditions
- Shortening movement at constant speed was imposed on the muscle during a tetanus
- Temperature, 0°C
- Weight of specimen, 22.3 mg

Data

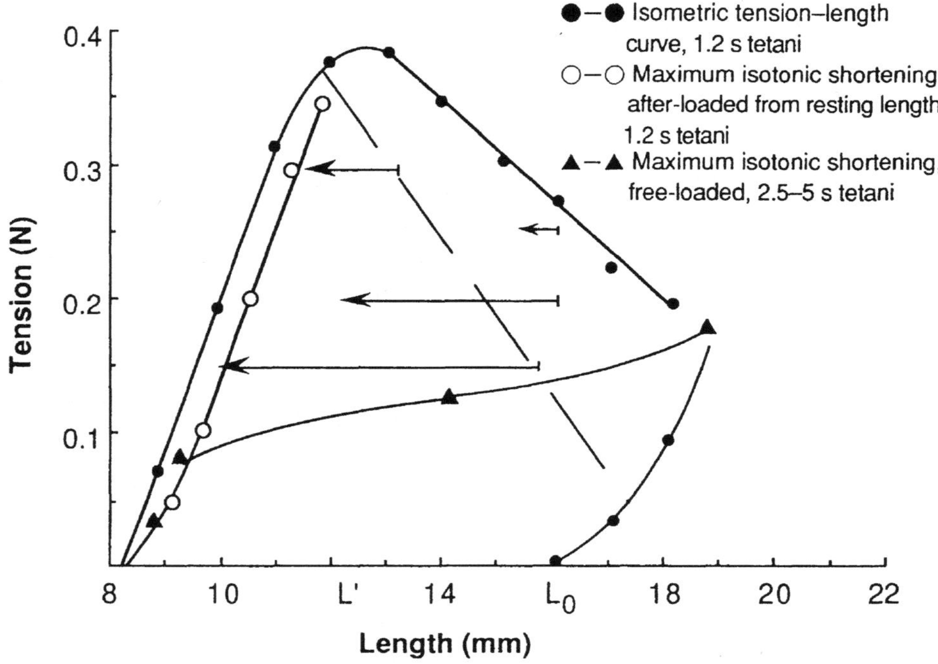

Comments
- The arrows show the maximum shortening for some other loads and initial lengths.
- The isometric and isotonic tension–length curves are approximately equivalent only if length and tension remained in the area to the left of the interrupted line.
- $P_0 L_0 / M = 0.33$ MN/m^2 (L_0, body length; P_0, isometric tension at L_0; M, volume of specimen).

Reference(s)
Deleze JB (1961) The mechanical properties of the semitendinosus muscle at lengths greater than its length in the body. J Physiol 158:154–164 (with permission)

Force–Length Relation (17)

• Tension–striation spacing •	• Frog • Striated muscle	• Isolated musc e fiber • Contracting state

Materials
- Frogs
- Striated muscle
- Single fiber

Testing Methods and Experimental Conditions
- Temperature, 3°–4°C
- Transverse tetanus stimulation
- Length of a region in an isolated muscle fiber, within which the striation spacing was sufficiently uniform, was measured with a photo-electric device
- Muscle fiber length was controlled with a servo system

Data

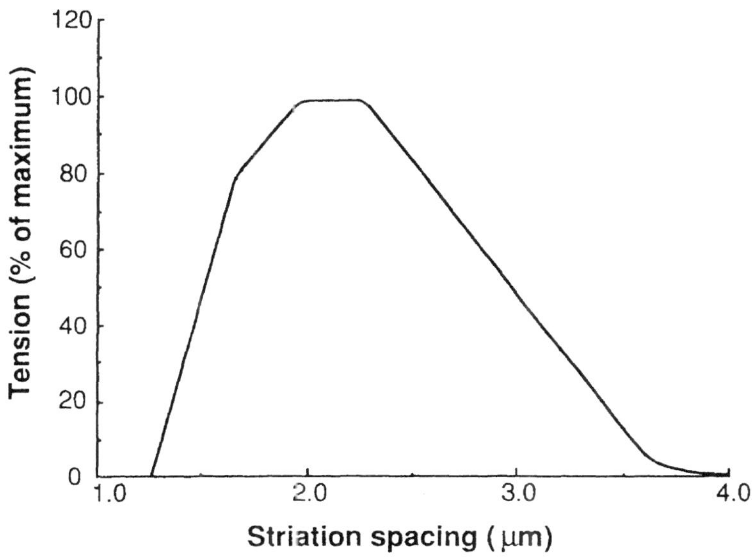

Comments
- Many features of the length–tension relation are simply explained by the sliding-filament theory.
- The peak of the curve consists of a plateau between sarcomere lengths of 2.05 and 2.2 μm.

Reference(s)
Gordon AM, Huxley AF, Julian FJ (1966) The variation in isometric tension with sarcomere length in vertebrate muscle fibres. J Physiol 184:170–192 (with permission)

Force–Velocity Relation (1)

• Force–power relation •	• Frog • Anterior tibialis muscle	• Isolated muscle fiber •

Materials
- Frogs, *Rana temporaria*
- Single living fiber from anterior tibialis muscle

Testing Methods and Experimental Conditions
- Shortening velocity was measured applying isotonic release
- Fibers were tetanized by electrical stimulation
- Experiments were carried out at 1.0°–2.5°C
- Force–velocity relation was determined for fibers in fatigue, which was produced by reducing the interval between isometric tetani to 15 or 30 s

Data

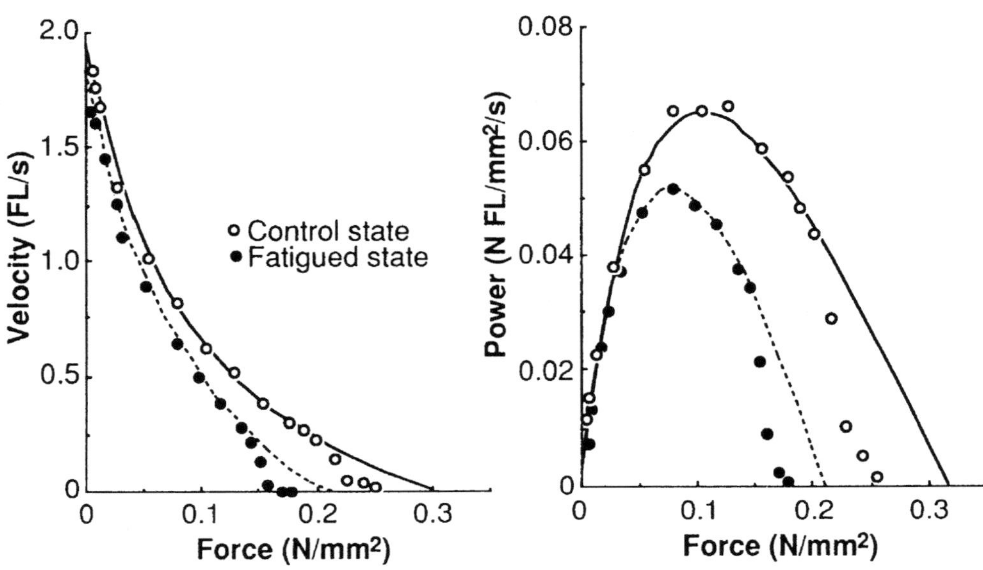

Comments
- Fatigue reduced isometric force (P_0) by $25.8 \pm 1.6\%$ (mean $\pm$ SE, $n = 13$) and depressed the maximum velocity of shortening by $10.2 \pm 2.2\%$. The force–velocity relation became less curved, a/P_0 (Hill's (1938) parameter) being increased by $29.5 \pm 8.8\%$. Thus, power was less affected than isometric force or V_{max}.

Reference(s)
Curtin NA, Edman KAP (1994) Force–velocity relation for frog muscle fibres: Effects of moderate fatigue and of intracellular acidification. J Physiol (London) 475:483–494 (with permission)

Force–Velocity Relation (2)

• Force–power relation •	• Human • Flexor muscle	• Training effect •

Materials
- Human
- Flexor muscle around elbow

Testing Methods and Experimental Conditions
- Force–velocity relationship at the maximal effort was measured in elbow flexion using different weights, and effects of training on the relationship were studied
- A subject of each group flexed the elbow with load of 0%, 30%, 60%, and 100% of P_0 (maximum tension) at maximal effort 10 times a day, 3 days a week for 12 weeks
- Force and velocity were calculated at the wrist

Data

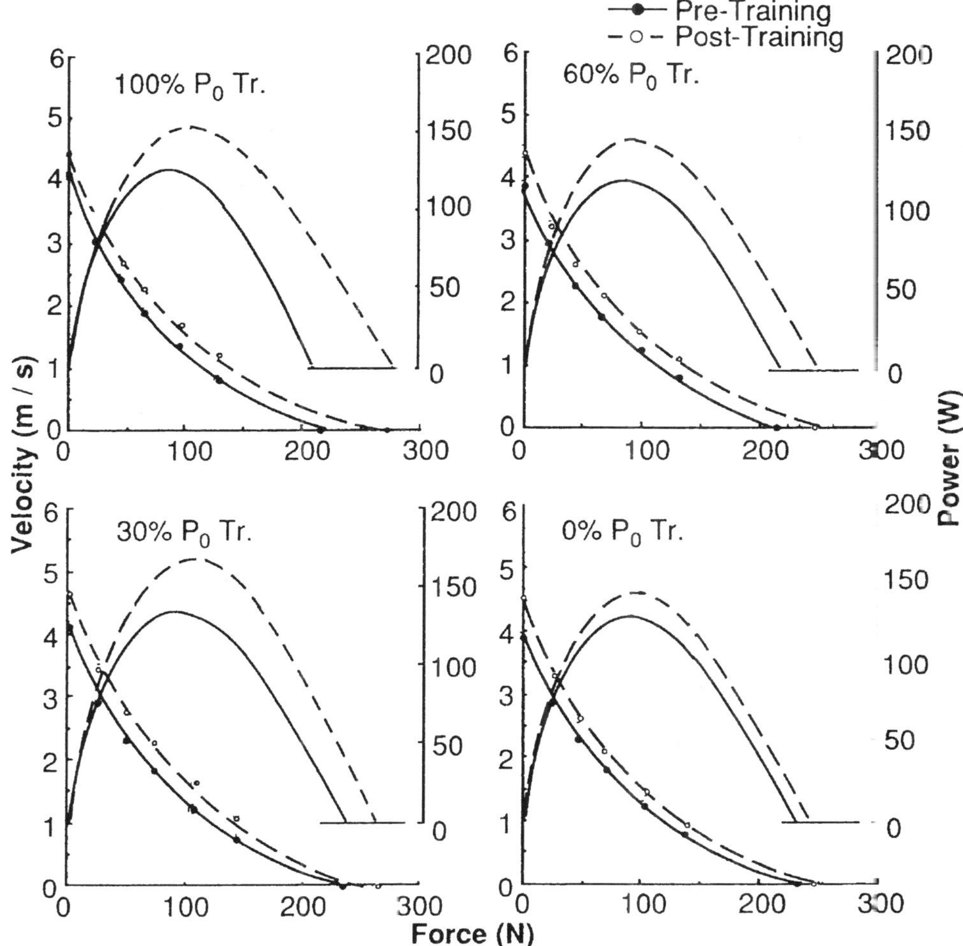

Comments
None.

Reference(s)

Kaneko M, Fuchimoto T, Toji H, Suei K (1981) Training effects on force, velocity and power relationship in human muscle. J Physical Fitness Jpn 30:86–93

Force–Velocity Relation (3)

• Force–power relation •	• Human • Leg extensor muscle	• •

Materials
- Human (non–athlete, skier, runner, football player)
- Leg extensor muscle

Testing Methods and Experimental Conditions
- Force–velocity relationship at the maximal effort was measured in knee extension using a lever-pulley with different weights
- Velocity is the tangential velocity of the ankle, and force is applied at the point 0.3 m from the knee

Data

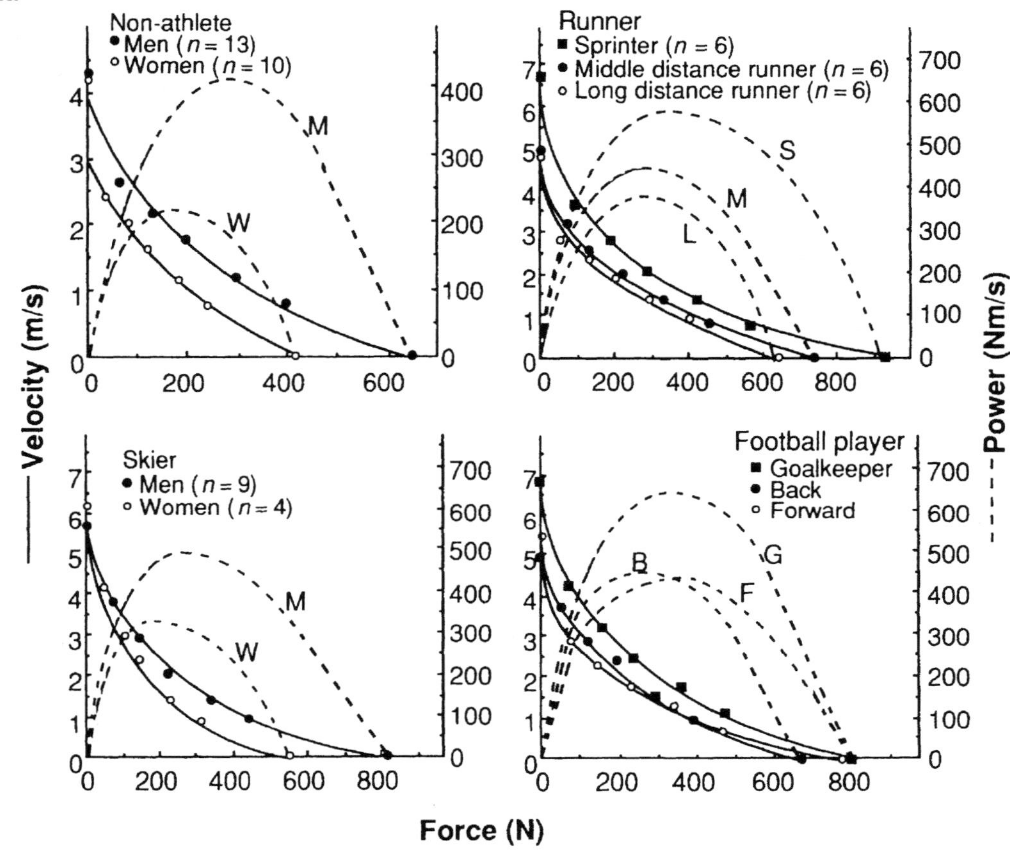

Comments
None.

Reference(s)
Kawahatsu K, Ikai M (1971) The development of the mechanical power and the force–velocity relation on the human leg extensor. Res J Physical Educ 16:223–232 (with permission)

Force–Velocity Relation (4)

• Force–velocity relation •	• Cat • Gastrocnemius muscle	• Intact whole muscle •

Materials
- Cats
- Gastrocnemius muscle

Testing Methods and Experimental Conditions
- Medial head of gastrocnemius muscle was exposed. The tendon was detached
  from its insertion and connected to a load through a velocity transducer
- Muscles were stimulated by sequentially supplying pulses to different subdivisions of
  the ventral nerve roots
- Recruitment order was set by varying the amplitude of the stimulus to the electrodes
  and the voltage to a pair of anodal block electrodes which were placed over the combined
  ventral root pool. Then isotonic shortening velocity was measured at different loads

Data

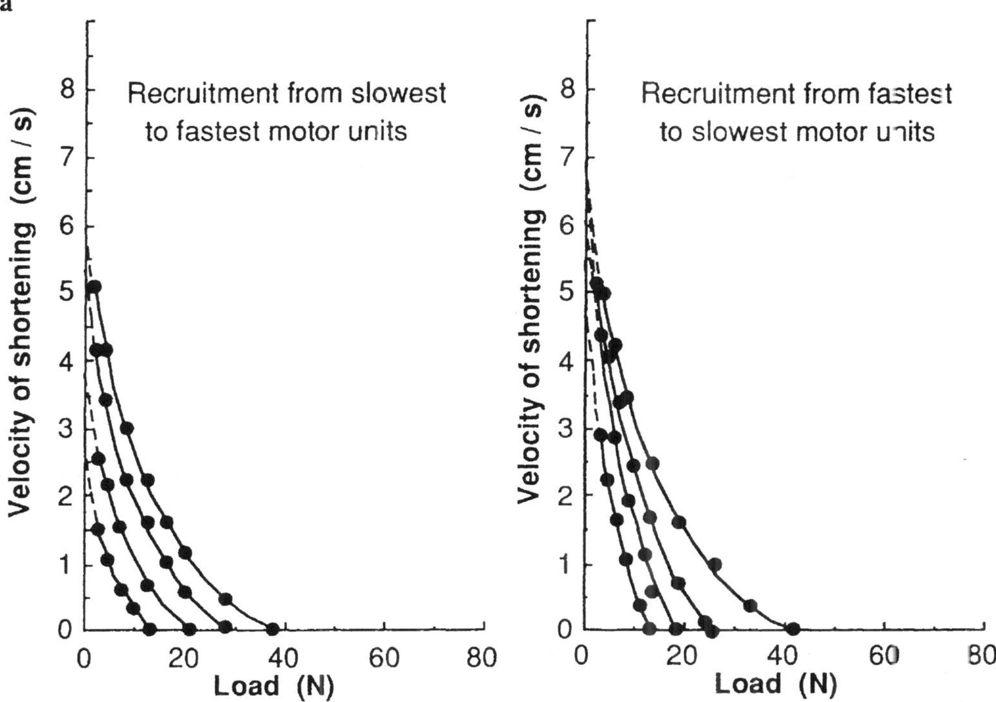

Comments
- The curves of the soleus muscle were similar to those of the gastrocnemius muscle,
  though the velocity of the former was slower than that of the latter.

Reference(s)
Petrofsky JS, Phillips CA (1980) The influence of recruitment order and fibre composition
on the force–velocity relationship and fatiguability of skeletal muscles in the cat. Med Biol
Eng Comput 18:381–390 (with permission)

Force–Velocity Relation (5)

• Force–velocity relation •	• Frog •	• Isolated muscle fiber • KCl concentration

Materials

- Frogs, *Rana pipiens*
- Single mechanically skinned fiber

Testing Methods and Experimental Conditions

- Shortening velocity was measured by using an isotonic release technique
- Fibers were activated by the addition of calcium
- pCa of the activating solution ranged from 5 to 6.7. KCl concentration ranged from zero to 280 mM
- Experiments were carried out at 5°–7°C

Data

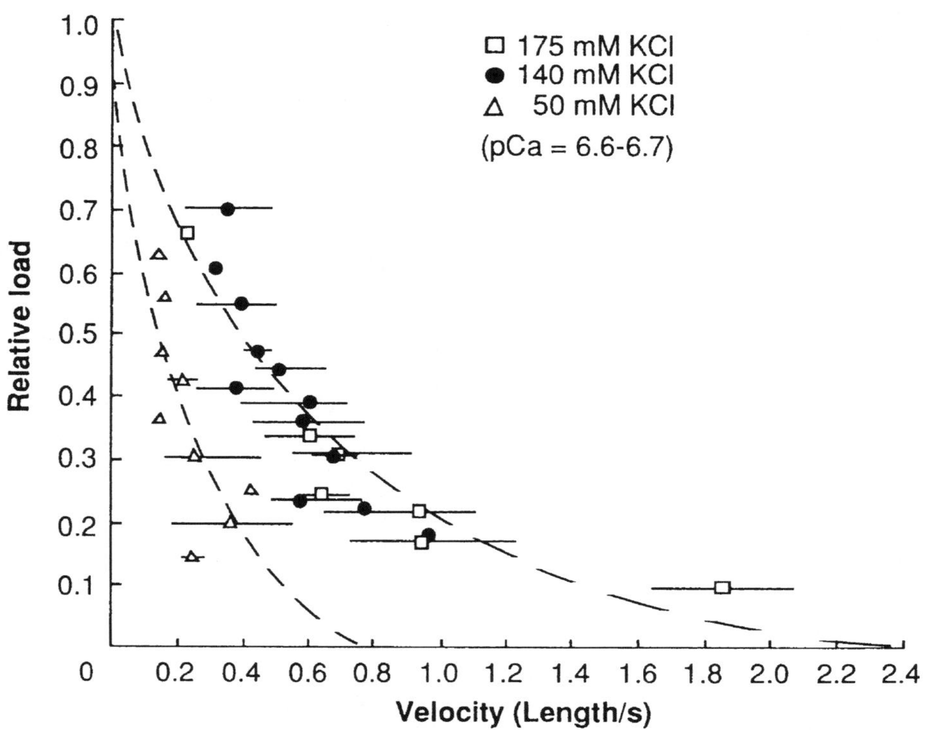

Comments

- The bars show velocity range.
- The shortening velocity at a given relative load was unaffected by the level of calcium activation at 140 mM KCl, and was independent of ionic strength when KCl concentration was increased from 140 to 280 mM. In contrast, the shortening velocity decreased as KCl concentration was reduced below 140 mM; the decrease in velocity was enhanced when fibers were only partially activated.

Reference(s)

Thames MD, Teichholtz LE, Podolsky RJ (1974) Ionic strength and the contraction kinetics of skinned fibers. J Gen Physiol 63:509–530 (with permission)

Force–Velocity Relation (6)

• Force–velocity relation •	• Frog • Anterior tibialis muscle	• Isolated muscle fiber •

Materials
- Frogs, *Rana temporaria*
- Single living fiber from anterior tibialis muscle

Testing Methods and Experimental Conditions
- Shortening velocity was measured applying isotonic release
- Fibers were tetanized by electrical stimulation
- Experiments were carried out at $1.0°–3.3°C$

Data

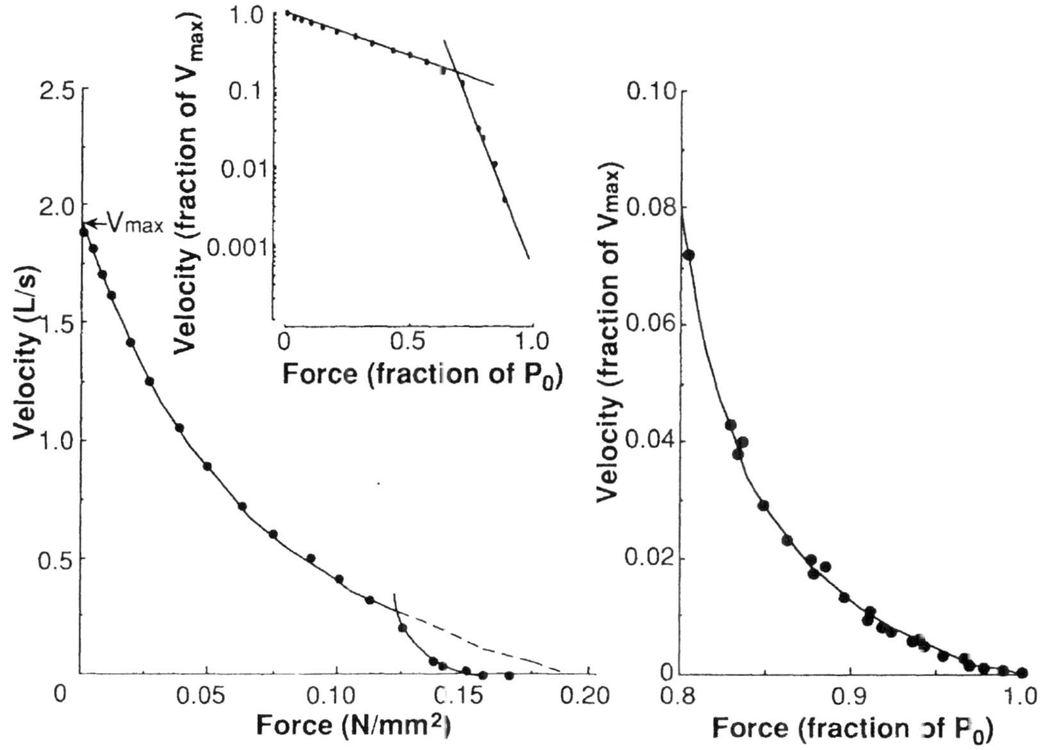

Comments
- The force–velocity relation had two distinct regions, each one exhibiting an upwards concave shape, that were located within the ranges 0%–78% and 78%–100% of the measured isometric force (P_0), respectively. The two portions of the force–velocity relation could be fitted well by hyperbolic functions or by single exponential functions.

Reference(s)
Edman KAP (1988) Double-hyperbolic force–velocity relation in frog muscle fibers. J Physiol (London) 404:301–321 (with permission)

Force–Velocity Relation (7)

· Force–velocity relation ·	· Frog · Anterior tibialis muscle	· Isolated muscle fiber ·

Materials
- Frogs, *Rana japonica*
- Single living fiber from anterior tibialis muscle

Testing Methods and Experimental Conditions
- Shortening velocity was measured under auxotonic conditions, where the distance shortened was proportional to the force exerted by the fiber
- Fibers were tetanized by electrical stimulation
- Experiments were carried out at 0°–2°C

Data

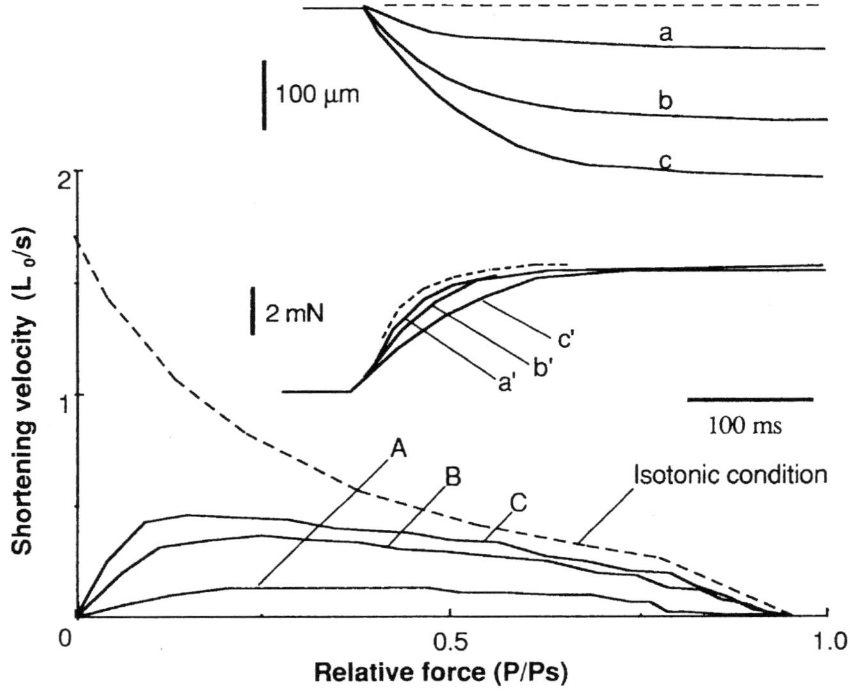

Comments
- The force (P)–velocity (V) curves (A, B, C) obtained in the auxotonic condition were convex upwards and always below the hyperbolic P–V curve obtained under the isotonic condition. Different curves were obtained depending on the compliance of auxotonic load. The shortening velocity for a given amount of load increased with increasing compliance.

Reference(s)
Iwamoto H, Sugaya R, Sugi H (1990) Force–velocity relation of frog skeletal muscle fibres shortening under continuously changing load. J Physiol (London) 422:185–202 (with permission)

Force–Velocity Relation (8)

• Force–velocity relation •	• Frog • Iliofibularis muscle	• Isolated muscle fiber •

Materials
- Frogs, *Xenopus laevis*
- Single living twitch and slow fibers from iliofibularis muscle

Testing Methods and Experimental Conditions
- Shortening velocity was measured by subjecting the fibers to isotonic release or afterloaded shortening
- Twitch fibers were tetanized by electrical stimulation. Slow fibers were activated by applying high K (30–75 mM) solutions
- Experiments were carried out at 5°–20°C for twitch fibers, and 21°–24°C for slow fibers

Data

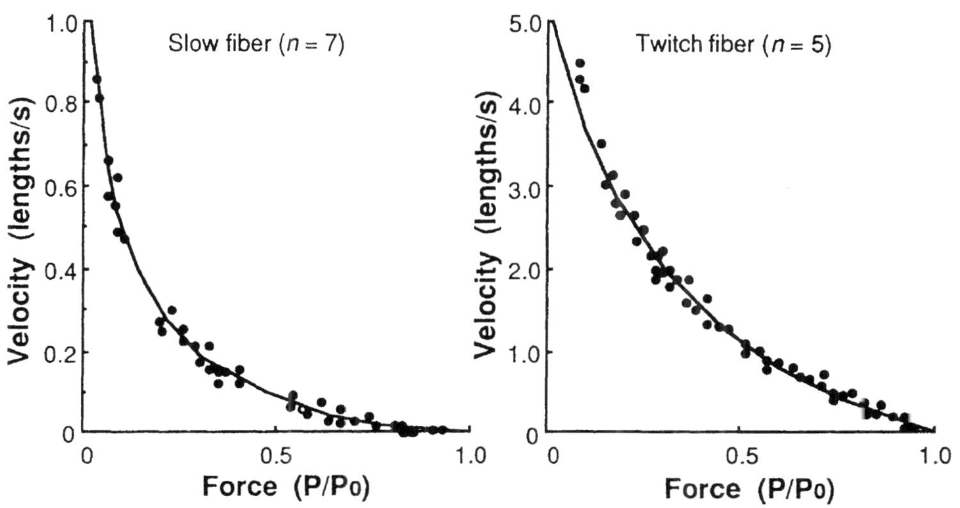

Comments
- Force–velocity data for slow fibers could be represented by a hyperbola with the constants $a = 0.10P_0$, $b = 0.11$ lengths/s; extrapolated V_{max} was 1.10 lengths/s.
- Force–velocity data for twitch fibers were reasonably well fitted by the hyperbola with the constants $a = 0.38P_0$, $b = 1.97$ lengths/s; extrapolated V_{max} was 5.20 lengths/s.
- Calculations based on AF Huxley's (1957) model for muscle contraction indicated that cross-bridge turnover rate was about 15 times lower in slow than in twitch fibers.

Reference(s)
Lännergren JI (1978) The force–velocity relation of isolated twitch and slow muscle fibres of *Xenopus laevis*. J Physiol (London) 283:501–521 (with permission)

Force–Velocity Relation (9)

• Force–velocity relation •	• Frog • Sartorius muscle	• Isolated muscle fiber •

Materials
- Frogs, *Rana temporaria* or *Rana esculenta*
- Whole sartorius muscle

Testing Methods and Experimental Conditions
- Shortening velocity was determined applying isotonic release
- Fibers were tetanized by electrical stimulation
- Experiments were carried out at 0°C

Data

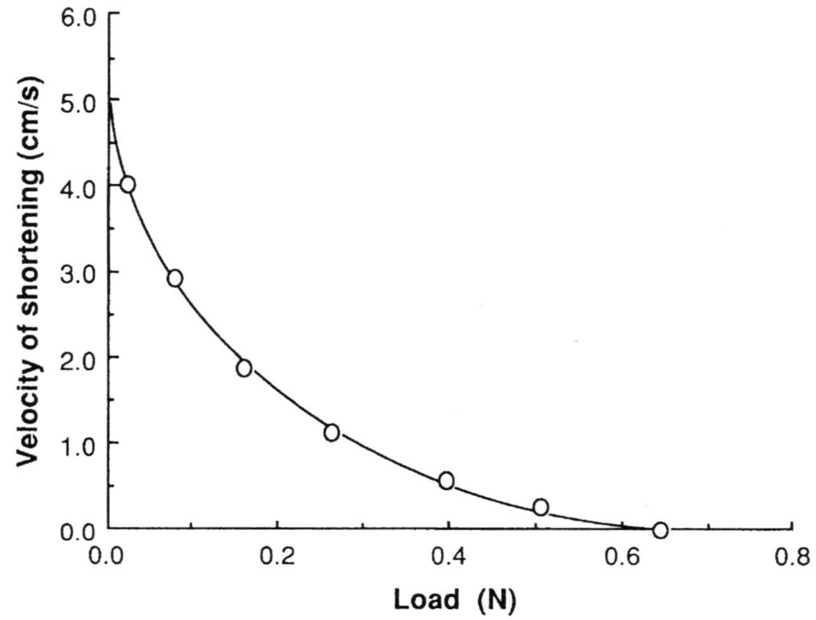

Comments
- Each circle denotes the mean of two observations in a series and reverse.
- The relation between speed of shortening, v, and load, P, is described by $(P + a)(v + b) =$ constant.

Reference(s)
Hill AV (1938) The heat of shortening and the dynamic constants of muscle.
Proc R Soc Lond B126:136–195 (with permission)

Force–Velocity Relation (10)

• Force–velocity relation •	• Frog • Sartorius muscle	• Isolated muscle fiber •

Materials

- Frogs, *Rana temporaria* or *Rana esculenta*
- Whole sartorius muscle

Testing Methods and Experimental Conditions

- Shortening velocity was determined applying isotonic release
- Loads greater than isometric tension were also applied
- Fibers were tetanized by electrical stimulation
- Experiments were carried out at 0°–20°C

Data

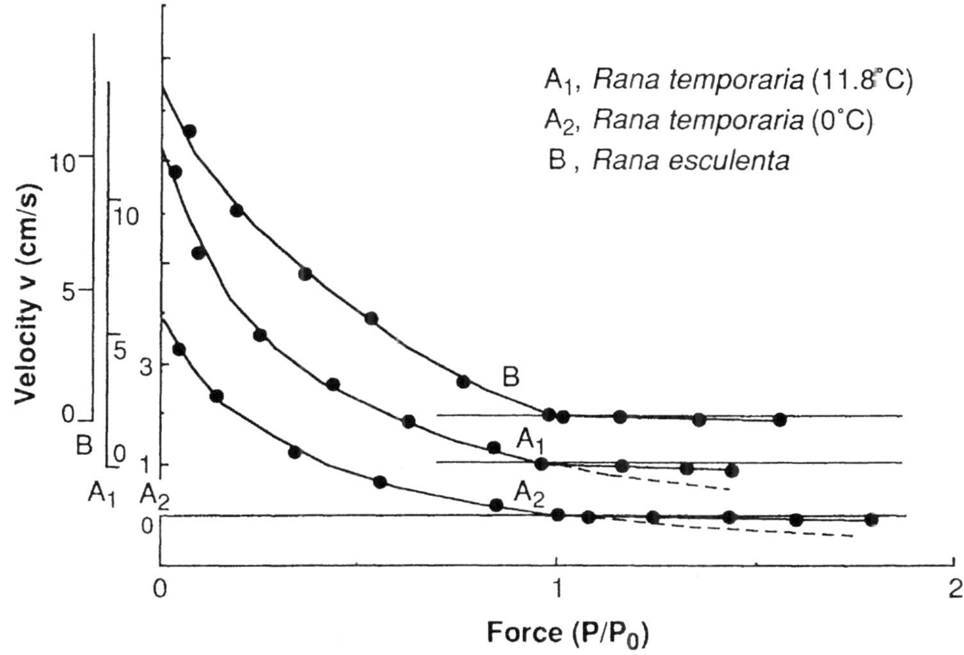

Comments

- The broken lines give the theoretical P–v relation for lengthening, using the constants a and b derived from the speeds of shortening.
- The observed relation between applied force and speed of isotonic shortening is satisfactorily described by Hill's hyperbolic equation $(P + a)(v + b) = constant$.
- The relation between a force greater than isometric and the resulting speed of "reversible" lengthening was determined. The observed velocities are considerably smaller than calculated from theory. If the applied force exceeds the isometric tension by 70–100%, the muscle "relaxes" rapidly.

Reference(s)

Katz B (1939) The relation between force and speed in muscular contraction. J Physiol (London) 96:45–64 (with permission)

Force–Velocity Relation (11)

• Force–velocity relation •	• Frog • Semitendinosus muscle	• Isolated muscle fiber •

Materials

- Frogs, *Rana nigromaculata*
- Semitendinosus muscle
- Small bundle of fast muscle fibers prepared from ventral caput

Testing Methods and Experimental Conditions

- Load–velocity relations were measured at different levels of contraction in Ringer's solution at 10°C

Data

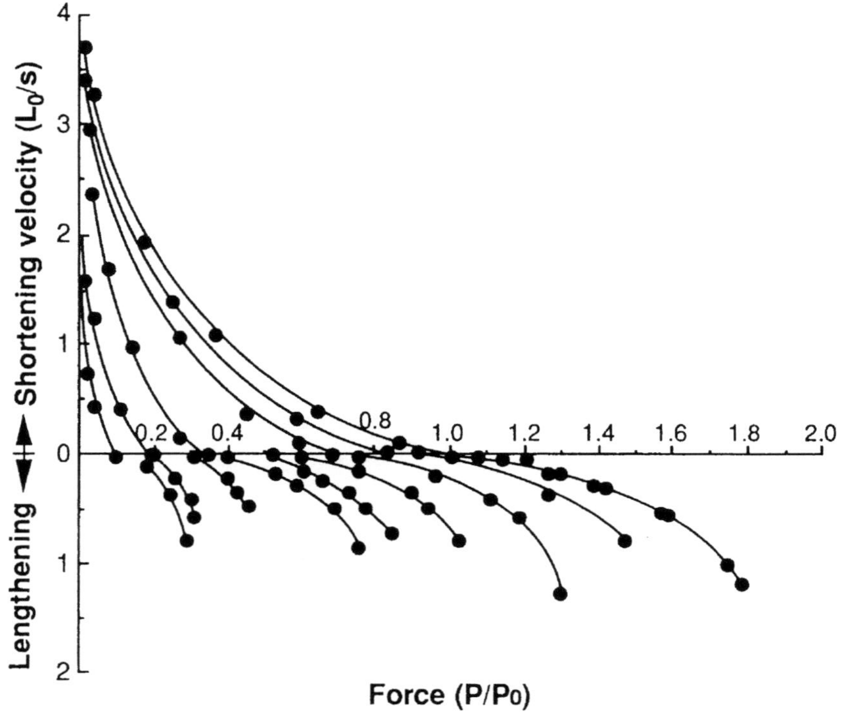

Comments

- Force–load–velocity relation (length region of $0.8 \sim 1.2\ L_0$):

  $(P + \alpha)(v + b) = b(F + \alpha)$ for shortening, where $\alpha = (a/P_0)F$, $a/P_0 = 0.25$, and $b = 0.9L_0/s$.

  $(P - 2F - \alpha')(v - b') = b'(F + \alpha')$ for lengthening, where $\alpha' = (a'/P_0)F$, $a'/P_0 = 0.4$, and $b' = 0.85L_0/s$.

  P_0, maximum isometric tension; F, contractile force (tension at $v = 0$); a and b, dynamic constants (shortening); a' and b', dynamic constants (lengthening); P, load; v, velocity (positive for shortening); L_0, standard length.

Reference(s)

Mashima H, Akazawa K, Kushima H, Fujii K (1972) The force–velocity relation and the viscous-like force in the frog skeletal muscle. Jpn J Physiol 22:103–120 (with permission)

Force–Velocity Relation (12)

• Force–velocity relation •	• Frog • Semitendinosus muscle	• Isolated muscle fiber •

Materials
- Frogs, *Rana temporaria*
- Single living fiber from semitendinosus muscle

Testing Methods and Experimental Conditions
- Shortening velocity was measured by applying isotonic release
- Fibers were tetanized by electrical stimulation
- Experiments were carried out at 0.5°–2.0°C
- Force–velocity relation was determined in normal (1.00 R), hypotonic (0.62 and 0.81 R) and hypertonic (1.22 R and 1.44 R) Ringer's solutions (the values refer to osmolality)

Data

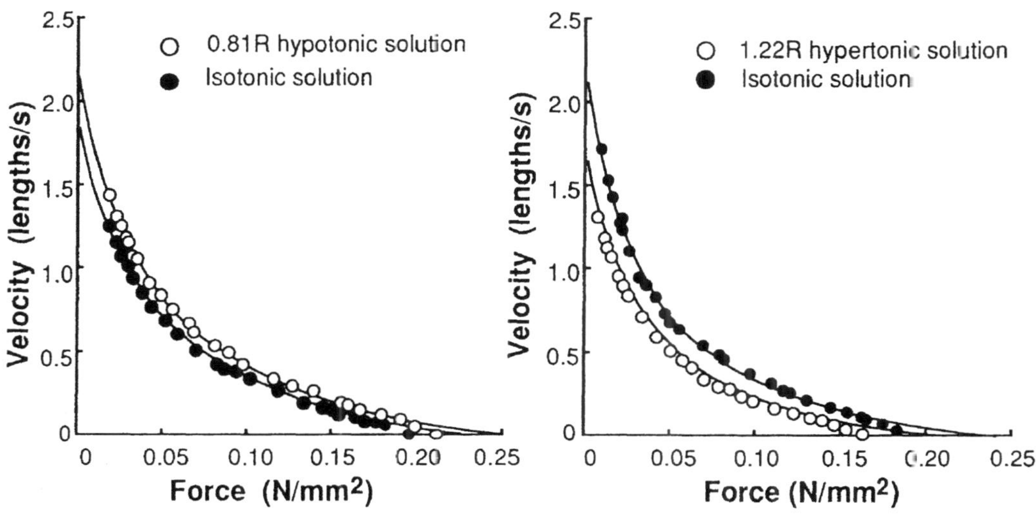

Comments
- Isometric force and maximum speed of shortening both changed inversely with the tonicity of the extracellular medium. Immersion of the fiber in 0.81 R hypotonic solution caused active tension and shortening velocity to increase by $10 \pm 1\%$ (mean $\pm$ SE, $n = 14$) and $12 \pm 1\%$, respectively. Conversely, force and shortening velocity decreased by $12 \pm 1\%$ ($n = 13$) and $22 \pm 2\%$, respectively, when normal Ringer was replaced by 1.22 R hypertonic solution. These changes doubled when the tonicity was altered from normal Ringer to 0.62 R and 1.44 R, respectively.

Reference(s)
Edman KAP, Hwang JC (1977) The force–velocity relationship in vertebrate muscle fibres at varied tonicity of the extracellular medium. J Physiol (London) 269:255–272 (with permission)

Force–Velocity Relation (13)

• Force–velocity relation •	• Frog • Semitendinosus muscle	• Isolated muscle fiber • Different calcium levels

Materials

- Frogs, *Rana pipiens*
- Briefly glycerinated single fiber from semitendinosus muscle

Testing Methods and Experimental Conditions

- Shortening velocity was measured by applying isotonic release (three successively decreasing loads in a single contraction)
- Fibers were activated by the addition of calcium
- Experiments were carried out at 4°C
- Force–velocity relation was determined for various calcium levels

Data

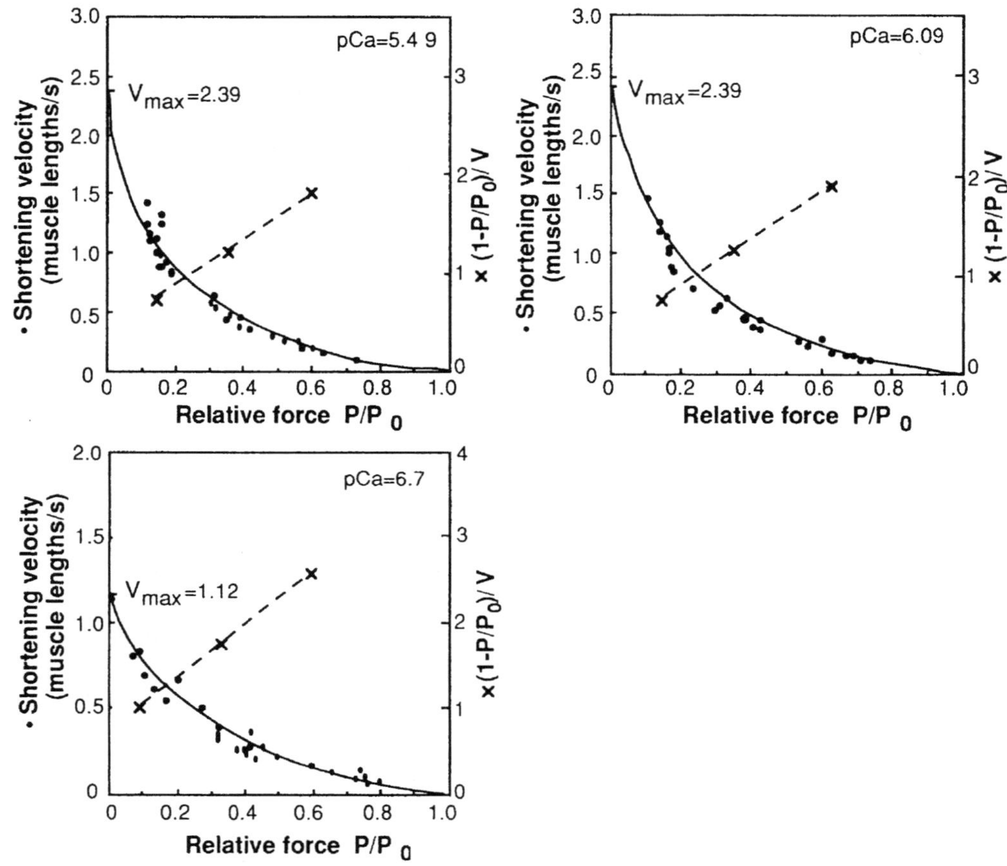

Comments

- The relative force–velocity relation was the same at pCa 6.09 and 5.49 where the steady force was at a maximum. The data points were well fitted by a hyperbola in which the extrapolated value for V_{max} was 2.39 muscle lengths/s.
- The relative force–velocity points obtained at higher pCa values at which the steady force was on average 37% of that developed at pCa 5.49 were fitted by a hyperbola. However, the extrapolated value for V_{max} was only 1.12 muscle lengths/s.

Reference(s)

Julian FJ (1971) The effect of calcium on the force–velocity relation of briefly glycerinated frog muscle fibres. J Physiol (London) 218:117–145 (with permission)

Force–Velocity Relation (14)

• Force–velocity relation •	• Human • Flexor muscle	• •

Materials
- Human
- Flexor muscle around elbow

Testing Methods and Experimental Conditions
- Force–velocity relationship at the maximal effort was measured in flexion of the elbow using different weights
- Force and velocity were measured at the hand

Data

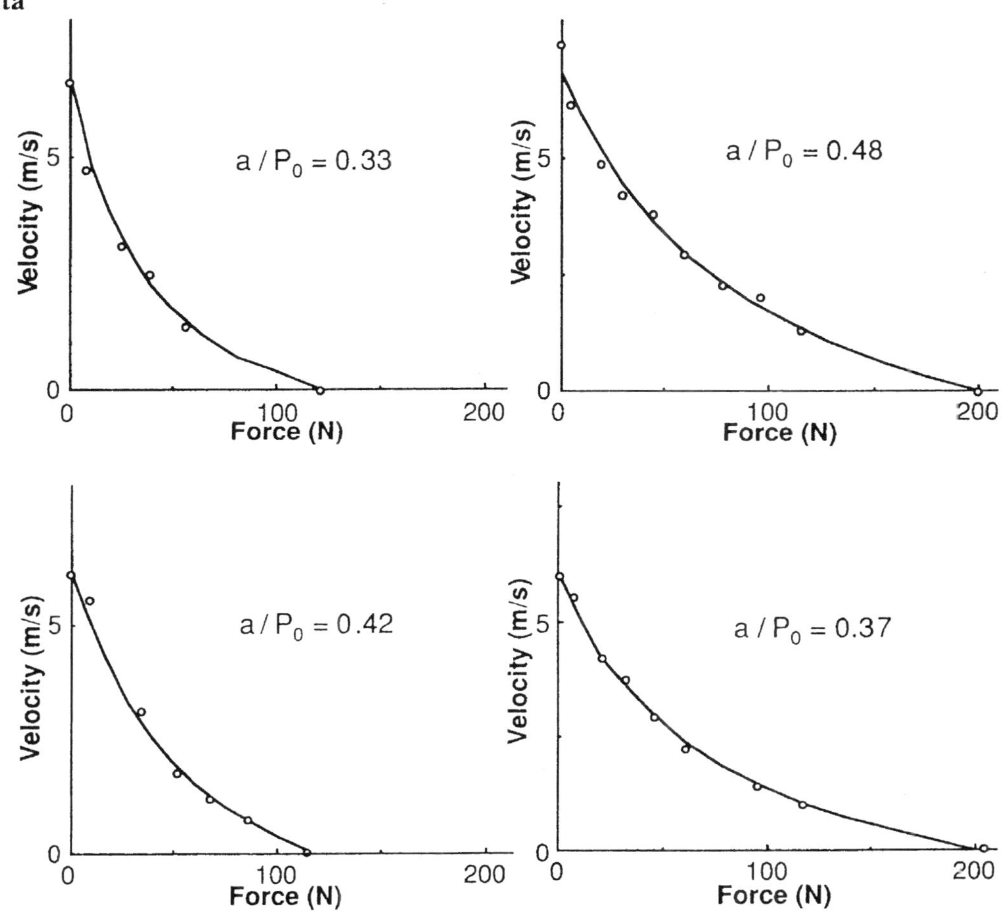

Comments
- P_0, force exerted at zero speed; a, constant of Hill's equation.
- Five muscles were involved in the movement of flexion of the elbow.

Reference(s)
Wilkie DR (1950) The relation between force and velocity in human muscles. J Physiol 110:249–280 (with permission)

Force–Velocity Relation (15)

| • Force–velocity relation | • Human | • |
| • | • Gastrocnemius muscle | • |

Materials
- Human
- Gastrocnemius muscle

Testing Methods and Experimental Conditions
- Force–velocity relationship at the submaximal effort (indicated by constant EMG) was measured in plantar flexion of the foot using a dynamometer
- Force–velocity curves were plotted at four different levels of submaximal excitation

Data

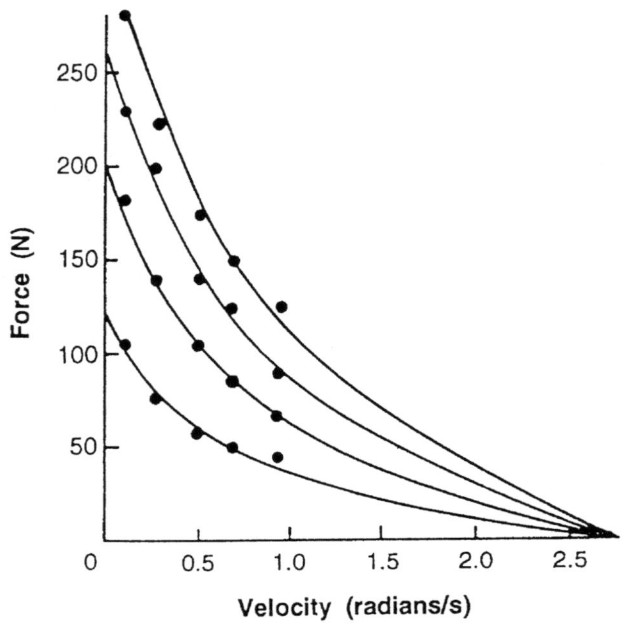

Comments
- Forces are approximately 1/10 of forces calculated in the tendon.

Reference(s)
Bigland B, Lippold OCJ (1954) The relation between force, velocity and integrated electrical activity in human muscles. J Physiol 123:214–224 (with permission)

Force–Velocity Relation (16)

• Force–velocity relation	• Rabbit	• Isolated muscle fiber
•	•	•

Materials
- Rabbits
- Myosin isolated from skeletal muscle

Testing Methods and Experimental Conditions
- In vitro force–movement assay study
- Myosin was coated on the tip of a flexible glass microneedle
- The myosin-coated glass microneedle was made to slide on the actin cables on the inner surface of the internodal cell of an alga, *Nitellopsis obtusa*
- Force and sliding were measured from the displacement of the tip of the elastic glass microneedle
- Experiments were carried out at 20°–23°C
- Force–velocity relations were obtained from the sliding movement with a sigmoidal time course and a nonsigmoidal time course

Data

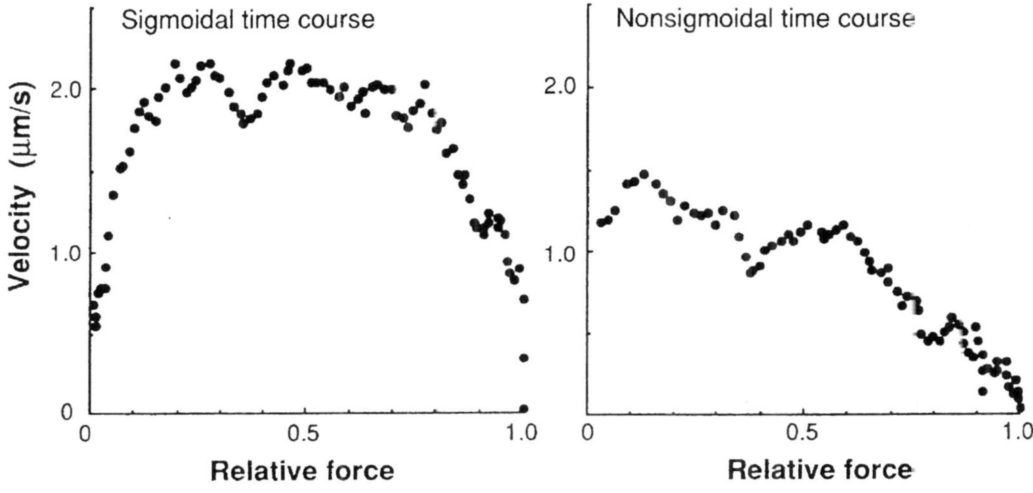

Comments
- The shape of the force–velocity curve was convex upwards, similar to that of the force–velocity curve of intact frog muscle fibers obtained under the auxotonic condition.

Reference(s)
Chaen S, Oiwa K, Shimmen T, Iwamoto H, Sugi H (1989) Simultaneous recordings of force and sliding movement between a myosin-coated glass microneedle and actin cables in vitro. Proc Natl Acad Sci USA 86:1510–1514 (with permission)

Force–Velocity Relation (17)

• Force–velocity relation •	• Rabbit •	• Isolated muscle fiber •

Materials
- Rabbits
- Myosin isolated from skeletal muscle

Testing Methods and Experimental Conditions
- In vitro force–movement assay study
- Myosin was coated on the surface of tosyl-activated polystyrene beads
- The myosin-coated polystyrene beads were made to slide on the actin cables on the inner surface of the internodal cell of an alga, *Nitellopsis obtusa*, mounted on the rotor of a centrifuge microscope
- The beads were made to slide under constant centrifugal force directed opposite to ("positive" loads) or in the direction of ("negative" loads) the bead movement
- Experiments were carried out at 20°–23°C

Data

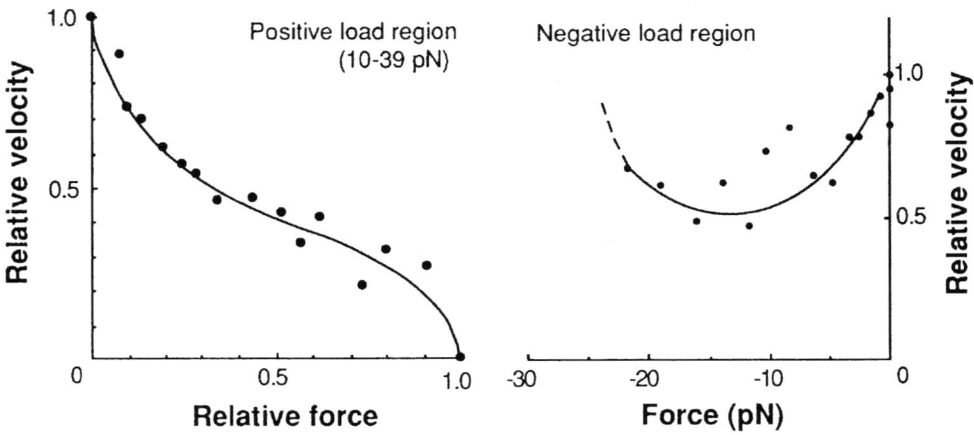

Comments
- The steady-state force–velocity curve determined for the positive load region was analogous to the double-hyperbolic force–velocity curve of single muscle fibers. The unloading velocity of bead movement was 1.6–3.6 μm/s.
- In the negative load region, the velocity of bead movement first decreased by 20%–60% with increasing negative loads and then increased towards the initial unloading velocity until the beads were eventually detached from the actin cables.

Reference(s)
Oiwa K, Chaen S, Kamitsubo E, Shimmen T, Sugi H (1990) Steady-state force–velocity relation in the ATP-dependent sliding movement of myosin-coated beads on actin cables in vitro studied with a centrifuge microscope. Proc Natl Acad Sci USA 87:7893–7897 (with permission)

Force–Velocity Relation (18)

• Force–velocity relation •	• Rabbit • Psoas muscle	• Isolated muscle fiber •

Materials
- Rabbits
- Single fiber from glycerinated psoas muscle

Testing Methods and Experimental Conditions
- Shortening velocity was measured using an isotonic release technique
- Fibers were activated by the addition of calcium
- Experiments were carried out at 25°C
- Force–velocity relation was determined after modification with paramagnetic probes (IASL [iodoacetamide spin label])

Data

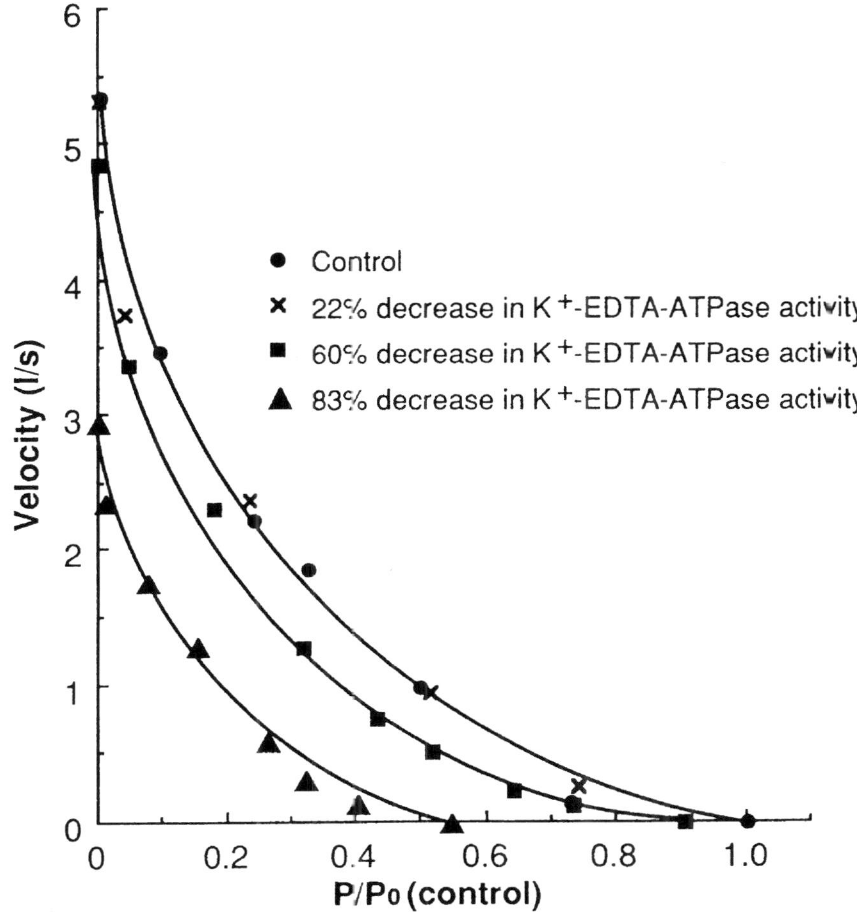

Comments
- In all cases, the velocities of contraction decreased in parallel with the decrease in tension, while the decrease in stiffness was less pronounced.

Reference(s)
Crowder MS, Cooke R (1984) The effect of myosin sulphydryl modification on the mechanics of fibre contraction. J Muscle Res Cell Motil 5:131–146 (with permission)

Force–Velocity Relation (19)

• Force–velocity relation •	• Rabbit • Psoas muscle	• Isolated muscle fiber • ATP concentration

Materials
- Rabbits
- Single fiber or bundle from glycerinated psoas muscle

Testing Methods and Experimental Conditions
- Shortening velocity was obtained using an isotonic release technique
- Fibers were activated by the addition of ATP (2.5 μM – 5 mM)
- Experiments were carried out at 10°C

Data

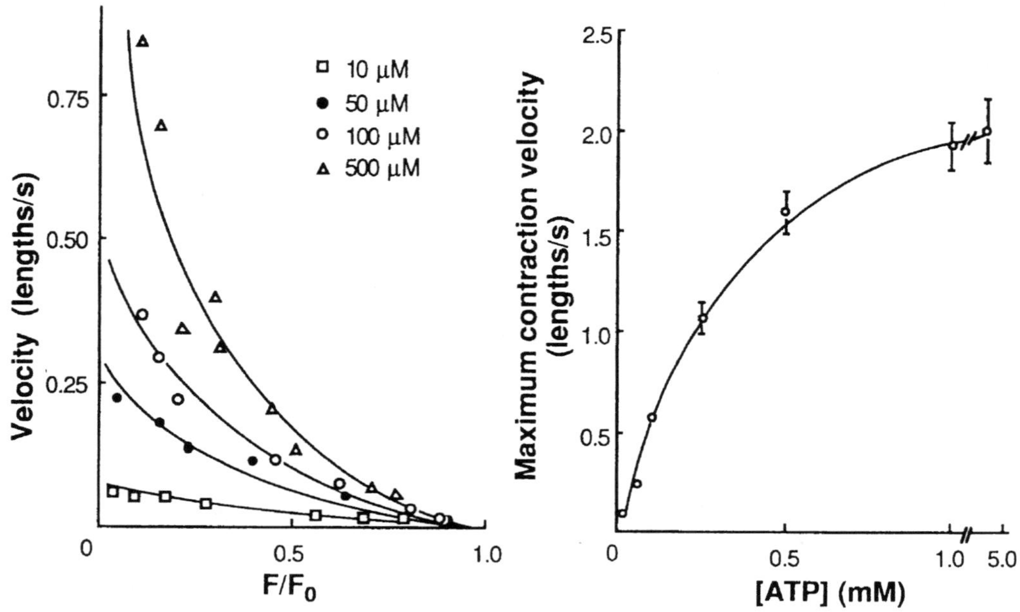

Comments
None.

Reference(s)
Cooke R, Bialek W (1979) Contraction of glycerinated muscle fibers as a function of the ATP concentration. Biophys J 28:241–258 (with permission)

Force–Velocity Relation (20)

• Force–velocity relation •	• Rabbit • Psoas muscle	• Isolated muscle fiber • MgATP concentration

Materials
- Rabbits
- Single fiber from glycerinated psoas muscle

Testing Methods and Experimental Conditions
- Shortening velocity was measured by using an isotonic release technique
- Fibers were activated by the addition of calcium
- Force–velocity relation was determined by using nonnucleotide ATP analogues 2-[(4-azido-2-nitrophenyl)amino]ethyl triphosphate (NANTP) and 2-[(4-azido-2-nitrophenyl)amino]propyl triphosphate (PrNANTP)) as substrates
- Experiments were carried out at 10°C

Data

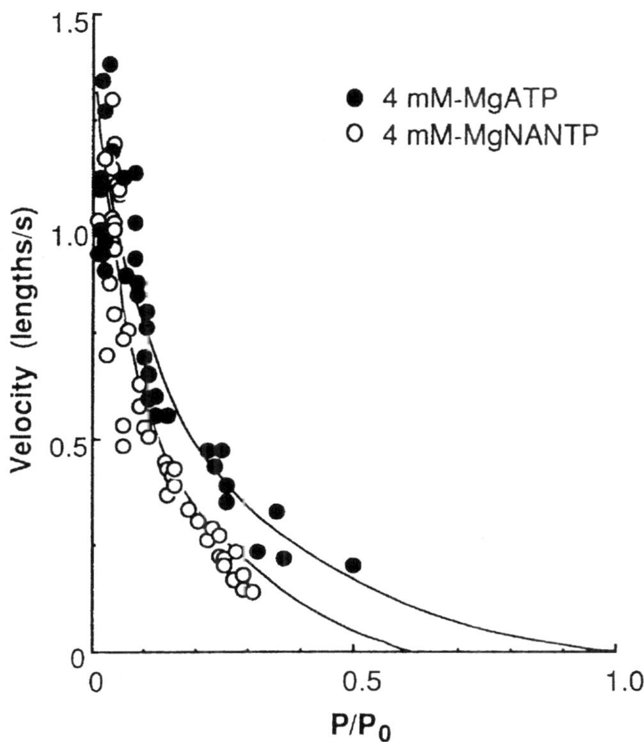

Comments
- At millimolar concentrations, in the absence of calcium, both analogues relaxed fibers. In the presence of calcium, MgNANTP produced isometric tension and stiffness. Maximum shortening velocity and triphosphatase rate were approximately the same for MgNANTP as for MgATP. In contrast, MgPrNANTP did not support isometric tension or active shortening in the presence of calcium.

Reference(s)
Pate E, Nakamaye KL, Franks-Skiba K, Yount RG, Cooke R (1991) Mechanics of glycerinated muscle fibers using nonnucleoside triphosphate substrates. Biophys J 59:598–605 (with permission)

Force–Velocity Relation (21)

• Force–velocity relation •	• Rat • Extensor digitorum longus	• Isolated muscle fiber •

Materials
- Rats (male)
- Extensor digitorum longus (EDL, fast-twitch muscle)

Testing Methods and Experimental Conditions
- Shortening velocity was measured by subjecting the fibers to isotonic releases or afterloaded shortening
- Twitch fibers were tetanized by electrical stimulation
- The sequence in which data were collected was 35°, 30°, 20°, and 35°C (closed triangles)

Data

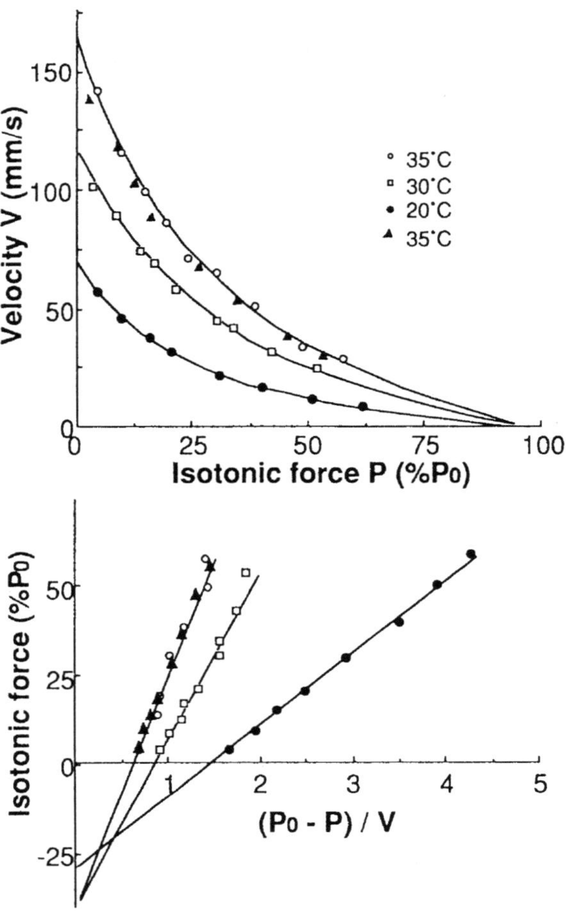

Comments
- The maximum velocity of shortening decreased with cooling in both EDL (fast twitch) and soleus (slow twitch, not illustrated) muscles. The calculated Arrhenius activation energy for maximum velocity of shortening was around 40–45 kJ in both muscles.

Reference(s)
Ranatunga KW (1982) Temperature-dependence of shortening velocity and rate of isometric tension development in rat skeletal muscle. J Physiol (London) 329:465–483 (with permission)

Force–Velocity Relation (22)

• Isometric stress • Shortening velocity	• Mouse, rat, rabbit, sheep • Extensor digitorum	• Isolated muscle fiber •

Materials
- Mice, rats, rabbits, sheep, and cows
- Chemically skinned single fiber from extensor digitorum longus (EDL, fast-twitch muscle) and soleus (slow-twitch muscle)

Testing Methods and Experimental Conditions
- Shortening velocity was measured using an isotonic release technique
- Fibers were activated by the addition of calcium
- Experiments were carried out at 5°–6°C

Data

Fiber type	Fiber diameter (μm)	Isometric stress (kPa)	Maximum velocity (μm/half-sarc. /s)	Maximum power (mW/g)	Maximum relative power (μm/half-sarc. /s)	Number of fibers
Mouse						
Fast	53.2 ± 9.9	153.3 ± 52.4	2.85 ± 0.46	34.9 ± 13.6	0.27 ± 0.033	11
Slow	31.1 ± 6.4	212.9 ± 20.8	1.61 ± 0.47	19.4 ± 5.6	0.11 ± 0.024	6
Rat						
Fast	72.7 ± 14.4	123.2 ± 41.2	1.73 ± 0.48	18.6 ± 9.1	0.18 ± 0.048	23
Slow	95.1 ± 13.9	99.9 ± 25.0	1.09 ± 0.47	6.67 ± 2.59	0.079 ± 0.019	10
Rabbit						
Fast	58.9 ± 12.2	123.1 ± 61.5	1.15 ± 0.22	14.3 ± 7.8	0.14 ± 0.017	19
Slow	53.4 ± 12.6	146.8 ± 54.1	0.84 ± 0.22	5.8 ± 3.2	0.051 ± 0.018	13
Sheep						
Fast	62.4 ± 16.1	159.4 ± 72.1	0.86 ± 0.14	9.1 ± 4.9	0.067 ± 0.007	49
Slow	59.3 ± 18.4	198.0 ± 105.6	0.62 ± 0.22	8.1 ± 8.3	0.04 ± 0.025	13
Cow 1						
Fast	52.2 ± 11.0	248.0 ± 112.3	0.77 ± 0.20	16.3 ± 9.3	0.08 ± 0.011	3
Slow	51.6 ± 14.1	232.7 ± 86.1	0.52 ± 0.12	4.4 ± 2.1	0.022 ± 0.003	9
Cow 2						
Fast	73.6 ± 15.3	87.8 ± 39.8	0.79 ± 0.17	6.0 ± 2.7	0.082 ± 0.007	11
Slow	77.1 ± 40.8	60.1 ± 62.4	0.38 ± 0.39	0.66 ± 0.72	0.013 ± 0.006	9

Comments
- While fiber diameter and isometric force showed no dependence on animal body size, maximum shortening velocity in both fast and slow fibers and maximum power output in fast fibers were found to vary with the $-1/8$ power of body size. For all sizes of animals, the average maximum velocity was 1.7 times faster in fast fibers than in slow fibers.

Reference(s)
Seow CY, Ford LE (1991) Shortening velocity and power output of skinned muscle fibers from mammals having a 25, 000-fold range of body mass. J Gen Physiol 97:541–560 (with permission)

Force–Velocity Relation (23)

• Maximal power •	• Human • Handgrip muscle	• Temperature effect •

Materials
- Human
- Handgrip muscle

Testing Methods and Experimental Conditions
- Force–velocity relationship of maximal contraction was measured in the temperature range 22°–38°C

Data

Parameter equation				Significance
$v_0 =$	0.99	+ 0.020	(t - 32.8)	$P < 0.001$
SE	0.02	0.004		
$F_0 =$	1.00	+ 0.001	(t - 32.8)	NS
SE	0.01	0.001		
$P_{max} =$	1.00	+ 0.033	(t - 32.8)	$P < 0.001$
SE	0.01	0.001		
$F_t / F_0 =$	1.00	+ 0.021	(t - 32.8)	$P < 0.001$
SE	0.01	0.002		
$a / F_0 =$	0.98	+ 0.086	(t - 32.8)	$P < 0.001$
SE	0.05	0.007		
$H =$	1.00	+ 0.020	(t - 32.8)	$P < 0.001$
SE	0.01	0.002		

t , muscle temperature; 32.8, resting muscle temperature; v_0 , the highest velocity with zero load (m/s); F_0 , the highest force with zero velocity (N); P_{max} , maximal power (w); F_t , the force at maximal power (N); a, constant of Hill's equation; H, distance between the origin and a force–velocity curve; NS, not significant.

Comments
None.

Reference(s)
Binkhorst RA, Hoofd L, Vissers ACA (1977) Temperature and force–velocity relationship of human muscles. J Appl Physiol 41:471–475 (with permission)

Force–Velocity Relation (24)

• Shortening velocity •	• Frog • Semitendinosus muscle	• Isolated muscle fiber • MgATP concentration

Materials
- Frogs, *Rana temporaria*
- Mechanically skinned single fiber from semitendinosus muscle

Testing Methods and Experimental Conditions
- Shortening velocity was measured by applying isotonic release
- Fibers were activated by the addition of calcium
- Experiments were carried out at 0°–5°C
- Force–velocity relation was determined in fully activated fibers at concentrations of ATP ranging from 10 µM to 10 mM

Data

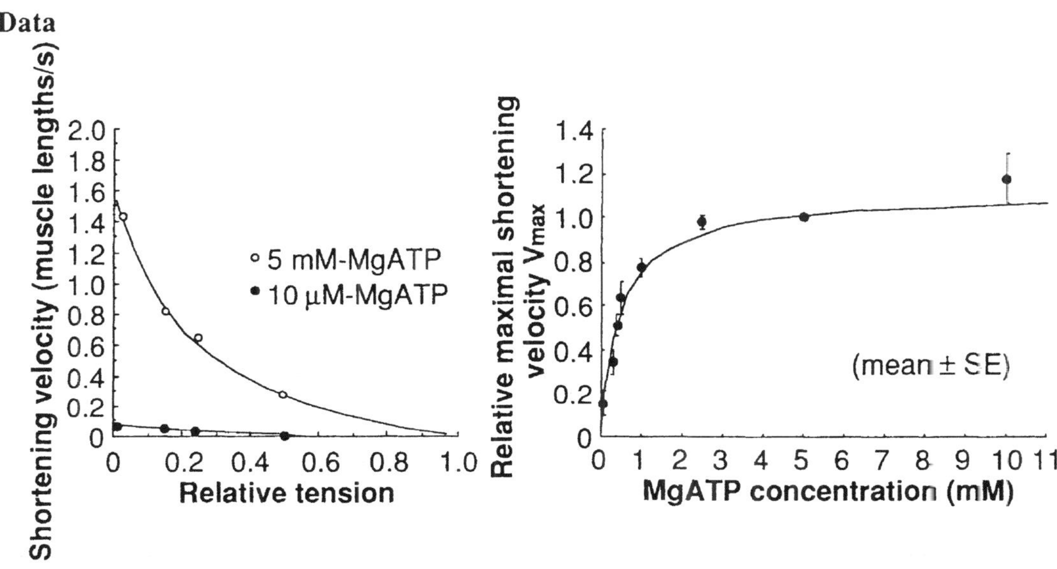

Comments
- V_{max} is expressed in values relative to the value obtained in the reference solution, 5 mM MgATP.
- V_{max} showed a roughly hyperbolic dependence on MgATP concentration, with Km (Michaelis constant) of 0.47 mM. At 5 mM MgATP, V_{max} was 2.16 muscle lengths per second, which is similar to that of intact fbers.
- The parameter a/P_0 of the Hill (1938) equation that is related to the curvature of the force–velocity relation showed a slight decrease with increasing MgATP concentration. Its value at 5 mM MgATP was 0.16, which is somewhat lower than found for intact fibers.

Reference(s)
Ferenczi MA, Goldman YE, Simmons RM (1984) The dependence of force and shortening velocity on substrate concentration in skinned muscle fibres from *Rana temporaria*. J Physiol (London) 350:519–543 (with permission)

Force–Velocity Relation (25)

• Tension–velocity relation •	• Cat • Soleus muscle	• Intact whole muscle •

Materials
- Cats
- Soleus muscle

Testing Methods and Experimental Conditions
- Soleus was exposed and dissected free from the surrounding structures, leaving it attached by only its origin, blood supply, and nerve supply
- The tendon was detached from its insertion and connected to a force transducer and a load
- By sequentially supplying pulses to different subdivisions of the ventral nerve roots, isotonic lengthening or shortening velocity was measured at different loads
- Velocity was measured 15 ms after step change of tension

Data

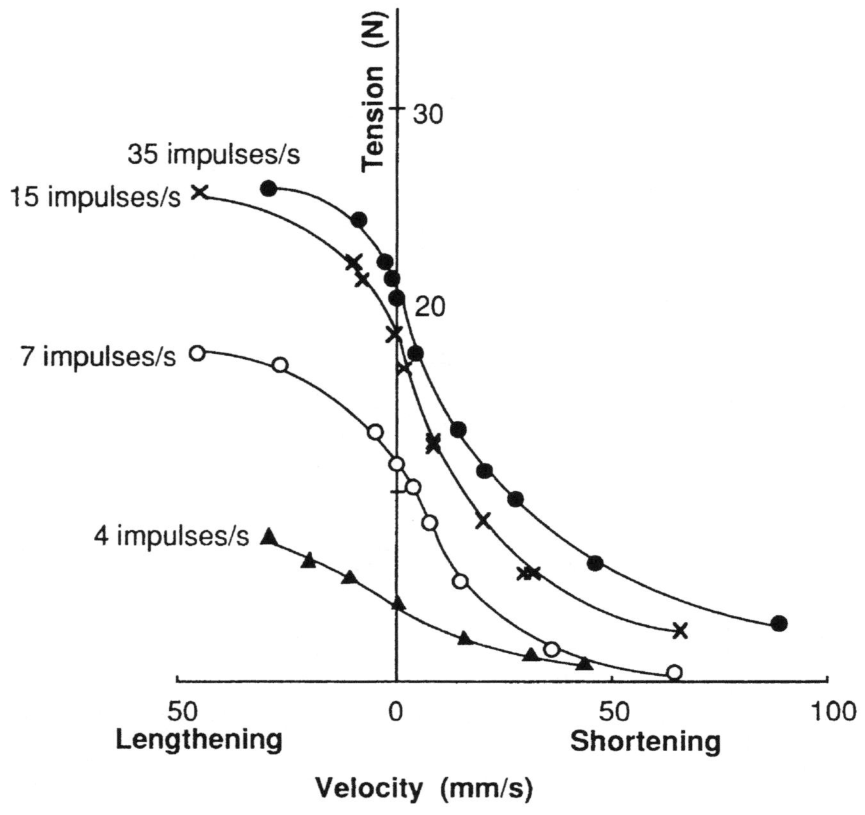

Comments
None.

Reference(s)
Joyce GC, Rack PMH (1969) Isotonic lengthening and shortening movements of cat soleus muscle. J Physiol 204:475–491 (with permission)

Force–Velocity Relation (26)

• Tension–velocity relation	• Frog	• Isolated musc e fiber
•	• Anterior tibialis muscle	•

Materials

- Frogs, *Rana esculenta*
- Single living fiber from anterior tibialis muscle

Testing Methods and Experimental Conditions

- Shortening velocity was determined applying controlled velocity release
- Fibers were activated by electrical stimulation
- Experiments were carried out at 3.7°–6°C
- The force–velocity relation was determined at different times (28–240 ms after the beginning of stimulation) during the development of isometric tetanus

Data

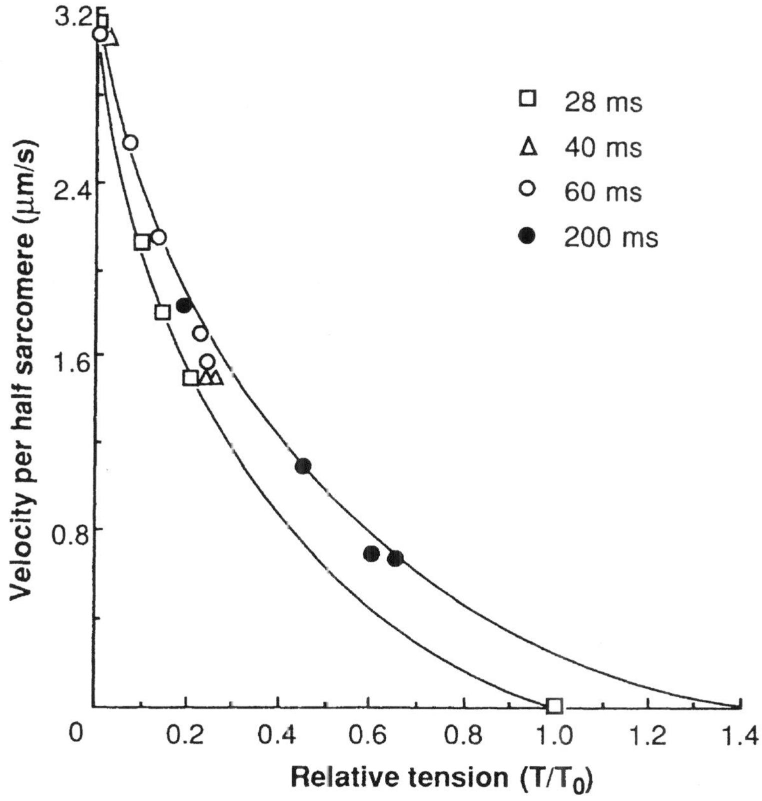

Comments

- During the rise of an isometric tetanus, the velocity of shortening at zero load remained constant.

Reference(s)

Ambrogi-Lorenzini C, Colomo F, Lombardi V (1983) Development of force–velocity relation, stiffness and isometric tension in frog single muscle fibres. J Muscle Res Cell Motil 4:177–189 (with permission)

Force–Velocity Relation (27)

• Tension–velocity relation	• Frog	• Isolated muscle fiber
•	• Semitendinosus muscle	•

Materials

- Frogs, *Rana esculenta*
- Single living fiber from semitendinosus muscle

Testing Methods and Experimental Conditions

- Shortening velocity was measured by applying controlled velocity releases
- Fibers were activated by electrical stimulation
- Force–velocity relation was also determined in the Ringer's solution in which 115 mM NaCl was replaced by $NaNO_3$

Data

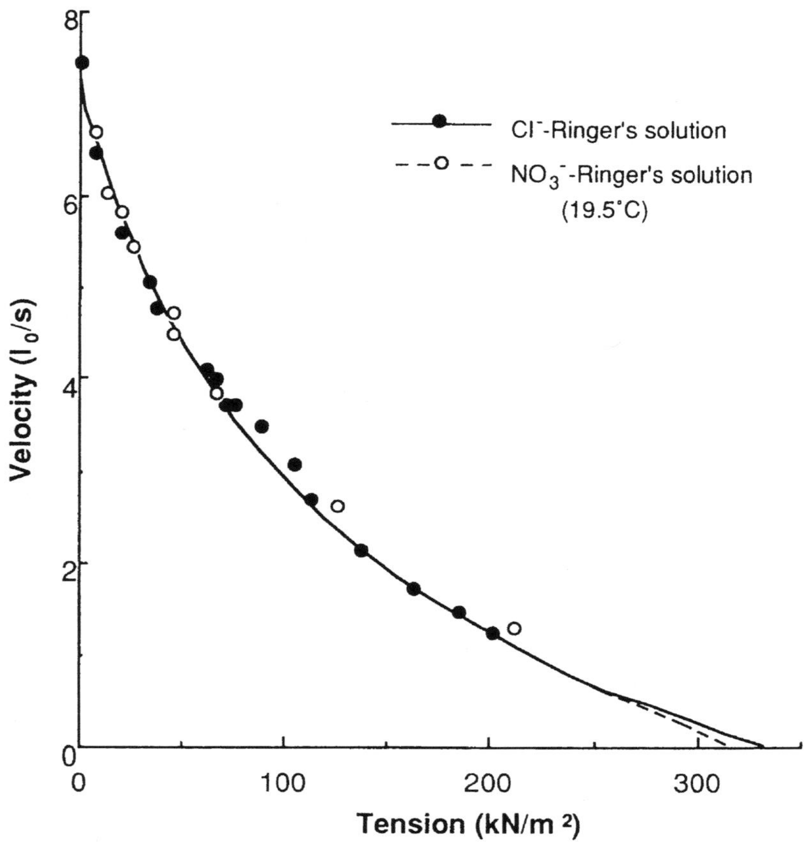

Comments

- NO_3^- ions did not affect the steady-state tension (P)–velocity (V) relation. The "relative" P–V relation appears to be independent of both the time after start of stimulation and the presence of NO_3^- ions in the bathing solution.

Reference(s)

Cecchi G, Colomo F, Lombardi V (1978) Force–velocity relation in normal and nitrate-treated frog single muscle fibres during rise of tension in an isometric tetanus. J Physiol (London) 285:257–273 (with permission)

Force–Velocity Relation (28)

• Tension–velocity relation •	• Rabbit • Psoas muscle	• Isolated muscle fiber •

Materials
- Rabbits
- Single fiber from glycerinated psoas muscle

Testing Methods and Experimental Conditions
- Shortening velocity was measured by using an isotonic release technique
- Fibers were activated by the addition of calcium
- Force–velocity relation was determined in the presence of 1–4 mM ADP or 12 mM inorganic phosphate (P_i)

Data

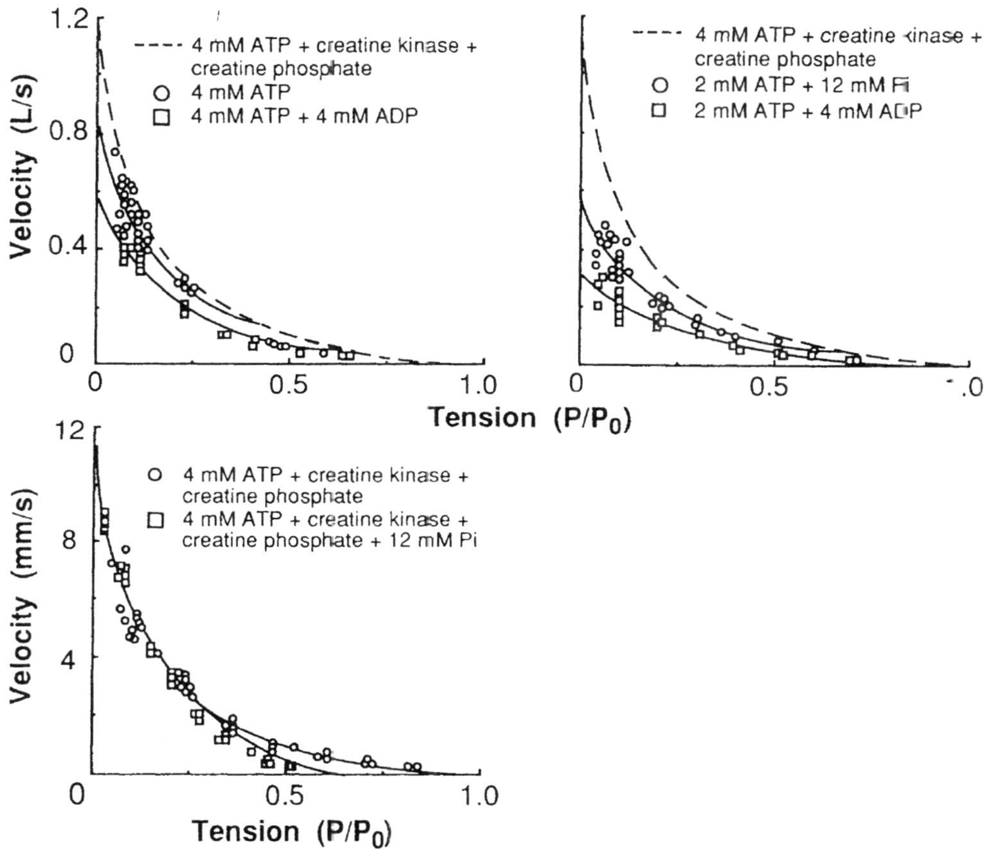

Comments
- Addition of phosphate decreased isometric force but did not affect the maximum velocity of shortening. As the [MgADP]/[MgATP] ratio in the fiber increases, the maximum velocity achieved by the fiber decreases while the isometric tension increases.

Reference(s)
Cooke R, Pate E (1985) The effects of ADP and phosphate on the contraction of muscle fibers. Biophys J 48:789–798 (with permission)

Force–Velocity Relation (29)

• Torque–velocity relation •	• Human • Extensor muscle	• •

Materials
- Human
- Knee extensor muscle

Testing Methods and Experimental Conditions
- Torque–velocity relationship at maximal contraction was measured in knee extension using an isokinetic loading dynamometer (Cybex)
- Force–velocity performances of 15 subjects were normalized with respect to the maximum torque

Data

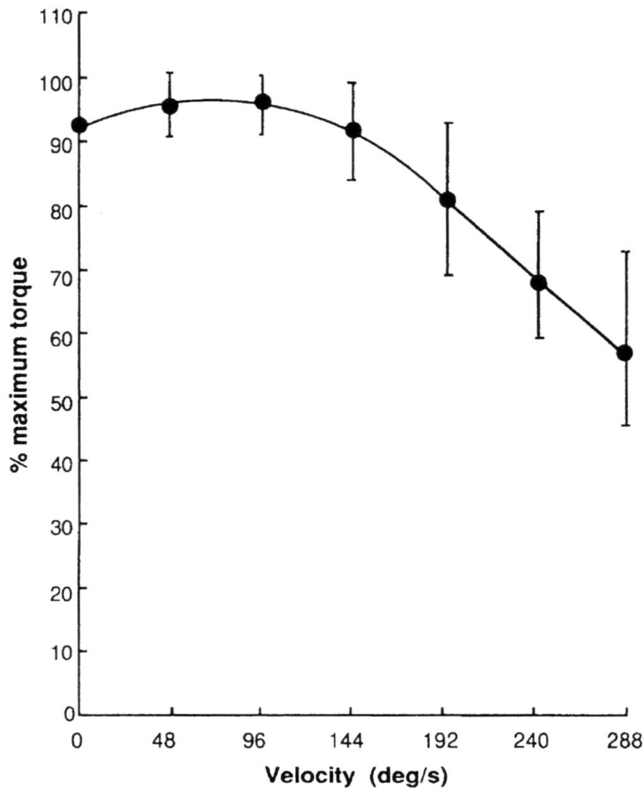

Comments
None.

Reference(s)
Perrine JJ, Edgerton VR (1978) Muscle force–velocity and power–velocity relationships under isokinetic loading. Med Sci Sports 10:159–166 (with permission)

Step–Stretch Response (1)

<table>
<tr><td>• Residue of tension
•</td><td>• Dogfish
• Jaw muscle</td><td>• Isolated musc.e fiber
•</td></tr>
</table>

Materials
- Small dogfish, *Scyllium canicula*
- Jaw muscle (striated muscle; coracomandibulars; weight, 740 mg)

Testing Methods and Experimental Conditions
- Stimulus frequency, 5 /s; stimulus duration, 30 s
- Temperature, 0°C
- Oxygenated salt solution, Pantin's (1946) solution
- The tension exerted by a fully active muscle during stretching and after the end of stretch is greater than in an isometric contraction. Influence of the length of muscle on th s residue of tension was examined
- Stretches of 5 mm were applied either on the rising limb of the tension–length curve (where the isometric tension increased with an increase of length), or around the top of the curve, or on the falling limb

Data

Preparation No.	R:sing limb	Top	Falling limb
1	2	–	24
2	2	–	13
3	11	–	20
4	16	–	52
5	11	19	–
6	0	22	–
7	–	20	–
8	–	17	–

Residue of tension above isometric during "after-stretch", following stretches of 5 mm over various parts (rising limb, top, and falling limb) of the isometric tension–length curve. Residue of the tension was expressed as percentage of the maximum isometric tension. In this table, for example, the muscle of preparation No. 1 showed that residue of the tension was 2% when the stretch was applied on the rising limb, and it was 24% when the stretch was on the falling limb.

Comments
- The isometric tension during tetanus varies with the length of a muscle; for example, 50 g at length 35 mm, 80 g at 44 mm, and 70 g at 50 mm (standard length was 37 mm).
- The residue of tension above isometric was greater when the stretch was applied around the peak or on the falling limb of the tension–length curve.
- As an analogy, the residue of tension could be attributed to the parallel elastic component of the contractile machinery.
- This would show that the elastic coefficient of the parallel elastic component depends on the region of the tension–length curve.

Reference(s)
Abbott BC, Aubert XM (1952) The force exerted by active striated muscle during and after change of length. J Physiol 117:77–86 (with permission)

Step–Stretch Response (2)

• Tension •	• Dogfish • Jaw muscle	• Isolated muscle fiber •

Materials

- Small dogfish (45 cm, *Scyllium canicula*)
- Jaw muscle (striated muscle; coracomandibulars; weight, 740 mg)

Testing Methods and Experimental Conditions

- Stimulus frequency, 5 /s; stimulus duration, 30 s
- Temperature, 0°C
- Oxygenated salt solution (Pantin's [1946] solution)
- Each muscle was brought to length 34 mm (standard length), both by stretching it from 29 to 34 mm and by shortening at the same speed of 0.6 mm/s from 39 to 34 mm by means of a Levin–Wyman ergometer
- Movement began after 3 s of stimulation

Data

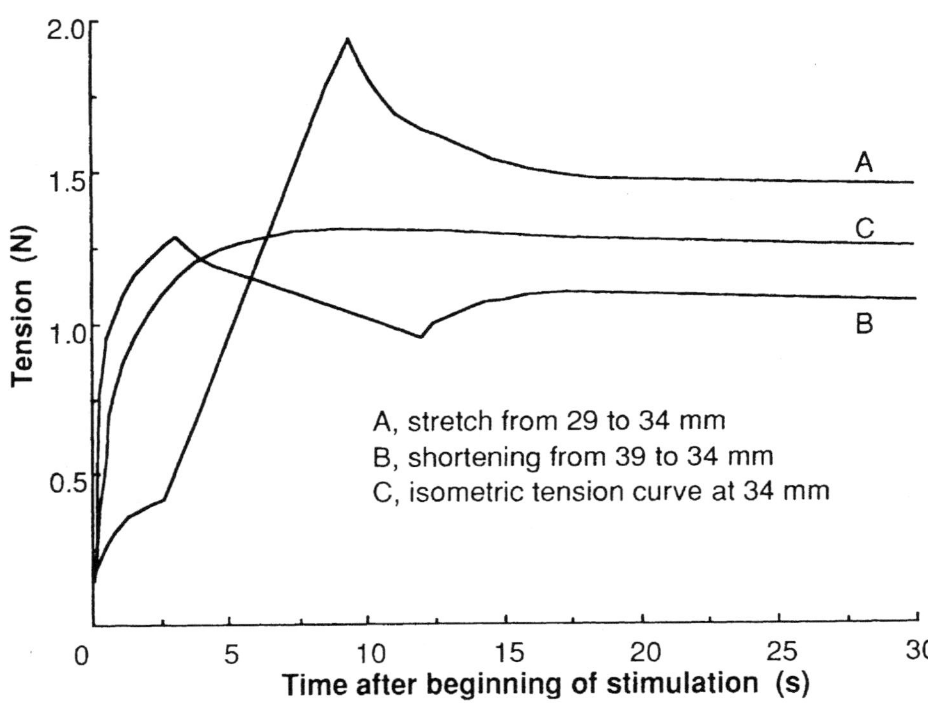

Comments

- After the end of stretch at constant speed, and provided that stimulation is continued, tension falls back towards, but remains considerably above, the isometric tension at the terminal length.
- Similarly, after a slow shortening at constant speed, tension remains considerably below the isometric value.
- This seems to show a spring-like behavior and may suggest the existence of a parallel elastic component.

Reference(s)

Abbott BC, Aubert XM (1952) The force exerted by active striated muscle during and after change of length. J Physiol 117:77–86 (with permission)

Step–Stretch Response (3)

• Tension •	• Toad • Sartorius muscle	• Isolated muscle fiber •

Materials
- Toads
- Sartorius muscle (33 mm, 45 mg)

Testing Methods and Experimental Conditions
- Stretching a muscle by 3 mm (from 30 to 33 mm) at various moments after maximal shock
- Latent period, 30 ms
- Stretch 3 mm at times 35, 64, 132, 250, 420, 740, and 1020 ms, and remain at the length for 15 – 20 ms

Data

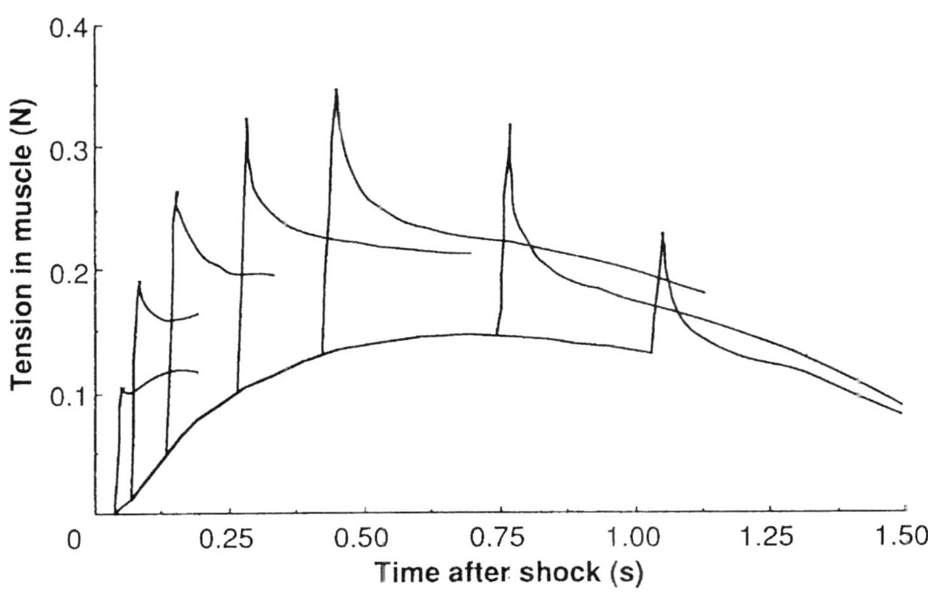

Comments
- 0.3 N tension is equivalent to 0.23 MN/m².
- Tension remains considerably above the isometric twitch tension at the terminal length
- This is an activation of the contractile machinery and seems to indicate the existence of a parallel elastic component of contracting muscle.

Reference(s)
Hill AV (1949) The abrupt transition from rest to activity in muscle. Proc R Soc B 184: 399–420 (with permission)

Stiffness (1)

• Compliance • Extension	• Rat • Gracilis anticus muscle	• Isolated muscle fiber •

Materials
- White Wistar rats (male; age, 50 days; weight, 140–165 g)
- Gracilis anticus muscle (rest length, 2.7 cm; weight, 60 mg)

Testing Methods and Experimental Conditions
- Krebs–Ringer solution, pH 7.3; temperature, 17.5°C
- Modified Wilkie's quick-release method, releasing from isometric condition to some fixed isometric load
- Muscle stimulation: current density = 0.08 amp/cm^2, train of 322 ms pulses with a pulse separation of 10.5 ms

Data

Muscle	Compliance (cm/dyne $\times 10^{-6}$)	Normalized compliance $(L_0/P_0 \times 10^2)$	Normalized extension at P_0
Cat papillary [a]			
L_0=1.4 cm	5 at P_0	3.6	0.1L_0
P_0=10 g	50 at 0.2P_0	36	
Rat anterior tibialis [a]			
L_0=2.5 cm	0.25 at P_0	4.4	0.05L_0
P_0=440 g	1.5 at 0.1P_0	26.4	
Rat gracilis anticus			
L_0=2.7 cm	1.1 at P_0	1.3	0.07L_0
P_0=30 g	17 at 0.1P_0	19.5	
Frog illiofibularis [a]			
L_0=2.5 cm	1.1 at 0.6P_0	2.0	0.03–0.05L_0
P_0=48 g	5 at 0.1P_0	9.6	
Frog sartorius[a]			
L_0=3 cm	1.6 at 0.7P_0	2.4	0.04L_0
P_0=45 g	20 at 0.02P_0	30.0	
Frog sartorius[a]			
L_0=2.8 cm	0.5 at P_0	2.0	0.04L_0
P_0=120 g	6 at 0.1P_0	26	

[a] A comparison of the series elastic components reported for different striated muscles.

Comments
- A basic similarity was observed among striated muscles both in normalized compliance and normalized total extension.
- There is little variability in the normalized results when different muscles are compared.

Reference(s)
Bahler AS (1967) Series elastic component of mammalian skeletal muscle. Am J Physiol 213:1560–1564 (with permission)

Stiffness (2)

• Inertia •	• Human • Hand	• •

Materials
- Human
- Hand

Testing Methods and Experimental Conditions
- While subjects maintained a given hand position, small external disturbance to the hand was applied by a manipulandum. The transient responses of the hand displacements and the reaction forces by the disturbance were measured, and the hand impedance was estimated using a second-order linear impedance

Data

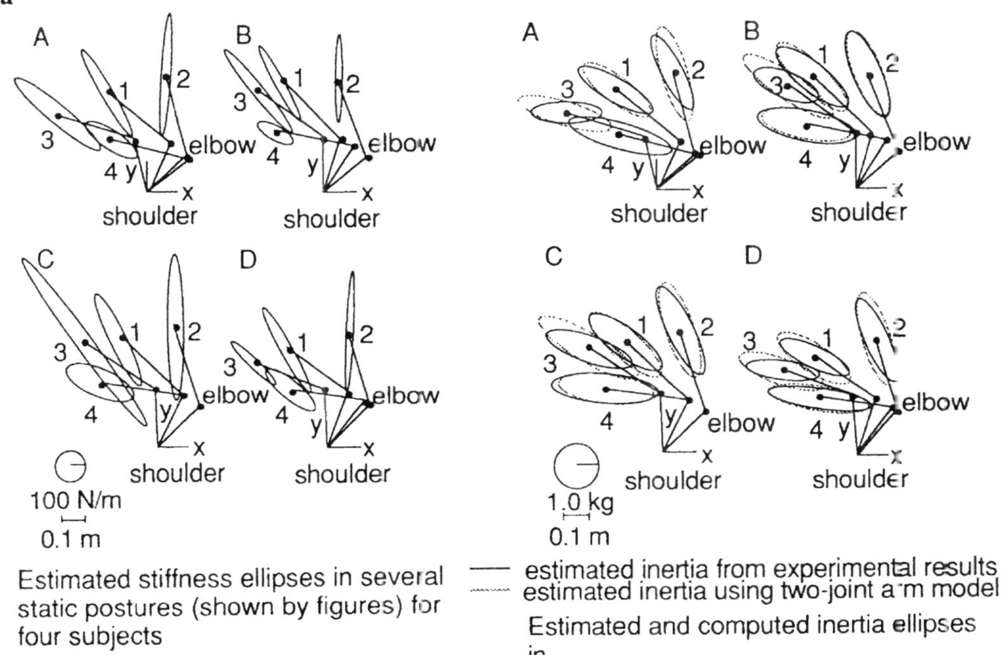

Estimated stiffness ellipses in several static postures (shown by figures) for four subjects

—— estimated inertia from experimental results
········ estimated inertia using two-joint arm model

Estimated and computed inertia ellipses in

Comments
- The estimated inertia agrees with the values computed from a two-joint arm model. The sizes of the ellipses are different in different subjects, but the directions of the ellipses are almost the same.

Reference(s)
Tsuji T, Goto K, Ito K, Nagamachi M (1994) Estimation of human hand impedance during maintenance of posture. Trans SICE 30:319–328 (with permission)

Stiffness (3)

• Stiffness •	• Cat • Soleus muscle	• Intact whole muscle •

Materials
- Cats decerebrated at the premammillary level
- Soleus muscle

Testing Methods and Experimental Conditions
- Muscle length perturbation amplitude of 1 mm
- Reflex effects extinguished by deep halothane anesthesia
- Operation force: stimulation of peripheral nerve with 30 – 100 Hz trains of negative pulses of 0.02 ms duration; brain stem stimulation with 30 – 50 Hz trains of 0.01 – 0.1 mA negative pulses of 0.5 ms duration

Data

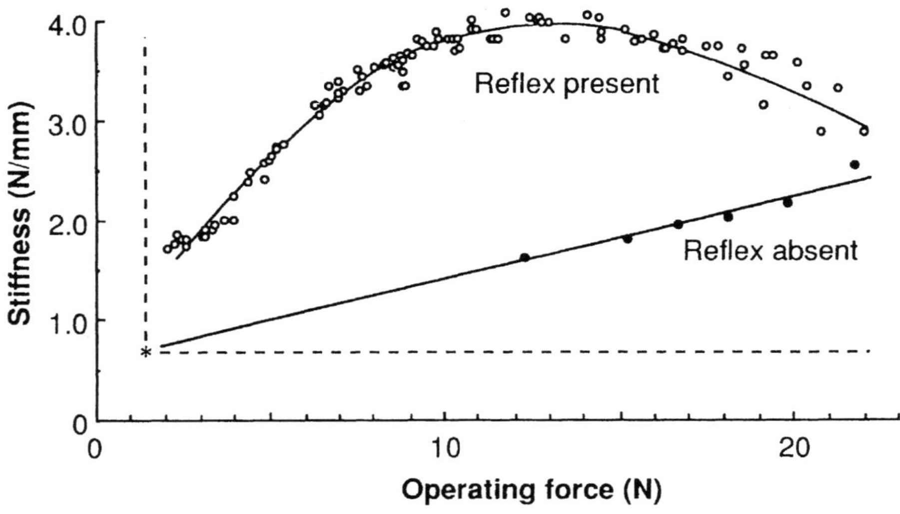

Comments
- Passive muscle stiffness due to elastic properties of noncontractile elements is shown by asterisk on the left and horizontal dotted line.
- Intrinsic stiffness of the muscle is linearly related to operating force.

Reference(s)
Hoffer JA, Andressen S (1981) Regulation of soleus muscle stiffness in premammillary cats. J Neurophysiol 45:267–285 (with permission)

Stiffness (4)

• Stiffness •	• Human • Flexor pollicis longus	• •

Materials
- Human
- Flexor pollicis longus muscle

Testing Methods and Experimental Conditions
- The thumb was stretched with a torque motor and the steady force was maintained in executing three types of task
- Both force and displacement were measured at the tip of the thumb
- Stretch-evoked stiffness was calculated as the stretch-evoked force per unit of stretch where stretch-evoked force was obtained by subtracting tonic force from the mean force computed over the interval 75–115 ms after the perturbation onset

Data

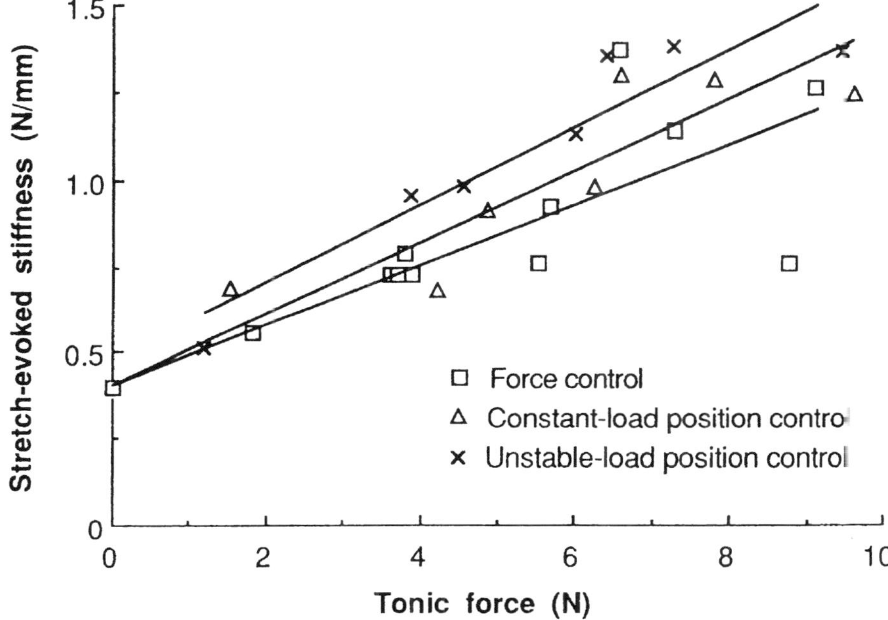

Comments
- The stretch-evoked stiffness around the joint increased with increasing flexing force, and with the level of contraction of flexor and extensor muscles for a given flexing force.

Reference(s)
Akazawa K, Milner TE, Stein RB (1983) Modulation of reflex EMG and stiffness in response to stretch of human finger muscle. J Neurophysiol 49:16–27 (with permission)

Stiffness (5)

• Stiffness •	• Rabbit • White muscle	• Myo-filament •

Materials

- Rabbits
- Skeletal white muscle
- Reconstituted single actin filament with and without tropomyosin

Testing Methods and Experimental Conditions

- Stiffness of single actin filaments was measured by using a technique for nanomanipulation of single actin filaments with a pair of glass microneedles under an optical microscope
- Temperature, 25°–27°C

Data

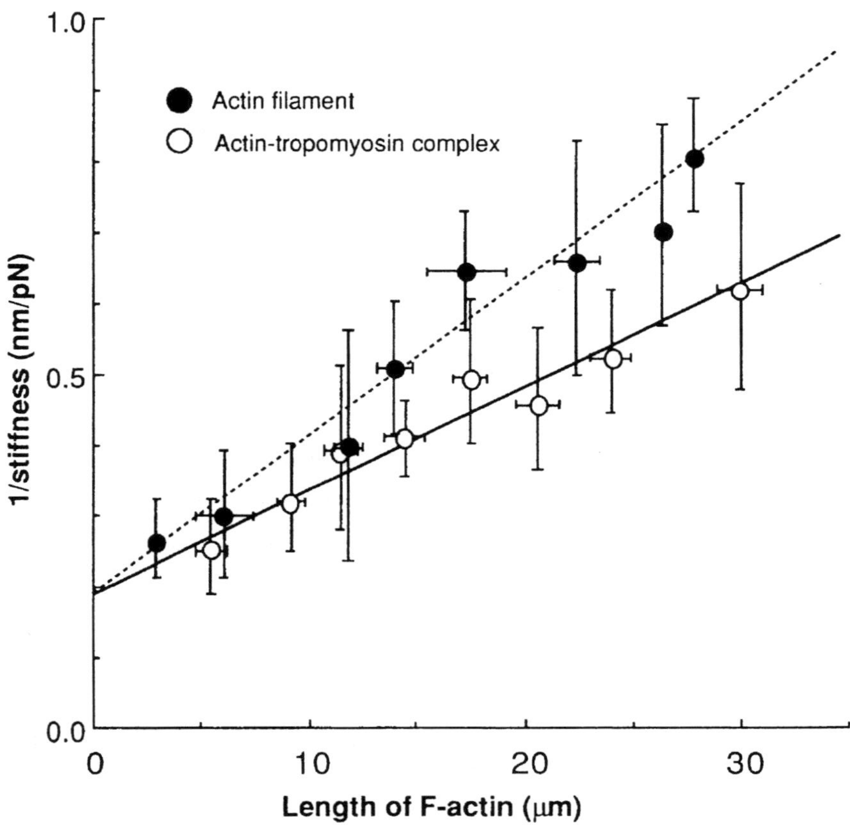

Comments

- Stiffness for 1-μm-long actin filaments with and without tropomyosin was 65.3 ± 6.3 and 43.7 ± 4.6 pN/nm, respectively.

Reference(s)

Kojima H, Ishijima A, Yanagida T (1994) Direct measurement of stiffness of single actin filaments with and without tropomyosin by in vitro nanomanipulation. Proc Natl Acad Sci 91:12962–12966 (with permission)

Stiffness (6)

| • Tension
• | • Frog
• Semitendinosus muscle | • Isolated muscle fiber
• |

Materials
- Frogs, *Rana temporaria* and *Rana esculenta*
- Semitendinosus muscle

Testing Methods and Experimental Conditions
- Tension and stiffness, measured simultaneously, are examined at rest, during single twitches, as tetanic contraction develops, and during the tetanic contraction itself
- Tension changes caused by periodic changes (100 Hz) in length are used for continual measurement of stiffness (Δtension/Δlength)
- Temperature, 15°C

Data

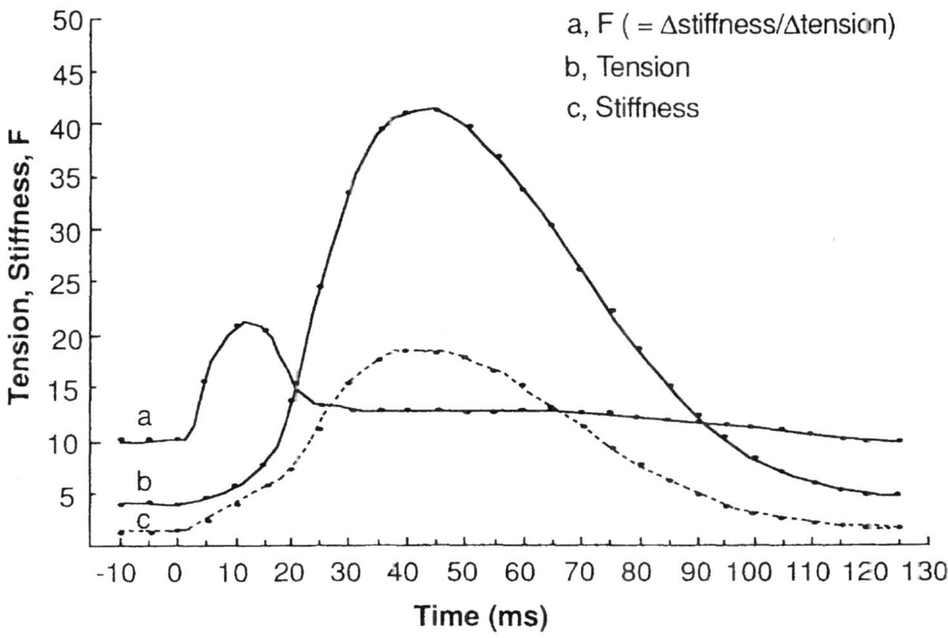

Comments
- Stimulation introduced at 0 ms.
- Tension and stiffness in arbitrary units.
- Dynamic stiffness varies proportionally with tension at rest and during contraction.

Reference(s)
Buchthal F, Kaiser E (1944) Factors determining tension development in skeletal muscle. Acta Physiol Scand 8:38–74 (with permission)

Active Tension (1)

• Contraction	• Bovine	• Adrenaline
•	• Mesenteric artery	• Quick release

Materials

- Bovine
- Mesenteric artery with an approximate diameter of 3–4 mm
- Immersed in ice-cold gassed Tyrode's solution until use

Testing Methods and Experimental Conditions

- Specimen of 12 mm in width, 10–12 mm in length and about 1 mm in thickness mounted in the isometric apparatus and immersed in an organ bath containing gassed Tyrode's solution at 37°C
- Length–tension relation determined in normal, monoiodoacetic acid-treated and adrenaline-treated specimen
- Active state tested by the quick-release method

Data

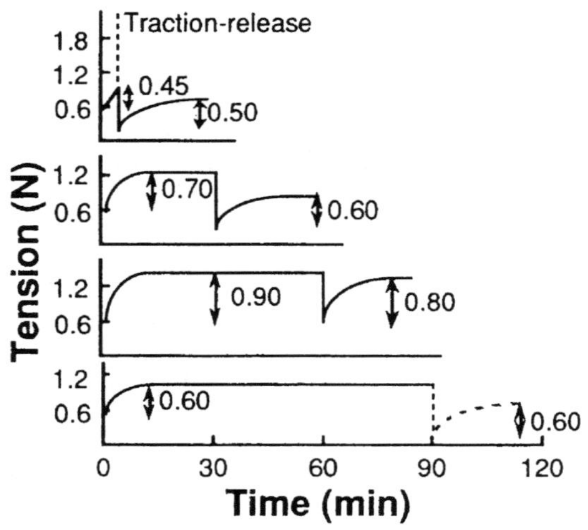

Comments

- Active state determined by traction–release. Adrenaline was added at time 0. After 6, 30, 60 and 90 min the specimen was extended 2 mm but so slowly that the tension increase never exceeded 2.0 N.
- The increase in tension after quick release was assumed to reflect the degree of active state of the muscle.

Reference(s)

Lundholm L, Lundholm EM (1966) Length at inactivated contractile elements, length–tension diagram, active state and tone of vascular smooth muscle. Acta Physiol Scand 68:347–359 (with permission)

Active Tension (2)

• Isometric force •	• Hog • Common carotid artery	• Correction for extracellular space

Materials
- Hogs weighing 80–90 kg for adult and 10–20 kg for weanling
- Common carotid artery

Testing Methods and Experimental Conditions
- Specimen having cross sections of 0.4–0.8 mm^2 and length of approximately 1 cm at their optimum length maintained in a bubbled physiological salt solution at 37°C
- Force-generating capacity of a smooth muscle evaluated
- Maximum active tension elicited equimolar substitution of KCl for NaCl and an increase in [CaCl$_2$] to 5 mM
- The fractional cell cross-sectional areas obtained using low-power electron micrographs

Data

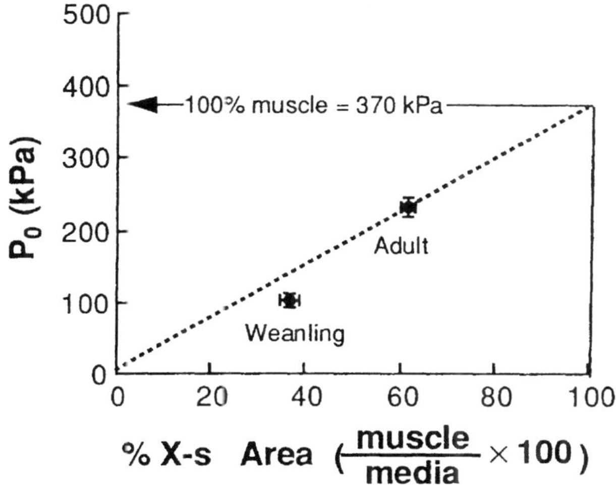

Comments
- Values are mean ± SE.
- The maximally stimulated arterial strips (60% smooth muscle cell by area of the cross section) developed an active force (220 kPa) comparable to that of skeletal muscle.
- Correction for extracellular space showed a value of 370 kPa for the smooth muscle component.
- The dashed line illustrates the graphical determination of force developed per square centimeter muscle cells in adult animals.

Reference(s)
Murphy RA, Herlihy JT, Megerman J (1974) Force-generating capacity and contractile protein content of arterial smooth muscle. J Gen Physiol 64:691–705 (with permission)

Active Tension (3)

• Twitch tension • Tetanus tension	• Rabbit • *Taenia coli*	• Quick stretch • Tetanus

Materials
- Female rabbits weighing 1.9–2.4 kg, New Zealand White strain
- *Taenia coli*

Testing Methods and Experimental Conditions
- Specimen of about 3 mm in width and 5 mm in length mounted in a Plexiglass chamber containing a flowing oxygenated Krebs' bicarbonate solution, allowed to equilibrate under 2 g tension for 2 h
- Active-state characteristics using the single-shock, quick-stretch technique of Hill
- DC shocks (6–9 V, 5 ms duration) for stimulation
- All experiments were done at 22°C in order to eliminate spontaneous activity

Data

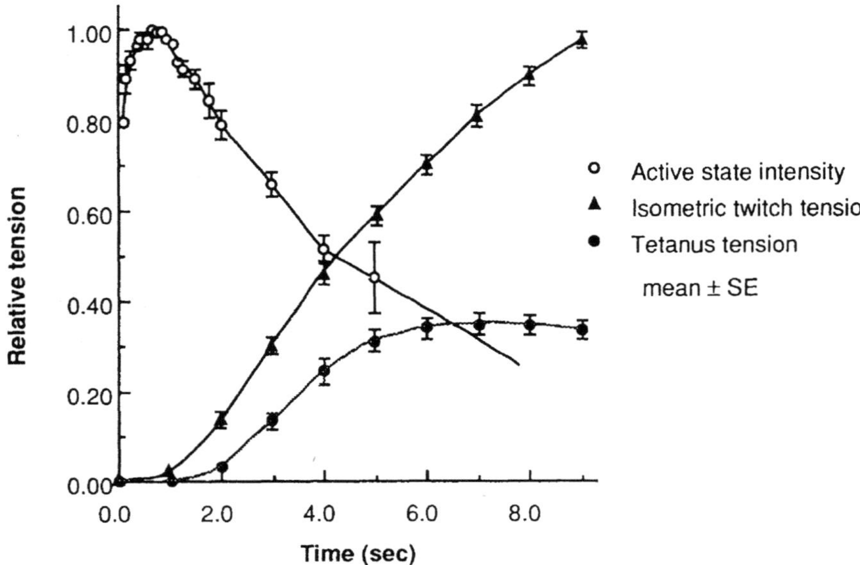

Comments
- The active state rises quickly after a shock reaching a peak at about 0.75 s.
- The tensions reached at 0.5 and 1.0 s following the stimulus were not significantly different ($P < 0.05$).

Reference(s)
Gordon AR, Siegman MJ (1971) Mechanical properties of smooth muscle. II. Active state. Am J Physiol 221:1250–1254 (with permission)

Force–Velocity Relation (1)

• Active force • Shortening velocity	• Rabbit • Urinary bladder	• Quick release • Hill's equation

Materials

- Rabbits weighing 2.5–3 kg
- Urinary bladder
- Transferred to oxygenated Krebs' solution, left in this medium for 1 h at 4°C, and filled with 10 ml Krebs solution

Testing Methods and Experimental Conditions

- A bundle of longitudinal muscle mounted in the bath
- Medium and temperature change to suppress spontaneous contractile activity: accommodated for 1 h at 37°C in a bubbled physiological salt solution, temperature lowered to 25°C, and medium changed to an oxygenated tris solution at 25°C
- Force–velocity relations obtained by the quick-release technique at different times during the twitch
- Single square wave pulses (5 ms, 100 V) at a frequency of 0.03–0.06 Hz to stimulate the muscle
- The results were fitted to Hill's equation: $V(P + a) = b(P_0 - P)$. P, tension in muscle; V, velocity of contraction; a, b, P_0, constants

Data

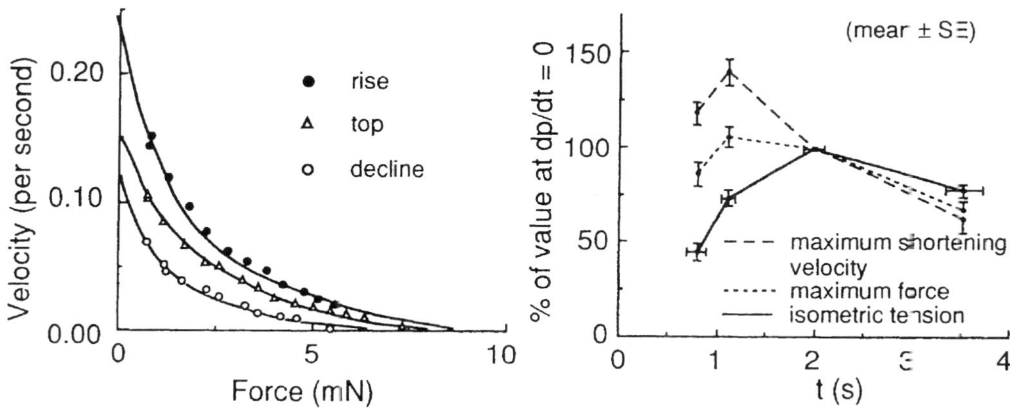

Comments

- Left diagram: Force–velocity relations. Full lines show the fits to Hill's equation.
- Right diagram: Maximum shortening velocity (V_{max}), extrapolated maximum tension (P_0) and isometric tension expressed as percent of their respective values at the peak.
- V_{max} and P_0 reach their highest values during the rising phase of the twitch.
- V_{max} declines rapidly with time and at the peak of the twitch it is only about 70% of the maximum value.

Reference(s)

Uvelius B (1979) Shortening velocity, active force and homogeneity of contraction during electrically evoked twitches in smooth muscle from rabbit urinary bladder. Acta Physiol Scand 106:481–486 (with permission)

Force–Velocity Relation (2)

• Isometric contraction • Length–tension relation	• Dog • Ureter	• Single twitch • Hill's equation

Materials
- Adult dogs weighing 25–35 kg
- Middle portions of ureter

Testing Methods and Experimental Conditions
- Cylindrical specimen removed to a aerated, modified Krebs' solution at 25°C
- Single twitch, length–tension relation, and force–velocity relation
- Before testing, preconditioned by repeated loading and unloading to stabilize its mechanical properties and stimulated by square waves during this period

Data

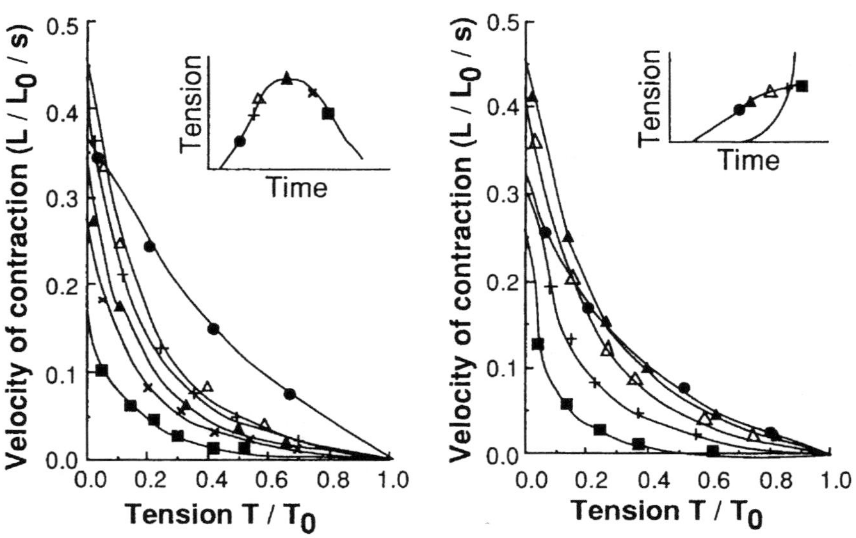

Comments
- Left diagram: Force–velocity relations for different times of release after stimulation.
- Right diagram: Force–velocity relations at a given time after stimulation for different levels of stretch.
- Measurements were analyzed using the hyperbolic Hill's equation.
- Maximum velocity of shortening occurred at 75–100 ms after release.
- Maximum velocity of contraction reached a peak for releases made in a region of stretch where the length was in the range of 0.85–0.90 L_{max}, where L_{max} is the length which the largest active tension can be developed.

Reference(s)
Zupkas PF, Fung YC (1985) Active contractions of ureteral segments. Trans ASME J Biomech Eng 107:62–67

Force–Velocity Relation (3)

• Isometric tension • Isotonic contraction	• Rat • Portal vein	• Quick release •

Materials
- Rats weighing 250–400 g, Sprague–Dawley strain
- Portal vein of 3–5 mm in length

Testing Methods and Experimental Conditions
- Force–velocity relation of longitudinal smooth muscle determined at different stages of highly uniform contractions induced by appropriate alteration of the ionic environment
- The experiments were performed by the method of isotonic quick release
- Temperature, 37°C

Data

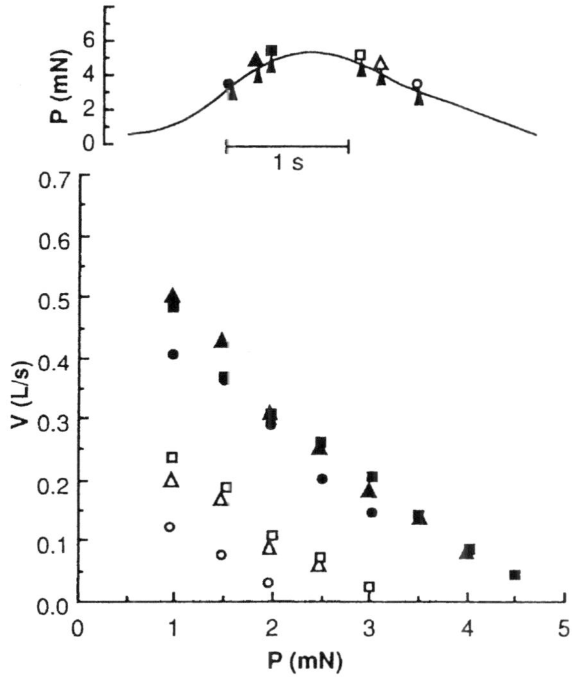

Comments
- Upper diagram: Isometric tension. Different symbols indicate time of release.
- Lower diagram: Force–velocity curves. Symbols corresponding to time of release as shown in upper diagram.
- During the rising phase the curves are not much different at the three force levels.

Reference(s)
Hellstrand P, Johansson B (1975) The force–velocity relation in phasic contractions of venous smooth muscle. Acta Physiol Scand 93:157–166 (with permission)

Force–Velocity Relation (4)

| • Isometric tension
• Maximum velocity | • Guinea pig
• *Taenia coli* | • Quick release
• Hill's equation |

Materials
- Guinea pigs weighing 0.25–0.5 kg
- *Taenia coli*

Testing Methods and Experimental Conditions
- Specimen mounted in the test chamber filled with aerated Krebs' solution at 36°–37°C, and restored to the in situ length
- Quick-release experiment
- Specimen stimulated with AC field at 50–100 Hz
- Force–velocity relation obtained

Data

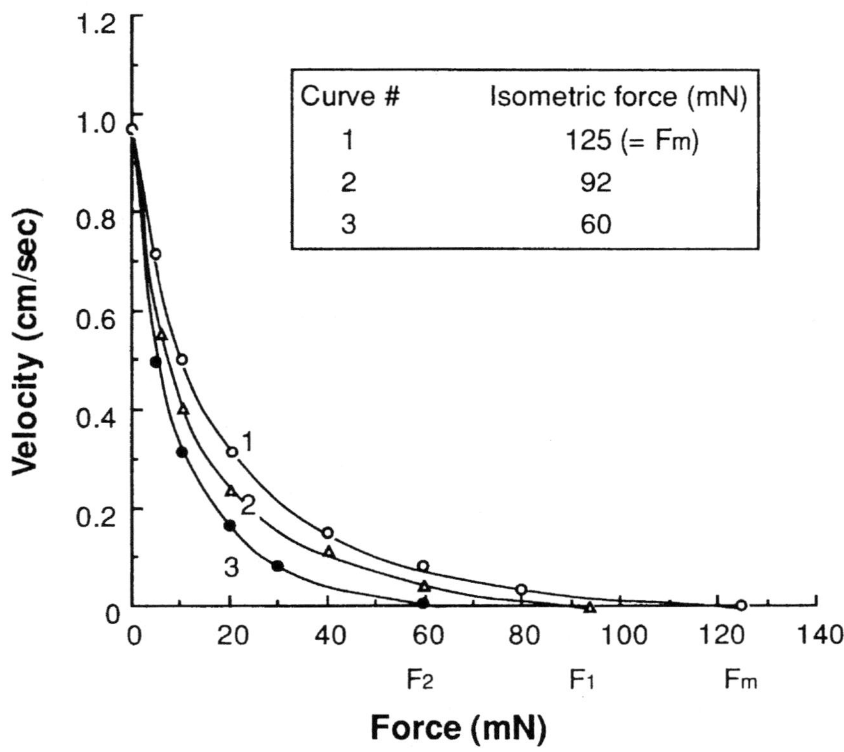

Comments
- The load–velocity curve obtained was hyperbolic, obeying Hill's equation:
  $$(P + a)(v + b) = b(F_m + a).$$
  P, load; v, shortening velocity; F_m, maximum isometric tension; a, heat constant; b, rate constant of energy liberation.
- The maximum velocity per unit muscle length was constant, irrespective of the initial muscle length.

Reference(s)
Mashima H, Okada T, Okuyama H (1979) The dynamics of contraction in the guinea pig *Taenia coli*. Jpn J Physiol 29:85–98

Force-Velocity Relation (5)

• Isotonic contraction • Maximal shortening velocity	• Rat • Mesenteric vein	• Influence of temperature • Hill's equation

Materials
- Rats
- Tetanized portal-anterior mesenteric vein

Testing Methods and Experimental Conditions
- Specimen suspended in a plexiglass chamber perfused with Tyrode's solution at 37°, 30°, or 25°C
- Force–velocity relations of afterloaded isotonic contractions studied at different temperatures
- Electric stimulus of 50 Hz, 6 V, sine wave parallel to the specimen
- The duration of the stimulus was chosen to stimulate until the peak tension was reached and to suppress the spontaneous activity

Data

	37°C	30°C	25°C
V_0 (ML/s)	0.51 ± 0.02	0.42 ± 0.05	$0.28 \pm 0.04***$
ΔL_0 (ML)	0.69 ± 0.07	0.73 ± 0.07	0.70 ± 0.06
dT/dt (mN/s)	8.47 ± 0.75	$5.97 \pm 0.55*$	$3.62 \pm 0.42***$
$T(dT/dt)$ (%)	42.7 ± 2.7	$35.0 \pm 2.3*$	$33.9 \pm 1.4*$
$t(dT/dt)$ (s)	0.90 ± 0.04	1.01 ± 0.05	$1.59 \pm 0.15***$
ΔT_0 (mN)	11.07 ± 1.22	11.51 ± 1.34	9.89 ± 1.26
Latency (s)	0.33 ± 0.04	0.34 ± 0.03	$0.56 \pm 0.07*$
a	0.35 ± 0.03	0.27 ± 0.04	0.34 ± 0.06
b (ML/s)	0.18 ± 0.02	$0.10 \pm 0.01**$	$0.09 \pm 0.02**$

$* \ P < 0.05; \ ** \ P < 0.005; \ *** \ P < 0.0005$ as compared with the respective values at 37°C.

Comments
- Parameters:
  V_0, extrapolated maximal velocity of shortening at zero load referred to muscle length (ML); ΔL_0, extent of isotonic shortening referred to ML; dT/dt, maximal rate of tension increase; T(dT/dt), tension development at dT/dt referred to the amplitude of isometric contraction; t(dT/dt), time to dT/dt; ΔT_0, amplitude of isometric contraction; Latency, period between onset of stimulation and beginning of contraction; a, force constant of Hill's equation referred to the peak total tension; b, velocity constant of Hill's equation referred to ML. Values are mean $\pm$ SE.
- Parameters of contraction velocity in vascular smooth muscle depend on temperature, whereas the extent of contraction is independent of temperature.

Reference(s)
Peiper U, Laven R, Ehl M (1975) Force velocity relationships in vascular smooth muscle. Pflügers Arch 356:33–45

Length–Tension Relation (1)

• Active tension •	• Rat • Urinary bladder	• •

Materials
- Male rats weighing about 350 g, Sprague–Dawley strain
- Normal and denervated urinary bladder

Testing Methods and Experimental Conditions
- Urinary bladders denervated, removed 10 days after the operation, and transferred to oxygenated Krebs' solution
- Longitudinal preparations of 3.5–5.1 mm in length mounted in the organ bath filled with a bubbled Krebs' solution at 37°C
- Length–tension relations measured under supramaximal stimulation with AC current

Data

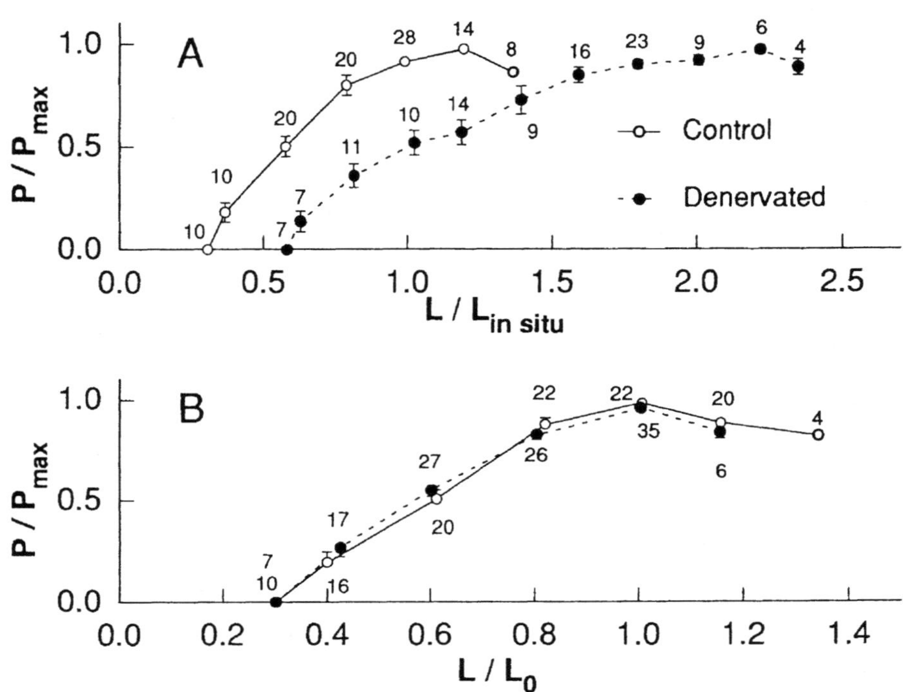

Comments
- Parameters and caption:
  (A) $L/L_{in\ situ}$, length relative to in situ length. (B) L/L_0, length relative to optimum length; P/P_{max}, force relative to maximum value. Values are mean ± SE. Number of observations is given for each point.
- The active length–tension relations of control bladders and denervated bladders were found to differ markedly from each other when comparisons were based on lengths relative to $L_{in\ situ}$.
- The relations were almost superimposed on each other when all lengths are expressed relative to L_0.

Reference(s)
Ekström J, Uvelius B (1981) Length–tension relations of smooth muscle from normal and denervated rat urinary bladders. Acta Physiol Scand 112:443–447 (with permission)

Length–Tension Relation (2)

• Active tension • Passive tension	• Dog • Mesenteric artery	• •

Materials

- Dogs
- Superior mesenteric artery of approximately 1 mm inside diameter
- The artery stored in Krebs' solution at 4°C until used

Testing Methods and Experimental Conditions

- Specimen cut helically to obtain a strip 10–25 mm long, 0.8–1 mm wide, and approximately 0.2 mm thick
- Strip mounted in a bath of aerated Krebs' solution at 38°C
- Effect of stretch on the resting tension and contractility studied isometrically
- Electric current or epinephrine for stimulation

Data

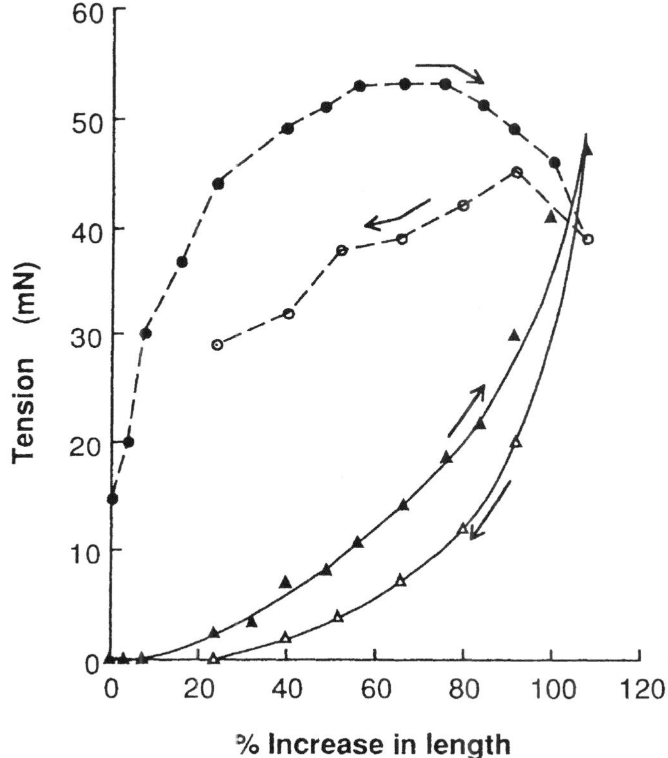

Comments

- Passive tension in response to stretch (▲) and release (△). Active tension developed in response to a standard stimulus in several increments of stretch (●) and decrements of release (○).
- After the strip has been stretched more than 50% of its initial length the tension rise is steep.
- With each increase in length there is a corresponding increase in magnitude of the response until an optimal length is reached.

Reference(s)

Sparks Jr HV, Bohr DF (1962) Effect of stretch on passive tension and contractility of isolated vascular smooth muscle. Am J Physiol 202:835–840 (with permission)

Length–Tension Relation (3)

| • Active tension
• Passive tension | • Rat
• Mesenteric artery | •
• |

Materials
- Wistar rats (3–5 months of age)
- Artery 0.7 mm long, 100–200 µm lumen diameter from mesenteric bed

Testing Methods and Experimental Conditions
- Active and passive isometric tension–length (internal circumference) relation
- Medium: Activating and relaxing solutions as for unique bubbled physiological salt solution (PSS) at 37°C
- Medium change at 12-min intervals

Data

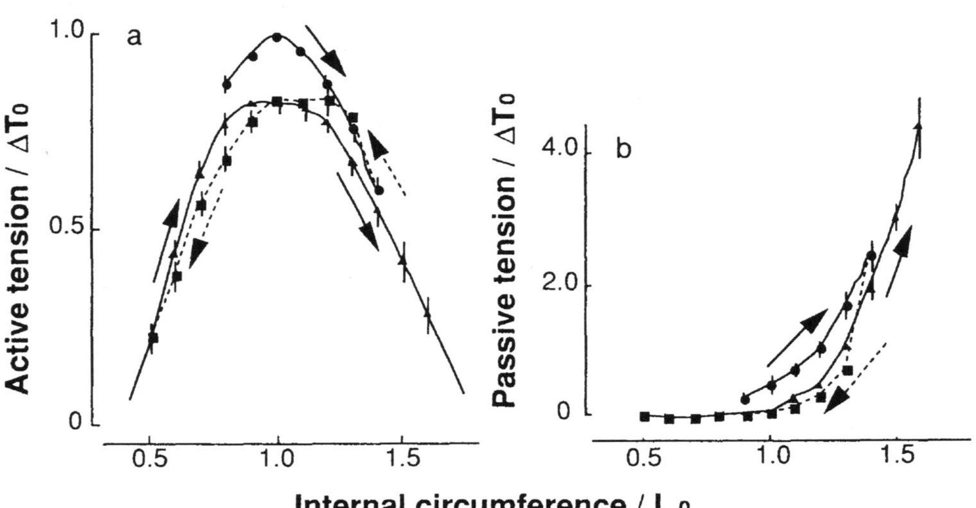

Comments
- Caption:
  L_0, length at which the active tension is maximum ($L_0/\pi = 208 \pm 9$ µm); ΔT_0, maximum tension ($\Delta T_0 = 3.08 \pm 0.38$ mN); ●, first ascending sequence; ■, descending sequence (dashed line); ▲, second ascending sequence. Values are mean ± SE.
- The specimen can develop active tension in the range from 0.38 L_0 to 1.82 L_0 (a).
- During the subsequent release sequence and stretch sequence for passive characteristic, the characteristic was shifted to the right (b).

Reference(s)
Mulvany MJ, Warshaw DM (1979) The active tension–length curve of vascular smooth muscle related to its cellular components. J Gen Physiol 74:85–104 (with permission)

Length–Tension Relation (4)

• Active tension • Resting tension	• Guinea pig • *Taenia coli*	• •

Materials
- Adult guinea pigs
- *Taenia coli* in about 5 cm lengths
- Kept moist with Krebs' solution at about 23°C until use

Testing Methods and Experimental Conditions
- The active and resting tension–length relations
- Medium: Aerated Krebs' solution, Ca-free solution for the resting tension
- Specimen activated in Krebs' solution by applying the AC stimulus for 15 s (at 23°C)

Data

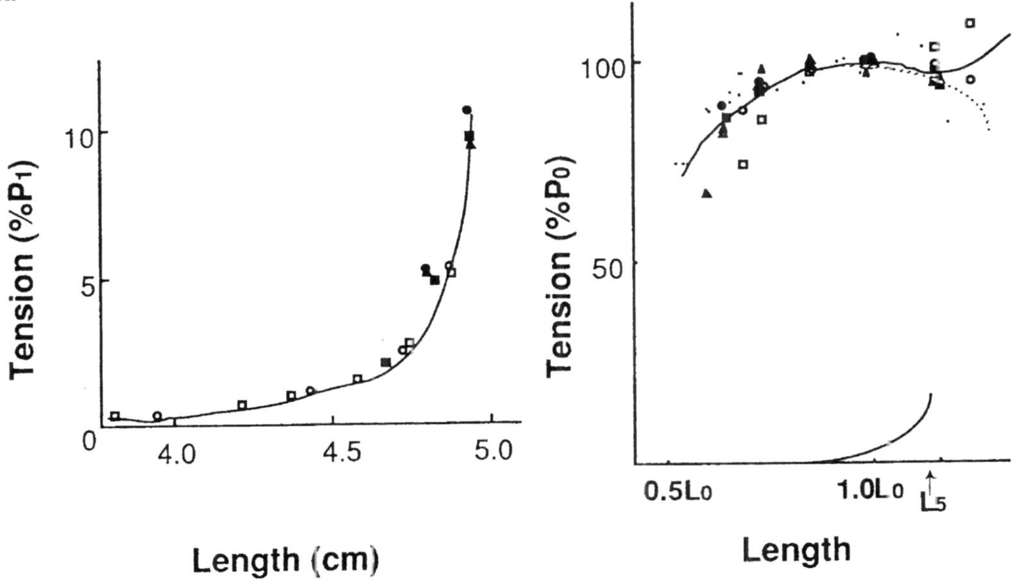

Comments
- Left diagram: Resting tension–length curve. Solid symbols indicate results while stretching and open symbols results releasing. P_1, active tension at which the resting tension was 5.0×10^{-4} N (8.4×10^{-2} N in this case).
- Right diagram: Total length–tension curve (top) and resting tension curve (bottom). L_0, length at which maximum active tension P_0 is obtained; L_5, length at which the resting tension is 5% of P_0.

Reference(s)

Lowy J, Mulvany MJ (1973) Mechanical properties of guinea pig *Taenia coli* muscles. Acta Physiol Scand 88:123–136 (with permission)

Length–Tension Relation (5)

• Active tension • Resting tension	• Guinea pig • *Taenia coli*	• •

Materials

- Guinea pigs
- *Taenia coli*

Testing Methods and Experimental Conditions

- Specimen mounted in a plastic chamber filled with aerated Krebs' solution at 37°C
- A uniform longitudinal AC field (50 Hz) to stimulate the specimen
- Adrenaline to stop spike discharge
- The chamber divided into three compartments by two partitions set 5 mm apart
- Relations between the muscle length and the resting tension by adrenaline, the maximum tension by electrical stimulus, the potentiated tension by successive stimuli, or the tension of spontaneous contraction

Data

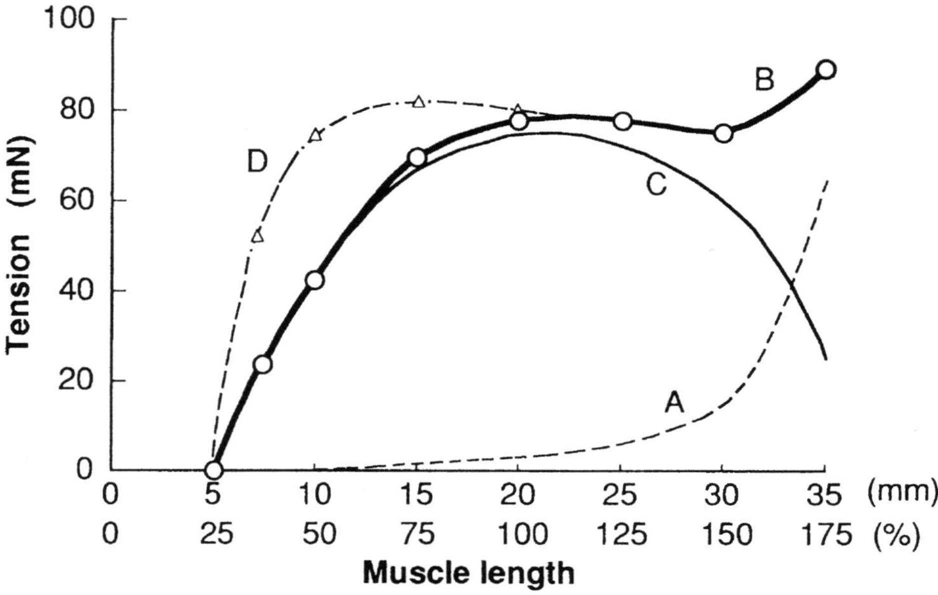

Comments

- Caption:
  A, resting tension; B, maximum tension produced by the optimum stimulus; C, active tension (= B − A); D, potentiated tension produced by successive stimuli.
- The slightest resting tension was produced at 75% of the in situ length (20 mm).
- The optimum length, at which the active tension became maximum, is between 100% and 125% of the in situ length.
- The active tension declines almost symmetrically on either side of the optimal length.
- The maximum tension potentiated by successive stimuli was attained at 75% of the in situ length.

Reference(s)

Mashima H, Yoshida T (1965) Effect of length on the development of tension in guinea-pigs *Taenia coli*. Jpn J Physiol 15:463–477

Length–Tension Relation (6)

• Active tension • Resting tension	• *Mytilus edulis* • Pedal retractor muscle	• Single cell measurement • Micropipette

Materials
- *Mytilus edulis* of 7–8 cm in shell length
- Smooth muscle cells isolated from strips of the pedal retractor muscle (PRM) by digestion with collagenase

Testing Methods and Experimental Conditions
- Isolated single cell held by suction at the cell ends with a pair of micropipettes
- Isometric tension measured by photoelectrically recording the elastic displacements of one of the pair of micropipettes
- A 0.1 ms rectangular current pulse for stimulation
- Supramaximal stimulation of about five times threshold intensity
- Tension in the resting cell measured 15 s after a stretch by 0.05–0.1 L_s at a rate of 1.0–1.5 L_s/s, where L_s is the length at which slack of the cell just disappears and tension begins to appear
- Medium: artificial sea water at 19°–24°C

Data

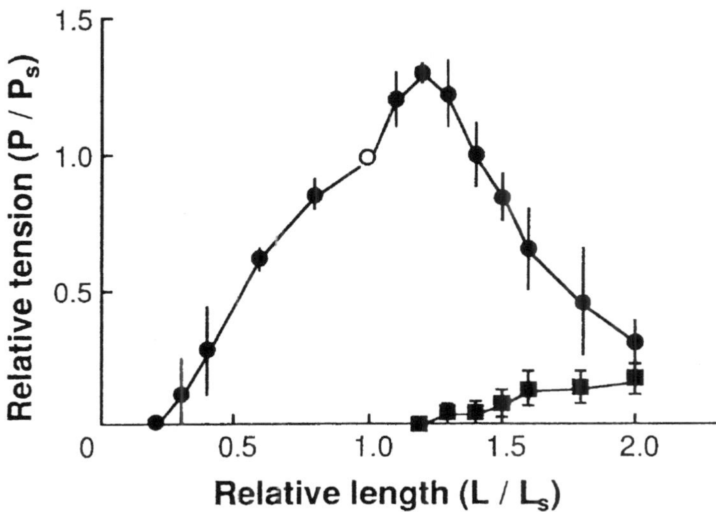

Comments
- Legend:
  P_s, tension at L_s; ●, twitch tension; ■, tension in the resting cell; ○, standard point for normalization.
- The isolated cell can generate tension at lengths ranging from 0.17 to more than 1.7 L_0, where L_0 is the optimal length for tension generation.
- The working range is much wider than that of the whole PRM and those of other muscle preparations.

Reference(s)
Ishii N, Takahashi K (1982) Length–tension relation of single smooth muscle cells isolated from the pedal retractor muscle of *Mytilus edulis*. J Muscle Res Cell Motil 3:25–38 (with permission)

Length–Tension Relation (7)

• Force–velocity curve •	• Rabbit • *Taenia coli*	• Hill's equation •

Materials
- Female rabbits weighing 1.9–2.4 kg, New Zealand White strain
- *Taenia coli*

Testing Methods and Experimental Conditions
- Specimen of about 3 mm in width and 5 mm in length mounted in a Plexiglass chamber containing a flowing oxygenated Krebs' bicarbonate solution, and allowed to equilibrate under 2 g tension for 2 h at 22°C
- Length–tension and force–velocity relations examined under conditions in which spontaneous mechanical activity was eliminated
- 10 V rms, 60 Hz, AC stimulation to tetanize the muscle

Data

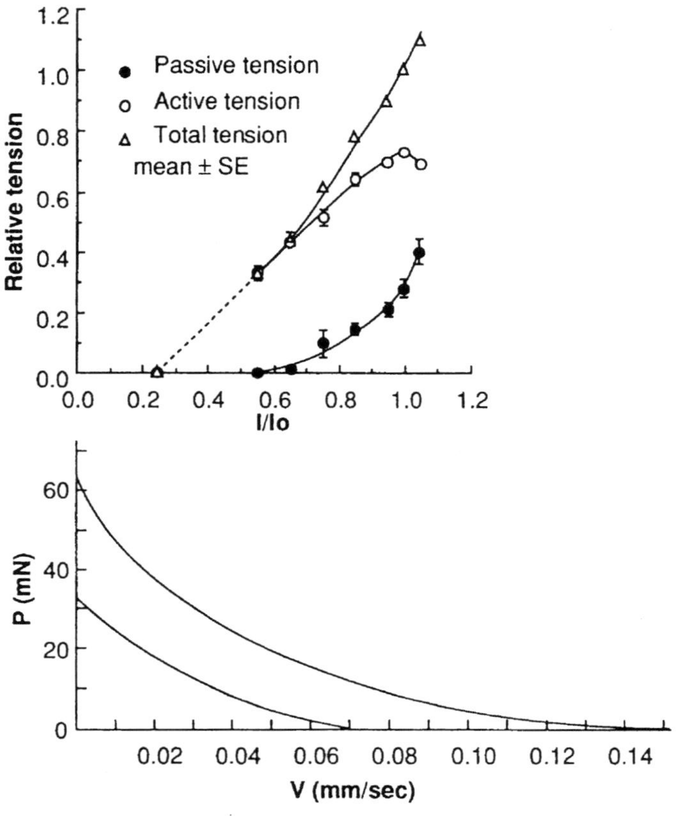

Comments
- Upper diagram: Length–tension curve; abscissa, muscle length normalized to optimum length l_0; ordinate, muscle tension normalized to total tension at l_0.
- Lower diagram: Force–velocity relations. Curves were generated from Hill's equation.
- At the optimal length, for tension development, appreciable passive tension was observed.
- The effect of decreasing the initial muscle length is a shift of the force–velocity curve toward the origin.

Reference(s)
Gordon AR, Siegman MJ (1971) Mechanical properties of smooth muscle. I. Length–tension and force–velocity relations. Am J Physiol 221:1243–1249 (with permission)

Length–Tension Relation (8)

• Stress relaxation •	• Cat • Ureter	• Hysteresis •

Materials

- Adult cats weighing 2.0–3.5 kg
- Upper and middle thirds of the ureter
- Placed in oxygenated modified Tyrode's solution at room temperature until ureteral segment of 8–11 mm in length was isolated

Testing Methods and Experimental Conditions

- Specimen mounted in a muscle chamber perfused at a rate of 6 ml/min with equilibrated Tyrode's solution at 37°C
- Length–tension relations studied
- Suprathreshold square-wave stimuli with duration of 40–100 ms and frequency of 4–6 / min applied transversely across the preparation

Data

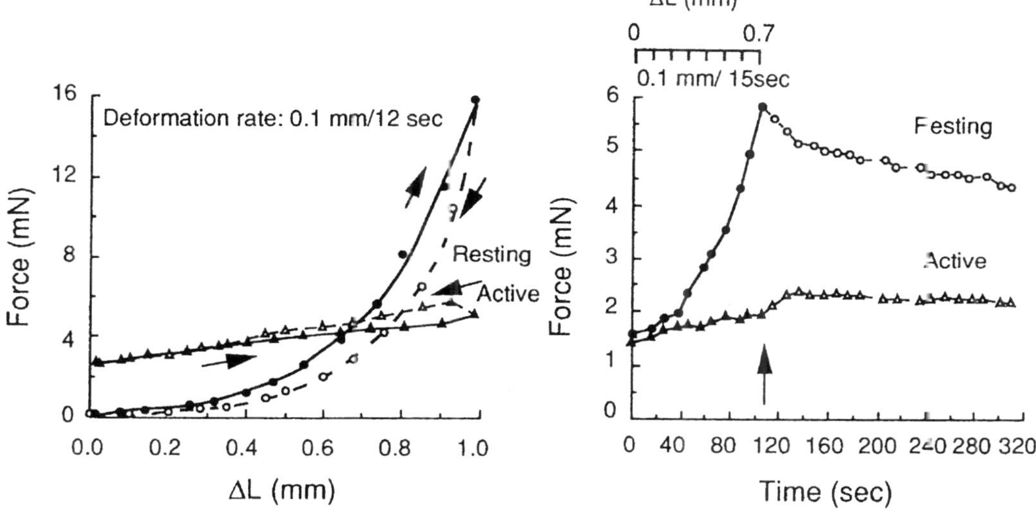

Comments

- Left diagram: Changes in resting and developed (active) force during lengthing–shortening cycle.
- Right diagram: Time courses of resting and developed (active) force during stretch of 0.7 mm at a constant rate (0.1 mm/15 s) and following cessation of the stretch (arrow).
- Changes in resting and developed tension are more readily apparent when the data are expressed in loop form.
- Resting tension declined with time while the muscle was maintained at a constant length.

Reference(s)

Weiss RM, Bassett AL, Hoffman BF (1972) Dynamic length–tension curves of cat ureter. Am J Physiol 222:388–393 (with permission)

Length–Tension Relation (9)

• Stress–strain curve • Strain rate	• Dog, rabbit, etc. • Ureter	• Stress relaxation •

Materials
- Dogs, rabbits, guinea pigs, human fetuses
- Ureter
- Specimens obtained from dog and human fetuses kept in oxygenated Krebs' solution at 10°C until test

Testing Methods and Experimental Conditions
- Specimens immersed in bubbled, modified Krebs' solution at 37 ± 0.25°C
- Simple elongation and stress–relaxation tests in the longitudinal direction
- Preconditioned for all tests

Data

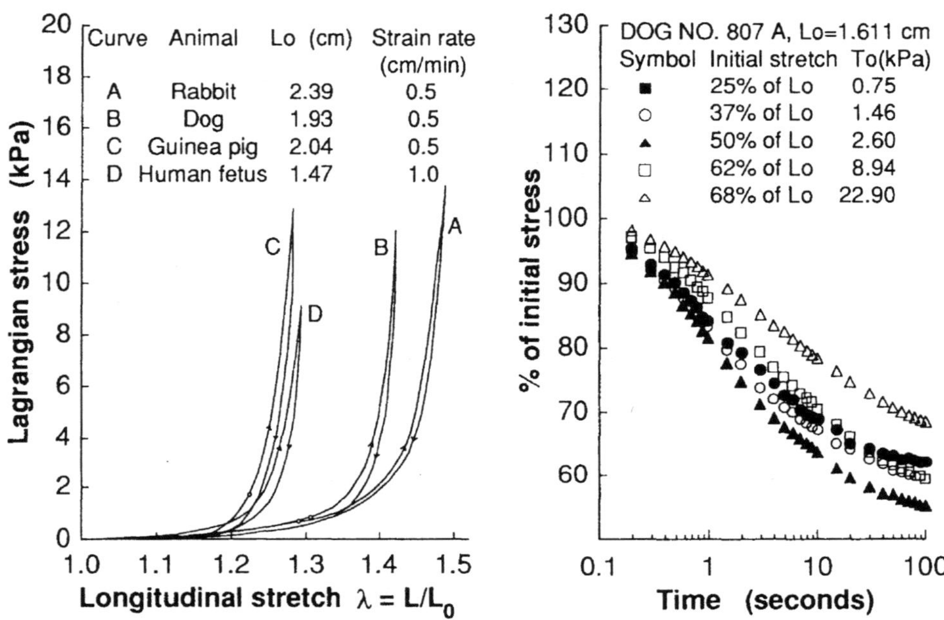

Comments
- Caption:

  L_o, preconditioned length or initial length; T_0, initial Lagrangian stress.
- These curves can be closely represented by exponential functions in the form

  $$T = (T^* + \beta)e^{\alpha(\lambda - \lambda^*)} - \beta.$$

  α, β, T^* and λ^*, constants; λ, stretch ratio.
- The time course of relaxation depends on amount of initial stretch.

Reference(s)
Yin FCP, Fung YC (1971) Mechanical properties of isolated mammalian ureteral segments. Am J Physiol 221:1484–1493 (with permission)

Length–Tension Relation (10)

• Stress–strain curve • Stress relaxation	• Guinea pig • *Taenia coli*	• Step–stretch • Spontaneous contraction

Materials
- Guinea pigs weighing 0.3–0.6 kg, Hartley strain
- Resting *Taenia coli*

Testing Methods and Experimental Conditions
- Specimen based in a bubbled physiological solution at 37°C at the in vivo length and equilibrated for 1 h before the experiments
- Epinephrine-treated solution to suppress spontaneous contraction
- Stress–relaxation characteristics before and after treatment with epinephrine
- Stress–strain relation in epinephrine at various strain rates
- Comparison with other kinds of muscle

Data

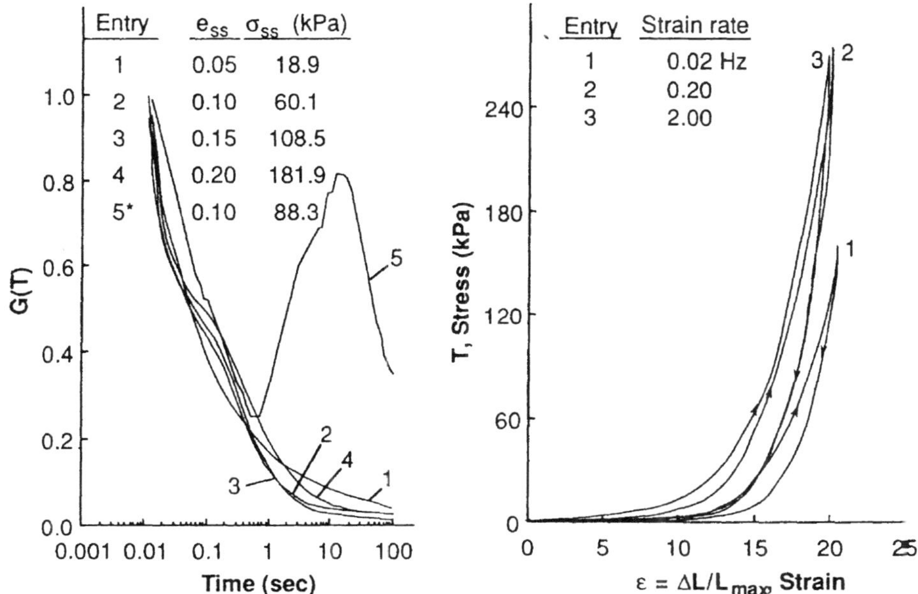

Comments
- Parameters:

  e_{ss}, strain at time (t_{ss}) from beginning of step stretch to attain a constant length; σ_{ss}, stress at t_{ss}; G(t), ratio σ/σ_{ss}; L_{max}, length at which the maximum isometric active tension is generated.
- Entry 5 with no epinephrine.
- σ_{ss} are lower after epinephrine injection, whereas the relaxation function G(t) continues to monotonically decrease in a fashion similar to that in the latent period of the spontaneous specimen.
- The stress–strain relation is nonlinear and varies with strain rate but approaches a limit at high strain rates.
- The relation is similar to that of the aorta, but different from that of the papillary muscle, mesentery, and ureter.

Reference(s)
Price JM, Patitucci PJ, Fung YC (1979) Mechanical properties of resting *Taenia coli* smooth muscle. Am J Physiol 236:C211–C220 (with permission)

Length–Tension Relation (11)

• Stress–strain curve • Time course of contraction	• Guinea pig • *Taenia coli*	• Isometric step stretch • Isotonic step loading

Materials
- Guinea pigs weighing 0.3–0.6 kg, Hartley strain
- *Taenia coli*

Testing Methods and Experimental Conditions
- Specimen mounted in the test chamber filled with bubbled physiological fluid at approximately the in vivo length L_{ph} and 37°C
- Isometric step stretch, isotonic step loading, length–tension, and stress–strain tests

Data

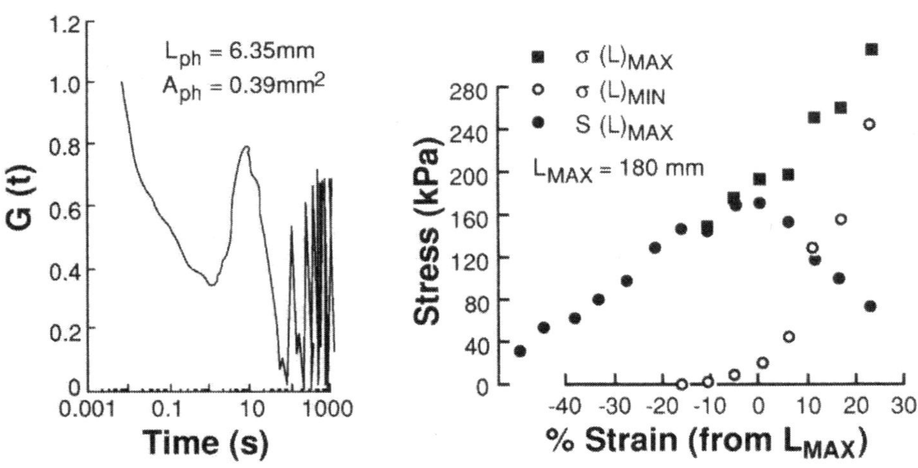

Comments
- Left diagram: Stress response to a 10% of L_{ph} step change in length. G(t), normalized stress response to an isometric step stretch; A_{ph}, the in vivo cross-sectional area.
- Right diagram: Functional dependence of σ (L)$_{max}$, σ (L)$_{min}$, and S(L)$_{max}$ on muscle length. σ (L)$_{max}$, maximum stress; σ (L)$_{min}$, minimum stress; S(L)$_{max}$, maximum active stress or σ (L)$_{max}$ − σ (L)$_{min}$; L_{max}, optimum length.
- The response can be divided into two phases — the short time latent period for t < 1 s and the long time minute rhythm for t > 1 s.

Reference(s)
Price JM, Patitucci P, Fung YC (1977) Mechanical properties of *Taenia coli* smooth muscle in spontaneous contraction. Am J Physiol 233:C47–C55 (with permission)

Stiffness (1)

<table>
<tr><td>• Isometric contraction
• Active force</td><td>• Toad
• Stomach</td><td>• Single cell measurement
• Latency</td></tr>
</table>

Materials

- Toads, *Bufo marinus*
- Single smooth muscle cell of the stomach

Testing Methods and Experimental Conditions

- Suspensions of the cells prepared by enzymatic digestion of the stomach muscularis
- Technique for isometrically measuring the force of contraction of a single cell
- Stiffness measured by applying small sinusoidal length oscillations (10–100 cycles/s, 0.5–2 μm amplitude)
- Temperature, 19°–22°C

Data

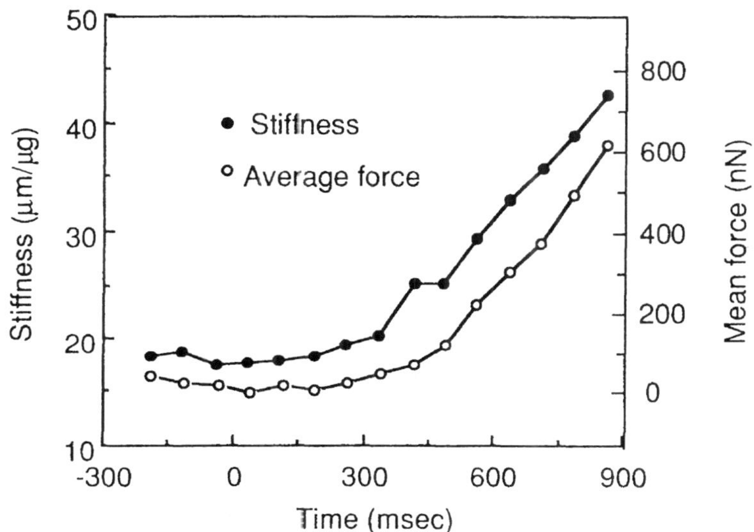

Comments

- Average force and stiffness as a function of time before and after electrical stimulation at time 0.
- There is a considerable delay between electrical activation of a smooth muscle cell and the onset of active force development.
- No increase in stiffness is measured until after active force is produced.
- The latency must reside in a step between the stimulus-induced potential change and the attachment of cross-bridges.

Reference(s)

Fay FS (1977) Isometric contractile properties of single isolated smooth muscle cells. Nature 265:553–556 (with permission)

Stiffness (2)

• Isotonic contraction • Isotonic relaxation	• Dog • Trachea	• Small force perturbation • Length dependence

Materials
- Dogs
- Cervical trachea
- Placed in a beaker of ice-cold, aerated Krebs–Henseleit solution until use
- A strip of muscle with an average length of 8 mm, an average weight of 2 mg, and a width of 0.5 mm

Testing Methods and Experimental Conditions
- Specimen fixed in the muscle bath containing aerated Krebs–Henseleit solution at 37°C and equilibrated for 2 h
- During this time, stimulated every 4 min electrically by 17 V, 60 Hz alternating current
- Stiffness of the series elastic component (SEC) in isotonic contraction and relaxation measured by applying small force perturbations and measuring the resulting length perturbations, and the effect of muscle length on the SEC stiffness by applying load clamps to the muscle at a fixed time (10 s after onset of electrical stimulation)
- Force perturbation: a train of rectangular force waves varying from 0 to 4 mN (10% of P_0, maximum isometric tension) with a frequency of 10 Hz

Data

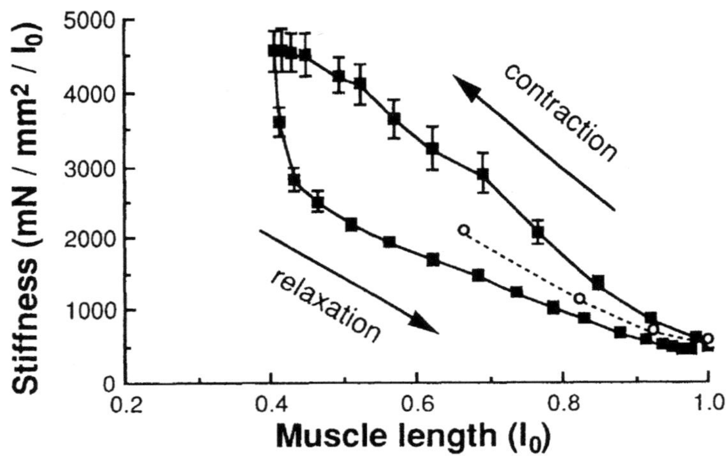

Comments
- The SEC stiffness was estimated by the ratio $\Delta P/\Delta L$; ΔP, small force perturbation with a constant amplitude; ΔL, change in the SEC length.
- Values are mean ± SE.
- SEC stiffness increased as muscle length decreased.
- This stiffness increase is not likely to be due to an increase in the number of attached cross bridges, but it is probably due to the gradual diminution of the SEC length itself during muscle shortening.

Reference(s)
Seow CY, Stephens NL (1989) Changes of tracheal smooth muscle stiffness during an isotonic contraction. Am J Physiol 256:C341–C350 (with permission)

Viscoelasticity (1)

• Series elastic component • Quick release	• Dog • Common carotid artery	• Maxwell model • Voigt model

Materials
- Dogs
- Common carotid artery

Testing Methods and Experimental Conditions
- Cylindrical segment mounted in a tissue bath filled with Krebs–Ringer dextrose solution at 36°–37°C and restored to in situ length (about 4 cm)
- Pressure–diameter test
- Inflation pressure: 10–15 mmHg steps to 300 mmHg
- Norepinephrine for activation and potassium cyanide for inactivation
- Series elastic element studied using a quick-release procedure
- Estimation of series elastic element with both Maxwell and Voigt models

Data

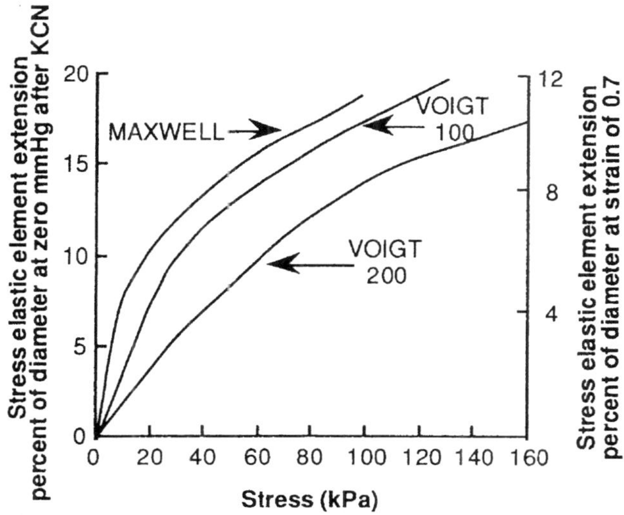

Comments
- Load–extension curves for the series elastic element of vessels excited isometrically at pressures between 50 and 150 mmHg.
- The computed series elastic element tension at an applied stress of 1.0×10^2 kPa was 18.9% for the Maxwell model and 14.7%–17.9% for the Voigt model.
- The values were reduced to 11.2% for the Maxwell model and 8.6%–10.5% for the Voigt model corresponding to the maximum active isometric tension.

Reference(s)
Dobrin PB, Canfield TR (1973) Series elastic and contractile elements in vascular smooth muscle. Circ Res 33:454–463 (with permission)

Viscoelasticity (2)

• Series elastic component • Stress–stiffness curve	• Dog • Common carotid artery	• Maxwell model • Voigt model

Materials

- Dogs
- Intact common carotid artery

Testing Methods and Experimental Conditions

- Cylindrical specimen of 10 cm in length mounted horizontally in a tissue bath filled with a Krebs–Ringer dextrose solution of $37 \pm 0.5°C$ at in situ length
- Pressure–diameter test to identify series elastic component (SEC)
- Static circumferential stress–strain curves and stress–stiffness curves to evaluate Maxwell and Voigt model elements at 33°, 36°, and 39°C
- Elastase or collagenase to digest the connective tissue component
- Activated with norepinephrine (NE), inactivated with potassium cyanide and sodium iodoacetate
- SEC stiffness computed for both models

Data

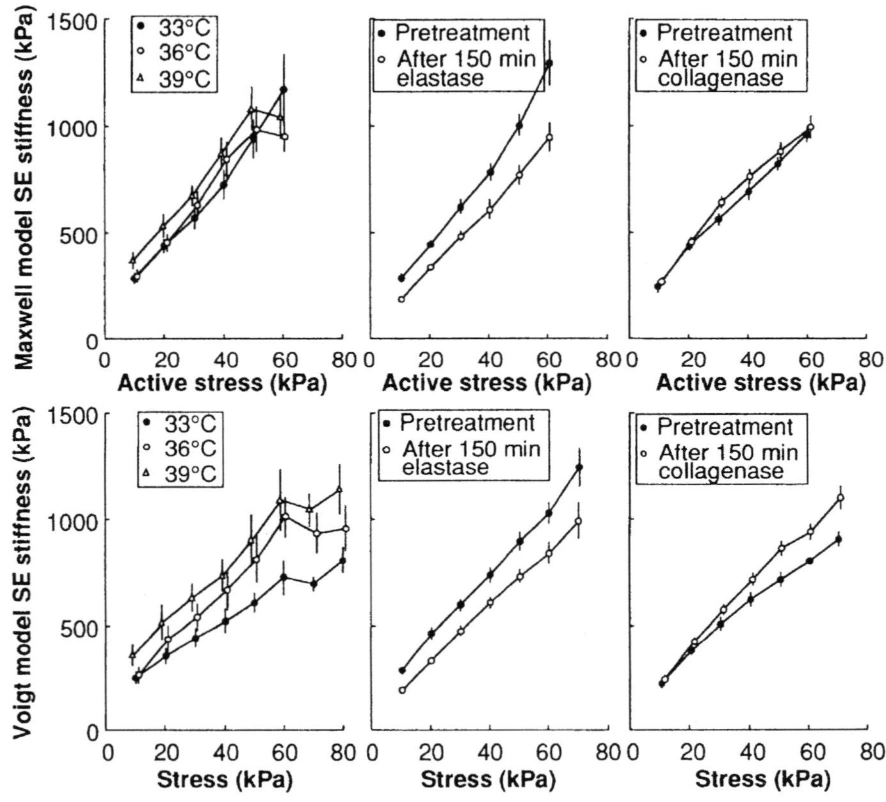

Comments

- Left diagrams, thermal variations; middle diagrams, elastase experiments; right diagrams, collagenase experiments. Values are mean ± SE.
- The Maxwell model is preferable to the Voigt model to describe the mechanical properties of the carotid artery wall.
- The SEC in intact carotid artery is primarily elastin.

Reference(s)

Dobrin PB, Canfield T (1977) Identification of smooth muscle series elastic component in intact carotid artery. Am J Physiol 232:H122–H130 (with permission)

Shear Property

• Stress–strain curve • Tangent modulus	• Rat • Periodontal ligament	• Strain rate •

Materials
- Wistar rats (age, 4 weeks)
- Periodontal ligament (PDL)

Testing Methods and Experimental Conditions
- Alveolar bone part of the transverse section of the mandible was clamped, and the tooth part was pushed by a metal rod
- Rod speed of 1, 10, 10^2, 10^3, and 10^4 mm/24 h
- Phosphate-buffered saline added with a mixture of antibiotics (room temperature)
- Shear stress was calculated by dividing load by the area of the PDL facing the root (cementum)
- Shear strain was approximated as deformation per average width of the lingual PDL

Data

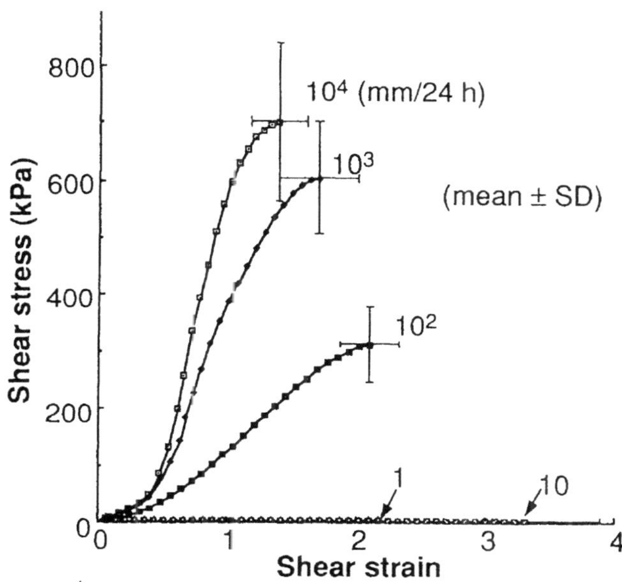

Velocity of loading (mm/24h)	Maximum shear stress (kPa)	Maximum shear strain	Tangent modulus (kPa)	Strain energy density (kPa)	Number of rats n
1	1.11 ± 0.39	2.20 ± 0.26	0.77 ± 0.19	1.46 ± 0.69	11
10	4.51 ± 0.98	3.29 ± 0.56	3.01 ± 1.38	7.20 ± 1.62	11
10^2	322 ± 67	2.02 ± 0.26	312 ± 77	269 ± 66	12
10^3	636 ± 99	1.67 ± 0.34	729 ± 224	474 ± 114	11
10^4	740 ± 156	1.36 ± 0.23	1115 ± 239	430 ± 111	10

All data are given as mean ± SD.

Comments
- The free surface of the PDL adhering to the cementum was rough and irregular after mechanical testing at higher velocities and rather smooth at lower velocities.

Reference(s)
Chiba M, Komatsu K (1993) Mechanical responses of the periodontal ligament in the transverse section of the rat mandibular incisor at various velocities of loading in vitro. J Biomech 26:561–570 (with permission)

Tensile Property (1)

• Junction strength •	• Rat • Medial collateral ligament	• Age effect • Sex

Materials

- Pregnant Sprague–Dawley rats (age, 15–730 days)
- Medial collateral ligament (MCL)

Testing Methods and Experimental Conditions

- Tensile test of femur–MCL–tibia complex
- Maintained at a temperature of 15°C
- Deformation rate of 0.24 mm/s

Data

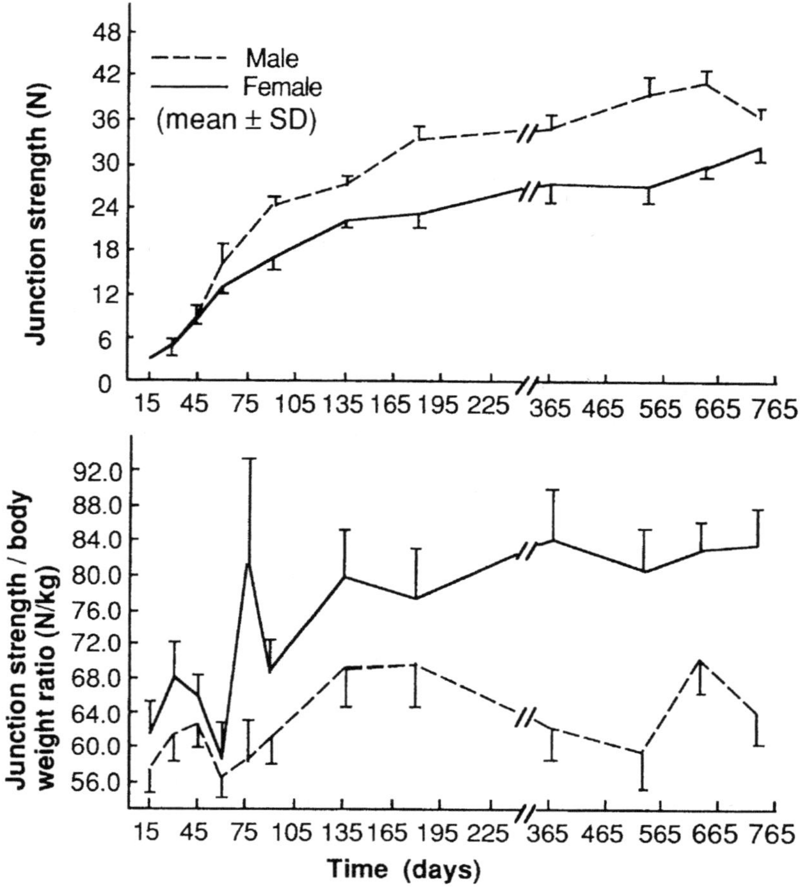

Comments

None.

Reference(s)

Tipton CM, Matthes RD, Martin RK (1978) Influence of age and sex on the strength of the bone–ligament junctions in knee joints of rats. J Bone Joint Surg 60A:230–234 (with permission)

Tensile Property (2)

• Load–deformation curve • Load–thickness curve	• Human • Spine ligament	• •

Materials
- Human (age, 43–85 years)
- Fresh frozen lumber spine motion segments (L3 and L4)
- Ligamentum flavum

Testing Methods and Experimental Conditions
- Lamina–ligament–lamina complex was fixed in a heavy vice and loaded with weights (9.8, 19.6, 39.2, and 78.4 N) on a string
- Elongation was determined from the distance between the metal pins inserted into the laminae
- Thickness was measured with a laser system
- Moistened with physiological saline

Data

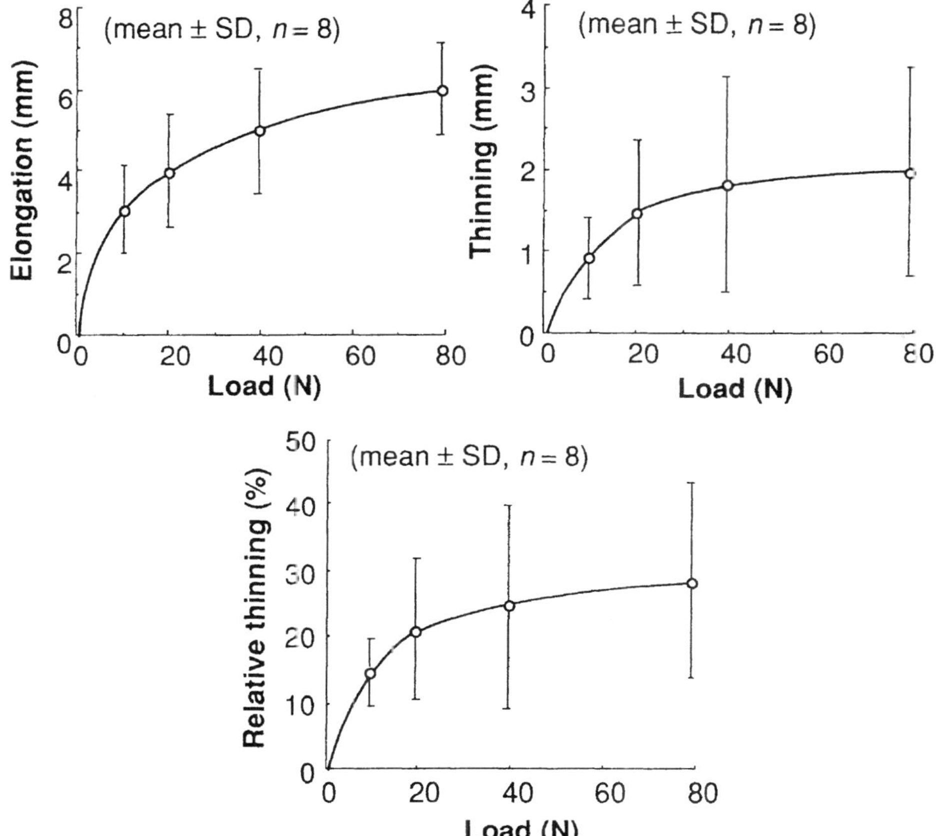

Comments
None.

Reference(s)

Schonstrom NR, Hansson TH (1991) Thickness of the human ligamentum flavum as a function of load: An in vitro experimental study. Clin Biomech 6:19–24 (with permission)

Tensile Property (3)

• Load–deformation curve •	• Rabbit • Anterior cruciate ligament	• Flexion angle • Loading direction

Materials
- New Zealand White rabbits (skeletally immature; weight, 3.2 ± 0.2 kg [mean ± SD])
- Anterior cruciate ligament (ACL)

Testing Methods and Experimental Conditions
- Tensile test of femur–ACL–tibia complex
- Directions of load were along the tibial axis for the left leg, and along the ligament axis for the right leg
- Crosshead speed of 200 mm/min, strain rate of 0.33 per second
- Moisture with saline solution

Data

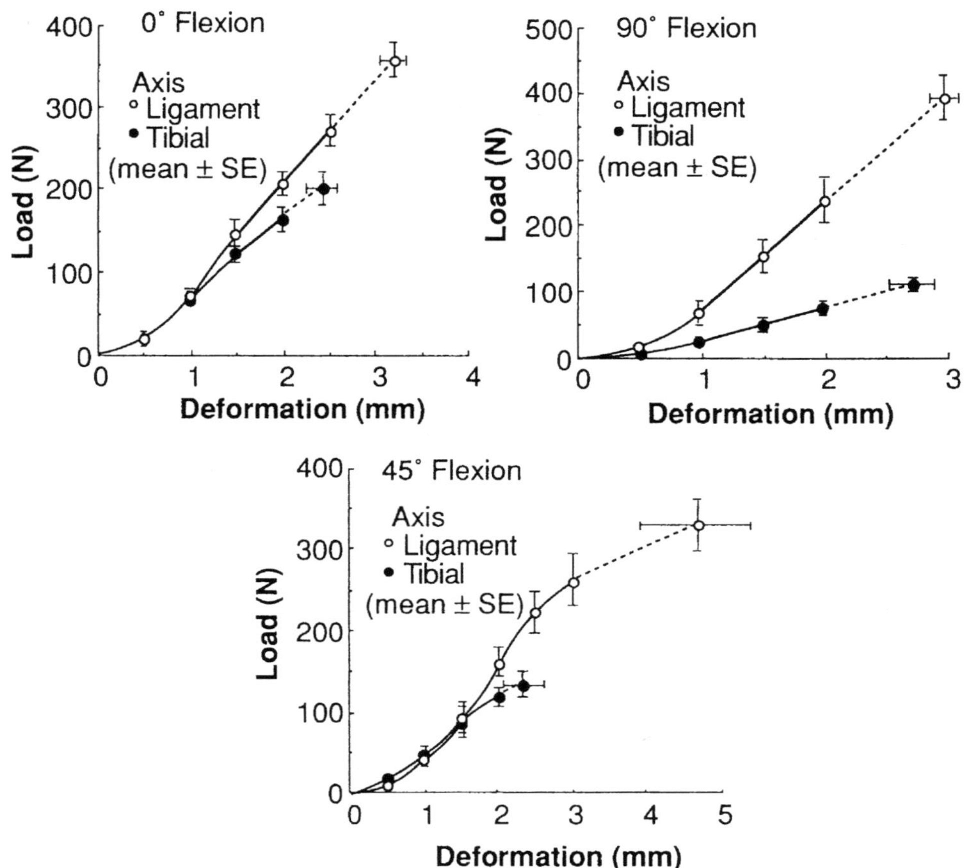

Comments
- Bony avulsion was the most common failure mode for the specimens tested along the ligament axis, while ligament "peeling off" from its tibial insertion was the most common failure mode for the specimens tested along the tibial axis.

Reference(s)

Woo SL-Y, Hollis JM, Roux RD, Gomez MA, Inoue M, Kleiner JB, Akeson WH (1987) Effects of knee flexion on the structural properties of the rabbit femur–anterior cruciate ligament–tibia complex (FATC). J Biomech 20:557–563 (with permission)

Tensile Property (4)

• Load–strain curve • Maximum load	• Human • Flexor tendon	• Locational dependence •

Materials
- Human (age, 45–84 [mean 72] years)
- Long flexor tendon (flexor digitorum superficialis [FDS], flexor digitorum profundus [FDP], flexor pollicis longus [FPL])

Testing Methods and Experimental Conditions
- Tensile test of isolated tendons
- Extension rate of 100 mm/min
- In room air

Data

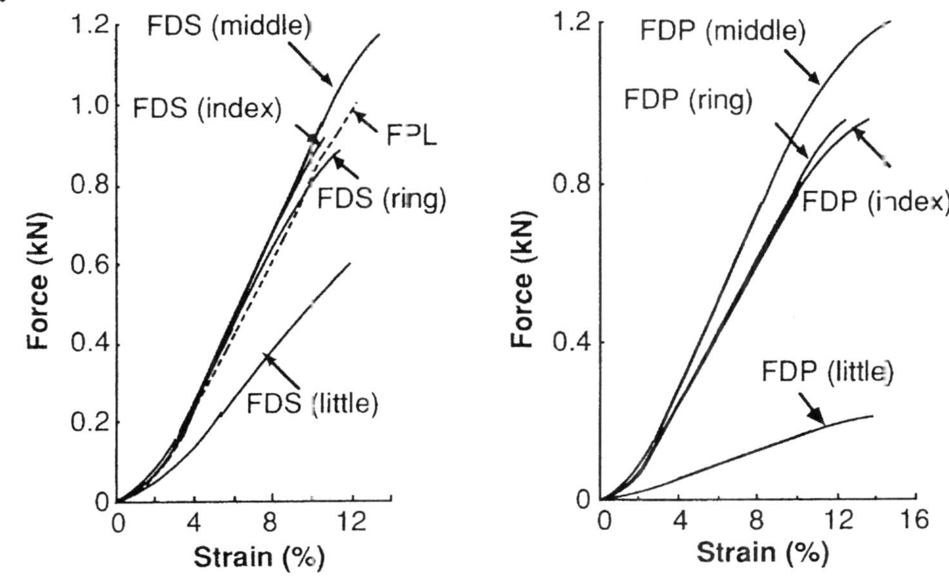

		Breaking strength (N)		Strain (%)	
		Max. - Min.	Mean ± SD	Max. - Min.	Mean ± SD
FDS	Index	1430 - 490	965 ± 209	21.5 - 10.0	13.4 ± 2.9
	Middle	1820 - 750	1252 ± 321	25.5 - 10.0	15.7 ± 5.3
	Ring	1115 - 620	955 ± 142	16.2 - 9.3	12.5 ± 2.1
	Little	345 - 110	212 ± 76	25.0 - 9.0	13.7 ± 4.7
FDP	Index	1385 - 660	935 ± 176	14.8 - 6.3	10.7 ± 2.1
	Middle	1520 - 780	1182 ± 213	21.8 - 7.4	12.9 ± 3.5
	Ring	1100 - 615	927 ± 149	15.3 - 8.9	11.5 ± 2.1
	Little	1205 - 460	623 ± 191	16.9 - 8.5	12.0 ± 2.8
FPL		1330 - 685	998 ± 180	15.5 - 9.3	12.1 ± 2.0

Comments
None.

Reference(s)
Pring DJ, Amis AA, Coombs RRH (1985) The mechanical properties of human flexor tendons in relation to artificial tendons. J Hand Surg 10B:331–336 (with permission)

Tensile Property (5)

• Load–deformation curve • Stress–strain curve	• Rabbit • Lateral collateral ligament	• •

Materials
- New Zealand White rabbits (male; age, 15 weeks; weight, 3.5 ± 0.2 kg)
- Lateral collateral ligment (LCL)

Testing Methods and Experimental Conditions
- Tensile test of LCL
- Crosshead speed of 20 mm/min
- Saline solution at 37°C
- Tensile load and strain were measured with a load cell and a video dimensional analyzer, respectively

Data

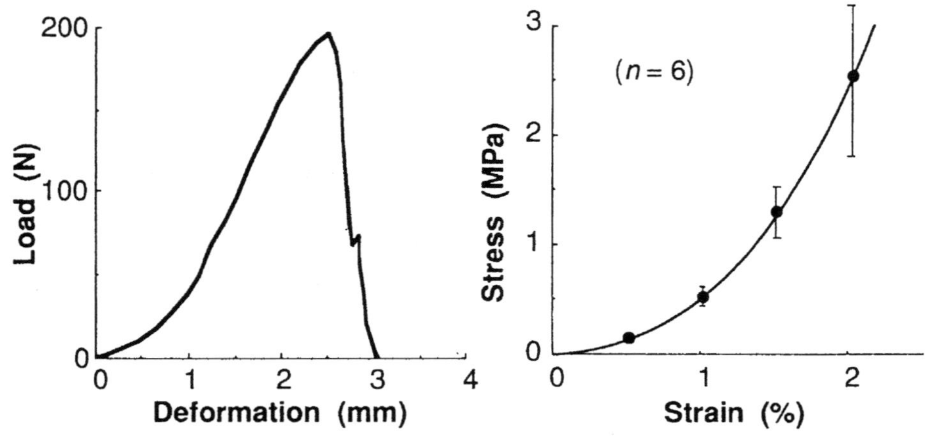

Comments
None.

Reference(s)

Amiel D, Woo SL-Y, Harwood FL, Akeson WH (1982) The effect of immobilization on collagen turnover in connective tissue: A biochemical–biomechanical correlation. Acta Orthop Scand 53:325–332 (with permission)

Tensile Property (6)

• Maximum load •	• Human • Ankle ligament	• •

Materials
- Human (age, 24–86 years)
- Ankle ligament (anterior talofibular ligament)

Testing Methods and Experimental Conditions
- Tensile test of fibula–anterior talofibular ligament–talus complex
- Displacement rate of 12.5 cm/min
- Saline solution

Data

Specimen no.	Sex	Age	Side (left or right)	Width/length (mm)	Tensile strength (N)	Location of rupture
1	M	62	R	8/12	196	Mid-substance
1	M	62	L	7/11	151	Mid-substance
2	M	62	R	6/13	156	Talus
3	F	69	R	7/10	263	Mid-substance
3	F	69	L	7/12	334	Mid-substance
4	F	27	L	6/15	89	Mid-substance
4	F	27	R	8/21	62	Mid-substance
5	F	64	R	7/9	80	Mid-substance
5	F	64	L	7/8	89	Mid-substance
6	F	73	L	7/11	71	Talus
7	M	65	R	10/12	267	Not clear
7	M	65	L	11/12	289	Not clear
8	F	79	R	7/11	129	Mid-substance
8	F	79	L	7/12	129	Mid-substance
9	M	75	L	8/12	80	Mid-substance
9	M	75	R	8/14	160	Mid-substance
10	M	69	R	7/10	58	Talus
10	M	69	L	7/12	160	Talus
11	M	72	R	6/10	214	Talus
11	M	72	L	6/10	174	Talus
12	M	51	R	8/12	138	Talus
12	M	51	L	8/13	44	Talus
13	F	66	R	9/13	245	Talus
13	F	66	L	10/14	312	Mid-substance
14	M	59	R	8/12	187	Talus
14	M	59	L	7/12	325	Talus
15	M	83	R	7/13	80	Talus
15	M	83	L	8/10	129	Talus
16	F	56	R	8/14	182	Talus
16	F	56	L	8/14	383	Talus
17	M	86	R	13/18	414	Mid-substance
17	M	86	L	13/18	102	Mid-substance
18	M	50	R	12/30	409	Mid-substance
18	M	50	L	15/30	374	Talus
19	M	52	R	10/14	405	Talus
19	M	52	L	9/14	556	Talus
Average		64		8.3/13.8	206	

Comments
- Grip slippage tended to occur, more or less, in every specimen, indicating that the displacement to rupture value was not meaningful.

Reference(s)
Pierre RKS, Rosen J, Whitesides TE, Szczukowski M, Fleming LL, Hutton WC (1983) The tensile strength of the anterior talofibular ligament. Foot Ankle 4:83–85 (with permission)

Tensile Property (7)

• Maximum load • Maximum deformation	• Dog • Anterior cruciate ligament	• Flexion angle •

Materials
- Beagles (male; age, about 18 months)
- Anterior cruciate ligament (ACL)

Testing Methods and Experimental Conditions
- Tensile test of femur–ACL–tibia complex
- Displacement rate of 51.0 cm/min
- Moisture with saline solution
- Deflection was determined from crosshead displacement

Data

Number of specimens	9	9	9
Bone to bone angle (degrees)	0	45	90
Deflection to ultimate load (cm)	0.787 ± 0.211	0.622 ± 1.49	0.813 ± 0.123
Total deflection (cm)	0.864 ± 0.225	0.864 ± 0.294	1.346 ± 0.282
Ultimate load (N)	1181 ± 276	454 ± 84	428 ± 77
Total energy absorbed (Nm)	9.76 ± 2.23	3.25 ± 0.64	5.01 ± 1.56

Comments
- Failures were categorized as types A1, A2, B, and C. Type A1 and A2 failures were purely osseous failures in the substrate adjacent to the ligamentous attachment; Type A1 failure occurred at the tibia and Type A2 at the femur. Type B failure was a ligamentous pull-off at the distal tibial insertion. Type C failure was primarily within the body of the ligament. The failures for the 0° tests were Type A1 (2/9), Type A2 (2/9), and Type B (5/9). The failures for the 45° tests were Type B (7/9) and Type A (2/7). The failures for the 90° tests were Type C (7/9) and combined Type B/C (2/9).

Reference(s)
Figgie III HE, Bahniuk EH, Helple KG, Davy DT (1986) The effects of tibial–femoral angle on the failure mechanics of the canine anterior cruciate ligament. J Biomech 19:89–91 (with permission)

Tensile Property (8)

• Maximum load • Stiffness	• Goat • Anterior cruciate ligament	• Freeze-dry • Ethylene oxide

Materials
- Spanish goats (adult)
- Normal and treated anterior cruciate ligaments (ACLs)
    Group 1, normal ACLs
    Group 2, ACLs treated with ethylene oxide (simulating clean procurement), then
        freeze-dried
    Group 3, ACLs freeze-dried only

Testing Methods and Experimental Conditions
- Tensile test of femur–ACL–tibia complex at room temperature
- Knee flexion of 30°
- Strain rate of 1 per second
- Change of the distance between bone ends was measured with a metal clip gage

Data

Group	n	Max. force (N)	Stiffness (N/mm)	Energy to max. force (Nm)	Elongation to max. force (mm)
1	12	2403 ± 133	692 ± 37	4.85 ± 0.29	4.85 ± 0.16
2	6	2059 ± 273	718 ± 60	3.67 ± 0.78	4.00 ± 0.30
3	6	2023 ± 214	680 ± 65	3.70 ± 0.50	4.30 ± 0.30

All data are given as mean ± SE.

Comments
- No significant differences between the groups were observed for any parameter.

Reference(s)
Jackson DW, Grood ES, Wilcox P, Butler DL, Simon TM, Holden JP (1988) The effects of processing techniques on the mechanical properties of bone–anterior cruciate ligament–bone allografts. An experimental study in goats. Am J Sports Med 16:101–105 (with permission)

Tensile Property (9)

| • Maximum load
• Stiffness | • Human
• Spine ligament | • Strain rate
• |

Materials
- Human (age, 46–88 [mean 70] years; height, 152–185 [mean 171.8] cm; weight, 41–109 [mean 66.6] kg)
- Anterior longitudinal ligament and ligamentum flavum

Testing Methods and Experimental Conditions
- Tensile test of vertebral body–ligament–vertebral body complex
- Electrohydraulic testing device
- Moistened by physiologic saline

Data

Preset loading rate (mm/s)	Actual loading rate (mm/s)	Ultimate load (N)	Elongation at failure (mm)	Stiffness[a] (N/mm)	Energy[b] (J)
Anterior longitudinal ligament					
8.89	8.71 ± 0.20	120.58 ± 13.76	7.48 ± 2.20	14.93 ± 2.20	0.54 ± 0.11
25.00	24.93 ± 0.18	122.36 ± 24.27	5.53 ± 1.13	36.20 ± 8.34	0.46 ± 0.11
250.00	240.60 ± 9.47	166.39 ± 31.34	6.40 ± 0.37	47.34 ± 9.61	0.65 ± 0.13
2500.00	2261.72 ± 78.38	349.48 ± 44.85	6.34 ± 0.59	82.71 ± 6.79	1.23 ± 0.26
Ligamentum flavum					
8.89	8.71 ± 0.20	130.64 ± 12.94	7.55 ± 0.49	21.93 ± 3.13	0.47 ± 0.04
25.00	24.93 ± 0.18	118.24 ± 12.83	5.65 ± 0.39	29.77 ± 4.34	0.35 ± 0.04
250.00	240.60 ± 9.47	181.52 ± 13.10	6.29 ± 0.82	62.29 ± 13.72	0.63 ± 0.11
2500.00	2261.72 ± 78.38	335.07 ± 28.01	7.95 ± 0.67	92.92 ± 12.57	1.32 ± 0.17

All data are given as mean ± SE.
[a] Stiffness was taken as the slope of the load–elongation response in its most linear region.
[b] Area under the load–elongation response up to failure.

Comments
- Ultimate tensile failure load, stiffness, and energy-absorbing capacity at failure increased with increasing loading rate for both tissues, although elongation at failure did not.

Reference(s)
Yoganandan N, Pintar F, Butler J, Reinartz J, Sances A, Larson SJ (1989) Dynamic response of human cervical spine ligaments. Spine 14:1102–1110 (with permission)

Tensile Property (10)

• Maximum load • Strain at failure	• Human • Ankle ligament	• •

Materials
- Human (age, 23–82 [mean 57.9] years)
- Ankle ligament (anterior talofibular, calcaneofibular, posterior talofibular, and tibiotalar ligament)

Testing Methods and Experimental Conditions
- Tensile test of bone–ligament–bone complex
- Deflection rate of 0.01–100 cm/s
- Moisture with saline solution at 21°C

Data

	Anterior talofibular ligament	Calcaneo-fibular ligament	Posterior talofibular ligament	Tibiotalar ligament
Length (cm)	1.05 ± 0.06	1.75 ± 0.07	1.53 ± 0.09	0.51 ± 0.01
Width (cm)	0.76 ± 0.04	0.68 ± 0.04	0.78 ± 0.04	2.35 ± 0.08
Number of specimens	12	16	4	6
Maximum load (N)	138.9 ± 23.5	345.7 ± 55.2	261.2 ± 32.4	713 8 ± 69.3
Deflection to failure (cm)	0.51 ± 0.05	0.63 ± 0.05	1.31 ± 0.16	1.05 ± 0.11
Strain to failure	0.53 ± 0.06	0.38 ± 0.03	1.00 ± 0.15	2.10 ± 0.23
Linear modulus of elasticity (N/cm)	399.9 ± 85.4	705.1 ± 69.0	397.5 ± 137.9	1288.2 ± 250.4
Rate of deflection (cm/s)	101 ± 7	106 ± 4	82 ± 13	80 ± 13
Energy to yield points (Ncm)	25.7 ± 3.6	135.8 ± 45.1	56.4 ± 24.0	74.7 ± 13.3
Failure type	8 midsubstance 4 talar avulsion	8 midsubstance 4 calcaneal avulsion 4 fibular avulsion	4 midsubstance	1 midsubstance 5 talar avulsion

All data are given as mean ± SE.

Comments
- Only tests that represent ligament failure rather than fracture or mounting breakage are included, e.g., only six tibiotalar ligaments are reported because the medial malleolus usually fractured before the ligament failed.

Reference(s)
Attarian DE, McCrackin HJ, DeVito DP, McElhaney JH, Garrett Jr WE (1985) Biomechanical characteristics of human ankle ligaments. Foot Ankle 6:54–58 (with permission)

Tensile Property (11)

• Maximum load • Strain at failure	• Monkey • Anterior cruciate ligament	• Strain rate •

Materials
- Rhesus monkeys (male; weight, 5.9–10.9 [mean 7.9] kg)
- Anterior cruciate ligament (ACL)

Testing Methods and Experimental Conditions
- Tensile test of bone–ACL–bone preparations at two strain rates
- The femur and tibia were fixed at 135° to each other
- Ringer's solution
- Extension rate of 8.467 mm/s for the fast-rate testing and 0.08476 mm/s for the slow-rate testing
- Crosshead velocity was monitored with an oscillograph

Data

	Fast	Slow
Linear load (N)*	892 ± 204	778 ± 163
Maximum load (N)**	997 ± 165	805 ± 175
Strain to linear load (%)	43.5 ± 12.2	42.7 ± 7.7
Strain to maximum load (%)**	51.5 ± 6.7	43.7 ± 7.0
Strain to failure (%)	57.1 ± 10.4	51.9 ± 10.1
Energy to failure (Nmm) **	4120 ± 1090	3010 ± 850
Slope at 10% strain (N/mm)	156.4 ± 30.9	151.3 ± 30.5
Slope at 20% strain (N/mm)	178.4 ± 44.1	167.5 ± 40.1
Slope at 40% strain (N/mm)	141.4 ± 46.0	127.4 ± 30.3

$*P < 0.05; **P < 0.005.$

Comments
- The major mode of specimen failure changed from a predominance of tibial avulsion fractures at the slow rate to ligament disruption at the fast rate.

Reference(s)
Noyes FR, Delucas JL, Torvik PJ (1974) Biomechanics of anterior cruciate ligament failure: An analysis of strain rate sensitivity and mechanism of failure in primates. J Bone Joint Surg 56A:236–253 (with permission)

Tensile Property (12)

• Maximum load • Tangent modulus	• Dog • Patellar tendon	• Age effect •

Materials
- Dogs (age, 6–180 months; weight, 26.7 ± 8.4 kg)
- Patellar tendon (PT)

Testing Methods and Experimental Conditions
- Tensile test of patella–PT–tibia complex
- Strain rate of 1 per second
- Moisture with saline solution at room temperature
- Strain was determined from grip-to-grip deformation

Data

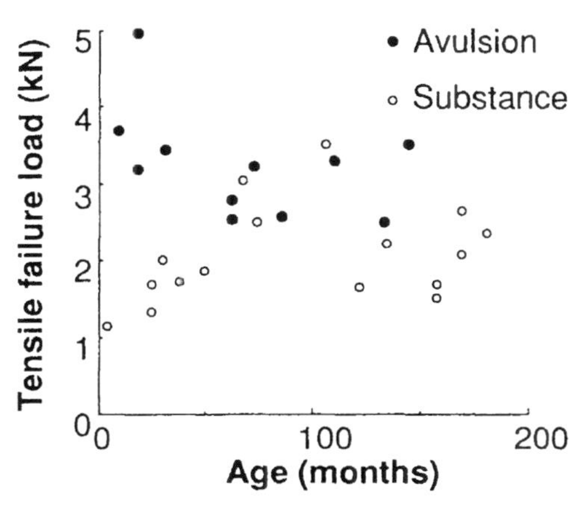

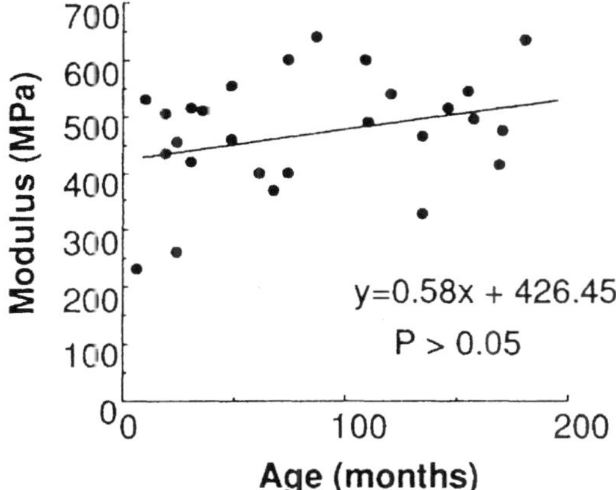

Comments
- Fifty-nine per cent (16/27) of the specimens failed by avulsion at the patella, but neither the failure load nor the mode of failure were a function of age.

Reference(s)
Haut RC, Lancaster RL, DeCamp CE (1992) Mechanical properties of the canine patellar tendon: Some correlations with age and the content of collagen. J Biomech 25:163–173 (with permission)

Tensile Property (13)

• Maximum load • Tensile strength	• Human • Ankle ligament	• •

Materials
- Human (12 females and 8 males; age, 67.8 ± 15.2 years; weight, 69.1 ± 15.1 kg; height, 1.71 ± 0.09 m)
- Ankle ligament (anterior fibulotalar ligament (AFTL), fibulocalcaneal ligament (FCL), posterior fibulotalar ligament (PFTL), tibionavicular ligament (TNL), tibiospring ligament (TSL), posterior tibiotalar ligament (PTTL))

Testing Methods and Experimental Conditions
- Tensile test of bone–ligament–bone complex
- Stretch rate of 0.32 cm/min, strain rate of 0.078 per minute
- Moisture with saline solution

Data
See next page.

Comments
- The ligaments obtained from male donors exhibited significantly higher yield and ultimate force values over those obtained from females. The cross-sectional areas of the ligaments from male donors were generally larger than those from females. The cross-sectional areas of the ligaments from heavier donors (> 67 kg) were generally larger than those from lighter donors. Anatomical mode of failure was not significantly related to the yield and ultimate point properties of ligaments.

Reference(s)
Siegler S, Block J, Schneck CD (1988) The mechanical characteristics of the collateral ligaments of the human ankle joint. Foot Ankle 8:234–242 (with permission)

	Lateral collateral			Medial collateral		
	AFTL	FCL	PFTL	TNL	TSL	PTTL
Initial length (cm)	1.781 ± 0.305	2.769 ± 0.330	2.116 ± 0.386	4.183 ± 0.493	1.859 ± 0.437	1.186 ± 0.396
Cross-sectional area (cm^2)	0.129 ± 0.077	0.097 ± 0.065	0.219 ± 0.181	0.071 ± 0.026	0.135 ± 0.071	0.452 ± 0.316
Ultimate load (N)	231 ± 129	307 ± 142	418 ± 191	120 ± 49	432 + 307	467 + 209
Ultimate elongation (cm)	0.246 ± 0.076	0.366 ± 0.071	0.348 ± 0.094	0.416 ± 0.127	0.648 ± 0.140	0.310 ± 0.081
Yield force (N)	222 ± 133	289 ± 138	400 ± 187	107 ± 49	351 ± 231	405 ± 218
Yield elongation (cm)	0.226 ± 0.081	0.343 ± 0.061	0.325 ± 0.086	0.361 ± 0.124	0.480 ± 0.117	0.251 ± 0.068
Stiffness constant x10^5 (N/m)	1.418 ± 0.793	1.266 ± 0.429	1.643 ± 0.555	0.391 ± 0.166	1.226 ± 0.669	2.343 ± 0.776
Ultimate stress (MPa)	24.20 ± 16.91	46.22 ± 36.62	25.95 ± 24.78	22.93 ± 16.82	33.97 ± 23.53	15.59 ± 15.07
Ultimate strain (cm/cm)	0.15 ± 0.06	0.13 ± 0.03	0.17 ± 0.05	0.10 ± 0.02	0.33 ± 0.14	0.30 ± 0.13
Yield stress (MPa)	22.59 ± 16.64	43.64 ± 35.85	25.00 ± 23.97	20.73 ± 15.03	26.97 ± 18.34	12.39 ± 8.46
Yield strain (cm/cm)	0.14 ± 0.07	0.13 ± 0.02	0.16 ± 0.04	0.09 ± 0.02	0.28 ± 0.12	0.25 ± 0.13
Elastic modulus (MPa)	255.5 ± 181.3	512.0 ± 333.5	216.5 ± 169.5	320.7 ± 268.5	184.5 ± 133.4	99.54 ± 79.32
Mode of failure (%)						
Avulsion	58	70	70	0	31	60
Tear	42	30	30	100	69	40

All data are given as mean ± SD.

Tensile Property (14)

• Maximum load • Tensile strength	• Human • Coracoacromial ligament	• •

Materials
- Human (age, 61–87 years)
- Lateral band of coracoacromial ligament from fresh frozen shoulder with and without rotator cuff tears

Testing Methods and Experimental Conditions
- Tensile test of bone–ligament–bone complex
- Strain rate of 0.001 per second
- Saline solution at 37°C
- Strain was determined using an optical system

Data

	Rotator cuff tears	No rotator cuff tears
Length (mm)	26.6 ± 4.8	34 ± 4.2
Thickness (mm)	1.21 ± 0.36	0.96 ± 0.33
Width (mm)	10.2 ± 2.8	8.0 ± 2.1
Area (mm^2)	12 ± 3.8	8 ± 3.8
Failure load (N)	305.4 ± 154.8	311.7 ± 127.4
Failure displacement (mm)	7.5 ± 2.3	7.8 ± 2.0
Stiffness (N/mm)	57.5 ± 37.3	59.3 ± 26.3
Failure stress (MPa)	25.3 ± 8.7	46.9 ± 30.7
Total specimen failure strain (%)	28.3 ± 8.6	23.7 ± 9.5
Ligamentous failure strain (%)	5.1 ± 1.2	5.6 ± 1.7
Total specimen modulus (MPa)	120.3 ± 38.9	291.6 ± 154
Ligamentous modulus (MPa)	658.4 ± 261.3	1174.4 ± 437.7

All data are given as mean ± SD, n=10.

Comments
- There was no description about failure mode.

Reference(s)
Soslowsky LJ, An CH, Johnson SP, Carpenter JE (1994) Geometric and mechanical properties of the coracoacromial ligament and their relationship to rotator cuff disease. Clin Orthop Relat Res 304:10–17 (with permission)

Tensile Property (15)

• Maximum load • Tensile strength	• Human • Patellar tendon	• Effect of twisting • Effect of width

Materials

- Human (age, 28 ± 7 years [mean ± SE])
- Patellar tendon (PT)

Testing Methods and Experimental Conditions

- Tensile test of bone–central third PT–bone composite (B–PT–B composite)
  - Group 1 = untwisted (10 mm versus 15 mm wide B–PT–B)
  - Group 2 = untwisted 10 mm wide versus 90° twisted 15 mm wide B–PT–B
  - Group 3 = 90° twisted versus 180° twisted 10 mm wide B–PT–B
  - Group 4 = untwisted 7 mm wide B–PT–B
- In room air
- Elongation rate of 5 cm/s
- Tissue elongation was determined from clamp-to-clamp displacement

Data

Group	Ultimate load (N)	Area (mm^2)	Stress (MPa)	Elongation (%)	Energy to failure (Nm)	Stiffness (kN/m)
1						
15 mm (n=5)	4389 ± 709	44.9 ± 8.6	97.5 ± 8.6	19.7 ± 1.8	26.5 ± 5.0	555.5 ± 67.1
10 mm (n=5)	3057 ± 351	32.3 ± 2.7	95.5 ± 16.8	17.9 ± 1.1	18.0 ± 1.1	455.4 ± 56.5
2						
10 mm (n=5)	2664 ± 395	34.9 ± 3.6	77.6 ± 17.6	15.8 ± 4.0	14.4 ± 3.7	424.1 ± 66.5
15 mm (n=5, 90°)	3397 ± 378	33.8 ± 3.4	101.3 ± 14.0	18.5 ± 3.2	21.7 ± 7.3	476.9 ± 12.4
3						
10 mm (n=5, 90°)	2542 ± 606	41.2 ± 11.9	64.5 ± 20.1	17.1 ± 4.0	16.0 ± 6.6	391.6 ± 34.5
10 mm (n=5, 180°)	2684 ± 684	38.4 ± 10.8	70.5 ± 19.3	17.2 ± 3.4	17.3 ± 5.5	423.1 ± 25.2
4						
7 mm (n=4)	2238 ± 316	24.6 ± 3.7	92.7 ± 20.1	18.7 ± 1.5	12.5 ± 2.4	325.8 ± 57.9

Comments

- The central third of PT is stronger than previously reported. Twisting to 90° increased the strength, although further twisting to 180° had no significant effect compared with 90° twisting.

Reference(s)

Cooper DE, Deng XH, Burstein AL, Warren RF (1993) The strength of the central third patellar tendon graft. Am J Sports Med 21:818–824 (with permission)

Tensile Property (16)

• Maximum load • Tensile strength	• Rat • Anterior cruciate ligament	• •

Materials

- Wistar strain-derived male albino rats (age, 14 weeks; weight, 393.4 ± 5.2 g [mean ± SE])
- Anterior cruciate ligament (ACL)

Testing Methods and Experimental Conditions

- Tensile test of femur–ACL–tibia complex
- Deformation rate was 3.34 mm/s

Data

Maximum linear load (N)	35.2 ± 1.4
Maximum load (N)	43.0 ± 2.2
Slope of linear load (N/mm)	62.9 ± 1.8
Ultimate stress (MPa)	49.2 ± 2.5
Modulus of elasticity (MPa)	250.7 ± 7.3
Strain to maximum linear load (%)	16.10 ± 0.59
Strain to maximum load (%)	23.83 ± 1.78
Strain at failure (%)	46.20 ± 2.61
Energy absorbed to failure (Nmm)	38.6 ± 4.3

All data are given as mean ± SE, $n = 29$.

Comments

- No description about experimental atmosphere.
- Control data obtained from non-exercised animals are shown above.

Reference(s)

Cabaud HE, Chatty A, Gildengorin V, Feltman RJ (1980) Exercise effect on the strength of the rat anterior cruciate ligament. Am J Sports Med 8:79–86 (with permission)

Tensile Property (17)

<table>
<tr><td>• Stiffness
• Maximum load</td><td>• Dog
• Anterior cruciate ligament</td><td>•
•</td></tr>
</table>

Materials
- German shepherd dogs (weight, 25–30 kg)
- Anterior cruciate ligament (ACL)

Testing Methods and Experimental Conditions
- Tensile test of femur–ACL–tibia complex
- Knee flexion of 45°
- Crosshead speed of 8.467 mm/s, strain rate of approximately 0.4 per second

Data

Number of specimens	10
Initial length (mm)	19.95 ± 2.8
Strain at maximum load (%)	37.2 ± 7.5
Stiffness (N/mm)	188 ± 36
Maximum load (N)	1210 ± 131
Energy to failure (Nmm)	7253 ± 1715

Comments
- No description on testing atmosphere and determination of elongation.

Reference(s)
Cabaud HE, Rodkey WG, Feagin JA (1979) Experimental studies of acute anterior cruciate ligament injury and repair. Am J Sports Med 7:18–22 (with permission)

Tensile Property (18)

• Stiffness • Maximum load	• Human • Anterior cruciate ligament	• Loading direction • Age effect

Materials

- Human younger group (age, 22–35 [mean 29] years); middle group (age, 40–50 [mean 45] years); older group (age, 60–97 [mean 75] years)
- Anterior cruciate ligament (ACL)

Testing Methods and Experimental Conditions

- Tensile test of femur–ACL–tibia complex
- One knee from each pair was oriented anatomically (anatomical orientation), and the contralateral knee was oriented with the tibia aligned vertically (tibial orientation)
- Crosshead speed of 200 mm/s
- Deformation was determined from crosshead displacement

Data

Age group	Specimen orientation	Stiffness (N/mm)	Ultimate load (N)	Energy absorbed (Nm)
Younger	Anatomical	242 ± 28	2160 ± 157	11.6 ± 1.7
	Tibial	218 ± 27	1602 ± 167	8.3 ± 2.0
Middle	Anatomical	220 ± 24	1503 ± 83	6.1 ± 0.5
	Tibial	192 ± 17	1160 ± 104	4.3 ± 0.5
Older	Anatomical	180 ± 25	658 ± 129	1.8 ± 0.5
	Tibial	124 ± 16	495 ± 85	1.4 ± 0.3

All data are given as mean $\pm$ SE, $n=9$.

Comments

- In the younger group, linear stiffness and ultimate load obtained from anatomically oriented ACLs were higher than those reported in the literature.

Reference(s)

Woo SL-Y, Hollis JM, Adams DJ, Lyon RM, Takai S (1991) Tensile properties of human femur–anterior cruciate ligament–tibia complex; the effects of specimen age and orientation. Am J Sports Med, 19:217–225 (with permission)

Tensile Property (19)

• Stiffness • Maximum load	• Monkey • Anterior cruciate ligament	• •

Materials
- Rhesus monkeys (weight, 7–15 kg)
- Anterior cruciate ligament (ACL)

Testing Methods and Experimental Conditions
- Tensile test of femur–ACL–tibia complex
- Knee flexion of 45°
- Crosshead speed of 8.467 mm/s, strain rate of approximately 0.6 per second

Data

Number of specimens	6
Initial length (mm)	13.1 ± 1.6
Strain at maximum load (%)	30.9 ± 8.8
Stiffness (N/mm)	180 ± 28
Maximum load (N)	466 ± 72
Energy to failure (Nmm)	2107 ± 171

Comments
- No description on testing atmosphere and determination of elongation.

Reference(s)
Cabaud HE, Rodkey WG, Feagin JA (1979) Experimental studies of acute anterior cruciate ligament injury and repair. Am J Sports Med 7:18–22 (with permission)

Tensile Property (20)

• Stiffness • Maximum load	• Pig • Anterior cruciate ligament	• Flexion angle • Loading direction

Materials

- Farm pigs (age, approximately 5 months; weight, 60 kg)
- Anterior cruciate ligament (ACL)

Testing Methods and Experimental Conditions

- Tensile test was performed with the knees at 30° and 90° of flexion with the loading direction along either the axis of the tibia (tibia axis) or the axis of the anterior cruciate ligament (ligament axis)
- Crosshead speed of 20 mm/min

Data

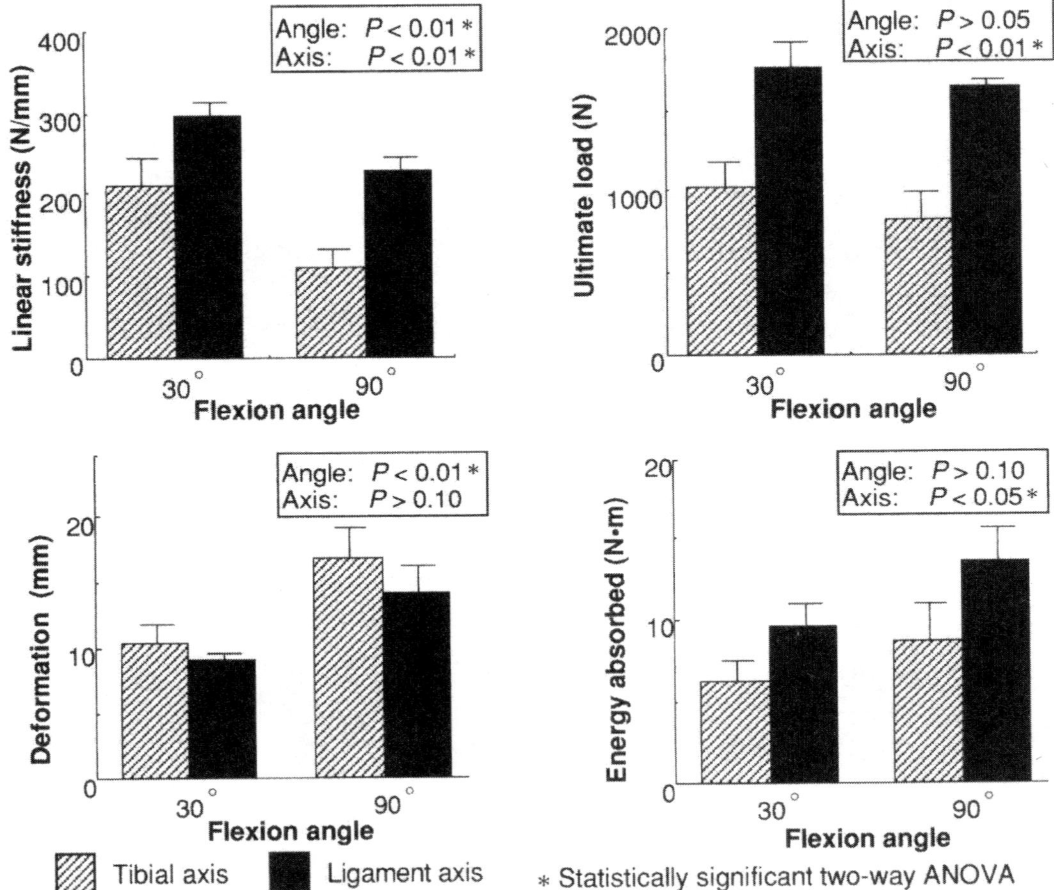

Comments

- Eighteen specimens (75%) tested along the ligament axis failed by bony avulsion while the remaining six specimens (25%) experienced substance failure. Twelve specimens (50%) tested along the tibial axis failed by the ACL "peeling off" the bone at the insertion site, ten (42%) failed by bony avulsion, and two (8%) failed by substance failure.

Reference(s)

Lyon RM, Woo SL-Y, Hollis JM, Marcin JP, Lee EB (1989) A new device to measure the structural properties of the femur–anterior cruciate ligament–tibia complex. Trans ASME J Biomech Eng 111:350–354

Tensile Property (21)

• Stiffness • Tangent modulus	• Dog • Flexor tendon	• Locational dependence •

Materials
- Dogs (mature)
- Flexor digitorum profundus tendon

Testing Methods and Experimental Conditions
- Tensile test of tendon
- Tendons were obtained from five zones (A–E). They range from fibrocartilaginous (zone B) to dense hypocellular collagen bundles (zone D), and from well vascularized (zone A) to avascullar (zone D). Zones of intermediate vascularity and histologic appearance are zones C and E
- Strain rate of 0.0066 per second
- Strain was determined using a photographic method

Data

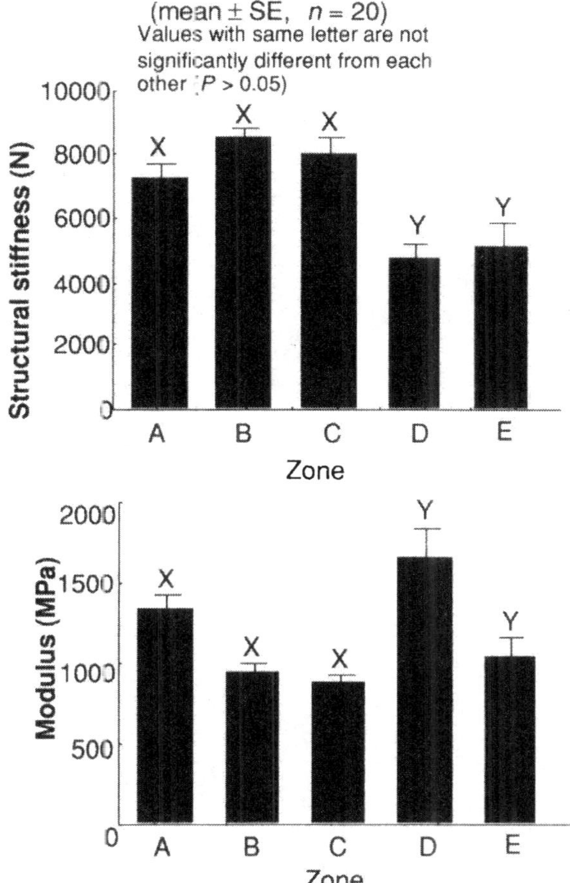

Comments
- Load to ultimate failure represented the strength of the tendon–claw interface because all tendons ruptured at this location.

Reference(s)
Amadio PC, Berglund LJ, An KN (1992) Biochemically discrete zones of canine flexor tendon: Evaluation of properties with a new photographic method. J Orthop Res 10:192–204 (with permission)

Tensile Property (22)

• Stiffness • Tangent modulus	• Goat • Patellar tendon	• Irradiation •

Materials
- Goats (skeletally mature; age, 4–6 years; weight, 50.6 kg)
- Patellar tendon (PT)

Testing Methods and Experimental Conditions
- ^{60}Co gamma irradiation of 2 and 3 Mrad
- Tensile test of bisected patella–PT–tibia complex
- Strain rate of 1 per second
- Moisture with saline solution at room temperature
- Strain was determined using a video dimension analyzer

Data

		Nonirrad.	Irrad. (2 Mrad)	Irrad. (3 Mrad)
Number of specimens		24	12	12
Length (mm)	Pre-irrad.	44.0 ± 3.9	43.8 ± 3.3	44.0 ± 4.6
	Post-irrad.	–	43.8 ± 3.4	43.6 ± 4.6
Area (mm^2)	Pre-irrad.	11.1 ± 2.5	11.3 ± 2.0	10.5 ± 1.6
	Post-irrad.	–	10.7 ± 1.8	9.7 ± 1.5
Width (mm)	Pre-irrad.	5.5 ± 0.8	5.6 ± 0.7	5.3 ± 0.6
	Post-irrad.	–	5.6 ± 0.6	5.4 ± 0.6
Thickness (mm)	Pre-irrad.	3.1 ± 0.4	3.2 ± 0.3	3.1 ± 0.3
	Post-irrad.	–	3.0 ± 0.4	2.9 ± 0.3
Stiffness (N/mm)		189.9 ± 33.6	179.3 ± 23.9	$158.2 \pm 14.3*$
Max. force (N)		1406.1 ± 363.8	1261.8 ± 252.1	$1026.2 \pm 101.2*$
Max. elongation (mm)		10.2 ± 1.5	9.9 ± 1.8	$8.9 \pm 1.3*$
Strain energy (Nm)		8.2 ± 3.2	7.2 ± 3.0	$4.9 \pm 1.1*$
Modulus (MPa)		1639.1 ± 435.9	1658.6 ± 432.2	1560.3 ± 262.0
Max. stress (MPa)		126.8 ± 20.8	117.7 ± 12.9	$107.8 \pm 17.5*$
Max. strain (%)		15.2 ± 3.9	14.0 ± 3.2	$12.3 \pm 2.9*$
Strain energy density (MPa)		11.6 ± 3.3	11.4 ± 2.9	$9.4 \pm 2.8*$
Mechanisms of failure				
Midsubstance		18 (75%)	11 (91.7%)	9 (75%)
Insertion site	Tibial	1 (4.2%)	0	1 (8.3%)
	Patellar	2 (8.3%)	0	1 (8.3%)
Bone avulsion	Tibial	1 (4.2%)	0	0
	Patellar	2 (8.3%)	1 (8.3%)	1 (8.3%)

All data are given as mean ± SD.
*$P < 0.05$ *vs* nonirradiated.

Comments
None.

Reference(s)
Gibbson MJ, Butler DL, Grood ES, Bylski-Austrow DI, Levy MS, Noyes FR (1991) Effects of gamma irradiation on the initial mechanical and material properties of goat bone–patellar tendon–bone allografts. J Orthop Res 9:209–218 (with permission)

Tensile Property (23)

<table>
<tr><td>• Stiffness
• Tangent modulus</td><td>• Human
• Wrist joint ligament</td><td>• Locational dependence
•</td></tr>
</table>

Materials
- Human (age, 63–78 [mean 68.7] years)
- Wrist joint ligament (radioscaphoid collateral [RSC], radiocapitate [RCP], radiolunate [RLP], lunatotriquetrum [LTP], triquetrocapitate [TCP], radiotriquetrum [RTD], triquetrotrapezium [TTD])

Testing Methods and Experimental Conditions
- Tensile test of ligament–bone complex
- Strain rate of 0.66 per second
- Moisture with Ringer's solution at room temperature
- Strain was determined from grip-to-grip deformation

Data

	Stiffness (N/mm)	Tangent modulus (MPa)
RSC	10.0 ± 7.6	22.6 ± 18.1
RCPp	19.3 ± 13.0	102.3 ± 38.0
RCPd	23.9 ± 16.6	67.4 ± 33.1
RCP	27.2 ± 10.8	83.0 ± 34.1
RTPp[a]	15.6 ± 9.0	26.9 ± 21.1
RTPd[a]	18.6 ± 10.0	51.2 ± 33.1
RLPp	32.7 ± 20.2	42.5 ± 19.4
RLPd	26.7 ± 18.0	49.5 ± 27.7
TCP	35.6 ± 17.4	28.9 ± 20.8
RTDp	24.9 ± 11.6	67.9 ± 37.5
RTDd	46.4 ± 24.0	119.3 ± 31.4
RTD	44.4 ± 20.9	93.1 ± 56.6
TTD	14.2 ± 6.1	47.5 ± 25.1

p, proximal; d, distal.

[a] RLP and LTP ligaments were connected in series. The complex was first tested, and its result is referred to by RTP. Subsequently, the LTP ligament was dissected and the RLP ligament was tested.

Comments
- The ligaments of the wrist joint can be divided into roughly two groups. The ligaments of the first group, containing the RTD and the RCP ligaments, have the highest tangent moduli of about 80–90 MPa. The ligaments in the second group, the RSC, RTPp, RTPd, RLPp, RLPd, TCP, and TTD ligament, have values for the tangent moduli ranging between approximately 25 and 50 MPa.

Reference(s)
Savelberg HHCM, Kooloos JGM, Huiskes R, Kauer JMG (1992) Stiffness of the ligaments of the human wrist joint. J Biomech 25:369–376 (with permission)

Tensile Property (24)

<table>
<tr><td>• Stress–strain curve
•</td><td>• Rabbit
• Anterior cruciate ligament</td><td>• Locational dependence
•</td></tr>
</table>

Materials

- Japanese White rabbits (female; weight, 3.8 ± 0.1 kg [mean ± SE])
- Anterior cruciate ligament (ACL)

Testing Methods and Experimental Conditions

- Tensile test of femur–lateral or medial ACL bundle–tibia complex
- ACL bundle was untwisted for uniform load in tension
- Crosshead speed of 20 mm/min
- Saline solution at 37°C
- Strain was determined using a video dimension analyzer

Data

	Medial bundle	Lateral bundle
Cross-sectional area (mm^2)	2.57 ± 0.25	3.92 ± 0.36

All data are given as mean ± SE, $n = 8$.

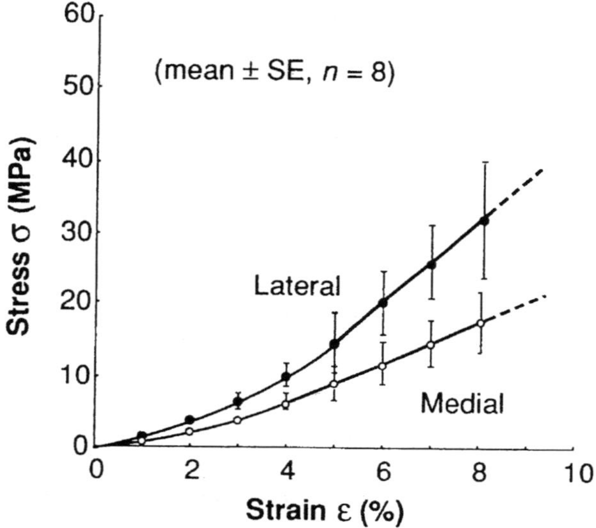

Comments

- Eighty-eight per cent (14/16) of the specimens failed at the femur or tibia insertion sites.
- There was a significant difference in the tangent modulus between the medial and lateral bundles.

Reference(s)

Yamamoto N, Hayashi F, Hayashi K (1992) Mechanical response of rabbit anterior cruciate ligament to overloading. In: Proc 7th Int Conf Biomed Eng, Singapore, 2–4 December, pp 110–112

Tensile Property (25)

• Stress–strain curve • Load–deformation curve	• Rabbit • Medial collateral ligament	• Age effect •

Materials
- New Zealand White rabbits (age, 1.5, 4–5, 6–7, and 12–15 months)
- Medial collateral ligament (MCL)

Testing Methods and Experimental Conditions
- Tensile test of femur–MCL–tibia complex
- Crosshead speed of 1 cm/min
- Saline solution at 37°C
- Strain was determined using a video dimension analyzer

Data

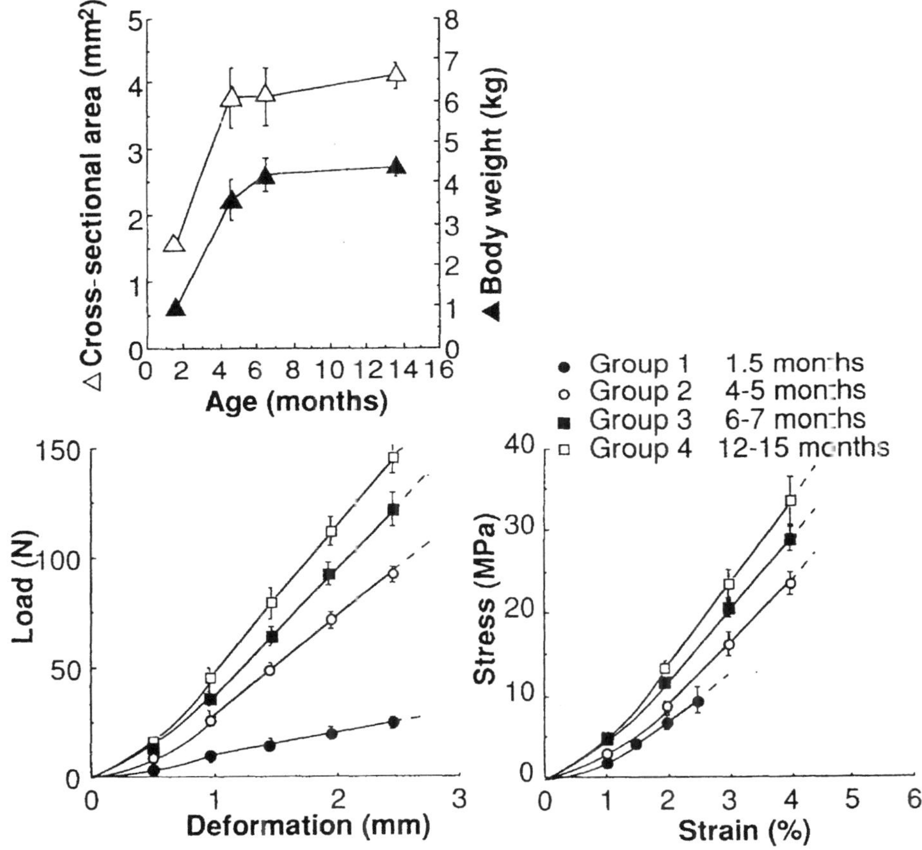

Comments
- For the younger animals with open epiphyses, 100% of the failures were induced by tibial avulsion. For the more mature animals with closed epiphyses, 62% were through the substance of the ligament, 12% were by tibial avulsion, and 26% failed at the femoral transfixing pin site.

Reference(s)
Woo SL-Y, Orlando CA, Gomez MA, Frank CB, Akeson WH (1986) Tensile properties of the medial collateral ligament as a function of age. J Orthop Res 4:133–141 (with permission)

Tensile Property (26)

• Stress–strain curve • Maximum load	• Pig • Flexor tendon	• •

Materials
- Miniature Yucatan swine (age, 2 years; weight, 77 ± 16 kg)
- Digital flexor tendon

Testing Methods and Experimental Conditions
- Tensile test of tendon–bone composite
- Crosshead speed of 2 cm/min
- Saline solution at 37°C
- Strain was determined using a video dimension analyzer
- Extension was determined from clamp to clamp deformation

Data

Cross-sectional area (mm^2)	17.3 ± 0.6
Load to failure (N)	1630 ± 70
Extension at failure (cm)	1.3 ± 0.1
Linear slope ($\times 10^5$ N/m)	2.76 ± 0.12

All data are given as mean ± SE, $n = 18$.

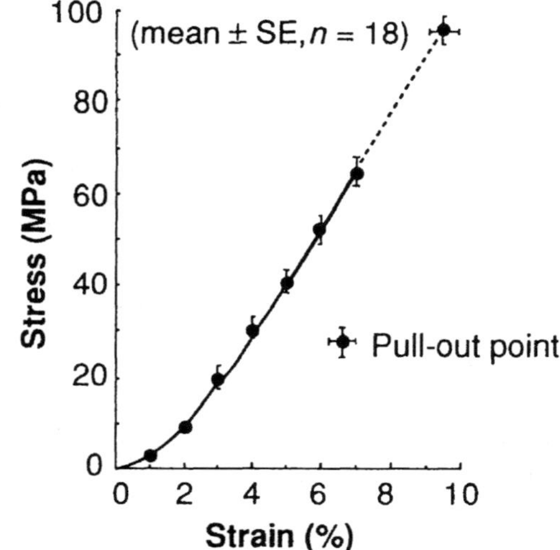

Comments
- All tensile failure occured at the tendon insertion to bone.
- Control data obtained from non-exercised animals are shown above.

Reference(s)
Woo SL-Y, Gomez MA, Amiel D, Ritter MA, Gelberman RH, Akeson WH (1981) The effects of exercise on the biomechanical and biochemical properties of swine digital flexor tendons. Trans ASME J Biomech Eng 103:51–56

Tensile Property (27)

• Stress–strain curve • Tangent modulus	• Human • Lateral collateral ligament	• •

Materials
• Human

Donor no.	Age	Sex	Height (m)	Mass (kg)	Cause of death
1	30	F	1.63	65.8	Drug overdose
2	30	M	1.73	71.7	Electrocution
3	21	F	1.72	83.9	Suicide

• Lateral collateral ligament (LCL)

Testing Methods and Experimental Conditions
• Tensile test of LCL fascicle–bone unit
• Strain rate of 1.0 per second
• Saline solution at 37°C
• Strain was determined from actuator displacement

Data

Donor no.	Number of fascicle	Initial length (mm)	Initial area (mm^2)
1	2	50.9 ± 2.4	1.54 ± 0.21
2	2	52.9 ± 1.9	2.98 ± 0.07
3	2	48.7 ± 0.6	3.73 ± 0.23

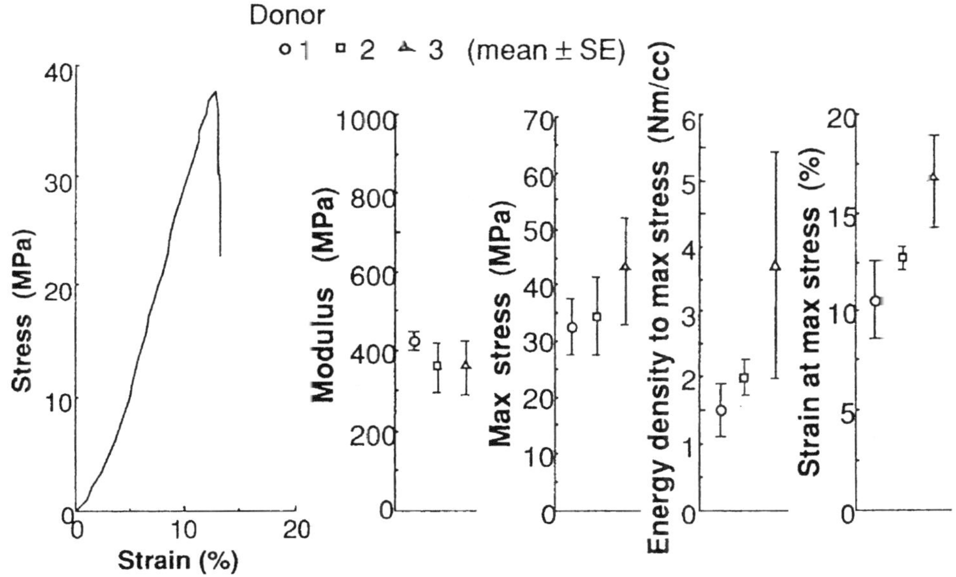

Comments
• The stress–strain curve is representative data.
• Each specimen appeared to fail through soft tissue with no evidence of bone avulsion.

Reference(s)
Butler DL, Kay MD, Stouffer DC (1986) Comparison of material properties in fascicle–bone units from human patellar tendon and knee ligaments. J Biomech 19:425–432 (with permission)

Tensile Property (28)

• Stress–strain curve • Tangent modulus	• Human • Anterior cruciate ligament	• •

Materials
• Human

Donor no.	Age	Sex	Height (m)	Mass (kg)	Cause of death
1	30	F	1.63	65.8	Drug overdose
2	30	M	1.73	71.7	Electrocution
3	21	F	1.72	83.9	Suicide

• Anterior cruciate ligament (ACL)

Testing Methods and Experimental Conditions
• Tensile test of ACL fascicle–bone unit
• Strain rate of 1.0 per second
• Saline solution at 37°C
• Strain was determined from actuator displacement

Data

Donor no.	Number of fascicle	Initial length (mm)	Initial area (mm^2)
1	3	25.4 ± 3.7	1.29 ± 0.13
2	3	31.7 ± 2.4	1.17 ± 0.39
3	3	29.1 ± 3.1	1.91 ± 0.09

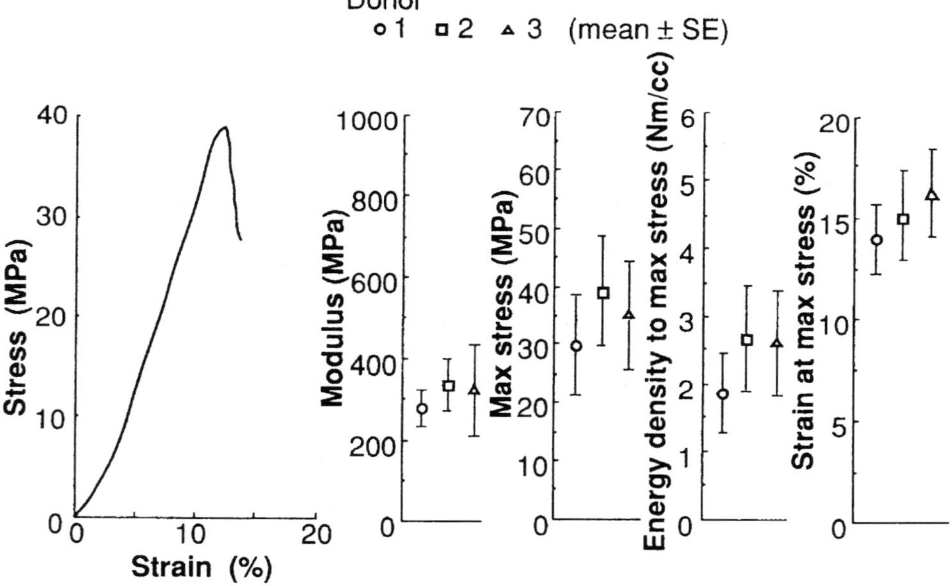

Comments
• The stress–strain curve is representative data.
• Each specimen appeared to fail through soft tissue with no evidence of bone avulsion.

Reference(s)
Butler DL, Kay MD, Stouffer DC (1986) Comparison of material properties in fascicle–bone units from human patellar tendon and knee ligaments. J Biomech 19:425–432 (with permission)

Tensile Property (29)

• Stress–strain curve • Tangent modulus	• Human • Patellar tendon	• •

Materials
• Human

Donor no.	Age	Sex	Height (m)	Mass (kg)	Cause of death
1	30	F	1.63	65.8	Drug overdose
2	30	M	1.73	71.7	Electrocution
3	21	F	1.72	83.9	Suicide

• Patellar tendon (PT)

Testing Methods and Experimental Conditions
• Tensile test of PT fascicle–bone unit
• Strain rate of 1.0 per second
• Saline solution at 37°C
• Strain was determined from actuator displacement

Data

Donor no.	Number of fascicle	Initial length (mm)	Initial area (mm^2)
1	5	44.5 ± 1.1	1.20 ± 0.28
2	8	57.8 ± 1.4	1.86 ± 0.31
3	6	55.6 ± 2.4	1.65 ± 0.21

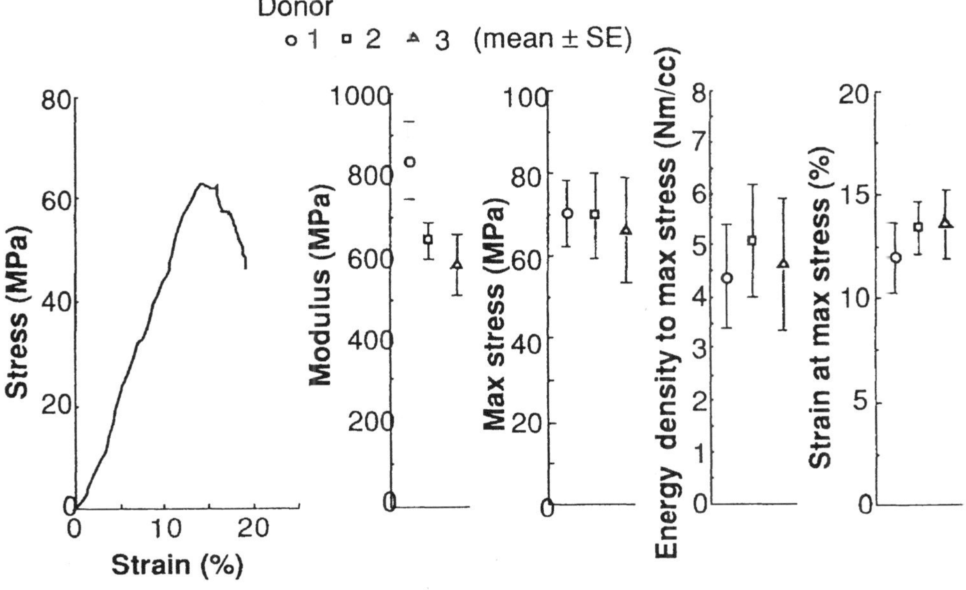

Comments
• The stress–strain curve is representative data.
• Each specimen appeared to fail through soft tissue with no evidence of bone avulsion.

Reference(s)
Butler DL, Kay MD, Stouffer DC (1986) Comparison of material properties in fascicle–bone units from human patellar tendon and knee ligaments. J Biomech 19:425–432 (with permission)

Tensile Property (30)

• Stress–strain curve • Tangent modulus	• Human • Patellar tendon	• Irradiation •

Materials
- Human (male; age, 34.7 ± 8.2 years)
- Patellar tendon (PT)

Testing Methods and Experimental Conditions
- Tensile test of patella–one half PT–tibia complex
- Strain rate of 1 per second
- 37°C bath of 0.15 M NaCl or moist by a drip of the saline solution
- Irradiation at 2.0 Mrad

Data

	Bath		Drip	
	Irrad.	Nonirrad.	Irrad.	Nonirrad.
Tangent modulus (MPa)	194 ± 22	307 ± 17	256 ± 21	191 ± 16
Peak failure stress (MPa)	34.5 ± 3.4	43.7 ± 3.9	37.6 ± 4.4	30.6 ± 2.6
Failure strain (%)	30.4 ± 2.0	23.4 ± 1.4	28.4 ± 2.2	26.6 ± 1.0

All data are given as mean ± SE, $n = 6$.

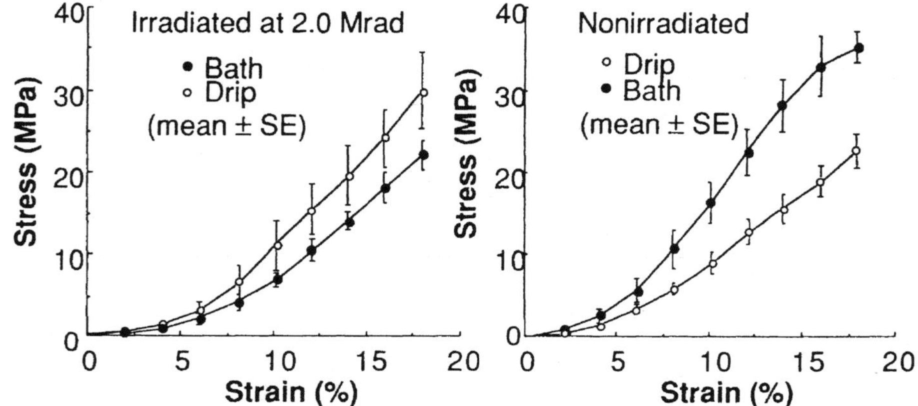

Comments
- The irradiated patellar tendons were visibly and histologically crimped, and the fascicles of collagen were separated.

Reference(s)
Haut RC, Powlison AC (1990) The effects of test environment and cyclic stretching on the failure properties of human patellar tendon. J Orthop Res 8:532–540 (with permission)

Tensile Property (31)

• Stress–strain curve • Tangent modulus	• Human • Posterior cruciate	• •

Materials
• Human

Donor no.	Age	Sex	Height (m)	Mass (kg)	Cause of death
1	30	F	1.63	65.8	Drug overdose
2	30	M	1.73	71.7	Electrocution
3	21	F	1.72	83.9	Suicide

• Posterior cruciate ligament (PCL)

Testing Methods and Experimental Conditions
• Tensile test of PCL fascicle–bone unit
• Strain rate of 1.0 per second
• Saline solution at 37°C
• Strain was determined from actuator displacement

Data

Donor no.	Number of fascicle	Initial length (mm)	Initial area (mm²)
1	3	32.0 ± 4.5	1.91 ± 0.24
2	2	31.4 ± 0.3	1.12 ± 0.45
3	3	29.5 ± 2.7	2.40 ± 0.24

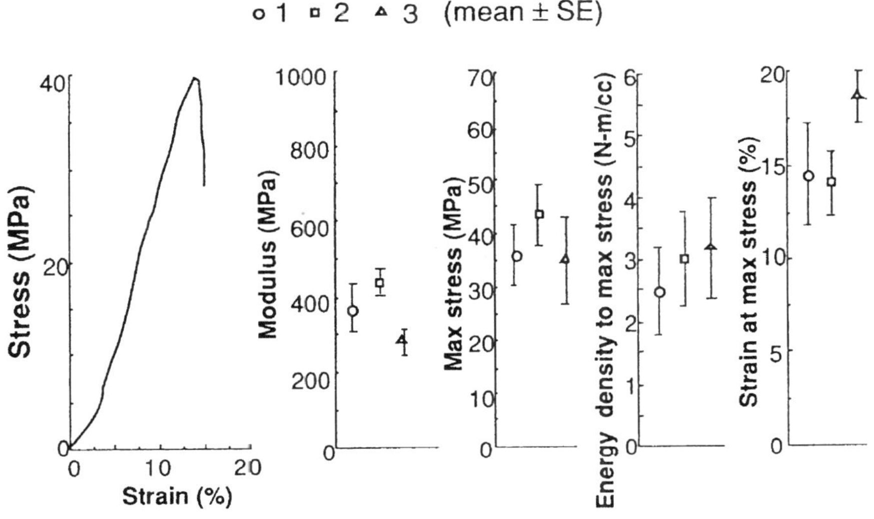

Comments
• The stress–strain curve is representative data.
• Each specimen appeared to fail through soft tissue with no evidence of bone avulsion.

Reference(s)
Butler DL, Kay MD, Stouffer DC (1986) Comparison of material properties in fascicle–bone units from human patellar tendon and knee ligaments. J Biomech 19:425–432 (with permission)

Tensile Property (32)

• Stress–strain curve • Tangent modulus	• Rabbit • Anterior cruciate ligament	• Strain rate •

Materials

- New Zealand White rabbits (male; skeletally mature [closed epiphyses]; age, 8.5 months; weight, 3.9 ± 0.1 kg [mean $\pm$ SE])
- Anterior cruciate ligament (ACL)

Testing Methods and Experimental Conditions

- Tensile test of femur–medial portion of ACL–tibia complex
- Extension rate of 0.003–113 mm/s, strain rate of 0.00016–3.81 per second
- Moisture with saline solution
- Strain was determined using a high-speed video motion analysis system

Data

Group	Cross-sectional area (mm^2)	Modulus (MPa)	AB[a] (MPa)	B[a]
Slow (0.003 mm/s)	3.4 ± 0.1	711 ± 18	125 ± 22	54 ± 5
Medium (0.33 mm/s)	3.7 ± 0.4	674 ± 62	68 ± 16	$81 \pm 1 2$
Fast (113 mm/s)	3.3 ± 0.2	930 ± 71	200 ± 55	61 ± 9

All data are given as mean $\pm$ SE, $n = 6$.
[a] Stress(σ) – strain(ε) curve was expressed by
$d\sigma/d\varepsilon = B\sigma + AB$ (AB, initial modulus; B, rate of change of the modulus).

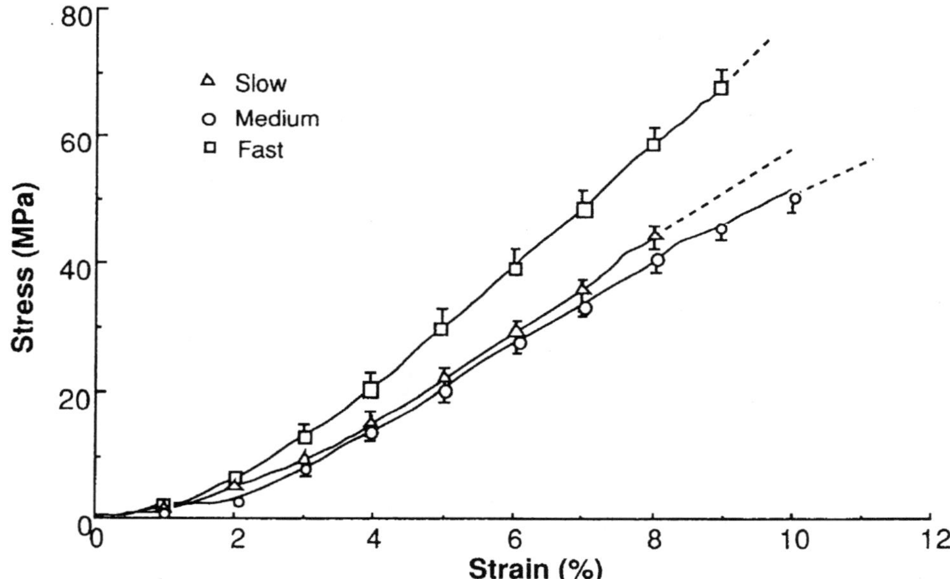

Comments

- Ninety percent of the ACL specimens failed by avulsion at the insertion site (72% at the tibial insertion and 18% at the femoral insertion), and 10% failed in the midsubstance.

Reference(s)

Danto MI, Woo SL-Y (1993) The mechanical properties of skeletally mature rabbit anterior cruciate ligament and patellar tendon over a range of strain rates. J Orthop Res 11:58–67 (with permission)

Tensile Property (33)

• Stress–strain curve • Tangent modulus	• Rabbit • Patellar tendon	• Strain rate •

Materials
- New Zealand White rabbits (male; skeletally mature [closed epiphyses]; age, 8.5 months; weight, 3.9 ± 0.1 kg)
- Patellar tendon (PT)

Testing Methods and Experimental Conditions
- Tensile test of quadriceps tendon–middle third of PT–tibia complex
- Extension rates of 0.008–113 mm/s, strain rates of 0.00016–1.35 per second
- Moisture with saline solution
- Strain was determined using a high-speed video motion analysis system

Data

Group	Cross-sectional area (mm^2)	Modulus (MPa)	AB [a] (MPa)	B [a]
low (0.008 mm/s)	4.4 ± 0.2	955 ± 97	159 ± 18	120 ± 6
medium (0.83 mm/s)	4.0 ± 0.2	1637 ± 132	187 ± 31	146 ± 5
ast (113 mm/s)	3.6 ± 0.1	1855 ± 77	384 ± 32	109 ± 3

All data are given as mean ± SE, $n = 6$.

Stress(σ)–strain(ε) curve was expressed by

$$d\sigma/d\varepsilon = B\sigma + AB \quad \text{(AB, initial modulus; B, rate of change of the modulus)}.$$

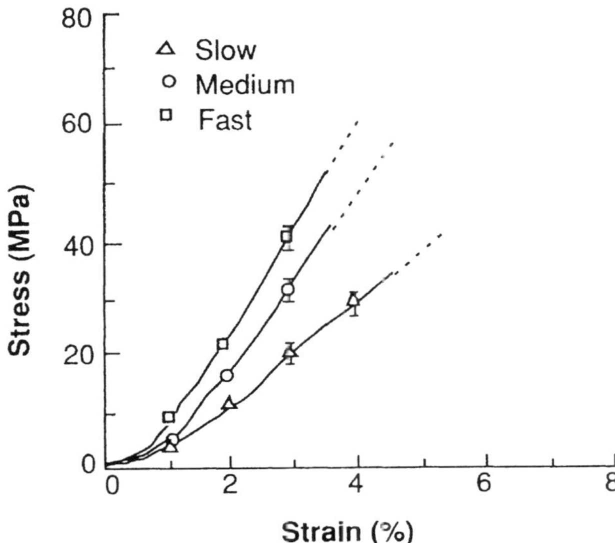

Comments
- All but one (94%) of the PT specimens failed by avulsion, the other specimen failed in the midsubstance of the PT.

Reference(s)

Danto MI, Woo SL-Y (1993) The mechanical properties of skeletally mature rabbit anterior cruciate ligament and patellar tendon over a range of strain rates. J Orthop Res 11:58–67 (with permission)

Tensile Property (34)

• Stress–strain curve • Tensile strength	• Dog • Anterior cruciate ligament	• •

Materials

- Mongrel dogs (adult, weight, 9.95 ± 1.81 kg [mean ± SD])
- Anterior cruciate ligament (ACL)

Testing Methods and Experimental Conditions

- Tensile test of femur–central one third ACL–tibia complex
- ACL bundle was untwisted for uniform load in tension
- Crosshead speed of 20 mm/min
- Moisture with saline solution at room temperature
- Strain was determined using a video dimension analyzer

Data

Tensile strength (MPa)	Tangent modulus (MPa)
150.3 ± 28.2	947.5 ± 228.6

All data are given as mean ± SD, $n = 6$.

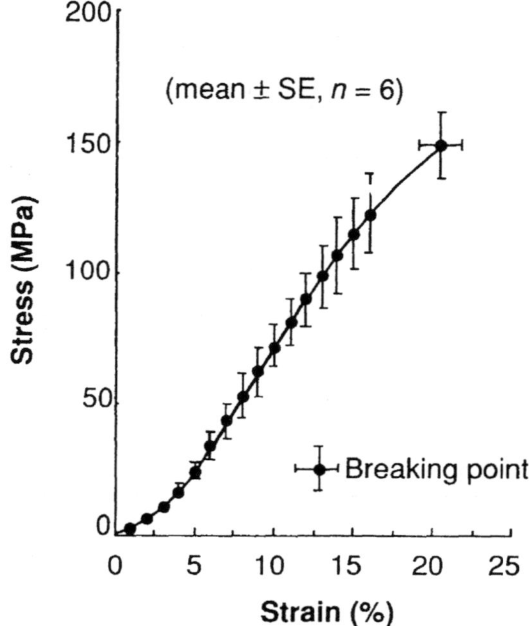

Comments

- All specimens failed in tendon substance.

Reference(s)

Keira M, Yasuda K, Kaneda K, Yamamoto N, Hayashi K (in press) Mechanical properties of the anterior cruciate ligament chronically relaxed by elevation of the tibial insertion. J Orthop Res (with permission)

Tensile Property (35)

• Stress–strain curve • Tensile strength	• Dog • Medial collateral ligament	• •

Materials

- Dogs (mature; weight, 22 ± 1 kg [mean ± SE])
- Medial collateral ligament (MCL)

Testing Methods and Experimental Conditions

- Tensile test of femur–MCL–tibia complex
- Strain rate of 0.24 per minute
- Saline solution at 37°C
- Strain was determined using a video dimension analyzer

Data

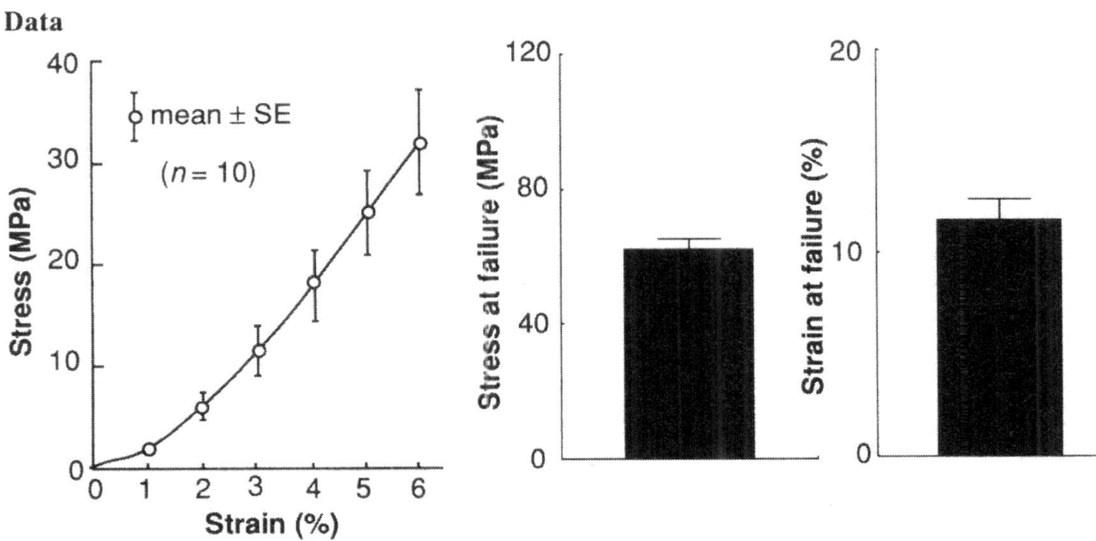

Comments

- All specimens failed at the mid-ligament substance.

Reference(s)

Woo SL-Y, Gomez MA, Seguchi Y, Endo CM, Akeson WH (1983) Measurement of mechanical properties of ligament substance from a bone–ligament–bone preparation. J Orthop Res 1:22–29 (with permission)

Tensile Property (36)

• Stress–strain curve • Tensile strength	• Pig • Extensor tendon	• •

Materials

- Miniature Yucatan swine (age, 2 years; weight, 77 ± 16 kg)
- Medial and lateral digital extensor tendons

Testing Methods and Experimental Conditions

- Tensile test of tendon
- Crosshead speed of 2 cm/min
- Saline solution at 37°C
- Strain was determined using a video dimension analyzer

Data

	Medial	Lateral
Maximum load (N)*	200 ± 20	290 ± 20
Maximum energy stored (Nm)*	0.76 ± 0.07	1.55 ± 0.22

All data are given as mean ± SE, $n = 9$.
*$P < 0.001$.

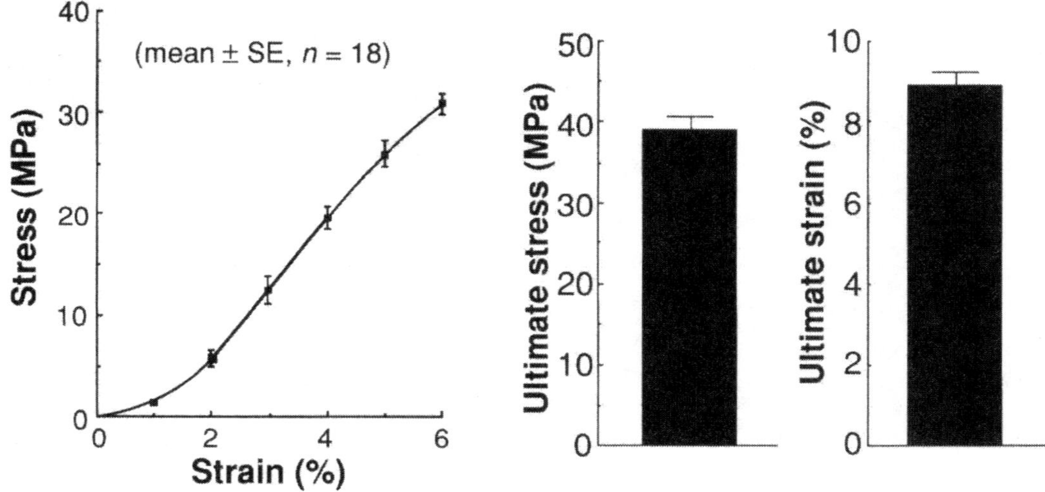

Comments

- There were no significant differences in the mechanical properties (ultimate stress and ultimate strain) between the medial and lateral tendons.
- Control data obtained from non-exercised animals are shown above.

Reference(s)

Woo SL-Y, Ritter MA, Amiel D, Sanders TM, Gomez MA, Kuei SC, Garfin SR, Akeson WH (1980) The biomechanical and biochemical properties of swine tendons – long-term effects of exercise on the digital extensors. Connect Tissue Res 7:177–183 (with permission)

Tensile Property (37)

• Stress–strain curve • Tensile strength	• Pig • Medial collateral ligament	• •

Materials
- Yucatan mini-swine (weight, 45 ± 2 kg [mean ± SE])
- Medial collateral ligament (MCL)

Testing Methods and Experimental Conditions
- Tensile test of femur–MCL–tibia complex
- Strain rate of 0.24 per minute
- Saline solution at 37°C
- Strain was determined using a video dimension analyzer

Data

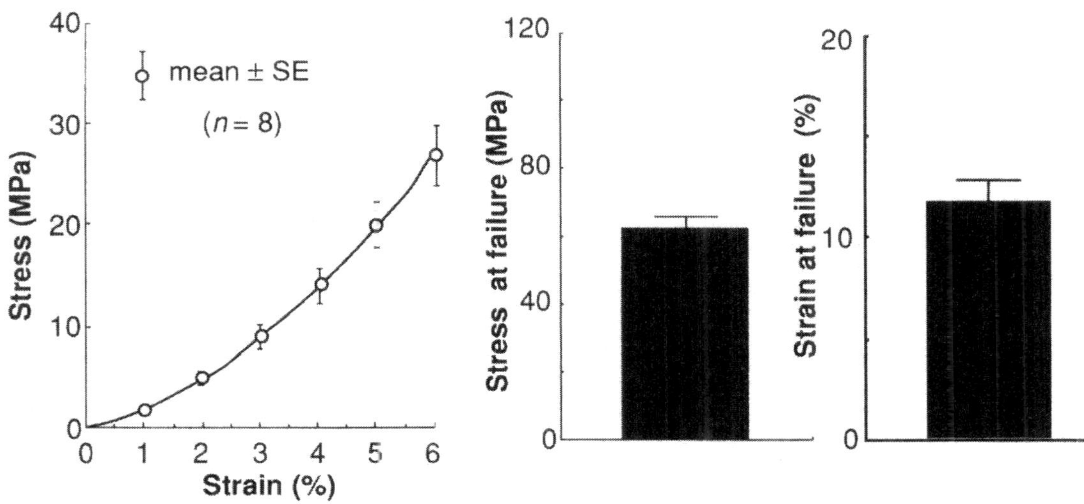

Comments
- One of eight specimens failed at the mid-ligament substance. The remaining specimens failed with a combination of ligament substance tear together with tibial avulsion.

Reference(s)
Woo SL-Y, Gomez MA, Seguchi Y, Endo CM, Akeson WH (1983) Measurement of mechanical properties of ligament substance from a bone–ligament–bone preparation. J Orthop Res 1:22–29 (with permission)

Tensile Property (38)

• Stress–strain curve • Tensile strength	• Rabbit • Achilles tendon	• Age effect •

Materials

- Japanese White rabbits (male; younger group (age, 8–10 month; weight, 4.1 ± 0.4 kg [mean ± SE]); older group (age, 4–5 years; weight, 5.1 ± 0.2 kg))
- Achilles tendon (lateral gastrocnemius tendon [LGT], medial gastrocnemius [MGT], and soleus tendon [ST])

Testing Methods and Experimental Conditions

- Tensile test of tendon
- Crosshead speed of 20 mm/min
- Moisture with saline solution at room temperature
- Strain was determined using a video dimension analyzer

Data

	Cross-sectional area (mm^2)	Tensile strength (MPa)	Elongation at failure (%)	Tangent modulus (MPa)
Younger				
LGT ($n = 4$)	5.3 ± 0.3*#	70.3 ± 3.5	15.7 ± 3.7	550.3 ± 113.9
MGT ($n = 4$)	4.1 ± 0.3	62.3 ± 13.3	9.1 ± 1.2*	791.6 ± 200.3*
ST ($n = 5$)	4.1 ± 0.3	68.9 ± 2.1	22.8 ± 5.0	533.4 ± 135.9
Older				
LGT ($n = 5$)	5.7 ± 0.3*#	64.5 ± 4.5	12.3 ± 3.2	709.7 ± 254.9
MGT ($n = 5$)	4.7 ± 0.2	65.5 ± 13.0	16.3 ± 1.8**	477.3 ± 131.8
ST ($n = 5$)	4.6 ± 0.2	73.0 ± 10.0	19.7 ± 3.4	432.6 ± 36.4

All data are given as mean ± SE.

*$P < 0.001$ *vs* ST; #$P < 0.001$ *vs* MGT; **$P < 0.001$ *vs* younger.

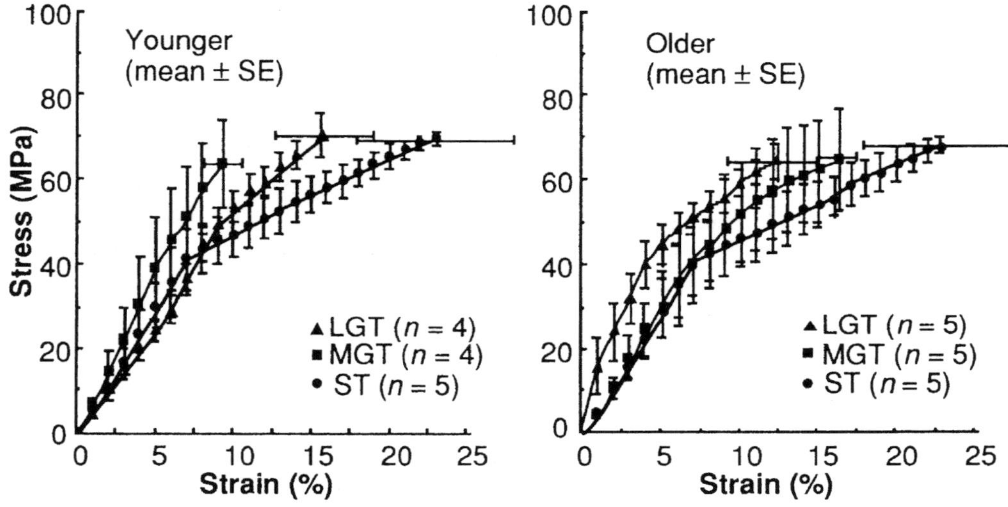

Comments

- All specimens failed in the midsubstance between the two markers drawn on the specimen surface for the strain measurement.

Reference(s)

Nakagawa Y, Hayashi K, Yamamoto N, Nagashima K (1994) The mechanical properties of the white and red muscle tendon in the rabbit Achilles tendon and aging effects. J Exercise Sport Physiol 1:41–46 (with permission)

Tensile Property (39)

• Stress–strain curve • Tensile strength	• Rabbit • Patellar tendon	• Age effect •

Materials
- Japanese White rabbits (female; age, 1, 2, and 6 months)
- Patellar tendon (PT)

Testing Methods and Experimental Conditions
- Tensile test of patella–central one third PT–tibia complex
- Strain rate from crosshead displacement, 0.0167 per second
- Saline solution at 37°C
- Strain was determined using a video dimension analyzer

Data

	1 month	2 months	6 months
Body weight (kg)*	0.6 ± 0.1	1.9 ± 0.2	4.1 ± 0.1
Whole cross-sectional area (mm^2)*	3.5 ± 0.2	8.8 ± 0.4	$12.1 = 0.7$
Length (mm)*	10.5 ± 0.3	14.7 ± 0.5	19.9 ± 0.2
Tangent modulus (MPa)*	285 ± 62	900 ± 80	1310 ± 33
Tensile strength (MPa)*	29.9 ± 1.7	40.8 ± 2.5	$58.6 = 3.7$
Strain at failure (%)	7.2 ± 0.8	5.1 ± 0.3	$5.3 = 0.3$

All data are given as mean $\pm$ SE, $n = 5$.
* $P < 0.05$ between two of three ages.

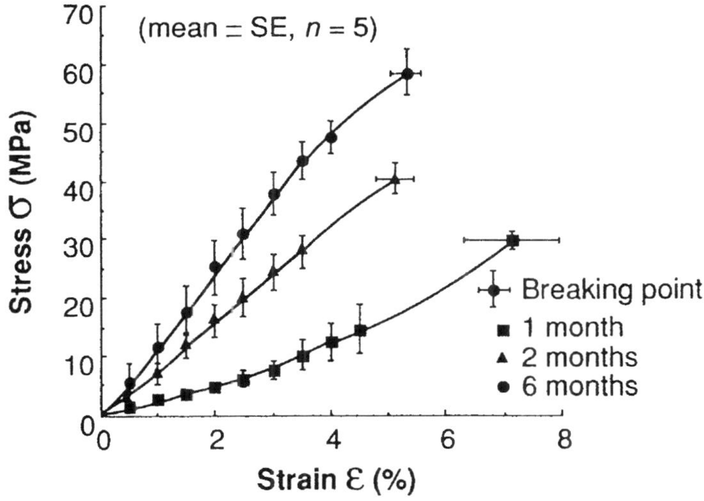

Comments
- In the 1 month group, 20% (1/5) of specimens failed in tendon substance, while 80% (4/5) failed at patellar insertion site. In the 2 months group, 60% (3/5) of specimens failed in tendon substance, while 40% (2/5) failed at patellar insertion site. In the 6 months group, all specimens failed in tendon substance.

Reference(s)
Murakami T, Yamamoto N, Hayashi K (1995) Effect of maturation on the in vivo tension and mechanical properties of the rabbit patellar tendon. In: Proc 4th Conf Biomech, Nagoya, 19–20 January, pp 15–16

Tensile Property (40)

• Stress–strain curve • Tensile strength	• Rabbit • Patellar tendon	• Collagen fascicle •

Materials

- Japanese White rabbits (female; weight, 3.8 ± 0.1 kg [mean ± SE])
- Patellar tendon (PT)

Testing Methods and Experimental Conditions

- PT was divided into 6 blocks (location number: 1, anterior-medial; 2, anterior-central; 3, anterior-lateral; 4, posterior-medial; 5, posterior-central; 6, posterior-lateral)
- Tensile testing of collagen fascicles dissected from each block (diameter, approximately 300 µm; length, 15 mm)
- Crosshead speed of 10 mm/min
- Saline solution at 37°C
- Strain was determined using a video dimension analyzer

Data

Location number	Tensile strength (MPa)	Strain at failure (%)	Tangent modulus (MPa)
1	16.5 ± 1.28	11.2 ± 0.18	206 ± 18
2	16.9 ± 1.46	10.0 ± 0.62	228 ± 27
3	16.0 ± 1.46	10.4 ± 0.91	211 ± 26
4	17.8 ± 1.74	10.9 ± 0.92	224 ± 35
5	19.3 ± 1.58	12.0 ± 0.51	222 ± 20
6	16.5 ± 1.86	10.8 ± 0.54	206 ± 29

All data are given as mean ± SE, $n = 7$.

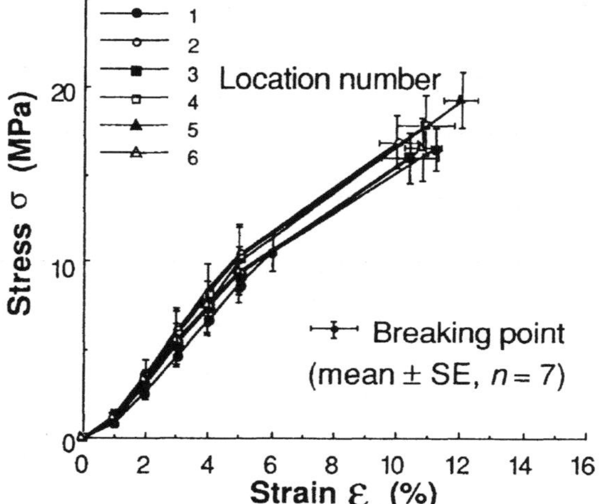

Comments

- There were no significant differences in the mechanical properties among the 6 locations.
- The mechanical properties of collagen fascicles obtained from the PTs were greatly different from those of the whole PTs.

Reference(s)

Yamamoto E, Hayashi K, Yamamoto N (1995) Mechanical properties of collagen fascicles of stress-shielded patellar tendons in the rabbit. In: Proc 1995 Summer Bioeng Conf, Beaver Creek, CO, 28 June–2 July, pp 199–200

Tensile Property (41)

• Stress–strain curve • Tensile strength	• Rabbit • Patellar tendon	• Locational dependence •

Materials
- Japanese White rabbits (female; weight, 4.0 ± 0.1 kg [mean $\pm$ SE])
- Patellar tendon (PT)

Testing Methods and Experimental Conditions
- Tensile test of patella–lateral, medial, or central 1/3 PT–tibia complex
- Crosshead speed of 20 mm/min
- Saline solution at 37°C
- Strain was determined using a video dimension analyzer

Data

	Lateral ($n = 7$)	Medial ($n = 7$)	Central ($n = 14$)
Cross-sectional area (mm^2)	4.92 ± 0.22	5.18 ± 0.30	4.27 ± 0.12
Length (mm)	22.7 ± 0.4	23.1 ± 0.4	20.6 ± 0.3
Tangent modulus (MPa)	1420 ± 97	1330 ± 130	1390 ± 53
Tensile strength (MPa)	66.4 ± 3.2*	55.5 ± 3.0	57.1 ± 2.5
Elongation at failure (%)	5.7 ± 1.0	5.1 ± 0.5	5.3 ± 0.2

All data are given as mean $\pm$ SE.
* $P < 0.05$ vs central.

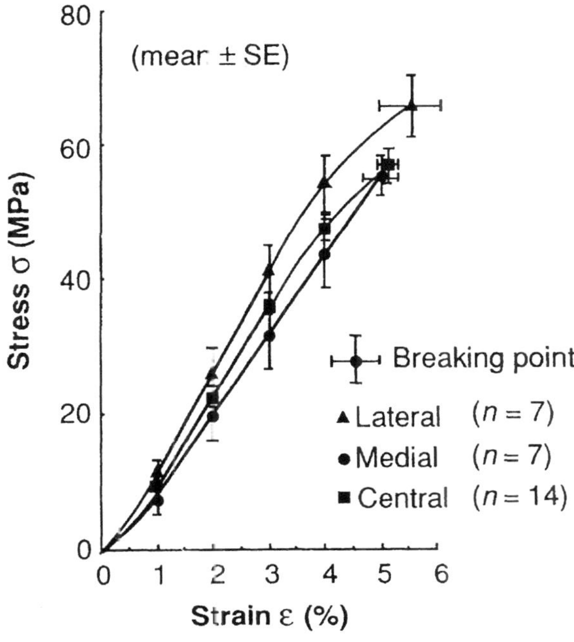

Comments
- Sixty-four per cent (9/14) of the central specimens failed in tendon substance, while 36% (5/14) failed at patella insertion sites. Similarly, 71% (5/7) of the medial and lateral specimens failed in tendon and 29% (2/7) at patella insertion sites.

Reference(s)
Yamamoto N, Hayashi K, Kuriyama H, Ohno K, Yasuda K, Kaneda K (1992) Mechanical properties of the rabbit patellar tendon. Trans ASME J Biomech Eng 114:332–337

Tensile Property (42)

• Stress–strain curve • Tensile strength	• Rabbit • Patellar tendon	• Strain rate •

Materials

- Japanese White rabbits (female; weight, 3.8 ± 0.1 kg [mean ± SE])
- Patellar tendon (PT)

Testing Methods and Experimental Conditions

- Tensile test of patella–central one third PT–tibia complex
- Extension rate of 0.33–560 mm/s, strain rate of 0.0057–13 per second
- Saline solution at 37°C
- Strain was determined using a high-speed video dimension analyzer

Data

Extension rate (mm/s)	Tensile strength* (MPa)	Strain at failure* (%)	Tangent modulus (MPa)
0.33	56.9 ± 4.2	5.2 ± 0.4	1340 ± 110
53	76.9 ± 2.3	7.4 ± 0.4	1050 ± 120
560	86.2 ± 3.6	9.2 ± 0.7	1040 ± 90

All data are given as mean ± SE, $n = 5$.
* $P < 0.05$.

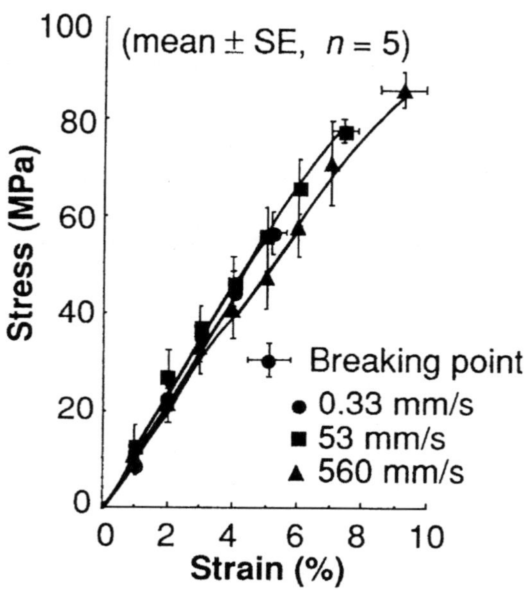

Comments

- In the case of extension rate of 0.33 mm/s, 60% of specimens failed in the tendon substance, while 40% failed at the patella insertion. At the highest extension rate (560 mm/s), 100% specimens failed in the tendon substance.

Reference(s)

Yamamoto N, Hayashi K (1993) Effects of strain rate on the mechanical properties of rabbit patellar tendon. In: Proc 3rd Conf Biomech, Tokyo, 20–22 January, pp 63–64

Tensile Property (43)

• Stress–strain relationship • Tangent modulus	• Human • Glenohumeral ligament	• Locational dependence •

Materials

- Human (9 male and 7 female; age, 56–87 years; 10 right and 6 left shoulders)
- Inferior glenohumeral ligament (superior band, anterior axillary pouch, posterior axillary pouch)

Testing Methods and Experimental Conditions

- Tensile test of bone–ligament–bone complex
- Elongation rate of 0.04 mm/s (~ 0.001 per second) or 0.004 mm/s (~0.0001 per second)
- Physiologic saline with protease inhibitors at room temperature
- Strain (total specimen strain) was calculated from the change in bone–ligament–bone specimen length measured from axial motion of actuator
- Strain was determined using a video dimension analyzer

Data

	Geometric data			Stress–strain data at failure		
Ligament region	Thickness (mm)	Width (mm)	Length (mm)	Stress (MPa)	Strain[a] (%)	Strain[b] (%)
Superior (n=16)	2.79 ± 0.49	13.33 ± 2.66	41.3 ± 4.5	5.2 ± 2.7	24.0 ± 6.2	8.3 ± 3.2
Anterior (n=16)	2.34 ± 0.43	12.61 ± 3.05	39.8 ± 5.6	5.5 ± 2.0	34.0 ± 10.5	15.1 ± 5.7
Posterior (n=16)	1.70 ± 0.55	10.86 ± 2.94	41.0 ± 4.2	5.6 ± 1.9	23.1 ± 4.6	9.9 ± 5.3
Average (n=48)	2.28 ± 0.66	12.27 ± 2.88	40.7 ± 4.7	5.5 ± 2.2	27.0 ± 8.9	10.9 ± 5.5

All data are given as mean ± SD.
[a] Total specimen strain.
[b] Video dimension analyzer.

Ligament region	A[a] (MPa)	B[a]	C[a] (MPa)	E[b] (MPa)
Superior (n=16)	0.43 ± 0.24	14.15 ± 4.17	5.50 ± 2.73	38.74 ± 18.09
Anterior (n=16)	0.55 ± 0.52	8.84 ± 3.03	4.38 ± 3.32	30.33 ± 10.58
Posterior (n=16)	0.92 ± 0.60	11.49 ± 4.99	8.42 ± 3.88	41.91 ± 12.50
Average (n=48)	0.64 ± 0.52	11.44 ± 4.60	6.11 ± 3.72	36.92 ± 14.55

All data are given as mean ± SD.
[a] Stress(σ)–strain(ε) relationship was expressed by $\sigma = A(e^{B\varepsilon} - 1)$, $d\sigma/d\varepsilon = B\sigma + C$.
[b] Young's modulus.
Total specimen strain data were used for the analysis.

Comments

- Three failure sites were seen for the inferior glenohumeral ligament: the glenoid insertion (40%), the ligament substance (35%), and the humeral insertion (25%).

Reference(s)

Bigliani LU, Pollock RG, Soslowsky LJ, Flatow EL, Pawluk RJ, Mow VC (1992) Tensile properties of the inferior glenohumeral ligament. J Orthop Res 10:187–197 (with permission)

Tensile Property (44)

• Tangent modulus •	• Human • Flexor tendon	• •

Materials
- Human
- Flexor digitorum profundus (FDP), extensor digitorum longus (EDL), palmaris longus, and plantaris tendon

Testing Methods and Experimental Conditions
- Cross-sectional area was measured with laser micrometer
- Tensile testing was carried out using universal testing machine
- Strain measurements were made with video dimension analyzer
- Testing was performed in 0.9% NaCl solution bath at 37°C
- Modulus was determined between 2.5% and 4.5% strain

Data

Tendon	n	Length (mm)	Area (mm^2)	Volume (mm^3)
Extensor digitorum longus	10	324.8 ± 12.6	3.3 ± 0.4	1005.6 ± 105.6
Palmaris longus	10	160.5 ± 4.7	3.1 ± 0.3	529.4 ± 51.0
Plantaris	11	334.1 ± 11.4	1.4 ± 0.3	557.3 ± 77.9

All data are given as mean ± SE.

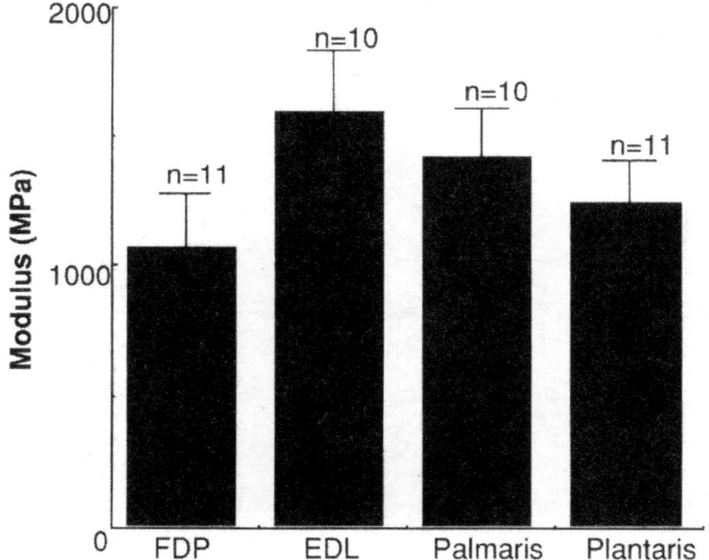

Comments
- No description on the dimensions of FDP tendon.
- There was no significant difference in the modulus among the four tendons.

Reference(s)
Carlson GD, Botte MJ, Josephan MS, Newton PO, Davis JLW, Woo SL-Y (1993) Morphologic and biomechanical comparison of tendons used as free grafts. J Hand Surg 18A:76–82 (with permission)

Tensile Property (45)

• Tangent modulus •	• Mammal • Flexor tendon	• Animal size •

Materials

- Mammals (white-tailed jackrabbit, grizzly bear, domestic cat, snow leopard, domestic dog, lion, etc.; 21 animals; weight, 470 g – 545 kg)
- Digital flexor tendon (plantaris, deep digital flexors), ankle extensor tendon (gastrocnemius), common digital extensor tendon from hind limbs

Testing Methods and Experimental Conditions

- Cyclic tensile test at a rate of 5–10 mm/min up to 5–6% strain
- Dripping 0.9% NaCl solution or in a saline-filled chamber for very thin tendons
- Extention measurements were made with a real-time video dimension analyzing system
- Tendon cross-sectional areas were calculated dividing the wet mass of a sample by its length and density

Data

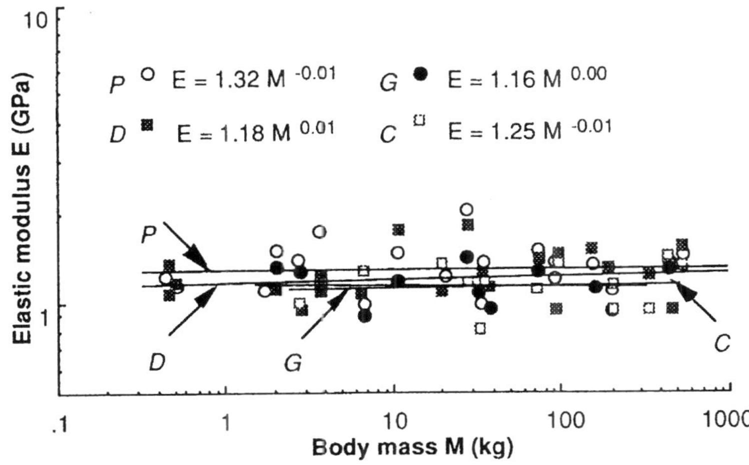

	n	a	b	95%CI	r
P (plantaris)	19	1.32	-0.01	0.04	-0.09
D (deep digital flexor)	21	1.18	0.01	0.04	0.09
G (gastrocnemius)	10	1.16	0.00	0.06	0.00
C (digital extensor)	17	1.25	-0.01	0.06	-0.11

The allometric equations were obtained by least-squares regression after transformation of the data to logarithms and were reported in the form $E = aM^b$. A Student's t-test and the standard error of the slope were used to assign 95% confidence limits (95% CI) to the allometric exponent, b. For all tendon types, the scaling exponent b for elastic modulus was not significantly different from zero.

Comments

- The elastic moduli of these anatomically and functionally distinct tendons are independent of species or body mass.

Reference(s)

Pollock CM, Shadwick RE (1994) Relationship between body mass and biomechanical properties of limb tendons in adult mammals. Am J Physiol 266:1016–1021 (with permission)

Tensile Property (46)

• Tangent modulus • Maximum load	• Human • Scapho-lunate ligament	• Locational dependence •

Materials
- Human
- Ligaments of scaphoid
- Scapho-lunate ligament

Testing Methods and Experimental Conditions
- Tensile test of ligaments
- Crosshead speed of 2 mm/s

Data

	Cross-sectional area (mm^2)	Young's modulus (MPa)	Ultimate load (N)
Ligaments of distal ligamentous complex of scaphoid	14	5.0	410
Lateral component	6		180
Medial component	8		230
Scapho-lunate ligament	6	25	150

Comments
- No description about experimental atmosphere.
- The ligaments of scaphoid consist of parallel regular, dense, wavy bundles. The scapho-lunate ligament is made of intermingled and dispersed fibers in its palmar part, forming a zone of weakness.

Reference(s)
Boabighi A, Kuhlmann AN, Kenesi C (1993) The distal ligamentous complex of the scaphoid and the scapho-lunate ligament. An anatomic, histological and biomechanical study. J Hand Surg 18B:65–69 (with permission)

Tensile Property (47)

• Tangent modulus • Tensile strength	• Dog • Diaphragmatic tendon	• Locational dependence •

Materials
- Mongrel dogs (weight, 15–25 kg)
- Diaphragmatic central tendon

Testing Methods and Experimental Conditions
- Specimens were cut in directions parallel to (Group 1) and perpendicular to (Group 2) the neighboring diaphragmatic muscles. Apex 1 ($n = 9$) and Apex 2 ($n = 8$) refer to Groups 1 and 2 specimens taken from central apical regions. Horn 1 ($n = 7$) and Horn 2 ($n = 8$) refer to Groups 1 and 2 specimens taken from horn-shaped regions
- Tensile test of tendon
- Crosshead speed of 10 mm/min, strain rate of approximately 0.02 per second
- Saline solution at 37°C
- Strain was determined using a video dimension analyzer

Data

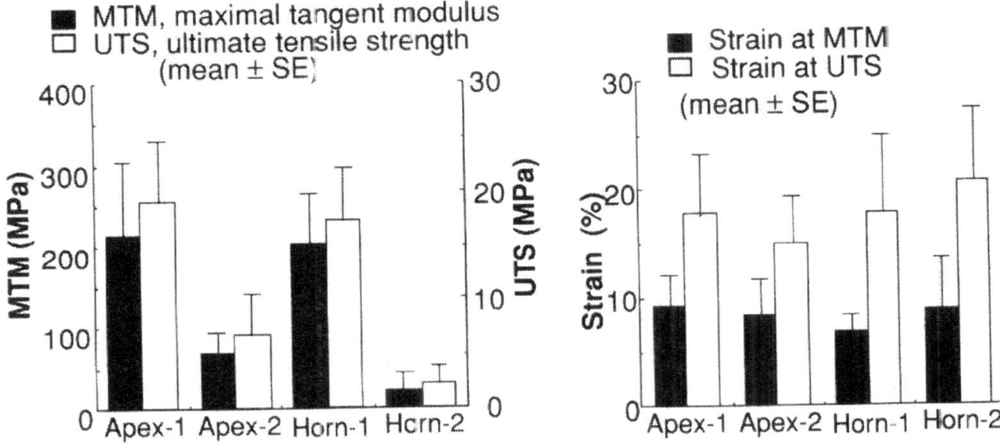

Comments
- Polarized microscopy showed that multiple sheets of collagen fiber bundles formed an orthogonal network in the tendon. Collagen fiber bundles along the Group 1 direction formed parallel trajectory lines connecting the neighboring costal and crural muscles; bundles along the Group 2 direction were observed to orient 90° away. At the central apex region, collagen bundles of Group 1 formed a fan-like trajectory pattern.

Reference(s)
Chuong CJ, Sacks MS, Johnson Jr RL, Reynolds R (1991) On the anisotropy of the canine diaphragmatic central tendon. J Biomech 24:563–576 (with permission)

Tensile Property (48)

• Tangent modulus • Tensile strength	• Dog • Flexor tendon	• Solvent preservation • Gamma irradiation

Materials

- Mongrel dogs (mature; weight, 8–12 kg)
- Flexor tendon

Testing Methods and Experimental Conditions

- Tendons in the solvent group were subjected to solvent preservation for sterilization and drying. Tendons in the solvent/gamma group were treated in the same manner and then sterilized by gamma irradiation. The same treatments were applied to the tendons in the gamma/solvent group but in reverse order
- Tensile test of middle section of tendon
- Crosshead speed of 20 mm/min
- Saline solution at 37°C
- Strain was determined using image processing equipment

Data

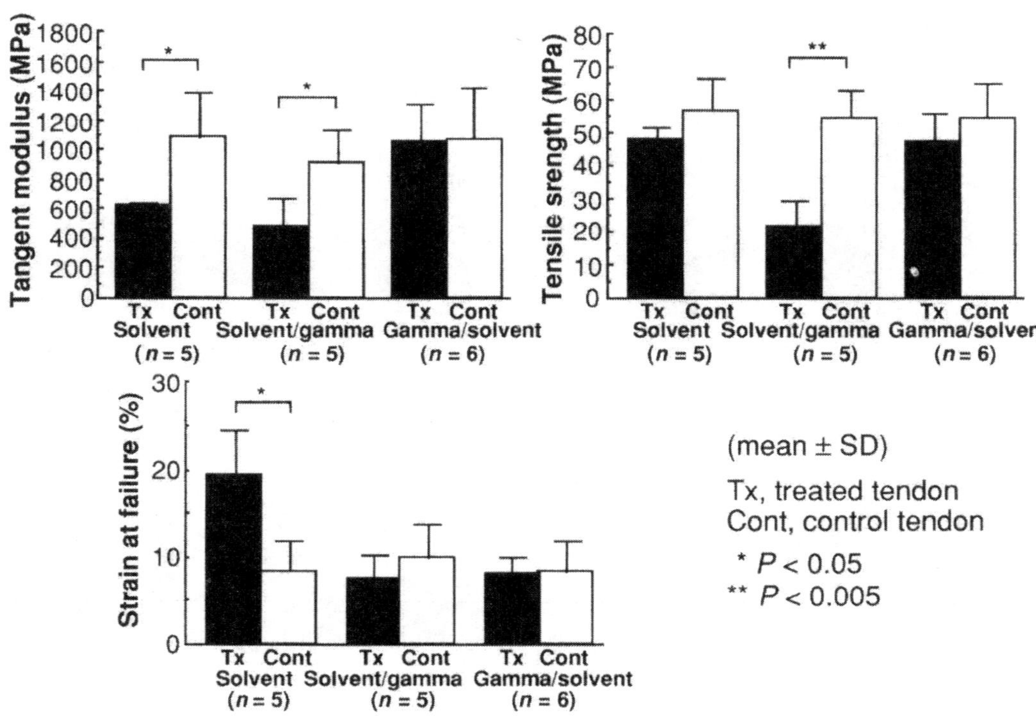

Comments

- Histologically, the tendons treated with solvent, with or without gamma irradiation, had a more prominent wavy pattern in the collagen fibers than the control tendons.

Reference(s)

Maeda A, Inoue M, Shino K, Nakata K, Nakamura H, Tanaka M, Seguchi Y, Ono K (1993) Effects of solvent preservation with or without gamma irradiation on the material properties of canine tendon allografts. J Orthop Res 11:181–189 (with permission)

Tensile Property (49)

• Tangent modulus • Tensile strength	• Human • Anterior cruciate ligament	• Locational dependence •

Materials
- Human (age, 26 ± 4 years [mean ± SD]; height, 1.64 ± 0.10 m; mass, 69 ± 7 kg)
- Anteror cruciate ligament (ACL)

Testing Methods and Experimental Conditions
- Tensile test of ACL fascicle–bone unit
- Strain rate of 1 per second
- Saline solution at 37°C
- Strain was determined from grip-to-grip displacement

Data

	Anteromedial bundle (AMB)	Anterolateral bundle (ALB)	Posterior bundle (PB)
Initial length (mm)	32.5 ± 4.7	28.6 ± 5.1	24.6 ± 4.5*
Initial area (mm^2)	2.92 ± 1.61	1.91 ± 1.02	4.57 ± 3.84
Modulus (MPa)	283.1 ± 114.4	285.9 ± 140.6	154.9 ± 119.5**
Maximum stress (MPa)	45.7 ± 19.5	30.6 ± 11.0	15.4 ± 9.5**
Strain energy density (Nm/cc)	3.3 ± 1.3	2.2 ± 0.6	1.1 ± 0.6***
Maximum strain (%)	19.1 ± 2.8	16.1 ± 3.9	15.2 ± 5.2

All data are given as mean ± SD.

* $P < 0.025$ *vs* AMB; ** $P < 0.05$ *vs* AMB and ALB; *** $P < 0.05$ *vs* ALB.

Comments
- All specimens failed in soft tissue, but the failure locations were distributed along the length. No failures occurred by tibial avulsion or bone fracture.

Reference(s)
Butler DL, Guan Y, Kay MD, Cummings JF, Feder SM, Levy MS (1992) Location-dependent variations in the material properties of the anterior cruciate ligament. J Biomech 25:511–518 (with permission)

Tensile Property (50)

• Tangent modulus • Tensile strength	• Human • Gracilis tendon	• •

Materials
- Human (age, 26 years)
- Gracilis tendon

Testing Methods and Experimental Conditions
- Tensile test of tendon
- Strain rate of 1 per second
- Strain was determined from film analysis and grip-to-grip motion

Data

Number of specimens	17
Length (mm)	41.9 ± 2.2
Area (mm^2)	7.6 ± 0.6
Volume (ml)	0.32 ± 0.02
Storage time (days)	60 ± 18
Modulus [a] (MPa)	612.8 ± 40.6
Maximum stress (MPa)	111.5 ± 4.0
Strain at maximum stress [a] (%)	26.7 ± 1.4
Strain at total failure [a] (%)	33.9 ± 1.5
Strain energy density to failure [a] (Nm/ml)	17.7 ± 1.7

All data are given as mean $\pm$ SE.

[a] Values obtained from grip-to-grip motions.

Number of specimens		5
Grip-to-grip strain at	Maximum stress	24.9 ± 3.1
	Total failure	32.5 ± 3.3
Local strain [a] at	Maximum stress	7.0 ± 1.9
	Total failure	7.8 ± 2.8
Local/grip [b] strain at	Maximum stress	29.9 ± 7.7
	Total failure	25.4 ± 6.8

All data are given as mean $\pm$ SE.

[a] Values obtained from film analysis.

[b] Percentages calculated from individual tests and then averaged.

Comments
- Some slippage and/or failure occurred in the grips.

Reference(s)

Butler DL, Grood ES, Noyes FR, Zernicke RF, Brackett K (1984) Effects of structure and strain measurement technique on the material properties of young human tendons and fascia. J Biomech 17:579–596 (with permission)

Tensile Property (51)

• Tangent modulus • Tensile strength	• Human • Patellar tendon	• •

Materials
- Human (age, 26 years)
- Patellar tendon (PT)

Testing Methods and Experimental Conditions
- Tensile test of patella–central or medial 1/3 PT–tibia complex
- Strain rate of 1 per second
- Strain was determined from film analysis and grip-to-grip motion

Data

	Central	Medial
Number of specimens	7	7
Length (mm)	48.7 ± 3.8	48.8 ± 2.8
Area (mm^2)	50.5 ± 3.8	49.9 ± 3.8
Volume (ml)	2.40 ± 0.03	2.46 ± 0.20
Width (mm)	13.8 ± 1.4	14.9 ± 1.1
Length to width ratio	3.5 ± 0.4	3.3 ± 0.2
Storage time (days)	11 ± 4	4 ± 1
Modulus[a] (MPa)	305.5 ± 59.0	361.3 ± 34.4
Maximum stress (MPa)	58.3 ± 6.1	56.7 ± 4.4
Strain at maximum stress [a] (%)	26.5 ± 2.9	24.2 ± 2.5
Strain at total failure [a] (%)	35.1 ± 4.4	30.7 ± 3.0
Strain energy density to failure[a] (Nm/ml)	9.9 ± 1.9	9.4 ± 1.6

All data are given as mean $\pm$ SE.
[a] Values obtained from grip-to-grip motions.

Number of specimens [a]		4
Grip-to-grip strain at	Maximum stress	30.2 ± 2.9
	Total failure	38.2 ± 2.9
Local strain[b] at	Maximum stress	12.0 ± 3.0
	Total failure	14.7 ± 4.6
Local/grip[c] strain at	Maximum stress	25.6 ± 1.4
	Total failure	27.5 ± 2.7

All data are given as mean $\pm$ SE.
[a] Consists of one central and three medial PT.
[b] Values obtained from film analysis.
[c] Percentages calculated from individual tests and then averaged.

Comments
- No bone-end failures were observed in the PT–bone complexes.

Reference(s)
Butler DL, Grood ES, Noyes FR, Zernicke RF, Brackett K (1984) Effects of structure and strain measurement technique on the material properties of young human tendons and fascia. J Biomech 17:579–596 (with permission)

Tensile Property (52)

• Tangent modulus • Tensile strength	• Human • Semitendinosus tendon	•

Materials

- Human (age, 26 years)
- Semitendinosus tendon

Testing Methods and Experimental Conditions

- Tensile test of tendon
- Strain rate of 1 per second
- Strain was determined from film analysis and grip-to-grip motion

Data

Number of specimens	11
Length (mm)	36.6 ± 3.1
Area (mm^2)	14.0 ± 0.5
Volume (ml)	0.51 ± 0.05
Storage time (days)	44 ± 23
Modulus [a] (MPa)	362.2 ± 21.6
Maximum stress (MPa)	88.5 ± 5.0
Strain at maximum stress [a] (%)	33.2 ± 1.8
Strain at total failure [a] (%)	52.4 ± 2.8
Strain energy density to failure [a] (Nm/ml)	23.4 ± 1.3

All data are given as mean $\pm$ SE.

[a] Values obtained from grip-to-grip motions.

Number of specimens		3
Grip-to-grip strain at	Maximum stress	27.4 ± 2.9
	Total failure	48.9 ± 1.6
Local strain [a] at	Maximum stress	8.2 ± 1.9
	Total failure	8.8 ± 1.7
Local/grip [b] strain at	Maximum stress	28.6 ± 7.1
	Total failure	17.1 ± 5.1

All data are given as mean $\pm$ SE.

[a] Values obtained from film analysis.

[b] Percentages calculated from individual tests and then averaged.

Comments

- Some slippage and/or failure occurred in the grips.

Reference(s)

Butler DL, Grood ES, Noyes FR, Zernicke RF, Brackett K (1984) Effects of structure and strain measurement technique on the material properties of young human tendons and fascia. J Biomech 17:579–596 (with permission)

Tensile Property (53)

• Tangent modulus • Tensile strength	• Human, monkey • Anterior cruciate ligament	• Age effect •

Materials
- Human
- Rhesus monkeys
- Anterior cruciate ligament (ACL)

Testing Methods and Experimental Conditions
- Tensile test of femur–ACL–tibia complex
- Instron biaxial electrohydraulic materials testing system for human specimens (strain rate: 1.0 per second)
- Instron material testing machine for rhesus monkey specimens (strain rate, 0.66 per second)
- The specimens were mounted in 45° of flexion
- The cross-sectional area of the ligament was measured with an area micrometer under blade pressure of 0.12 MPa
- The mean length of the ligament was determined from the length of the posterior and anterior fibers, which were measured with calipers

Data

	Older human (48–86 years)	Younger human (16–26 years)	Rhesus monkey
No. of specimens	20	6	25
Elastic modulus (MPa)	$65.3 \pm 24.0^{*}$	$111 \pm 26^{\#}$	186 ± 26
Linear stress (MPa)	11.3 ± 5.1	$25.5 \pm 14.0^{\#\#}$	56.2 ± 7.6
Maximum stress (MPa)	$13.3 \pm 5.0^{***}$	$37.8 \pm 9.3^{\#}$	66.1 ± 8.4
Strain energy to failure (Nm/ml)	$3.1 \pm 1.5^{**}$	$10.3 \pm 3.1^{\#}$	19.4 ± 3.8
Strain at linear stress (%)	21.6 ± 6.9	$25.5 \pm 8.1^{\#\#\#}$	38 ± 5.9
Strain at maximum stress (%)	$30 \pm 10.0^{*}$	44.3 ± 8.5	47 ± 5.6
Strain at failure (%)	$48.5 \pm 11.9^{*}$	60.25 ± 6.78	60 ± 6.3
Stiffness (kN/m)	129 ± 39	182 ± 56	194 ± 28
Linear force (kN)	0.622 ± 0.283	1.17 ± 0.75	0.71 ± 0.12
Maximum force (kN)	0.734 ± 0.266	1.73 ± 0.66	0.83 ± 0.11
Energy to failure (Nm)	4.89 ± 2.36	12.8 ± 5.5	3.0 ± 0.6

All data are given as mean $\pm$ SD.

$^{*}P < 0.01$; $^{**}P < 0.005$; $^{***}P < 0.001$ *vs* younger human.

$^{\#}P < 0.001$; $^{\#\#}P < 0.005$; $^{\#\#\#}P < 0.025$ *vs* rhesus monkey.

Comments
None.

Reference(s)
Noyes FR, Grood ES (1976) The strength of the anterior cruciate ligament in humans and rhesus monkeys. J Bone Joint Surg 58-B:1074–1082 (with permission)

Tensile Property (54)

• Tensile strength • Maximum load	• Human • Posterior cruciate	• Locational dependence •

Materials
- Human (age, 75 [range 53–98] years)
- Posterior cruciate ligament (PCL) (anterolateral [aPC] and posteromedial [pPC] bundles)

Testing Methods and Experimental Conditions
- Tensile test of femur–PCL bundle–tibia complex
- Crosshead speed of 1000 mm/min, strain rate of approximately 0.5 per second
- Strain was determined from crosshead displacement

Data

	pPC	aPC	P (paired t)
Length (mm)	33.8 ± 3.2	35.3 ± 3.4	> 0.1
Area (mm^2)	10.0 ± 1.3	43.0 ± 11.3	< 0.001
Max. load (N)	258 ± 83	1620 ± 500	< 0.001
Stiffness (N/mm)	77 ± 32	347 ± 140	< 0.001
Linear strain (%)	12.9 ± 2.9	12.3 ± 5.7	> 0.1
Linear stress(MPa)	18.7 ± 10.3	28.7 ± 17.0	0.086
Modulus (MPa)	145 ± 69	248 ± 119	0.011
Max. strain (%)	19.5 ± 5.4	18.0 ± 5.3	> 0.1
Max. stress (MPa)	24.4 ± 10.0	35.9 ± 15.2	0.027
Energy density at failure (J/cc)	2.06 ± 0.73	3.12 ± 1.4	0.045

All data are given as mean $\pm$ SD.

Comments
- In all specimens, interstitial fiber failure occurred.

Reference(s)
Race A, Amis AA (1994) The mechanical properties of the two bundles of the human posterior cruciate ligament. J Biomech 27:13–24 (with permission)

Tensile Property (55)

• Tensile strength • Maximum load	• Rabbit • Medial collateral ligament	• Freezing •

Materials
- New Zealand White rabbits (younger rabbits (open epiphysis; weight, 3.7 ± 0.1 kg) and older rabbits (closed epiphysis; weight, 4.3 ± 0.2 kg))
- Medial collateral ligament (MCL)

Testing Methods and Experimental Conditions
- In stored group, each hind limb was double wrapped in saline-soaked gauze, sealed in a plastic bag, and then stored airtight at $-20°C$ for 1.5 or 3 months. In fresh group, no postmortem freezing storage was performed
- Tensile test of femur–MCL–tibia complex
- Crosshead speed of 1 cm/min, strain rate of 0.004 per second
- Saline solution at 37°C
- Strain was determined using a video dimension analyzer

Data

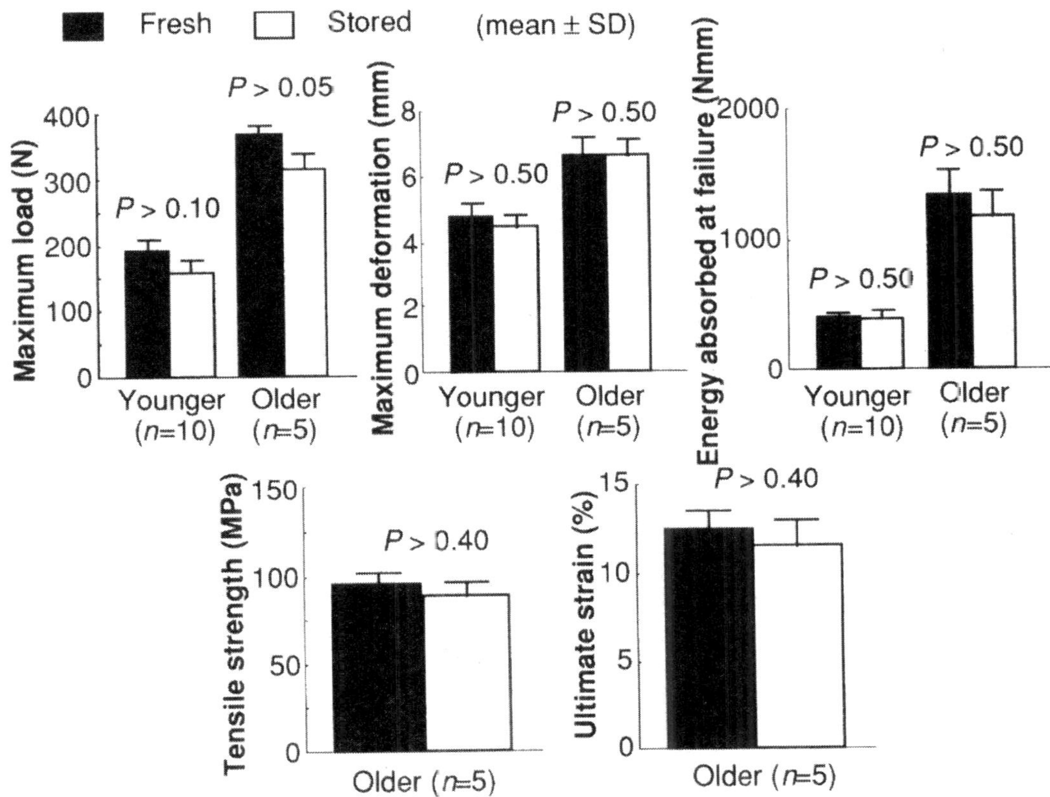

Comments
- For the younger rabbits, nineteen out of twenty samples failed by avulsion at the tibial junction. There was one case of midsubstance failure. For the older rabbits, only five paired samples had midsubstance failures, while the remaining four pairs of limbs had either one or both sides fail by tibial avulsion. The data of the older rabbits were obtained from the five paired samples.

Reference(s)
Woo SL-Y, Orlando CA, Camp JF, Akeson WH (1986) Effects of postmortem storage by freezing on ligament tensile behavior. J Biomech 19:399–404 (with permission)

Tensile Property (56)

• Tensile strength • Tangent modulus	• Human • Spine ligament	• •

Materials
• Human
• Lumbar spine anterior longitudinal ligament (ALL)

Testing Methods and Experimental Conditions
• Tensile test of bone–ALL–bone complex
• Crosshead speed of 2.5 mm/s, strain rate of approximately 0.01 per second
• Strain was determined from digitized video recordings of markers attached to the ALL at 12 sites and crosshead displacement

Data

Age (years)		21	21	21	29	43	43
Level		T12–L1	L2–L3	L4–L5	L1–L2	T12–L1	L2–L3
MSH (mm)		77.0	76.0	73.8	78.4	73.0	83.0
DHA (mm)		6.5	9.8	10.0	8.5	9.5	9.0
DHP (mm)		5.3	8.4	8.0	7.2	9.0	9.0
LA (mm^2)		36.5	39.8	43.5	33.5	39.5	36.5
Load (N)		1060	1163	1607	890	885	756
Difference (mm)		8.0	20.1	11.1	10.5	14.2	9.8
Stress (MPa)		29.0	29.2	36.9	26.6	21.7	20.7
Peak strain (%)	ε_{MSH}	14.0	32.0	18.3	17.1	23.2	13.3
	ε_{max}	11.4	12.1	12.5	14.7	8.0	13.8
	$\varepsilon_{1\text{-}4}$	4.26	2.65	6.94	5.70	4.35	5.79
	ε_S	7.30	7.72	10.1	10.3	5.01	8.72
	ε_A	4.56	5.04	3.15	4.74	3.69	3.46
	ε_B	1.61	6.02	5.62	4.12	5.24	6.75
Tensile moduli (MPa)	$E_{1\text{-}4}$	863	1367	505	713	680	424
	E_S	615	640	379	724	835	286
	E_A	811	871	1200	565	481	551
	E_B	894	447	469	502	481	349

MSH, anterior motion segment height.
DHA and DHP, anterior and posterior disc heights.
LA, cross-sectional area.
ε_{MSH}, peak strain from crosshead displacement.
ε_{max}, maximum peak strain from digitized video recordings of 12 markers.
$\varepsilon_{1\text{-}4}$, ε_S, ε_A, ε_B, peak strain for "overall ligament", substance, insertion A, insertion B.
$E_{1\text{-}4}$, E_S, E_A, E_B, tensile modulus for "overall ligament", substance, insertion A, insertion B.

Comments
• Failure consistently occurred in the ligament mid-substance, and ultimate strain at the ligament failure site was 12.1 ± 2.3% (mean ± SD).

Reference(s)
Neumann P, Keller TS, Ekstrom L, Perry L, Hansson TH, Spengler DM (1992) Mechanical properties of the human lumbar anterior longitudinal ligament. J Biomech 25: 1185–1194 (with permission)

Tensile Property (57)

• Tensile strength • Tangent modulus	• Rabbit • Medial collateral ligament	• Strain rate • Age effect

Materials

- New Zealand White rabbits (skeletally immature (open epiphyses; age, 3.5 months; weight, 2.8 ± 0.1 kg [mean $\pm$ SE]) and skeletally mature (closed epiphyses; age, 8.5 months; weight, 4.2 ± 0.1 kg))
- Medial collateral ligament (MCL)

Testing Methods and Experimental Conditions

- Tensile test of femur–MCL–tibia complex
- Extension rate of 0.008–113 mm/s, strain rate of 0.00011–2.22 per second
- Deformation was determined from crosshead displacement
- Strain was determined using a high-speed video motion analysis system

Data

Extension rate (mm/s)	0.008	0.1	1.0	10.0	113.0
Immature					
Linear stiffness (N/mm)	24.4 ± 1.2	40.0 ± 2.4	40.0 ± 1.9	41.2 ± 3.3	38.1 ± 1.4
Load at failure (N)	54.3 ± 3.0	84.5 ± 8.5	88.4 ± 8.6	105.7 ± 8.9	123.7 ± 6.7
Deformation at failure (mm)	2.6 ± 0.1	2.7 ± 0.1	2.9 ± 0.2	3.2 ± 0.2	3.5 ± 0.2
Energy absorbed at failure (Nm)	0.07 ± 0.01	0.09 ± 0.01	0.12 ± 0.02	0.16 ± 0.02	0.20 ± 0.02
Cross-sectional area (mm^2)	2.9 ± 0.1	3.0 ± 0.1	3.2 ± 0.1	3.1 ± 0.1	2.9 ± 0.1
Modulus (MPa)	610 ± 100	620 ± 120	700 ± 60	860 ± 70	970 ± 190
Mature					
Linear stiffness (N/mm)	52.8 ± 1.6	54.8 ± 1.4	54.3 ± 2.5	59.6 ± 2.8	60.7 ± 1.7
Load at failure (N)	311.5 ± 12.1	295.8 ± 28.1	337.5 ± 23.7	347.1 ± 13.7	403.7 ± 7.5
Deformation at failure (mm)	6.4 ± 0.2	5.6 ± 0.4	6.6 ± 0.2	6.3 ± 0.2	6.6 ± 0.1
Energy absorbed at failure (Nm)	0.98 ± 0.08	0.83 ± 0.14	1.10 ± 0.11	1.10 ± 0.09	1.35 ± 0.06
Cross-sectional area (mm^2)	4.1 ± 0.1	3.6 ± 0.1	4.0 ± 0.2	4.0 ± 0.2	3.8 ± 0.2
Tensile strength (MPa)	75.2 ± 2.4	81.4 ± 7.4	85.7 ± 7.7	87.2 ± 6.0	106.7 ± 6.8
Strain at failure (%)	9.5 ± 0.5	10.0 ± 0.5	11.0 ± 0.5	11.5 ± 1.0	13.0 ± 1.0
Modulus (MPa)	700 ± 70	840 ± 150	800 ± 60	910 ± 50	760 ± 160

All data are given as mean $\pm$ SE, $n = 6$.

Comments

- All specimens from the skeletally immature rabbits failed by tibial avulsion, whereas specimens from the skeletally mature rabbits failed within the ligament substance.

Reference(s)

Woo SL-Y, Peterson RH, Ohland KJ, Sites TJ, Danto MI (1990) The effects of strain rate on the properties of the medial collateral ligament in skeletally immature and mature rabbits: A biomechanical and histological study. J Orthop Res 8:712–721 (with permission)

Tensile Property (58)

• Tensile strength • Tangent modulus	• Rat • Tail tendon	• •

Materials

- Fischer rats (age, 9 months)
- Tail tendon

Testing Methods and Experimental Conditions

- Tensile test of tendon
- Initial lengths of test specimens were 100, 50, 20, and 10 mm
- Strain was determined from grip-to-grip length
- Strain rate of 0.01 per second
- Saline solution at 37°C

Data

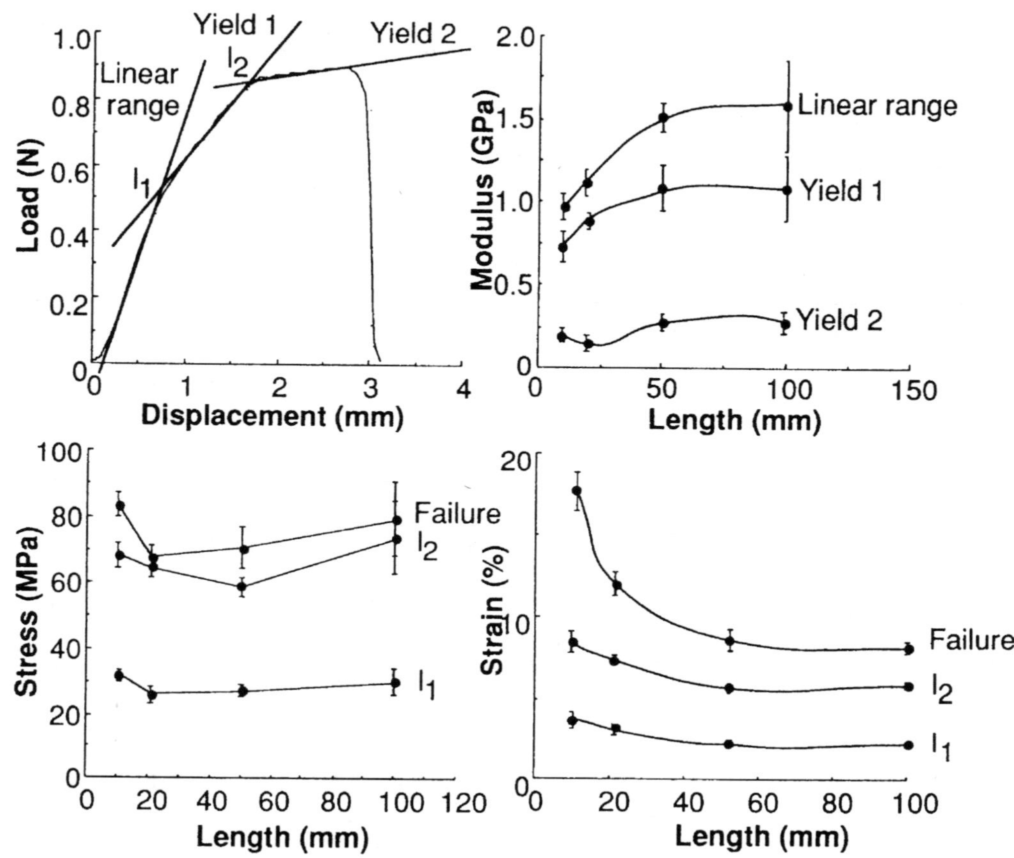

Comments

- All specimens failed in the midsection well away from the grips.

Reference(s)

Haut RC (1986) The influence of specimen length on the tensile failure properties of tendon collagen. J Biomech 19:951–955 (with permission)

Tensile property (59)

• Tensile strength • Tangent modulus	• Sheep • Anterior cruciate ligament	• Loading direction •

Materials

- Sheep (mature; age, 1–5 years; weight, 40–60 kg)
- Anterior cruciate ligament (ACL)

Testing Methods and Experimental Conditions

- Tensile test of femur–ACL–tibia complex
- Femur and tibia were mounted in the following four configurations:
  (1) Tibia mounted horizontally and the femur at 45° to the vertical (load applied in anterior draw)
  (2) Both the tibia and the femur mounted at 22.5° to the load (ACL axis at approximately 25° to load)
  (3) Both the tibia and femur mounted vertically, placing the knee in hyperextension (load along tibial axis)
  (4) Both the tibia and femur mounted horizontally, placing the knee in hyperextension (load in anterior draw)
- Crosshead speed of 500 mm/min
- Strain was determined using a digital video dimension analyzer (DVDA) and from crosshead displacement

Data

	(1)[a]	(1)[b]	(2)[b]	(3)[b]	(4)[b]
Cross-sectional area (mm^2)	19 ± 2	21 ± 3	24 ± 3	20 ± 4	21 ± 1
Length (mm)	18 ± 1	18 ± 1	18 ± 1	18 ± 1	19 ± 1
Ultimate tensile load (N)	2370 ± 220	2580 ± 500	1910 ± 150	1200 ± 190	2470 ± 330
Ultimate tensile stress (MPa)	124 ± 5	123 ± 15	80 ± 10	60 ± 3	117 ± 17
Ultimate specific extension (%)	44 ± 6	69 ± 6	45 ± 4	37 ± 7	93 ± 20
Young's modulus (MPa)	437 ± 122	204 ± 28	243 ± 39	227 ± 13	180 ± 34
Rupture energy (MJ/m^3)	20 ± 6	38 ± 7	14 ± 2	10 ± 3	47 ± 16

All data are given as mean ± SD.

[a] Specific extension measured by DVDA near insertion sites.

[b] Specific extension from crosshead displacement.

Comments

- The orientation and degree of flexion of the bone–ligament–bone complex significantly alter the apparent ultimate tensile properties.

Reference(s)

Rogers GJ, Milthorpe BK, Muratore A, Schindhelm K (1990) Measurement of the mechanical properties of the ovine anterior cruciate ligament bone–ligament–bone complex: A basis for prosthetic evaluation. Biomaterials 11:89–96 (with permission)

Tensile Property (60)

• Tensile strength • Tangent modulus	• Sheep • Posterior cruciate	• •

Materials

- German sheep (female; skeletally mature; age, 2 years)
- Posterior cruciate ligament (PCL)

Testing Methods and Experimental Conditions

- Tensile test of femur–PCL–tibia complex at 84° of knee flexion with the loading direction along the PCL axis
- Crosshead speed of 800 mm/min
- Deformation was determined from grip-to-grip motion

Data

Length (mm)	29.5 ± 1.4
Cross-sectional area (mm^2)	22.4 ± 1.7
Maximum load (N)	925 ± 70
Deformation (mm)	9.1 ± 0.9
Stiffness (N/mm)	130 ± 11
Maximum stress (MPa)	41.3 ± 1.9
Elastic modulus (MPa)	172 ± 14

All data are given as mean $\pm$ SD, $n = 6$.

Comments

- All specimens failed in their ligamentous portion.

Reference(s)

Bosch U, Kasperczyk WJ (1992) Healing of patellar tendon autograft after posterior cruciate ligament reconstruction — a process of ligamentization? An experimental study in a sheep model. Am J Sports Med 20:558–565 (with permission)

Viscoelastic Property (1)

• Cyclic load relaxation •	• Rabbit • Medial collateral ligament	• Water content •

Materials
- New Zealand White rabbits (immature; age, 3 months)
- Medial collateral ligament (MCL)

Testing Methods and Experimental Conditions
- Femur–MCL–tibia complexes were soaked in test solution for 1 h prior to cycling. Only test specimens were cycled; control specimens remained in the environmental solution until cycling of test specimens was completed
- Extension between 0 and 0.68 mm at rate of 10 mm/min
- Phosphate-buffered saline (PBS, 0.9%) and 2, 10, and 25% sucrose solutions (SUC); all solutions were kept at 35°C, pH 7.4
- Cyclic relaxation is represented by the ratio of the peak load of the nth cycle normalized against the peak load of the first cycle

Data

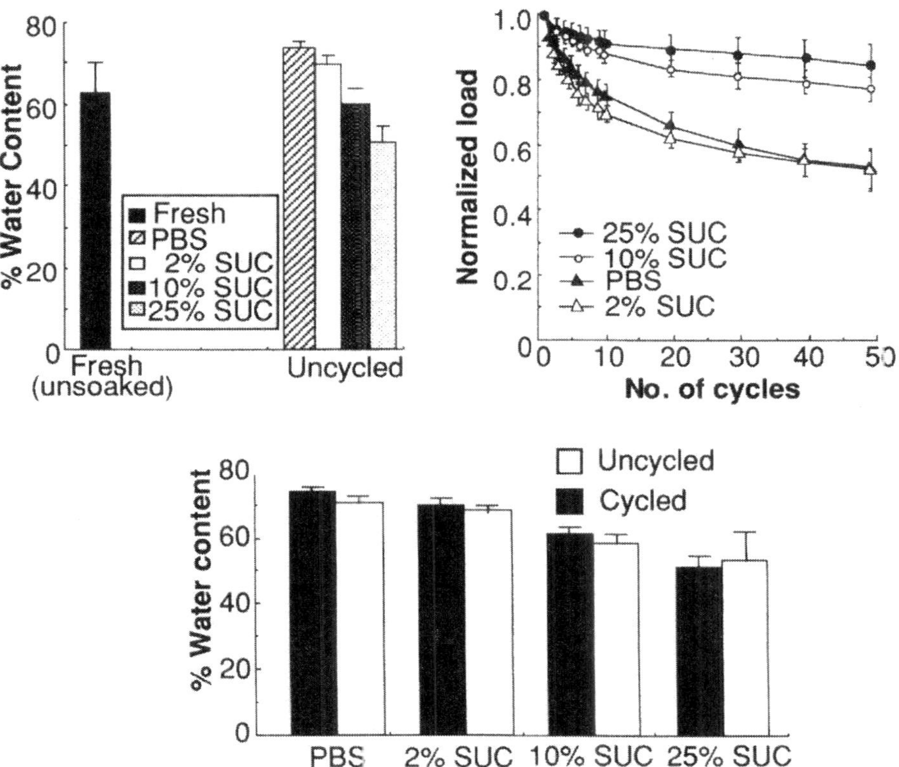

Comments
None.

Reference(s)
Chimich D, Shrive N, Frank C, Marchuk L, Bray R (1992) Water content alters viscoelastic behaviour of the normal adolescent rabbit medial collateral ligament. J Biomech 25:831–837 (with permission)

Viscoelastic Property (2)

• Cyclic load relaxation • Static load relaxation	• Rabbit • Medial collateral ligament	• Age effect •

Materials
- New Zealand White rabbits (age, 3, 6, 9, and 12 months)
- Medial collateral ligament (MCL)

Testing Methods and Experimental Conditions
- Static and cyclic relaxation test of femur–MCL–tibia complex
- Specimen was subject to 30 cycles of uniaxial tension between 0 and 0.68 mm extension at 10 mm/min. Immediately after the cyclic test, the specimen was held at a constant 0.68 mm extension for 1200 s
- Phosphate-buffered saline at 35°C

Data

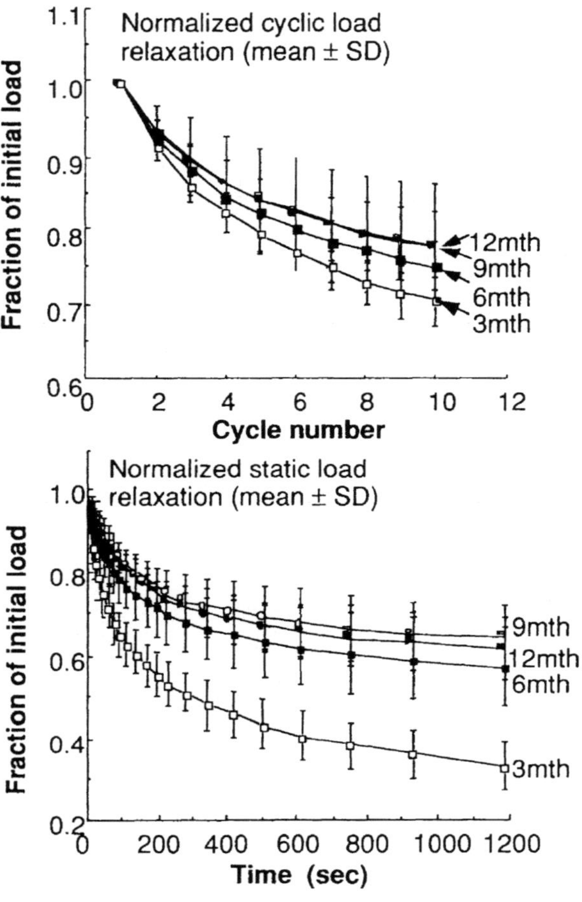

Comments
- In both the static and the cyclic tests, specimens from 3-month-old animals relaxed significantly more than specimens at 6, 9, or 12 months of age, whereas those 6, 9, and 12 months old were not significantly different from each other.

Reference(s)
Lam TC, Frank CB, Shrive NG (1993) Changes in the cyclic and static relaxations of the rabbit medial collateral ligament complex during maturation. J Biomech 26:9–17 (with permission)

Viscoelastic Property (3)

| • Hysteresis
• | • Mammal
• Flexor tendon | • Animal size
• |

Materials
- Mammals (white-tailed jackrabbit, grizzly bear, domestic cat, snow leopard, domestic dog, Lion, etc., 21 animals; weight, 470 g – 545 kg)
- Digital flexor tendon (plantaris, deep digital flexors), ankle extensor tendon (gastrocnemius), common digital extensor tendon from hind limbs

Testing Methods and Experimental Conditions
- Cyclic tensile test at a rate of 5–10 mm/min up to 5–6% strain
- Dripping 0.9% NaCl solution or in a saline-filled chamber for very thin tendons
- Extention measurements were made with a real-time video dimension analyzing system
- Tendon cross-sectional areas were calculated dividing the wet mass of a sample by its length and density

Data

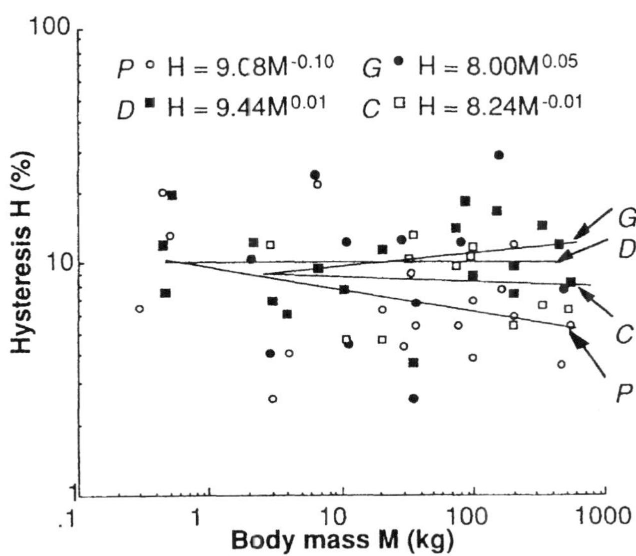

	n	a	b	95% CI	r
P (plantaris)	18	9.08	-0.10	0.14	-0.36
D (deep digital flexor)	21	9.44	0.01	0.08	0.03
G (gastrocnemius)	10	8.00	0.05	0.34	0.11
C (digital extensor)	17	8.24	-0.01	0.13	-0.06

The allometric equations were obtained by least-squares regression after transformation of the data to logarithms and were reported in the form $E = aM^b$. A Student's t-test and the standard error of the slope were used to assign 95% confidence limits (95% CI) to the allometric exponent, b. For all tendon types, the scaling exponent b for elastic modulus was not significantly different from zero.

Comments
- The hysteresis of these anatomically and functionally distinct tendons are independent of species or body mass.

Reference(s)
Pollock CM, Shadwick RE (1994) Relationship between body mass and biomechanical properties of limb tendons in adult mammals. Am J Physiol 266:1016–1021 (with permission)

Viscoelastic Property (4)

• Hysteresis • Cyclic load relaxation	• Dog • Medial collateral ligament	• Temperature dependence •

Materials
- Mongrel dogs (adult; weight, 29.1 ± 1.5 kg)
- Medial collateral ligament (MCL)

Testing Methods and Experimental Conditions
- Cyclic test of femur–MCL–tibia complex
- Extension rate of 2 cm/min
- First cycle from 0 to 2 mm extension was used for hysteresis measurement, and the following ten cycles between 1 and 2 mm extension were used to determine the temperature dependence
- Saline solution at 2–37°C

Data

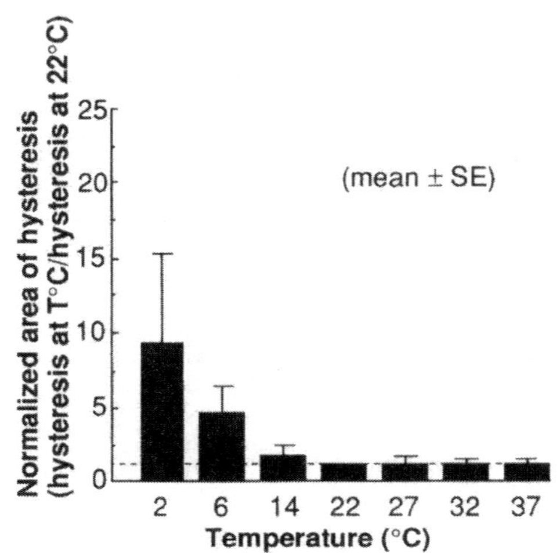

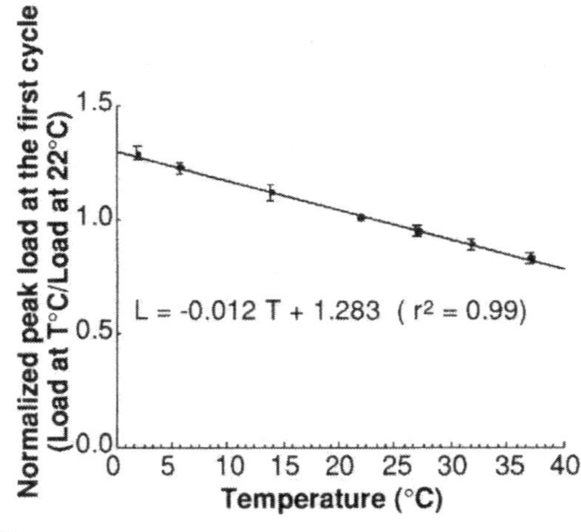

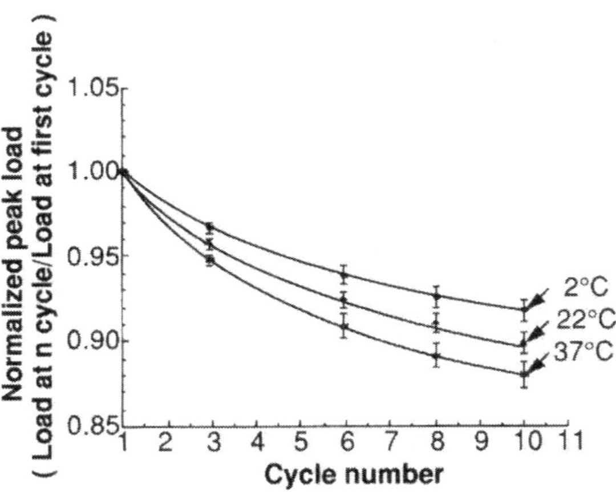

Comments
None.

Reference(s)

Woo SL-Y, Lee TQ, Gomez MA, Sato S, Field FP (1987) Temperature dependent behavior of the canine medical collateral ligament. Trans ASME J Biomech Eng 109:68–71

Viscoelastic Property (5)

• Stress relaxation • Stress–strain curve	• Human • Patellar tendon	• Age effect •

Materials
- Human (younger group (age, 29–50 years); older group (age, 64–93 years))
- Patellar tendon (PT)

Testing Methods and Experimental Conditions
- Stress–relaxation test, cyclic stress–relaxation test, and tensile test of central third of patella–PT–tibia complex
- In the first stress–relaxation test, specimens were stretched at 100 mm/min to a stress level of 1 MPa and were held at the resulting elongation for 15 min. The second stress–relaxation test followed the same procedure as the first, except that the stress level was increased to 4 MPa. A 30-min rest period followed each test. Next, the specimens were cyclically extended between the two elongations obtained in the stress–relaxation tests (15 cycles at a rate of 3 cycles/min). After another 30–min rest period, the specimens were stretched at an elongation rate of 200 mm/min until failure occurred
- Saline solution at 37°C
- Strain was determined using a video dimension analyzer (VDA)

Data

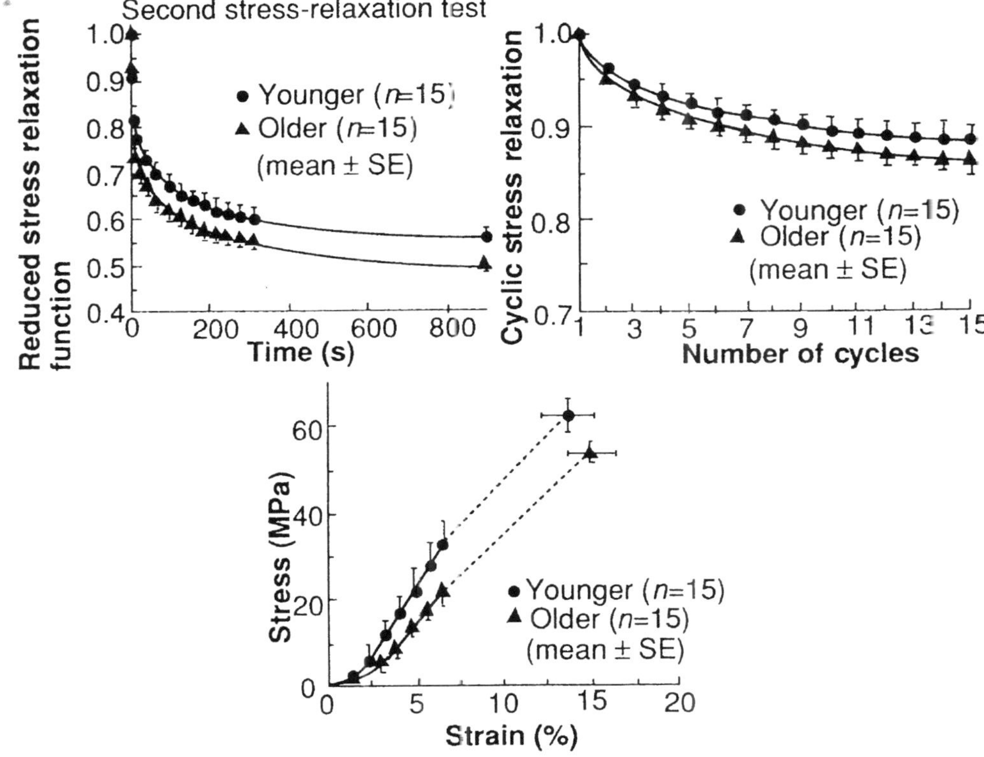

Comments
- All specimens failed within the tendon substance near the tibial insertion.

Reference(s)
Johnson GA, Tramaglini DM, Levine RE, Ohno K, Choi NY, Woo SL-Y (1994) Tensile and viscoelastic properties of human patellar tendon. J Orthop Res 12: 796–803 (with permission)

Coefficient of Friction (1)

• Pendulum test •	• Dog, bovine, pig • Ankle, elbow, and hip	• •

Materials
- Dogs, bovine, pigs
- Articular cartilage from ankle, metatarsal-phalangeal, elbow, hip, and shoulder joints

Testing Methods and Experimental Conditions
- Friction test by using pendulum apparatus
- Coefficient of friction was calculated from a decay in amplitude

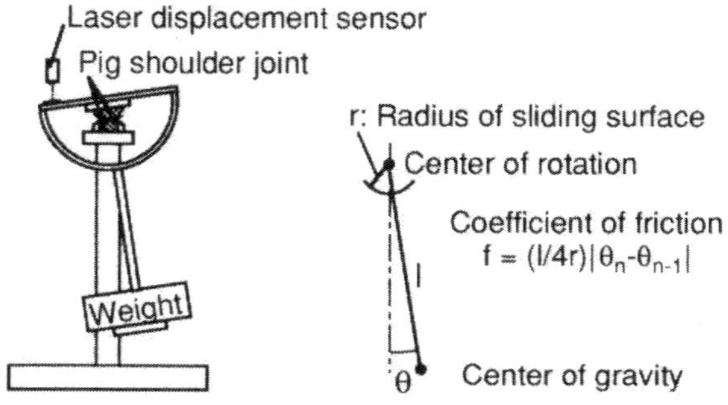

Data

Animal joint	Total weight of pendulum (N)	Lubricant	Coefficient of friction
Canine ankle joints [a]	3.9 – 10.8	Canine synovial fluid	0.017 – 0.029
Bovine metatarsal-phalangeal joints [b]	980 – 4900	Bovine synovial fluid	0.006 – 0.011
Canine elbow joints [c]	39	Canine synovial fluid	0.009 – 0.02
Canine hip joints [d]	100 – 175	Canine synovial fluid	0.00563 – 0.012
Canine hip joints [e]	88 – 118	Canine synovial fluid	0.009 ± 0.003
Pig shoulder joints [f]	100	Saline solution of sodium hyaluronate	0.003 – 0.011

Comments
- The value for coefficient of friction estimated from the amplitude decay in the pendulum test represents the average frictional property in swinging motion.

Reference(s)

[a]Barnett CH, Cobbold AF (1962) Lubrication within living joints. J Bone Joint Surg 44B:662–674

[b]Radin EL, Paul IL, Pollock D (1970) Animal joint behaviour under excessive loading. Nature 226:554–555

[c]Chikama H (1985) The role of the protein and the hyaluronic acid in the synovial fluid in animal joint lubrication. J Jpn Orthop Assoc 59:559–572

[d]Sasada T, Tsukamoto Y, Mabuchi K (1988) Biotribology, Sangyou-tosyo, Tokyo, p 79

[e]Mabuchi K, Tsukamoto Y, Obara T, Yamaguchi T (1993) Improvement in the lubricating property of animal joints by the injection of hyaluronic acid. J Jpn Soc Biomaterials 11:20–26

[f]Higaki H, Murakami T, Nakanishi Y (1995) Boundary lubricating ability of protein and phospholipid in natural synovial joints. Trans Jpn Soc Mech Engrs 61:3396–3401

Coefficient of Friction (2)

• Pendulum test	• Dog	• Washing
•	• Hip joint	•

Materials
- Mongrel dogs (skeletally mature)
- Hip joint
- Articular cartilage from femoral head and acetabulum cup

Testing Methods and Experimental Conditions
- Pendulum-type friction test for whole hip joint at room temperature (23°–25°C)
- Direction of hip joint loading aligned to that in the natural posture of dog
- Compression joint load was the body weight of individual donor
- Distance between the fulcrum and the center of gravity was 60–70 cm
- Joints were tested (1) intact, (2) after washing joint surface by saline solution, and (3) after adding 1 ml of 1% hyaluronic acid (HA) which was originated from a culture broth of *streptococcus zooepidemicus* (molecular weight, 2×10^6 dalton)

Data

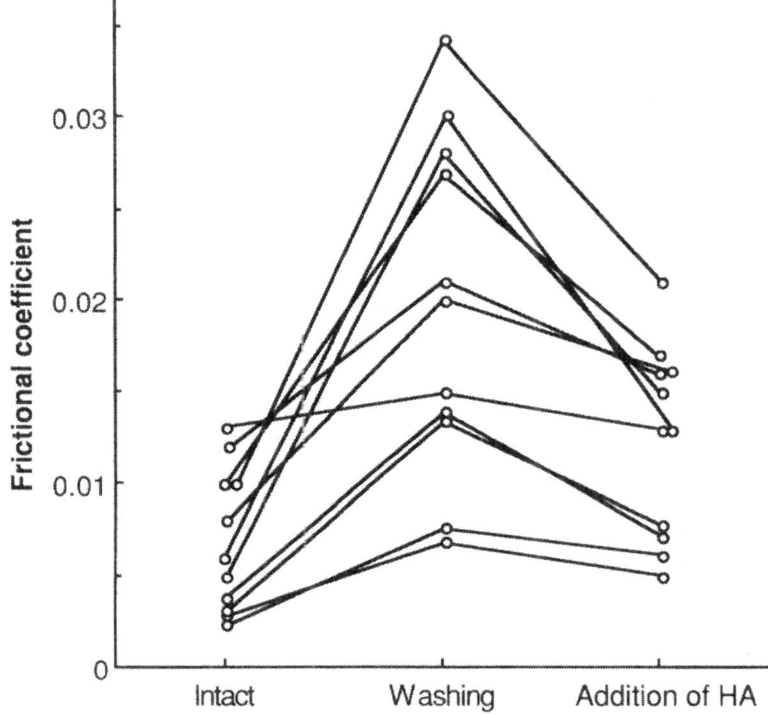

Comments
- Frictional coefficients were (1) 0.007 ± 0.004 [mean $\pm$ SD], (2) 0.020 ± 0.009, and (3) 0.013 ± 0.005, with statistically significant differences among three conditions $(P < 0.01)$.
- Viscosity was 22 Pa•s at the shear rate of 1.4 per second and 0.4 Pa•s at 270 per second for the HA used, and was 0.001 Pa•s for the saline solution.

Reference(s)
Mabuchi K, Tsukamoto Y, Obara T, Yamaguchi T (1994) The effect of additive hyaluronic acid on animal joints with experimentally reduced lubricating ability. J Biomed Mat Res 28: 865–870 (with permission)

Coefficient of Friction (3)

• Pendulum test •	• Human • Finger and knee joint	• •

Materials

- Human
- Middle finger and knee joint in vivo

Testing Methods and Experimental Conditions

- Friction test by using pendulum system with living joints
- Initial angular amplitude of 15° for finger joints and 11.5° for knee joints
- Loading conditions from 3.9 N to 10.8 N for finger joints and 216 N for knee joints

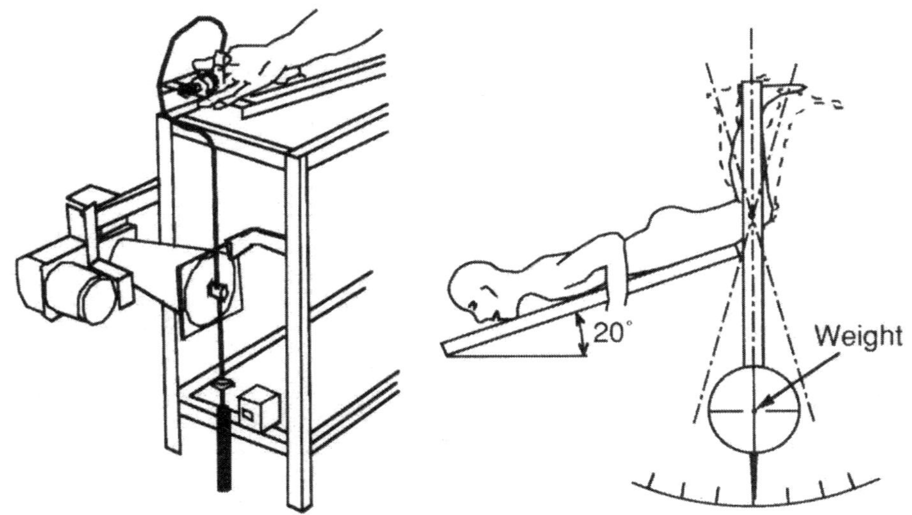

Schematic representations of pendulum friction systems for middle finger [a] and knee joint [b]

Data

Joint	Coefficient of friction
Finger [a]	Decreased from 0.018 to 0.0075 with increasing load
Knee [b]	Less than 0.01 in the range of small oscillating amplitude

Comments

- In the friction tests for living joints, soft tissues generated considerable resistance even under conditions where the muscle activity is suppressed.
- Frictional property caused by articular surfaces should be evaluated based on the values measured in the range of small amplitude.

Reference(s)

[a]Barnett CH, Cobbold AF (1962) Lubrication within living joints. J Bone Joint Surg 44B:662–674 (with permission)

[b]Sasada T, Maezawa H (1973) A measurement of the friction in the living human knee joint. J Jpn Soc Lubrication Engrs 18:901–906 (with permission)

Coefficient of Friction (4)

• Pendulum test •	• Human, bovine • Hip joint	• Digestive enzyme •

Materials
- Human and bovine
- Articular cartilage from hip joint

Testing Methods and Experimental Conditions
- Friction test by using pendulum apparatus
- Coefficient of friction was directly measured using a load transducer during swinging motion
- Lubricating condition with human and animal synovial fluid
- Instantaneous coefficients of friction at the same sliding speed of 7 mm/s were compared under different load conditions
- Effect of digestive enzyme on friction was examined for several hip joints to evaluate the lubricating ability of hyaluronic acid and proteins

Data

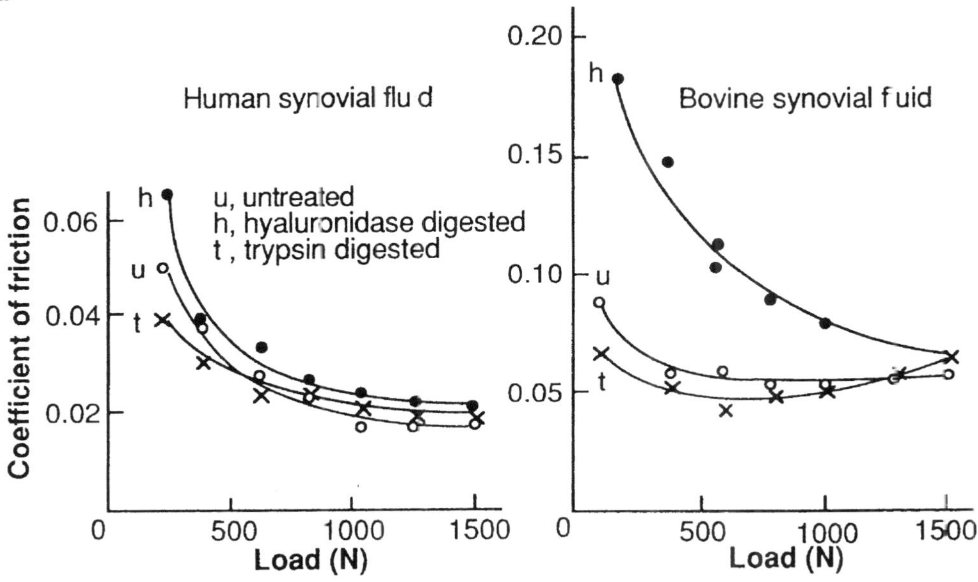

Comments
- The effect of digestive enzyme such as hyaluronidase and trypsin on friction should be evaluated from the view point of lubrication modes, which depend on lubricant viscosity, loading and speed conditions, and cartilage surface properties. In most of pendulum tests, friction has a tendency to decrease with increasing load.

Reference(s)

O'Kelly J, Unsworth A, Dowson D, Hall D, Wright V (1978) A study of the role of synovial fluid and its constituents in the friction and lubrication of human hip joints. Eng Med 7:73–83 (with permission)
Unsworth A, Dowson D, Wright V (1975) The frictional behavior of human synovial joints-Part 1 natural joints. Trans ASME J Lub Tech 97:369–376

Coefficient of Friction (5)

• Pendulum test •	• Human • Hip joint	• Surface wipe •

Materials
- Human
- Articular cartilage from hip joint

Testing Methods and Experimental Conditions
- Friction test by using pendulum apparatus
- Coefficient of friction was calculated from the decay of amplitude
- Lubricated with human synovial fluid

Data

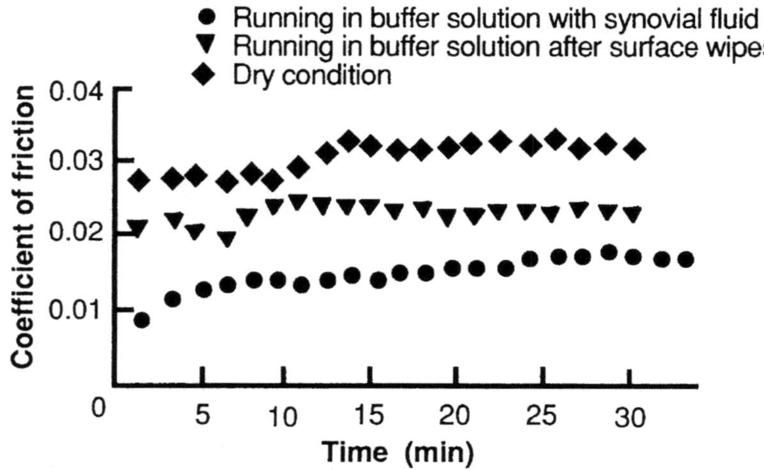

Comments
- The increase in friction with loading time appears to be due to cartilage deformation losses such as ploughing.

Reference(s)
Clarke IC, Contini R, Kenedi RM (1975) Friction and wear studies of articular cartilage: a scanning electron microscope study. Trans ASME J Lub Tech 97:358–368

Coefficient of Friction (6)

• Reciprocating test •	• Pig • Shoulder joint	• •

Materials
- Pigs
- Articular cartilage from caput humeri (head of humerus) in shoulder joint

Testing Methods and Experimental Conditions
- Friction test by using reciprocating apparatus with a stroke of 30 mm and frequency of 1.0 Hz
- Load, 2.0–19.6 N (mean contact pressure, 0.17–1.02 MPa)
- Sliding pair of spherical articular cartilage as stationary specimen and flat glass plate (center line average height, 0.02 μm) as reciprocating specimen
- Lubricants, pig synovial fluid in knee joint and water solution of sodium hyaluronate (molecular weight, 1.02×10^6)
- Room temperature

Data

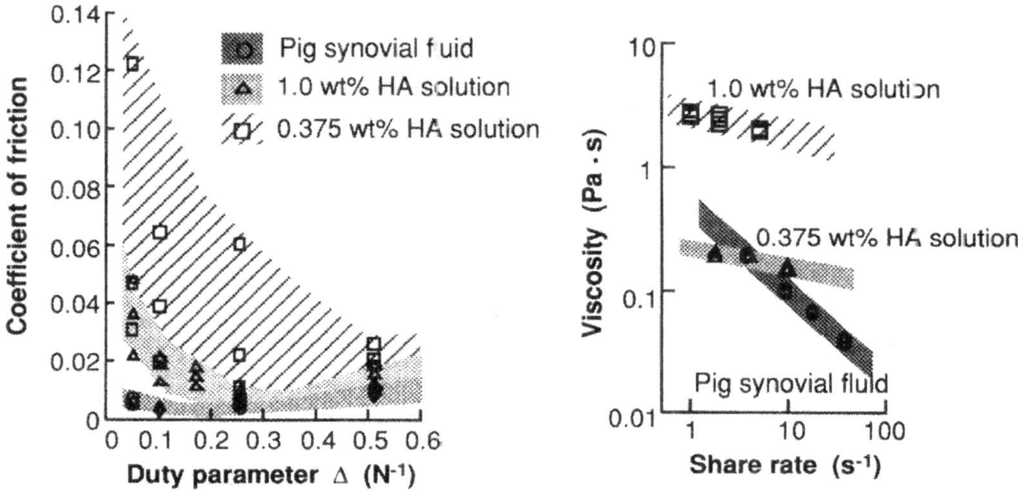

Comments
- Frictional behavior of articular cartilage significantly depends on lubricants, mating materials, and operating conditions. Synovial fluid exhibits excellent low-friction characteristics compared with sodium hyaluronate solution of higher viscosity.

Reference(s)
Higaki H, Murakami T (1994) Role of constituents in synovial fluid and surface layer of articular cartilage in joint lubrication (Part 1) — Experimental Study in Application of Enzyme Digestion. J Jpn Soc Tribol 39:625–632

Coefficient of Friction (7)

• Pendulum test •	• Pig • Shoulder joint	• Detergent •

Materials

- Pigs
- Articular cartilage from shoulder joint

Testing Methods and Experimental Conditions

- Pendulum test of articular cartilage surfaces untreated and treated with detergent
- Total weight of pendulum, 100 N
- Initial angle, 0.1 rad
- Room temperature
- Treatment with detergent, ultrasonic cleaning in 10 wt% saline solution of polyoxyethylene p-t-octylphenyl ether (Triton X-100) for 30 min and then in saline solution for 30 min
- Lubricant, saline solution of sodium hyaluronate (molecular weight, 1.02×10^6; concentration, 0.5 and 0.2 wt%)

Data

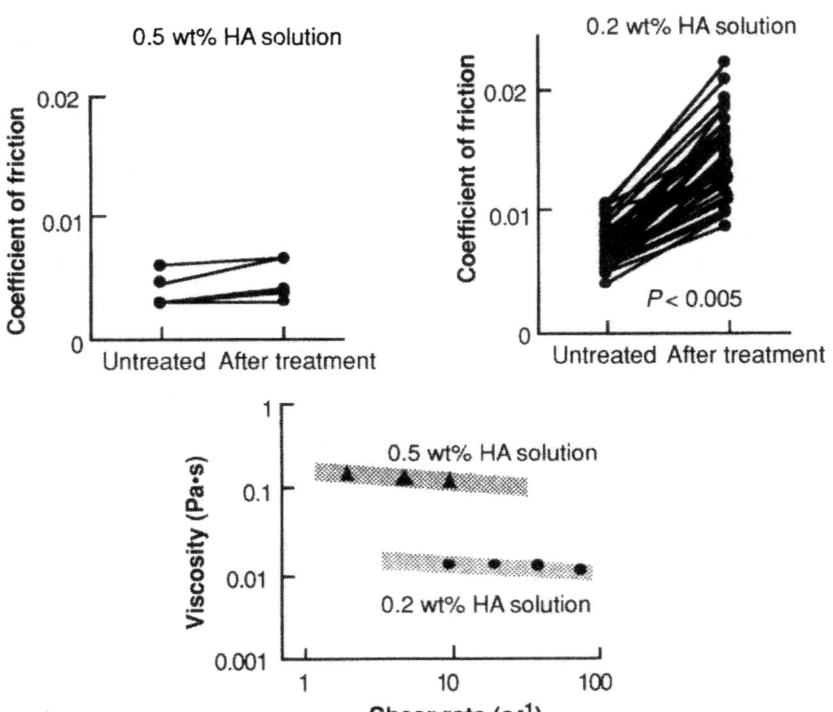

Comments

- Friction was increased by the treatment with detergent only in the lower viscosity lubricant, which suggests the boundary lubricating ability induced by adsorbed film plays a major role under thin fluid film condition.

Reference(s)

Higaki H, Murakami T, Nakanishi Y (1995) Boundary lubricating ability of protein and phospholipid in natural synovial joints. Trans Jpn Soc Mech Eng 61:3396–3401

Coefficient of Friction (8)

• Reciprocating test •	• Pig • Shoulder joint	• Material dependence •

Materials
- Pigs
- Articular cartilage from caput humeri (head of humerus) in shoulder joint

Testing Methods and Experimental Conditions
- Reciprocating friction tester; stroke, 30 mm; frequency, 1.0 Hz
- Loading condition, 3.9 to 19.6 N (mean contact pressure, 0.22 to 1.02 MPa)
- Stationary specimen, spherical articular cartilage
- Reciprocating specimen, flat plates (glass, zirconia, SUS316, ultrahigh molecular weight polyethylene [UHMWPE], polytetrafluoroethylene [PTFE])
- Lubricants, pig synovial fluid from knee joint and 1.0 wt% water solution of sodium hyaluronate (molecular weight, 1.02×10^6)
- Room temperature

Data

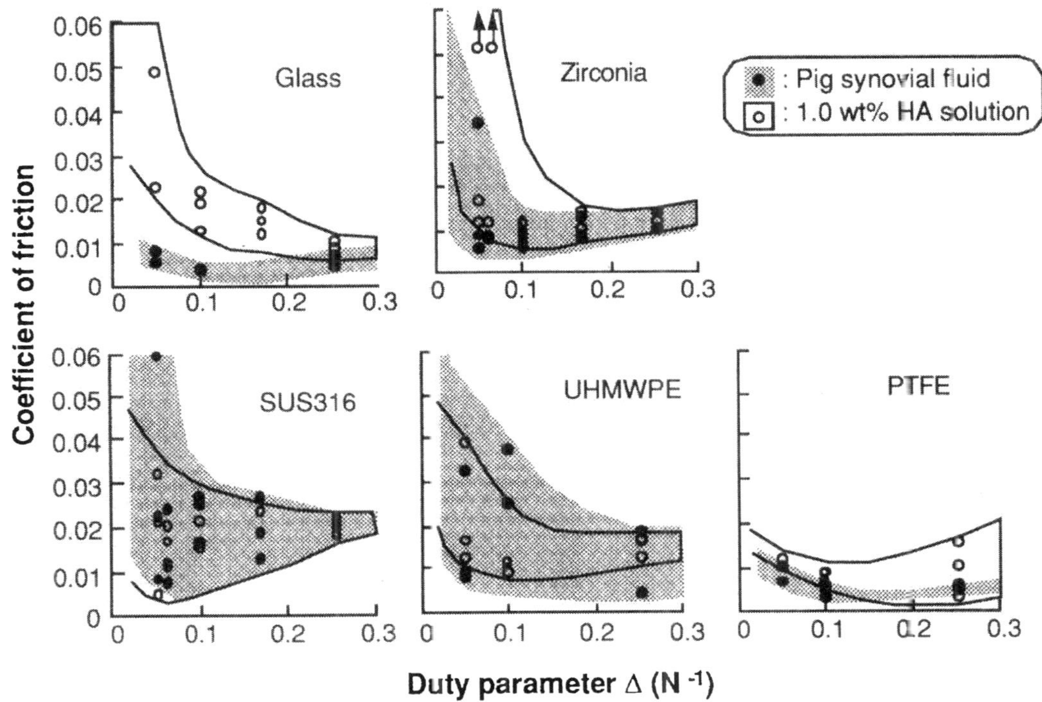

Comments
- Frictional behavior of articular cartilage significantly depends on lubricants, mating materials, and operating conditions. Under lubricated conditions with synovial fluid, hydrophilic materials tend to exhibit lower friction among glass, ceramics, and metal. Hydrophobic PTFE shows low friction for both synovial fluid and sodium hyaluronate solution.

Reference(s)
Higaki H, Murakami T (1995) Role of constituents in synovial fluid and surface layer of articular cartilage in joint lubrication (Part 2) — boundary lubrication by proteins. J Jpn Soc Tribol 40:598–604

Coefficient of Friction (9)

• Walking simulation •	• Human • Hip joint	• •

Materials

• Human
• Articular cartilage from hip joint

Testing Methods and Experimental Conditions

• Friction test by using simulator system for walking motion with bovine and human synovial fluid
• Loading conditions were up to 2489 N for bovine synovial fluid and from 210 N to 1390 N for human synovial fluid
• Relative speed was up to 20 mm/s for bovine synovial fluid and 38 mm/s for human synovial fluid

Data

	Coefficient of friction
Bovine synovial fluid [a]	< 0.09
Human synovial fluid [b]	0.005 – 0.1

Comments

• In simulator tests for walking condition, the dynamic loading at high loads enhance fluid film formation even with low viscosity synovial fluid.

Reference(s)

[a]O'Kelly J, Unsworth A, Dowson D, Hall DA, Wright V (1978) A study of the role of synovial fluid and its constituents in the friction and lubrication of human hip joints. Eng Med 7:73–83
[b]Roberts BJ, Unsworth A, Mian N (1982) Modes of lubrication in human hip joints. Ann Rheum Dis 41:217–224

Compressive Property

• Stress–strain curve • Modulus	• Bovine • Knee joint	• Chemical stress • NaCl concentration

Materials

- Cattle (age, 2 years)
- Articular cartilage from intact knee joint
- Plug of cartilage and subchondral bone from the medial facet of the femoropatellar groove

Testing Methods and Experimental Conditions

- Uniaxial confined compression test
- Plane cartilage disk (diameter, 6.4 mm; thickness, 600 µm) microtomed from plug
- NaCl concentration range, 0.005–1.0 M
- Temperature of $20° \pm 0.5°C$

Data

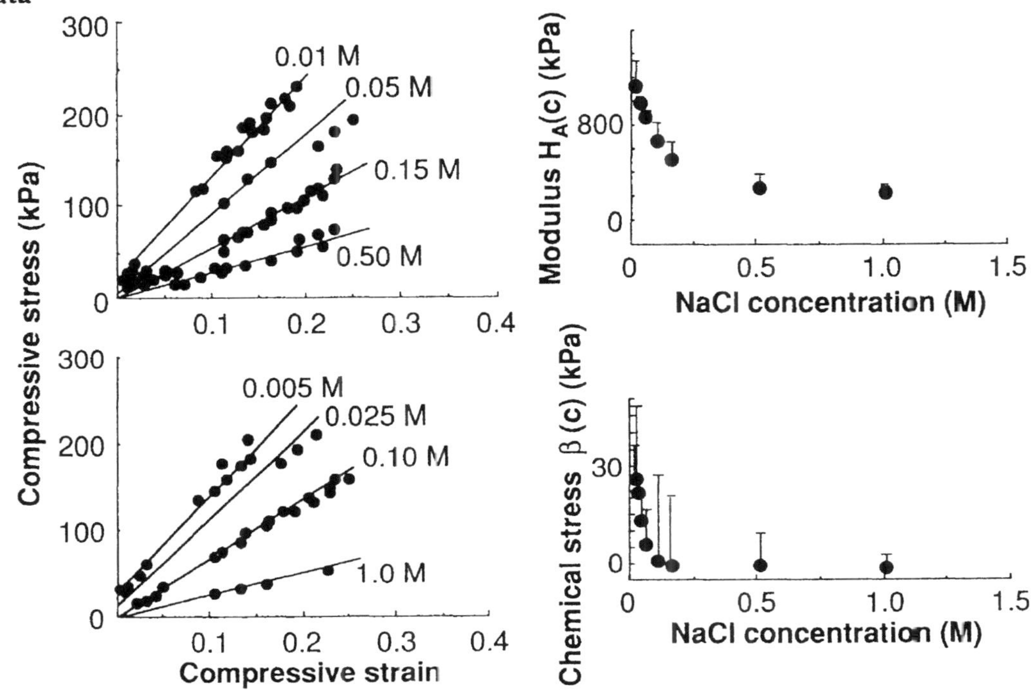

Comments

- The minimal free swelling thickness of the sample was taken as the reference of strain.
- The equilibrium stress–strain constitutive law was represented in terms of the concentration-dependent intrinsic material parameters $H_A(c)$ and $\beta(c)$ as $\sigma = H_A(c)\,\varepsilon - \beta(c)$, where σ and ε are compressive stress and strain, respectively.

Reference(s)

Eisenberg SR, Grodzinsky AJ (1985) Swelling of articular cartilage and other connective tissues: electromechanochemical forces. J Orthop Res 3:148–159 (with permission)

Creep (1)

• Aggregate modulus •	• Human • Elbow	• Locational dependence •

Materials
- Human (age, 28–82 [mean 65.5] years)
- Articular cartilage from elbow

Testing Methods and Experimental Conditions
- Indentation creep test with a 2.0 mm diameter, flat-ended, cylindrical, rigid, porous indenter tip
- Positions of the specimen: anterior (AN), medial (ME), posterior (PO), lateral (LA), and central (CE) for the radial head, and superolateral (SL), superomedial (SM), midlateral (ML), midmedial (MM), posterolateral (PL), and posteromedial (PM) for the capitellum
- Aggregate modulus was determined using the linear biphasic theory

Data

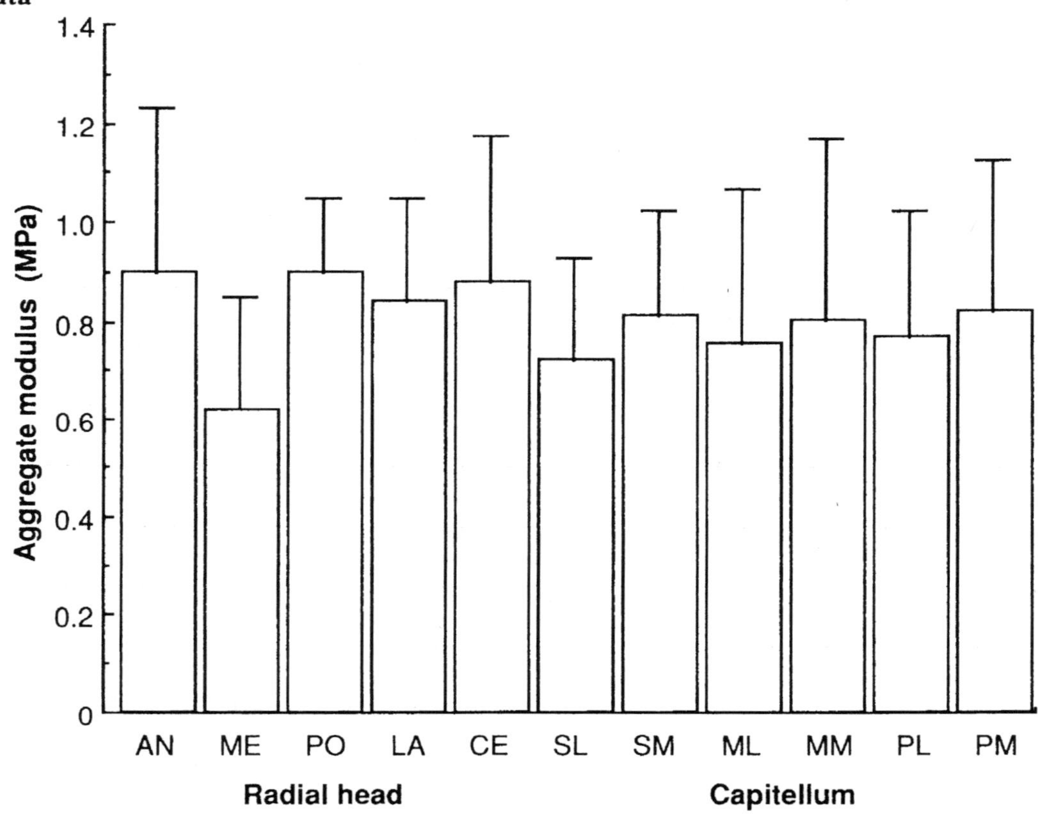

Comments
- Significant differences exist in the mechanical properties and thickness of cartilage topographically in the capitellum and radial head, and between the two surfaces.

Reference(s)
Schenck Jr RC, Athanasiou KA, Constantinides G, Gomez E (1994) A biomechanical analysis of articular cartilage of the human elbow and a potential relationship to osteochondritis dissecans. Clin Orthop Relat Res 299:305–312 (with permission)

Creep (2)

• Aggregate modulus • Permeability	• Bovine • Femoral condyle	• •

Materials
- Bovine
- Articular cartilage from femoral condyle

Testing Methods and Experimental Conditions
- Compression creep test; circular plug specimens were suddenly compressed at a constant load (0.10 MPa) with a permeable loading block (plug diameter, 6.35 mm; average thickness, 1.78 mm)
- Radial flow of the interstitial fluid and bulk lateral motion of the tissue were restricted

Data

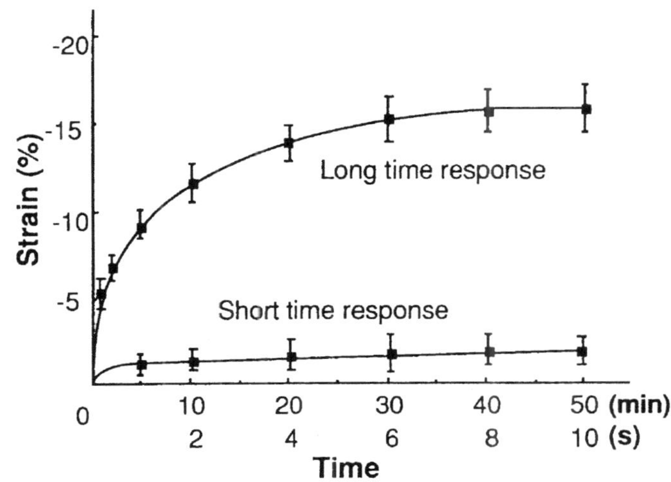

Specimen number	Permeability ($\times 10^{-14}$ m^4/Ns)	Aggregate modulus (MPa)	Thickness (mm)
1	0.38	0.73	1.81
2	0.55	0.68	1.75
3	0.30	0.64	1.61
4	0.97	0.69	1.74
5	1.33	0.87	1.91
6	0.30	0.80	1.54
7	1.19	0.58	1.95
8	0.98	0.62	1.88
9	1.22	0.79	1.88
10	0.39	0.65	1.69
mean ± SD	0.76 ± 0.42	0.70 ± 0.09	1.78

Comments
- As the tissue is compressed, the permeability drops appreciably.
- The articular cartilage does not possess an instantaneous response subsequent to a sudden application of load during confined compression.

Reference(s)
Mow VC, Kuei SC, Lai WM, Armstrong CG (1980) Biphasic creep and stress relaxation of articular cartilage: theory and experiments. Trans ASME J Biomech Eng 102:73–84

Creep (3)

• Aggregate modulus • Poisson's ratio	• Human, bovine, monkey • Femur	• Locational dependence •

Materials
- Human, bovine, Cynomolgus monkeys, greyhound dogs, and New Zealand rabbits
- Articular cartilage from distal femur

Testing Methods and Experimental Conditions
- Biphasic indentation creep test; test area, intermittent low-weight-bearing areas for patellar groove and high-weight-bearing areas for femoral condyles; indenter, porous-permeable tip (1.0 mm in diameter for rabbits, 1.5 mm for other); tare load, 0.0343–0.049 N for 15 min; test load, 0.098 N for rabbit, monkey, and canine, 0.1961 N for human and bovine
- Material coefficients (aggregate modulus, Poisson's ratio, and permeability) were determined by the linear biphasic theory

Data

	Poisson's ratio	Aggregate modulus (MPa)	Permeability ($\times 10^{-15}$ m^4/Ns)	Tissue thickness (mm)
Lateral condyle				
Human ($n = 4$)	0.098 ± 0.069	0.701 ± 0.228	1.182 ± 0.207	2.31 ± 0.53
Bovine ($n = 10$)	0.396 ± 0.023	0.894 ± 0.293	0.426 ± 0.197	0.94 ± 0.17
Dog ($n = 6$)	0.300 ± 0.075	0.603 ± 0.237	0.774 ± 0.563	0.58 ± 0.20
Monkey ($n = 6$)	0.236 ± 0.044	0.778 ± 0.176	4.187 ± 1.545	0.57 ± 0.12
Rabbit ($n = 6$)	0.337 ± 0.092	0.537 ± 0.258	1.806 ± 1.049	0.25 ± 0.06
Medial condyle				
Human ($n = 6$)	0.074 ± 0.084	0.588 ± 0.114	1.137 ± 0.160	2.21 ± 0.59
Bovine ($n = 10$)	0.383 ± 0.047	0.899 ± 0.427	0.455 ± 0.332	1.19 ± 0.24
Dog ($n = 5$)	0.372 ± 0.050	0.904 ± 0.218	0.804 ± 0.776	0.90 ± 0.15
Monkey ($n = 6$)	0.236 ± 0.055	0.815 ± 0.180	2.442 ± 1.129	0.72 ± 0.09
Rabbit ($n = 6$)	0.197 ± 0.094	0.741 ± 0.101	2.019 ± 1.621	0.41 ± 0.10
Patellar groove				
Human ($n = 4$)	0.000 ± 0.000	0.530 ± 0.094	2.173 ± 0.730	3.57 ± 1.12
Bovine ($n = 10$)	0.245 ± 0.065	0.472 ± 0.147	1.422 ± 0.580	1.38 ± 0.19
Dog ($n = 6$)	0.093 ± 0.067	0.555 ± 0.144	0.927 ± 0.844	0.52 ± 0.12
Monkey ($n = 6$)	0.197 ± 0.123	0.522 ± 0.159	4.737 ± 2.289	0.41 ± 0.05
Rabbit ($n = 6$)	0.206 ± 0.126	0.516 ± 0.202	3.842 ± 3.260	0.20 ± 0.04

All data are given as mean ± SD.

Comments
- Significant differences exist in some of these material properties among species and sites.
- Patellar groove cartilage can undergo greater and faster compression.

Reference(s)
Athanasiou KA, Rosenwasser MP, Buckwalter JA, Malinin TI, Mow YC (1991) Interspecies comparisons of in situ intrinsic mechanical properties of distal femoral cartilage. J Orthop Res 9:330–340 (with permission)

Creep (4)

• Aggregate modulus • Poisson's ratio	• Human • Hip joint	• Locational dependence •

Materials
- Human (age, 24, 28, 29, 47, and 50 years)
- Articular cartilage from hip joint

Testing Methods and Experimental Conditions
- Biphasic creep indentation test; test area, femoral head and acetabulum; indenter, porous-permeable tip (2.0 mm in diameter); tare load, 0.0687 N; test load, 0.438 N
- Intrinsic material coefficients (aggregate modulus, Poisson's ratio, and permeability) were calculated based on the linear biphasic theory

Data

	Aggregate modulus (MPa)	Poisson's ratio	Permeability ($\times 10^{-15}$ m^4/Ns)	Thickness (mm)
Femoral head				
Anteromedial	1.198 ± 0.245	0.055 ± 0.073	0.906 ± 0.497	1.84 ± 0.17
Anterolateral	1.067 ± 0.561	0.055 ± 0.069	0.940 ± 0.409	1.23 ± 0.28
Superomedial	1.816 ± 0.868	0.058 ± 0.074	1.002 ± 0.576	1.66 ± 0.13
Superolateral	1.054 ± 0.742	0.058 ± 0.069	0.814 ± 0.262	1.11 ± 0.23
Posteromedial	1.555 ± 0.603	0.013 ± 0.028	1.101 ± 0.610	1.79 ± 0.24
Posterolateral	1.169 ± 0.628	0.039 ± 0.060	0.781 ± 0.450	1.11 ± 0.23
Inferomedial	0.948 ± 0.238	0.047 ± 0.048	0.880 ± 0.470	1.42 ± 0.27
Inferolateral	0.679 ± 0.162	0.045 ± 0.064	0.899 ± 0.444	1.03 ± 0.16
Acetabulum				
Anteromedial	1.340 ± 0.739	0.045 ± 0.051	0.826 ± 0.365	1.22 ± 0.19
Anterolateral	1.147 ± 0.495	0.034 ± 0.054	0.710 ± 0.362	1.26 ± 0.15
Superomedial	1.284 ± 0.650	0.055 ± 0.075	1.133 ± 1.114	1.11 ± 0.34
Superolateral	1.072 ± 0.546	0.011 ± 0.022	0.983 ± 0.609	1.83 ± 0.45
Posteromedial	1.424 ± 0.602	0.097 ± 0.077	0.737 ± 0.588	1.06 ± 0.24
Posterolateral	1.147 ± 0.444	0.019 ± 0.025	0.819 ± 0.440	1.09 ± 0.23

All data are given as mean $\pm$ SD, $n = 10$.

Comments
- There are significant differences between these properties regionally in the acetabulum and femoral head and between the two anatomical structures.
- The superomedial aspect of the femoral head has the greatest aggregate modulus within the hip joint.

Reference(s)
Athanasiou KA, Agarwal A, Dzida FJ (1994) Comparative study of the intrinsic mechanical properties of the human acetabular and femoral head cartilage. J Orthop Res 12:340–349 (with permission)

Creep (5)

• Equilibrium modulus • Permeability	• Human • Patella	• Water content •

Materials

- Human (age, 16–85 [mean 56.4] years; $n = 103$)
- Articular cartilage from patella

Testing Methods and Experimental Conditions

- Confined compression creep test; cylindrical plug specimen from the lateral facet of the patella (6.35 mm in diameter, 1.69–5.17 mm in thickness) was mounted in an impermeable confining chamber
- Cartilage surface was suddenly loaded at 0.1 MPa with a rigid, highly permeable filter
- Water content of the cartilage was determined from the weight

Data

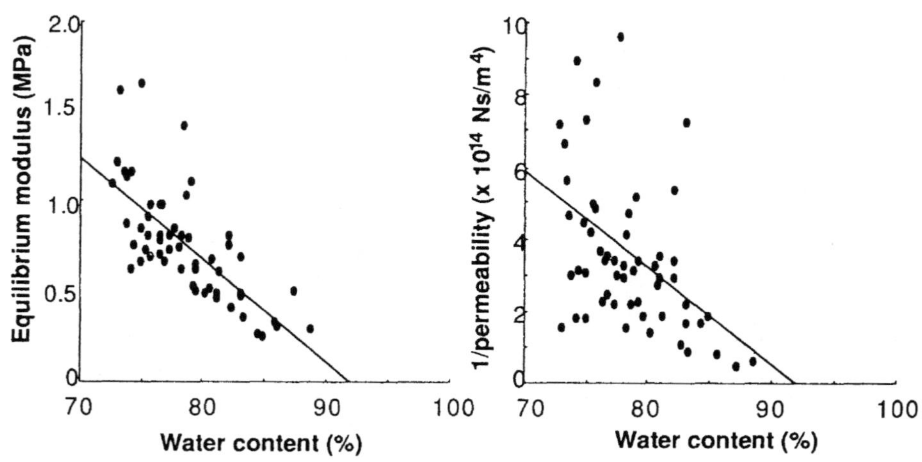

Comments

- The equilibrium modulus (H) and water content (W) are related by the regression equation $H = 5.29 - 0.058\,W$; $r = -0.73$, $n = 58$, $P < 0.0001$.
- The permeability (k) and water content are related by the regression equation $1/k = 24.8 \times 10^{14} - 0.27 \times 10^{14}\,W$; $r = 0.50$, $n = 58$, $P < 0.0001$.
- The intrinsic equilibrium modulus and the permeability both were highly correlated with the water content of the tissue; as water content increases, the matrix of the tissue becomes softer and more permeable.

Reference(s)

Armstrong CG, Mow VC (1982) Variations in the intrinsic mechanical properties of human articular cartilage with age, degeneration, and water content. J Bone Joint Surg 64A:88–94 (with permission)

Creep (6)

• Shear modulus •	• Rabbit • Femoral condyle	• Storage •

Materials
- New Zealand white rabbits (four young adult males; weight, 3.46–3.69 kg)
- Articular cartilage from medial femoral condyle

Testing Methods and Experimental Conditions
- Indentation test for whole condyle
- Tested in sterile Hank's balanced salt solution at 37.0° ± 0.5°C
- Stabilized under 5 g preload
- Instantaneous load of 25 and 40 g at a rate of 16.3 N/s for unrelaxed shear modulus
- Creep test at 25 g sustained load for relaxed shear modulus
- Tests repeated at days 1, 4, 7, and 12 for each specimen stored in tissue culture media

Data

Day of storage	Preload deformation (μm)	Unrelaxed shear modulus ($\times 10^5$ Pa)	Relaxed shear modulus ($\times 10^5$ Pa)
0	25.2 ± 2.5	3.46 ± 0.24	0.62 ± 0.04
	$P < 0.03$	$P < 0.07$	NS
1	29.6 ± 2.8	3.87 ± 0.25	0.63 ± 0.05
	$P < 0.02$	$P < 0.01$	NS
4	40.4 ± 4.7	5.04 ± 0.32	0.57 ± 0.05
	$P < 0.07$	$P < 0.02$	NS
7	46.1 ± 5.9	6.29 ± 0.50	0.57 ± 0.04
	NS	NS	NS
12	44.3 ± 4.8	6.19 ± 0.52	0.55 ± 0.05

All data are given as mean ± SE. Each *P* value is a comparison of the values between two sequential periods (i.e., day 0 *vs* day 1, day 1 *vs* day 4 , etc.). NS, not significant.

Comments
- Shear modulus was calculated from G = P (1-ω) /4 ak, where G is the shear modulus; P, applied force; ω, deformation; a, radius of indentation; k(a/h, υ), geometric scaling function with cartilage thickness h and Poisson's ratio υ.
- Preload deformation showed a significant and progressive increase through day 4 while the relaxed modulus remained unchanged.
- Morphologically, lacunar enlargement and chondrocyte swelling were noted in stored cartilage.

Reference(s)
Thomas VJ, Jimenez SA, Brighton CT, Brown N (1984) Sequential change in the mechanical properties of viable articular cartilage stored in vitro. J Orthop Res 2:55–60 (with permission)

Dynamic Shear Modulus (1)

• Storage modulus • Damping factor	• Bovine, dogfish • Scapula, rostrum	• •

Materials

- *Bos taurus* (cow) 1. Cartilage from central part of right scapula (perpendicular to vertebral border)
  2. Tendon of medial extensor of right forefoot
- *Squalus acanthias* (dogfish), cartilage from rostrum

Testing Methods and Experimental Conditions

- Prismatic specimens with gauged length of 21 mm and cross section of 5 × 2 mm
- Torsional test by a free-oscillating torsion technique
- At temperatures between 110 K and room temperature
- Load frequency of 1 Hz

Data

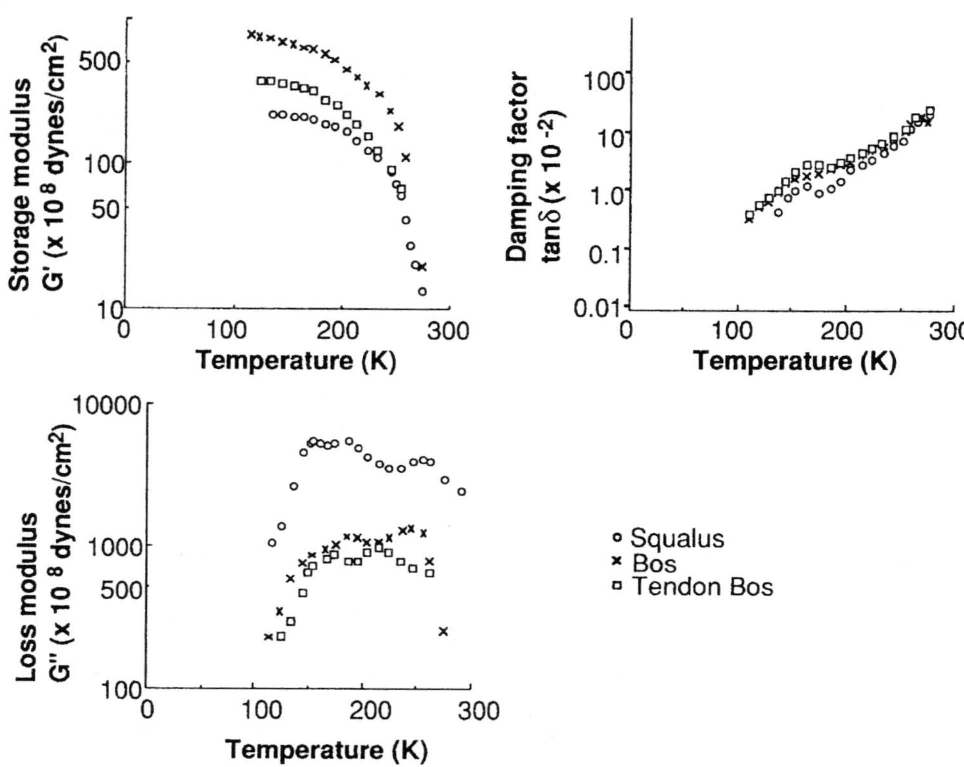

Comments

- The curves of storage modulus (G', dynes/cm²) show a sharp decrease at 200 K, and reach a very low value. The curves of damping factor (tan δ = G''/G') are very similar, and show damping maxima at 150 K and at 250 K and a short plateau around 190 K.
- Dynamic shear modulus: G* = G' + iG''

Reference(s)

Ramaekers JGM (1979) The rheological behaviour of skeletal material originating from several classes of vertebrates. Neth J Zool 29:166–176 (with permission)

Dynamic Shear Modulus (2)

• Viscoelasticity • Anisotropic	• Bovine • Meniscus	• •

Materials
- Bovine (age, 12–15 months), meniscus

Testing Methods and Experimental Conditions
- Cyclic shear deformation over a physiological range of frequencies (1–100 rad/s)
- Shear deformation for infinitesimal strains (0.005–0.05 rad)
- Disc shaped specimens (6 mm diameter, 1.0 mm thickness) were oriented radially, axially, and circumferentially

Data

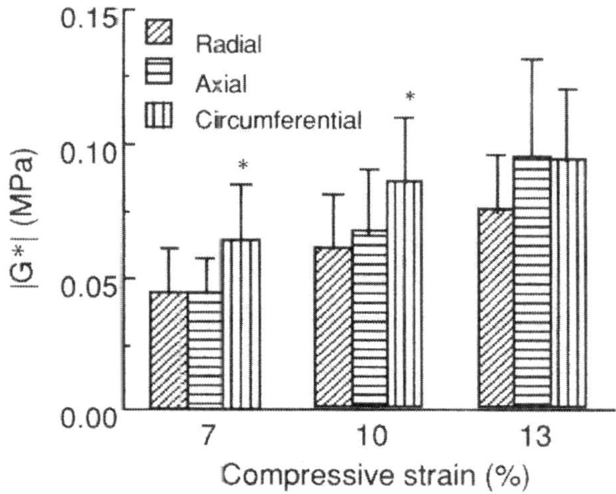

Comments
- *Statistically significant difference was detected.
- Viscoelastic behavior in shear depends on frequency, shear strain and compressive strain.
- Anisotropic shear properties are related to the collagen fiber orientation.

Reference(s)
Zhu W, Chern KY, Mow VC (1994) Anisotropic viscoelastic shear properties of bovine meniscus. Clin Orthop Relat Res 306:34–45 (with permission)

Stress–Relaxation (1)

• Permeability • Equilibrium modulus	• Human • Femoral	• •

Materials
- Human
- Articular cartilage from proximal femoral chondroepiphysis

Testing Methods and Experimental Conditions
- Ramp / stress–relaxation test in compression
- 2.5 mm diameter cylindrical specimens from 2 mm thick transversely sectioned slabs
- Specimen axis was in the weight-bearing longitudinal direction
- Two strain rates of 0.0042 per second (fast) and 0.00042 per second (slow)
- 2% prestrain by 0.008 N preload, and 8% additional strain at the peak

Data

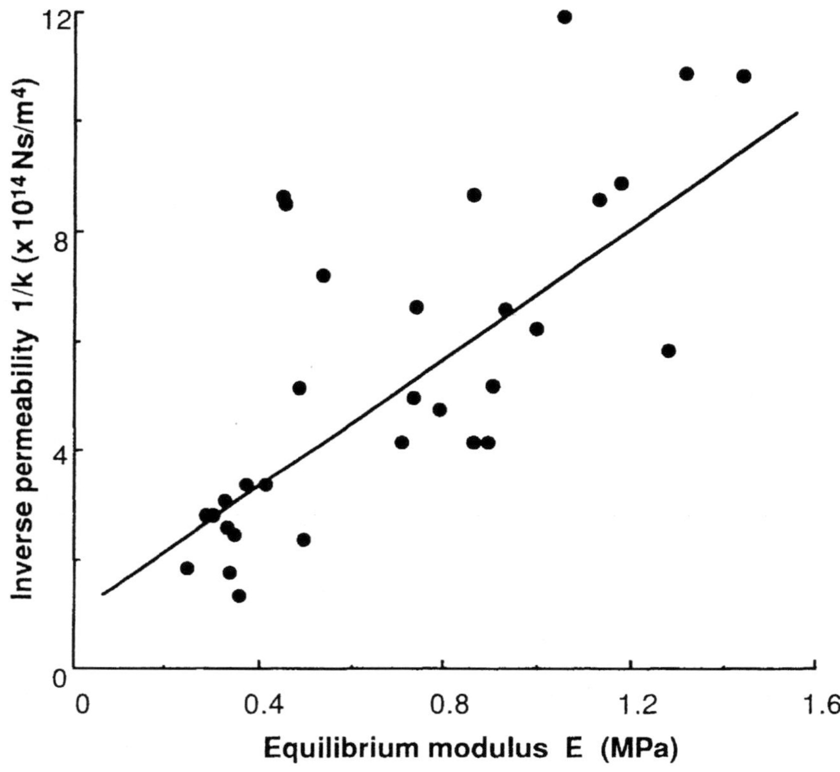

Comments
- The linear regression curve (r = 0.736) is given by
  $$E = 0.1616 \times 10^{-14}(1/k) - 0.193$$

Reference(s)
Brown TD, Singerman RJ (1986) Experimental determination of the linear biphasic constitutive coefficients of human fetal proximal femoral condroepiphysis. J Biomech 19:597–605 (with permission)

Stress–Relaxation (2)

• Relaxation function •	• Bovine • Humeral head	• •

Materials
- Bovine (age, 15–18 months)
- Articular cartilage from humeral head

Testing Methods and Experimental Conditions
- Uniaxial tensile stress relaxation test using dumbbell shape specimen
  ($n = 25$, h = 250–325 μm)
- Immersed in normal saline solution at 37°C
- Relaxation functon G(t) was defined as $\sigma(t) / \sigma(t = \sim 250$ ms) in place of $\sigma(t) / c(t = 0)$
- λ was defined as stretched length/initial length

Data

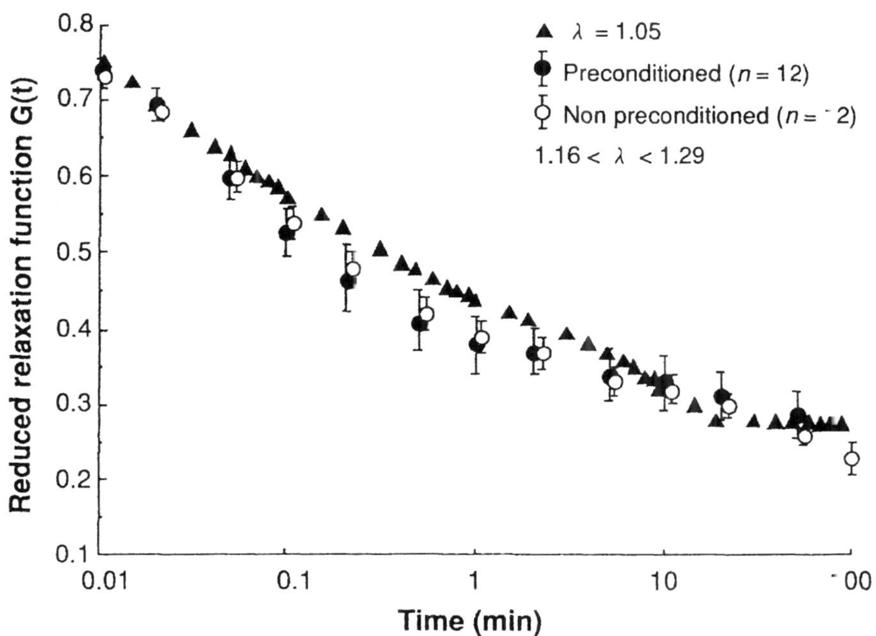

Comments
- The slope of stress relaxation curve is independent of stretch ratio.
- Preconditioning and non-preconditioning showed no effect on the slope of stress relaxation.

Reference(s)
Woo SL-Y, Simon BR, Kuei SC, Akeson WH (1980) Quasi-linear viscoelastic properties of normal articular cartilage. Trans ASME J Biomech Eng 102:85–90

Stress–Relaxation (3)

• Relaxation function • Equilibrium stress	• Bovine • Tibial plateau	• •

Materials

- Cows (mature)
- Stifle joint
- Articular cartilage from medial side of normal tibial plateau

Testing Methods and Experimental Conditions

- Full-thickness cylindrical plugs of 3 mm in diameter with approximately 0.2 mm of underlying subcondral bone ($n = 7$)
- Stored in Hanks balanced salt solution (BSS) at 4°C for 12–18 h before testing
- Shear stress relaxation test in Hanks BSS at 37° ± 2°C
- Four different shear strains ranging from 3.6% to 15.2% by using a fast ramp function (rate, 25 mm/s)

Data

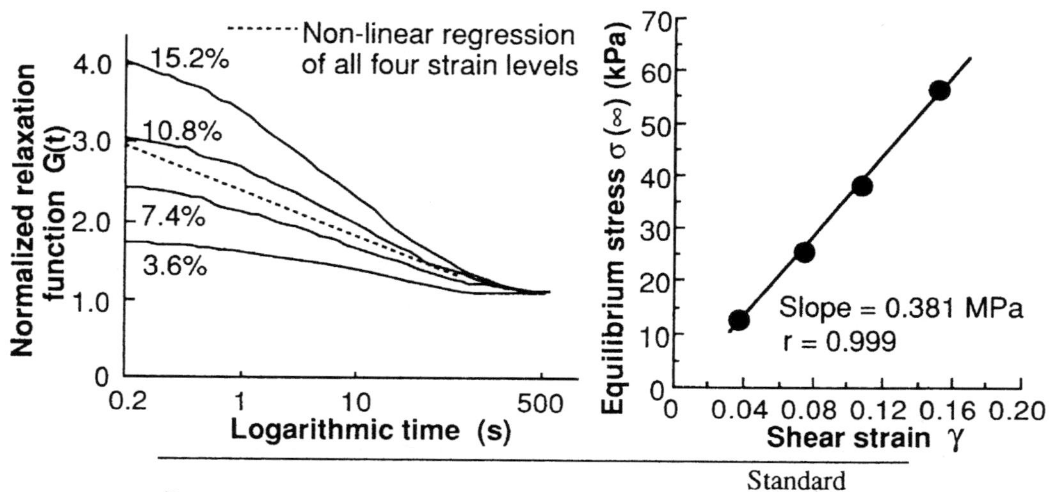

Parameter	Range	Mean	Standard deviation
A	1.74 to 4.43	3.15	0.89
B	-0.010 to 0.058	0.020	0.025
τ_1 (s)	0.0004 to 0.1646	0.0427	0.0572
τ_2 (s)	42.0 to 378	201	121
b_1	0.342 to 1.26	0.702	0.313
b_2	-5.64 to 27.2	5.54	11.8
b_3	-46.2 to 63.6	20.2	38.6

Comments

- Normalized relaxation function is defined as $G(t) = \sigma(t) / \sigma(\infty)$.
- For the linear portion in the semilogarithmic plot of $G(t)$, the material parameters A, B, τ_1, and τ_2 were determined for a normalized relaxation model of

$$g(\gamma, t) = 1 + \int_{\tau_1}^{\tau_2} [(A\gamma + B) / \tau] \exp t(-t / \tau) \, d\tau$$

- The instantaneous stress data were fitted to a polynomial of the form
$$\sigma(t = 0) = b_1\gamma + b_2\gamma^2 + b_3\gamma^3.$$

Reference(s)

Spit AD, Mak AF, Wassel RP (1989) Nonlinear viscoelastic properties of articular cartilage in shear. J Orthop Res 7:43–49 (with permission)

Tensile Property

• Tensile fracture stress •	• Human • Femoral head, ankle joint	• Age effect •

Materials

- Human (age, 7–90 years)
- Articular cartilage from femoral head and ankle joint

Testing Methods and Experimental Conditions

- Tensile fracture test
- Specimen, parallel-sided central gauge section (3.0 × 1.0 × 0.2 mm)
- Sampling site, superficial and mid-depth zones of the femoral head, and talus of the ankle
- Environment, Ringer's solution
- Extension rate of 5 mm/min

Data

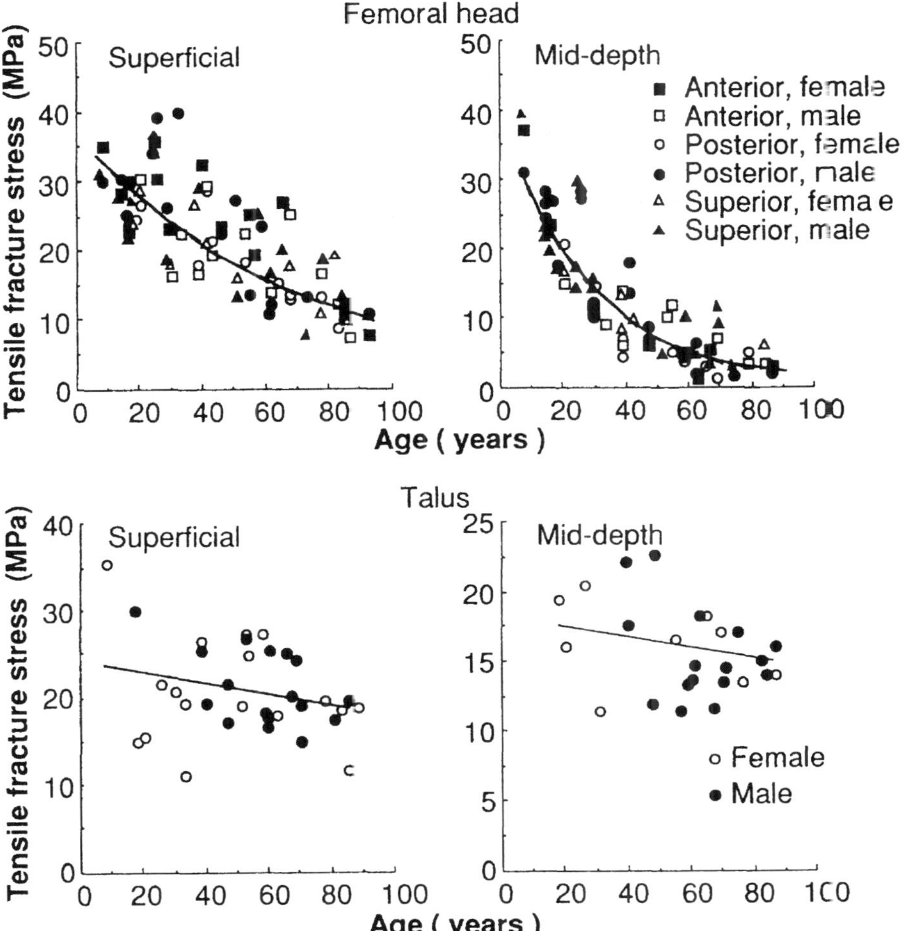

Comments

- The tensile fracture stress of cartilage from both the superficial and mid-depth zones of the femoral head decreased considerably with age. In contrast, the fracture stress of both levels of cartilage from the talus of the ankle showed no significant decrease.

Reference(s)

Kempson GE (1991) Age-related changes in the tensile properties of human articular cartilage: a comparative study between the femoral head of the hip joint and the talus of the ankle joint. Biochimica et Biophysica Acta 1075:223–230 (with permission)

Force–Elongation Relation

• Extension • Elastic modulus	• Rabbit • Mesentery	• Constitutive equation •

Materials
- Rabbits
- Mesentery

Testing Methods and Experimental Conditions
- Tensile tests
- Stretch rate, 0.254 cm/min
- In a physiological solution at room temperature

Data

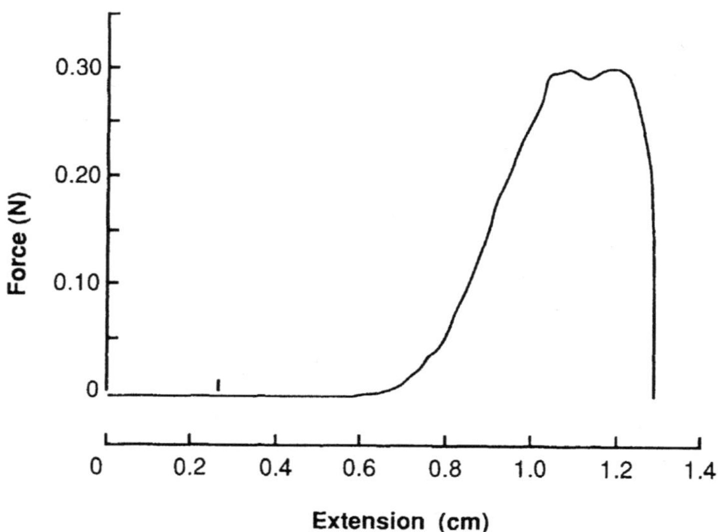

Comments
- The specimen failed by tearing at some unpredictable points.
- $dT/d\lambda = aT(1 - bT)$.

  T, tension; λ, extension ratio; a, b, constants.
- One typical example, $a = 12.4$, $b = -7.29 \times 10^{-3}$ Pa^{-1}.

Reference(s)
Fung YC (1967) Elasticity of soft tissue in simple elongation. Am J Physiol 213:1532–1544
(with permission)

Anisotropy

<table>
<tr><td>• Elastic modulus
•</td><td>• Human
• Valve leaflets</td><td>• Fresh
•</td></tr>
</table>

Materials
- Human aortic and mitral valve leaflets
- Specimens from either the radial or oblique direction of the leaflets

Testing Methods and Experimental Conditions
- Tensile testing
- Room temperature
- Excised 24 to 48 hours after death
- Each strip 1.0 to 1.5 cm long and 0.5 cm wide
- Crosshead speed, 1.27 mm/min

Data

	Mitral valve leaflets	Aortic valve leaflets	
Parameter	Fresh	Circumferential	Radial
Pretransition modulus (kPa)	11.07 ± 4.70	19.47 ± 6.78	11.03 ± 4.51
Stress at transition (kPa)	3.45 ± 0.98	6.87 ± 2.37	7.37 ± 3.22
Elongation at transition (%)	14.32 ± 3.09	12.8 ± 5.6	23.85 ± 8.275
Post-transition modulus (MPa)	2.91 ± 1.37	5.86 ± 2.95	1.71 ± 0.68
Elastic recovery (%)	98.1 ± 2.9	99.1 ± 0.545	98.57 ± 3.36
Thickness (mm)	0.90 ± 0.09	0.60 ± 0.07	0.62 ± 0.06

All data are given as mean $\pm$ SD, $n=6$

Comments
- The uniaxial behavior of all tissue tested was anisotropic and was characterized by an extremely low modulus of elasticity.

Reference(s)
Clark RE (1973) Stress–strain characteristics of fresh and frozen human aortic and mitral leaflets and chordae tendineae — Implications for clinical use. J Thorac Cardiovasc Surg 66: 202–208 (with permission)

Stress–Strain Relation

• Elastic modulus •	• Human • Chordae tendineae	• Fresh • Frozen

Materials

- Human chordae tendineae
- Fresh specimens were preserved at 4°C or frozen at –70°C

Testing Methods and Experimental Conditions

- Tensile testing
- Room temperature
- Excised 24 to 48 hours after death
- Crosshead speed, 1.27 mm/min
- Mechanical properties were tested within 1 week of harvest

Data

	Chordae tendineae	
	Fresh	Frozen
Pretransition modulus (Pa)	29.37 ± 16.12	40.87 ± 21.39
Stress at transition (Pa)	7.43 ± 3.28	4.70 ± 1.99
Elongation at transition (%)	12.143 ± 5.309	8.10 ± 1.687
Post-transition modulus (MPa)	8.82 ± 4.66	7.73 ± 3.83
Elastic recovery (%)	99.125 ± 0.895	98.8 ± 0.92
Thickness (mm)	0.653 ± 0.19	0.54 ± 0.083

All data are given as mean $\pm$ SD, $n = 6$

Comments

- The chordae tendineae data demonstrated the nonlinear behavior of valvular tissue.
- Chordae tendineae were affected by freezing.

Reference(s)

Clark RE (1973) Stress–strain characteristics of fresh and frozen human aortic and mitral leaflets and chordae tendineae — Implications for clinical use. J Thorac Cardiovasc Surg 66: 202–208 (with permission)

Force–Elongation Relation

• Extension ratio • Stiffness	• Rabbit • Skin	• Anisotropy •

Materials
- Albino male rabbits
- Rectangular specimen from an area exterior to the nipple line

Testing Methods and Experimental Conditions
- Biaxial stretch tests: first stretched in the x–direction then stretched in the y–direction, with the transverse stretch ratio kept at 1
- Uniaxial tension tests
- Stretch rate, 0.2 mm/s

Data

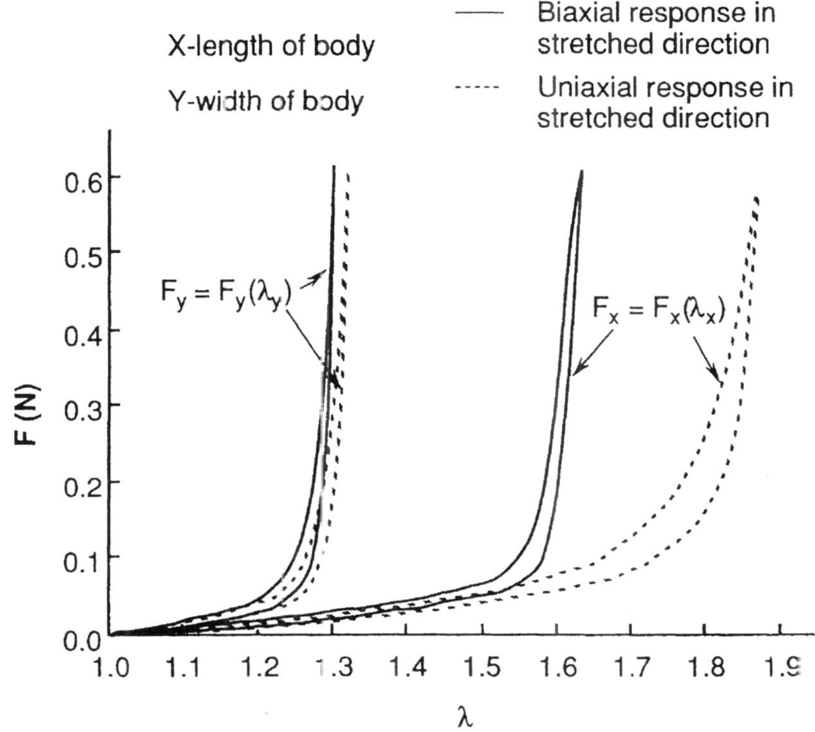

Comments
None.

Reference(s)

Lanir Y, Fung YC (1974) Two-dimensional mechanical properties of rabbit skin — II. Experimental results. J Biomech 7:171–182 (with permission)

Stress–Strain Relation

• Tensile strength • Extensibility	• Rat • Skin	• Collagen • Gelation

Materials

- Rat dorsal skin
- Collagen film, which is obtained by gelation of collagen prepared from rat skin, followed by drying

Testing Methods and Experimental Conditions

- Strips of collagen films (2 mm wide) and strips of rat dorsal skin (2 mm wide) were immersed in Ringer's solution (pH 7.4, 22°C)
- Stress values for strip speciments were calculated as load values normalized to the measured cross sectional area
- Strain values were obtained by expressing deformation values in units of original specimen length

Data

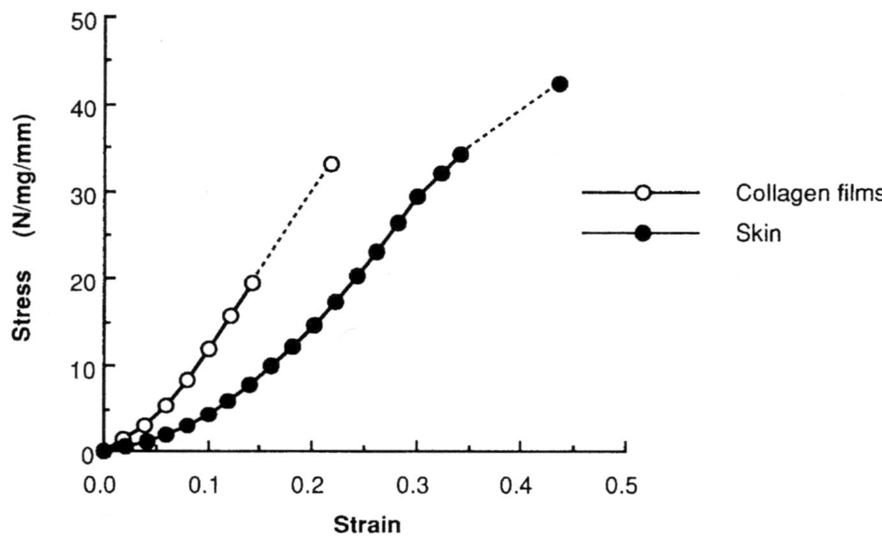

Comments

- The mechanical properties of the collagen films differed from those of the rat skin from which the film collagen was extracted, in being stiffer but with less strength and extensibility.

Reference(s)

Oxlund H, Andreassen TT (1980) The role of hyaluronic acid, collagen and elastin in the mechanical properties of connective tissues. J Anat 131:611–620 (with permission)

Force–Elongation Relation

• Tensile strength • Elastic stiffness	• Rabbit • Abdominal wall	• Position (Location) •

Materials

- Adult male albino rabbits
- Specimens excised from abdominal wall
- Specimen location: six from transverse plane, two from musculus rectus abdominis, and eight from linea alba

Testing Methods and Experimental Conditions

- Tensile tests
- Dumbbell-shaped and strip-shaped specimens
- Tests less than 4 hours after the animal was killed
- Temperature, 21°C ; relative humidity, 65%
- Crosshead speed, 20 mm/min

Data

Position	Breaking strength (N)	Energy absorption (N)	Maximum strain	Elastic stiffness (N)
I	2.90 ± 0.28	144.61 ± 22.85	0.80 ± 0.24	8.65 ± 4.75
II	3.32 ± 1.57	171.54 ± 49.82	0.69 ± 0.07	8.82 ± 4.18
III	2.95 ± 0.38	126.00 ± 44.24	0.70 ± 0.10	8.07 ± 1.58
IV	3.12 ± 0.78	145.72 ± 45.24	0.74 ± 0.10	7.83 ± 2.37
V	2.93 ± 0.71	150.55 ± 42.53	0.78 ± 0.13	7.90 ± 3.38
VI	2.82 ± 0.99	163.23 ± 69.77	0.76 ± 0.16	6.50 ± 2.31
Rectus	11.64 ± 2.77	7861.31 ± 372.07	1.50 ± 0.38	10.41 ± 5.21

All data are given as mean ± SD.
Positions I–VI correspond to the locations from lateral to central sides in the abdominal wall.

Position	Breaking strength (N)	Energy absorption (N)	Maximum strain	Elastic stiffness (N)
I	10.39 ± 1.66	431.74 ± 119.34	0.75 ± 0.16	21.77 ± 7.41
II	12.59 ± 3.89	533.84 ± 159.30	0.75 ± 0.17	26.22 ± 10.00
III	14.78 ± 5.55	557.61 ± 223.81	0.67 ± 0.14	32.27 ± 11.41
IV	16.95 ± 4.59	755.08 ± 367.07	0.72 ± 0.18	37.27 ± 11.41
V	17.13 ± 3.69	603.42 ± 224.17	0.56 ± 0.12	51.00 ± 17.83
VI	16.30 ± 4.55	476.85 ± 102.24	0.47 ± 0.08	55.95 ± 22.17
VII	17.32 ± 4.93	614.23 ± 269.09	0.58 ± 0.20	49.22 ± 15.66
VIII	19.95 ± 6.39	763.50 ± 411.50	0.59 ± 0.20	60.06 ± 24.41

All data are given as mean ± SD.
Position I–VIII correspond to the locations from cranial to caudal sides of the linea alba.

Comments

- Mechanical properties depended on the original locations and on the fiber direction.
- The results will be related to problems of wound healing.

Reference(s)

Nilsson T (1982) Biomechanical studies of rabbit abdominal wall. Part I. — The mechanical properties of specimens from different anatomical positions. J Biomech 15:123–129 (with permission)

Pressure–Thickness Relation

• Elasticity •	• Cats • Lung	• Pulmonary capillary bed •

Materials
- Male mongrel cats
- Pulmonary alveolar

Testing Methods and Experimental Conditions
- The silicone elastomer microvascular casting method
- Measurement of sheet thickness, h, for fixed lungs
- Transmural pressure (alveolar-capillary) pressure, Δp, for fixing

Data

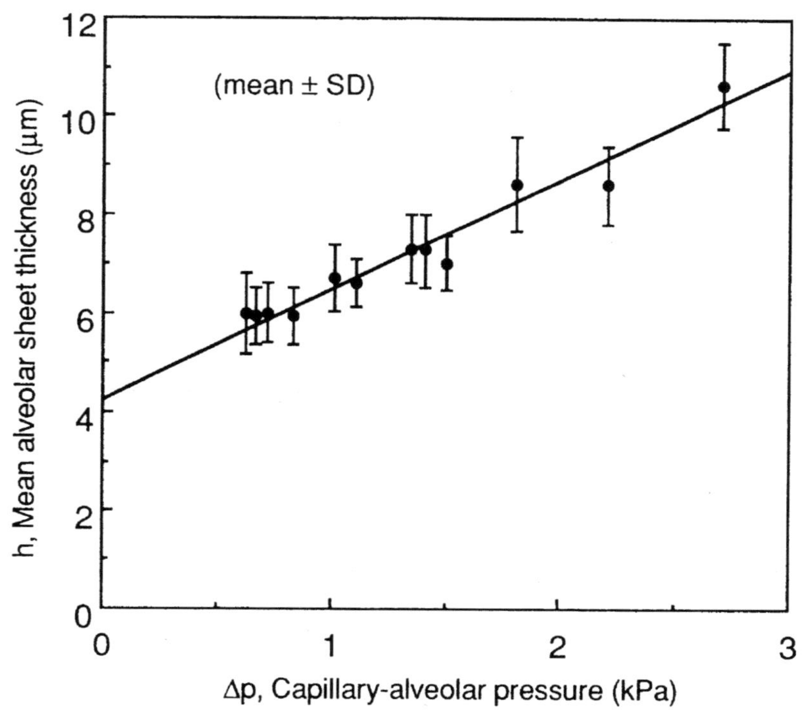

Comments
- $h = 4.28 + 2.2342\,\Delta p$.

Reference(s)
Sobin SS, Fung YC, Tremer HM, Rosenquist TH (1972) Elasticity of the pulmonary
alveolar microvascular sheet in the cat. Circ Res 30:440–450 (with permission)

Elastic Modulus

<table>
<tr><td>• Tensile strength
• Stiffness</td><td>• Mice
• Nerve</td><td>• Injured
•</td></tr>
</table>

Materials
- Adult mice
- Sciatic nerves

Testing Methods and Experimental Conditions
- Tensile tests
- Stretch rate, 1.22 mm/s
- Nerve crush by the application for 20 s with mosquito forceps at a point midway between the sciatic notch and patella
- 2, 6, 12, and 24 days after crush
- Specimens divided into proximal, medial, and distal segments
- Specimens divided into ipsilateral and contralateral sides

Data

Treatment group	Y (MPa)	σ_{pl} (MPa)	$(\Delta l/l_0)_{pl}$	μ (gm/cm)	A (cm^2)
			Biomechanical variables		
Control	7.0	3.2	0.43	0.0031	0.0025
Experimental					
Day 2	4.6	2.6	0.63	0.0036	0.0028
Day 6	7.4	3.0	0.55	0.0036	0.0030
Day 12	9.4	4.2	0.51	0.0033	0.0025
Day 24	8.8	3.8	0.40	0.0033	0.0032
Proximal	5.9	3.0	0.58	0.0031	0.0027
Medial	8.7	3.4	0.45	0.0035	0.0028
Distal	8.0	3.2	0.54	0.0035	0.0031
Ipsilateral	7.4	3.2	0.51	0.0040	0.0031
Contralateral	7.6	3.5	0.53	0.0029	0.0026

Y, elastic moduli; σ_{pl}, stress at the proportional limit; $(\Delta l/l_0)_{pl}$, strain at the proportional limit; μ, mass per unit length; A, cross-sectional area

Comments
- Strength and stiffness increased, and elasticity decreased time dependently by injury.
- Changes in mechanical properties appeared to be related to the epineurium.

Reference(s)
Beel JA, Groswald DE, Luttges MW (1984) Alterations in the mechanical properties of peripheral nerve following crush injury. J Biomech 17:185–193 (with permission)

3. Hard Tissues

Bending Property (1)

• Elastic modulus	• Human	• Dry and wet
•	• Mandible	•

Materials
- Human (male; adult)
- Ramus region of mandible
- Buccal cortical bone, dry and embalmed by formalin

Testing Methods and Experimental Conditions
- Small specimens (width, 0.965 mm; height, 0.425 mm)
- Divided into nine portions oriented at 15° intervals to the mandibular plane
- Three-point bending test
- Room temperature

Data

	Elastic modulus	
Angle (Deg)	Dry bone (GPa)	Wet bone (GPa)
0	17.0 ± 3.2	15.7 ± 2.5
15	16.3 ± 1.8	13.6 ± 4.0
30	12.8 ± 2.1	10.1 ± 1.8
45	11.3 ± 0.8	8.5 ± 0.7
60	11.3 ± 2.2	8.1 ± 1.9
75	13.0 ± 2.8	8.7 ± 1.3
90	15.9 ± 1.8	10.5 ± 3.9
105	17.7 ± 1.6	12.7 ± 2.8
120	20.5 ± 2.9	15.9 ± 0.9

All data are given as mean ± SD.

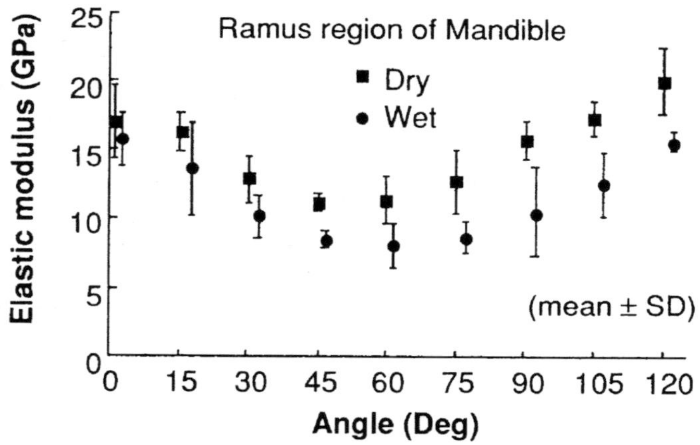

Comments
- The elastic modulus of the ramus region varied with the orientation. The two maximum values in dry bone were 20.5 GPa at 120°, and 17.0 GPa at 0°. The minimum values were 11.3 GPa at 45° and 60°. In wet bone, high values were observed at 120° (15.9 GPa) and 0° (15.7 GPa). The minimum was 8.1 GPa at 60°.

Reference(s)
Tamatsu Y (1994) A measurement of local elastic modulus of buccal compact bone of the human mandible. Jpn J Oral Biol 36:306–329

Bending Property (2)

• Elastic modulus •	• Human • Mandible	• Dry and wet •

Materials
- Human (male; adult)
- Molar region of mandible
- Buccal cortical bone, dry and embalmed by formalin

Testing Methods and Experimental Conditions
- Small specimens (width, 0.965 mm; height, 0.425 mm)
- Divided into seven portions oriented at 15° intervals to the mandibular plane
- Three-point bending test
- Room temperature

Data

Angle (Deg)	Elastic modulus	
	Dry bone (GPa)	Wet bone (GPa)
0	20.0 ± 0.9	17.6 ± 2.3
15	20.0 ± 0.7	17.7 ± 2.9
30	18.2 ± 1.2	16.1 ± 1.8
45	17.9 ± 2.2	14.2 ± 0.9
60	13.2 ± 1.0	12.7 ± 1.3
75	11.8 ± 1.0	8.0 ± 1.3
90	11.1 ± 0.5	7.4 ± 0.2

All data are given as mean ± SD.

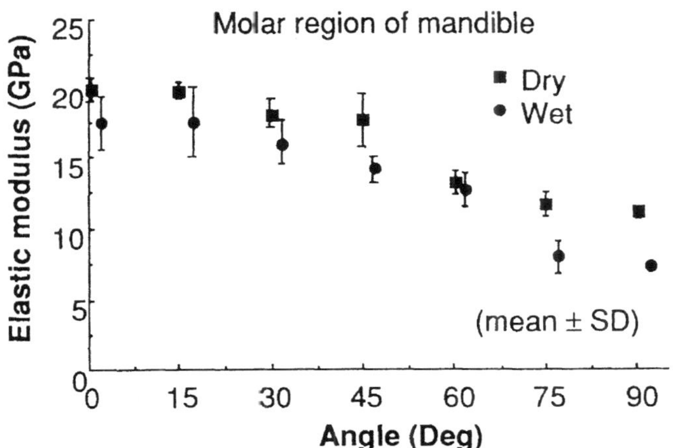

Comments
- The elastic modulus of the molar region varied with the orientation. The values of the molar region were higher than those of the incisal and premolar regions at angles ranging from 30° to 60°.

Reference(s)
Tamatsu Y (1994) A measurement of local elastic modulus of buccal compact bone of the human mandible. Jpn J Oral Biol 36:306–329

Bending Property (3)

· Elastic modulus ·	· Human · Mandible	· Dry and wet ·

Materials
- Human (male; adult)
- Premolar region of mandible
- Buccal cortical bone, dry and embalmed by formalin

Testing Methods and Experimental Conditions
- Small specimens (width, 0.965 mm; height, 0.425 mm)
- Divided into seven portions oriented at 15° intervals to the mandibular plane
- Three-point bending test
- Room temperature

Data

	Elastic modulus	
Angle (Deg)	Dry bone (GPa)	Wet bone (GPa)
0	20.9 ± 1.7	17.2 ± 1.7
15	20.0 ± 1.8	18.0 ± 1.2
30	17.4 ± 1.3	15.4 ± 1.3
45	13.8 ± 2.2	11.1 ± 2.0
60	13.4 ± 1.2	10.9 ± 2.1
75	11.8 ± 1.6	8.6 ± 2.0
90	10.1 ± 1.5	9.2 ± 1.1

All data are given as mean ± SD.

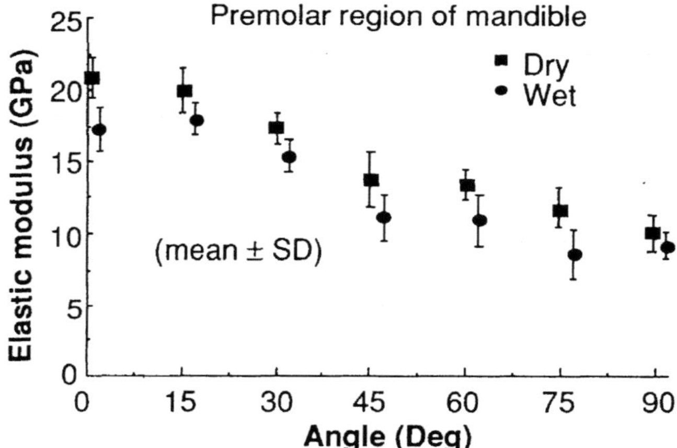

Comments
- The elastic modulus of the premolar region varied with the orientation. These characteristics closely approximate theoretical values of unidirectional fiber-reinforced composite materials, assuming that the reinforced major axis parallels the mandibular plane.

Reference(s)
Tamatsu Y (1994) A measurement of local elastic modulus of buccal compact bone of the human mandible. Jpn J Oral Biol 36:306–329

Bending Property (4)

• Elastic modulus •	• Human • Mandible	• Dry and wet •

Materials
- Human (male; adult)
- Incisal region of mandible
- Buccal cortical bone, dry and embalmed by formalin

Testing Methods and Experimental Conditions
- Small specimens (width, 0.965 mm; height, 0.425 mm)
- Divided into seven portions oriented at 15° intervals to the mandibular plane
- Three-point bending test
- Room temperature

Data

Angle (Deg)	Elastic modulus	
	Dry bone (GPa)	Wet bone (GPa)
0	20.2 ± 3.2	16.7 ± 2.2
15	18.6 ± 1.9	17.1 ± 1.7
30	15.5 ± 1.3	11.7 ± 3.4
45	14.6 ± 2.3	9.9 ± 2.5
60	12.0 ± 2.7	8.5 ± 2.6
75	10.4 ± 2.5	9.7 ± 1.0
90	10.2 ± 2.1	9.2 ± 1.6

All data are given as mean ± SD.

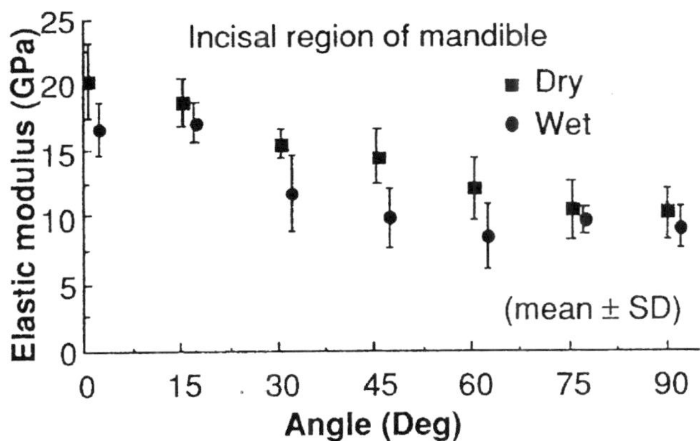

Comments
- The elastic modulus of the incisal region varied with the orientation. These characteristics closely approximate theoretical values of unidirectional fiber-reinforced composite materials, assuming that the reinforced major axis parallels the mandibular plane.

Reference(s)
Tamatsu Y (1994) A measurement of local elastic modulus of buccal compact bone of the human mandible. Jpn J Oral Biol 36:306–329

Bending Property (5)

• Elastic modulus •	• Human • Mandible	• Dry •

Materials
- Human (child; deciduous periods)
- Ramus region of mandible
- Buccal cortical bone, dry bone

Testing Methods and Experimental Conditions
- Small specimens (width, 0.965 mm; height, 0.425 mm)
- Divided into seven portions oriented at 30° intervals to the mandibular plane
- Three-point bending test
- Room temperature

Data

Angle (Deg)	Elastic modulus (GPa)
0	9.4 ± 3.0
30	6.6 ± 0.7
60	4.9 ± 1.1
90	7.4 ± 1.5
120	11.9 ± 2.3
150	11.3 ± 3.4
180	8.6 ± 4.7

All data are given as mean ± SD.

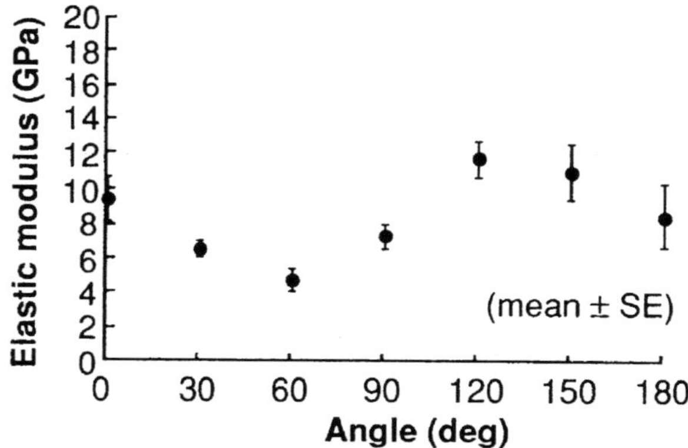

Comments
- Characteristics closely approximate theoretical values of unidirectional fiber-reinforced composite material in which the reinforced major axis parallels the posterior margin of ramus.

Reference(s)
Tamatsu Y (1994) A measurement of local elastic modulus of buccal compact bone of the human mandible. Jpn J Oral Biol 36: 306–329

Bending Property (6)

• Elastic modulus • Bending strength	• Human • Tibial diaphysis	• Cortical bone •

Materials

- Human (3 females [age, 55, 63, and 67 years] ; 4 males [age, 29, 55, 59, and 73 years])
- Mid-diaphysis of tibia
- Cortical bone

Testing Methods and Experimental Conditions

- Three-point bending test; sample size, $2 \times 2 \times 40$ mm
- Displacement rate, 5 mm/min; strain rate, 0.001 per second
- Span length, 30 mm
- Attenuation coefficient was determined from computed tomography (CT)

Data

Parameter	Average	Standard deviation	Range
Attenuation coefficient μ (HU)	1858	117	1624 – 2068
Modulus elasticity E (GPa)	17.5	1.62	14.3 – 21.1
Bending strength S (MPa)	214	21.1	178.2 – 250.2
Energy EA(J/m^2)	0.09	0.02	0.05 – 0.15
Apparent density d_a (g/cm^2)	1.861	0.057	1.748 – 1.952
Ash fraction Asf	0.658	0.015	0.607 – 0.685

	Pearson correlation coefficients r between variables for linear fit of data				Pearson correlation coefficients r between variables for power fit of data			
	μ (HU)	E (GPa)	S (MPa)	EA (J/m^2)	μ (HU)	E (GPa)	S (MPa)	EA (J/m^2)
μ (HU)	-	0.55	0.50	0.00	-	0.55	0.52	0.07
S (MPa)	0.50	0.72	-	0.37	0.52	0.73	-	0.37
d_a (g/cm^3)	0.65	0.74	0.67	0.17	0.65	0.75	0.68	0.0
Asf	0.46	0.48	0.41	-0.14	0.45	0.49	0.42	0.0

Linear	Power
$E = 70.4 \times 10^5 \mu$	$E = 10^{7.81} \mu^{0.74}$
$E = -23.5 \times 10^9 + (21.91 \times 10^9) d_a$	$E = 10^{9.59} d_a^{2.39}$
$E = 49.5 \times 10^9$ Asf	$E = 10^{10.58} Asf^{1.88}$
$S = 71.28 + 0.077 \mu$	$S = 10^{0.086} \mu^{0.69}$
$S = -189.9 + 217.1 d_a$	$S = 10^{1.81} d_a^{1.93}$
$S = 461.7 Asf$	$S = 10^{2.59} Asf^{1.42}$
$\mu = 1425.8 d_a$	$\mu = 10^{2.87} d_a^{1.45}$
$\mu = 3480.9 Asf$	$\mu = 10^{3.49} Asf^{1.23}$

Comments

- Clinical CT equipment used in this study was not sufficient to accurately estimate the mechanical properties of cortical bone.

Reference(s)

Snyder SM, Schneider E (1991) Estimation of mechanical properties of cortical bone by computed tomography. J Orthop Res 9:422–431 (with permission)

Bending Property (7)

• Elastic modulus • Maximum elastic stress	• Rat • Femur	• Whole bone •

Materials
- Wistar rats (20 males and 25 females)
- Femur

Testing Methods and Experimental Conditions
- Three-point bending test: whole bones, placed horizontally with the anterior aspect facing down on two supports separated by a constant 13 mm distance in a wet condition, were centrally loaded at a constant low loading rate (10 N/min) until fracture
- Bone segments between supports are assumed as hollow, elliptically shaped cylinders

Data

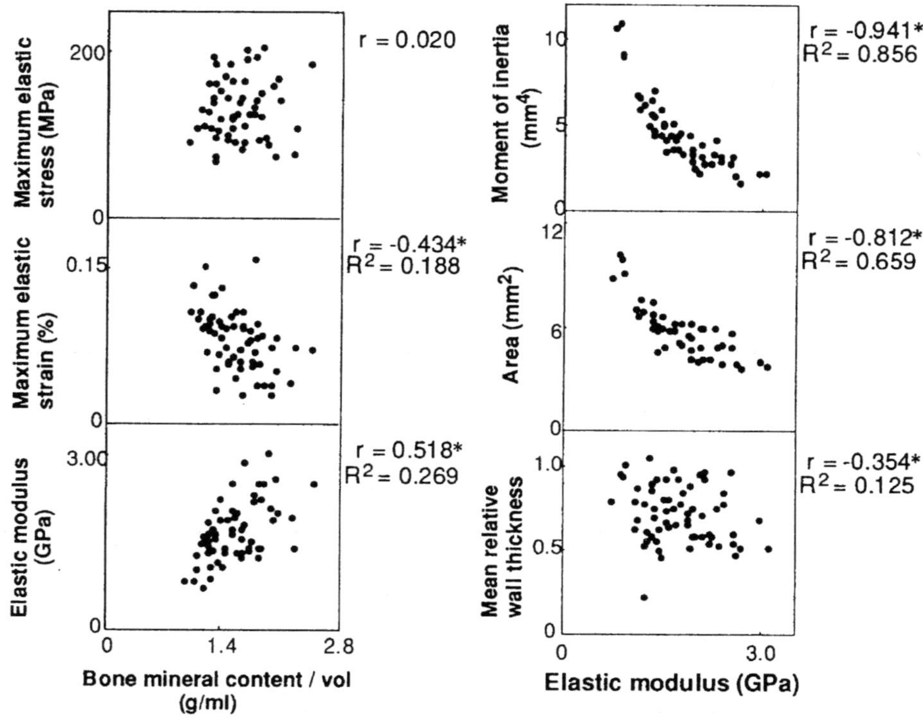

*Statistical significance ($P < 0.001$)

Comments
- Relative wall thickness = $[(B-b)/b+(H-h)/h])/2$, where H and B are horizontal and vertical external diameters of the fracture cross sections, respectively, and h and b are horizontal and vertical internal diameters, respectively.
- Negative correlation between the modeling-dependent cross-sectional architecture (moment of inertia) and the mineral-dependent stiffness (elastic modulus).
- Strength and stiffness of the integrated diaphyses depend on both cross-sectional inertia and body weight, but not on bone mineral density.

Reference(s)
Ferretti JL, Capozza RF, Mondelo N, Zanchetta JR (1993) Interrelationships between densitometric, geometric, and mechanical properties of rat femora: inferences concerning mechanical regulation of bone modeling. J Bone Miner Res 8:1389–1396 (with permission)

Bending Property (8)

• Fracture moment • Bending stiffness	• Dog • Tibia, radius, ulna	• Cortical bone • Mineral content

Materials

- Canine
- 12 tibiae, 21 radii, and 23 ulnae
- Cortical bone

Testing Methods and Experimental Conditions

- Three-point and four-point bending tests
- Strain rate of 0.15–0.0005 per second

Data

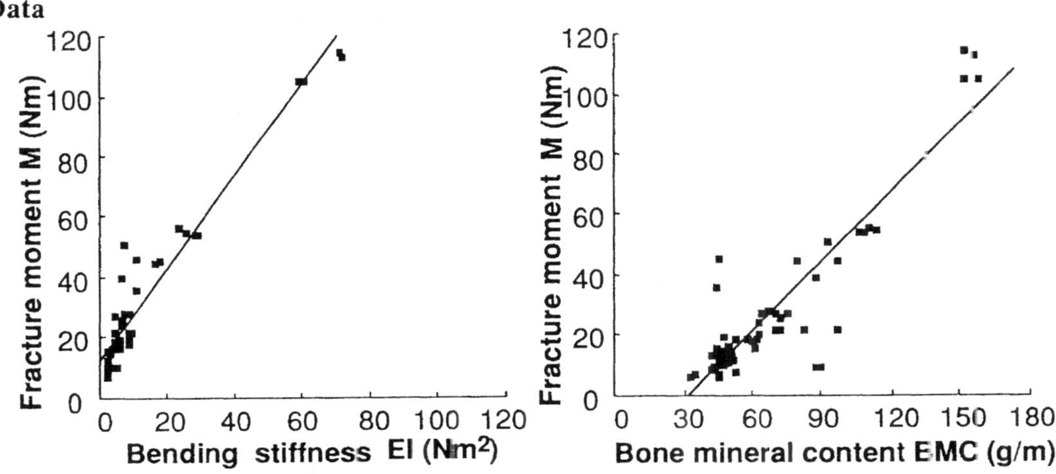

Comments

- M=1.40 + 1.238 EI + 0.183 BMC + 13.7 RATE (r = 0.967)

  M, fracture moment; EI, bending stiffness; BMC, bone mineral content; RATE, strain rate.

Reference(s)

Borders S, Petersen KR, Orne D (1977) Prediction of bending strength of long bones from measurements of bending stiffness and bone mineral content. Trans ASME J Biomech Eng 99:40–44

Bending Property (9)

• Maximum bending stress • Ultimate load	• Rat • Femoral diaphysis	• Age effect •

Materials
• Wistar rats (24 males; age, 6, 12, and 52 weeks)
• Femoral diaphysis

Testing Methods and Experimental Conditions
• Three-point bending test at deflecting speed of 10 mm/min until fracture

Data

Age (weeks)	6	12	52
Peak strain [a] (μstrain)	270 (181–347)	329 (219–392)	230 (183–256)
Body weight/in vivo strain ratio [b]	6 (5–7)	9 (7–11)	21 (17–24)
Load/deflection ratio [b]	202 (183–206)	434 (357–505)	715 (600–803)
Maximum bending stress (MPa)	121 (104–130)	199 (165–232)	220 (188–248)
Ultimate load (N)	95 (86–103)	195 (184–224)	389 (337–425)

[a] Peak strain (maximum deformation in vivo) during running on the seventh postoperative day.

[b] Stiffness expressed as the body weight/in vivo strain ratio (N/μstrain $\times 0.001$) and the conventional load/deflection ratio (N/mm).

Comments
• Maximum bending stress increased from 6-12 weeks of age, but then there was no further increase. Ultimate load increased steadily over the entire 52-week period.

Reference(s)
Indrekvam K , Husby OS , Gjerdet NR, Engester LB, Langeland N (1991) Age-dependent mechanical properties of rat femur: measured in vivo and in vitro. Acta Orthop Scand 62:248–252 (with permission)

Bending Property (10)

• Maximum moment • Bending stiffness	• Dog • Tibia, femur, humerus	• Age effect •

Materials
- Canine
- Tibia, femur, humerus, radius, and ulna (100 long bones)

Testing Methods and Experimental Conditions
- Three- and four-point bending tests
- Room temperature

Data

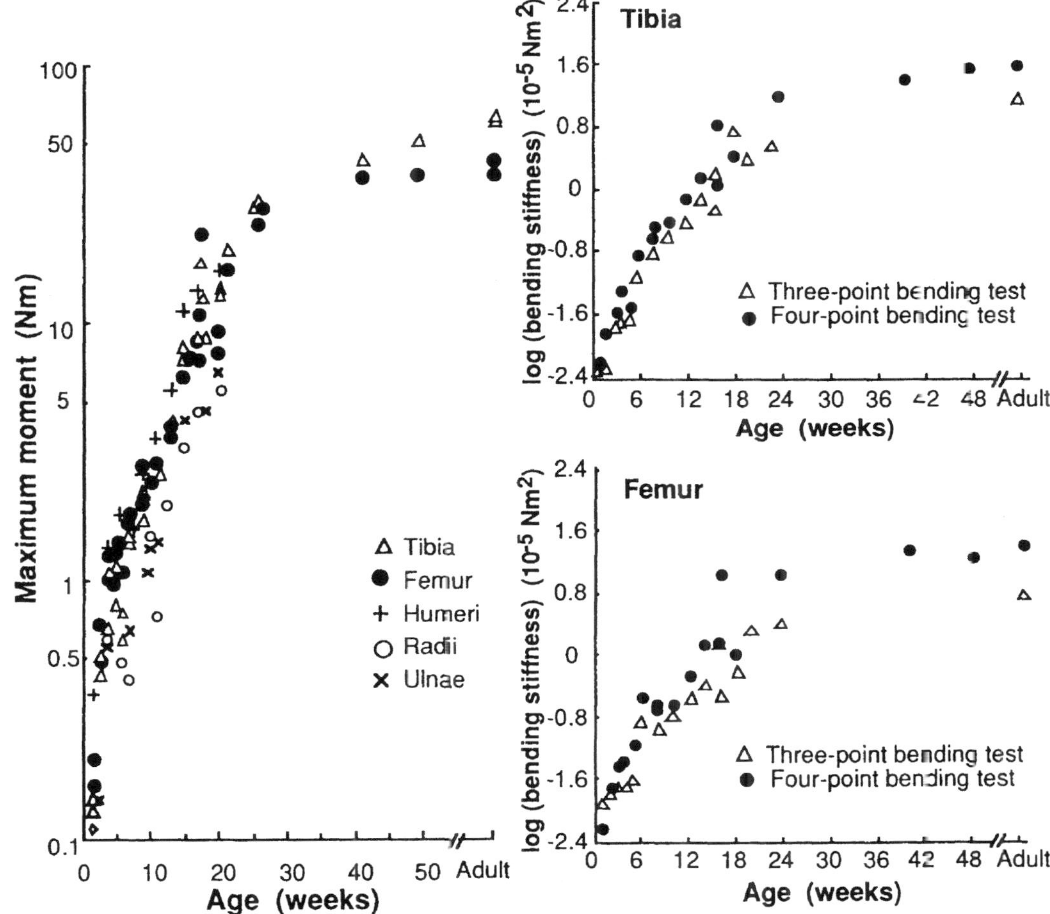

Comments
- Mechanical and geometric properties were found to follow a biphasic growth process, with a rapid increase in bending strength and moment of inertia from 1 to 24 weeks of age and a substantially decreased rate thereafter to maturity.

Reference(s)
Torzilli PA, Takebe K, Burstein AH, Heiple KG (1981) Structural properties of immature canine bone. Trans ASME J Biomech Eng 103:232–238

Bending Property (11)

• Maximum strength •	• Human • Femur	• Mineral content •

Materials

- Human (18 males and 18 females; age, 57–87 years)
- Femur

Testing Methods and Experimental Conditions

- Based on the CT measurements of density and area, the mass of a transverse slice of the femur was estimated
- Femur was mounted on an Instron device (TTMM 5 Ton, Instron, High Wycombe, UK) with the condyles standing on the platform and the cross head lying on the femoral head
- A vertical bending load was applied at a speed of 5 mm/min and the load deflection curve registered using a plotter

Data

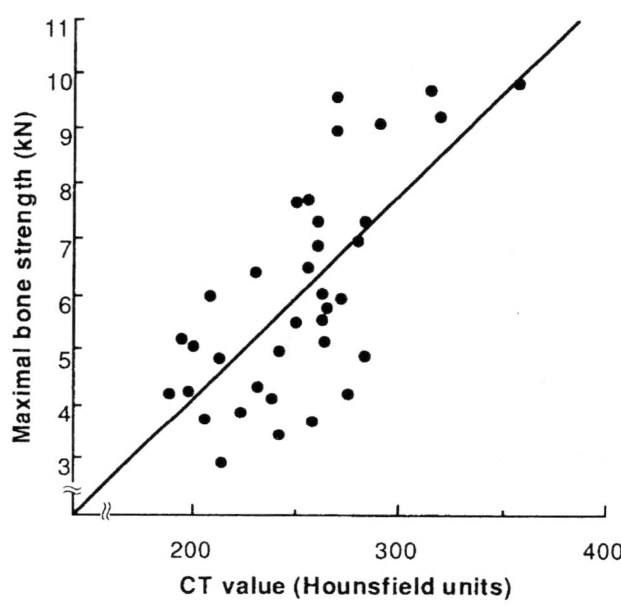

Comments

- Highly significant correlations were demonstrated between strength and cancellous bone density. Even higher correlations were revealed when the bone masses of the proximal and distal femoral areas were calculated.

Reference(s)

Alho A, Husby T, Hoiseth A (1988) Bone mineral content and mechanical strength. Clin Orthop Relat Res 227:292–297 (with permission)

Bending Property (12)

• Modulus • Maximum stress	• Human • Femur	• Density •

Materials
- Human (age, 28–90 years)
- Proximal femur

Testing Methods and Experimental Conditions
- Rectangular flat plate specimens
- Three-point bending tests
- All bone samples were kept moist with normal saline

Data

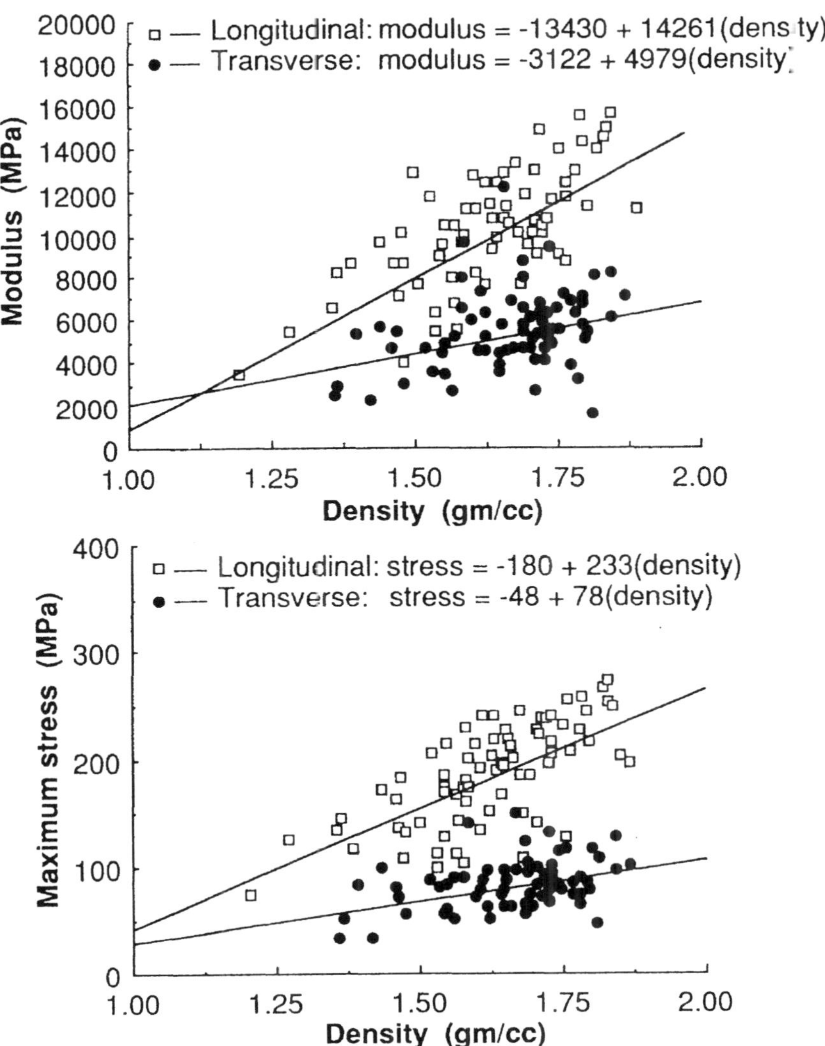

Comments
None.

Reference(s)
Lotz JC, Gerhart TN, Hayes WC (1991) Mechanical properties of metaphyseal bone in the proximal femur. J Biomech 24:317–329 (with permission)

Compressive Property (1)

• Anisotropy • Elastic modulus	• Dog • Femoral bone	• Cancellous bone • Locational dependence

Materials

- Large mongrel dogs (mature; weight, greater than 25 kg)
- Proximal femur
- Cancellous bone

Testing Methods and Experimental Conditions

- Uniaxial compression tests of cubic specimens (5 mm on a side)
- Strain rate of 0.001 per second
- Elastic constants (E) in the anterior-posterior (AP) direction, the medial-lateral (ML) direction and the proximal-distal (PD) direction were measured. Then yield stresses (σ_y) and ultimate compressive stresses (σ_M) in P-D direction were determined

Data

Position	E_{PD} (GPa)	E_{ML} (GPa)	E_{AP} (GPa)	σ_y (MPa)	σ_M (MPa)
1	0.468	0.367	0.434	13.3	14.8
	0.524	0.329	0.358	12.4	13.6
2	0.517	0.640	0.630	15.7	18.3
	0.670	0.465	0.549	16.8	18.9
	0.521	0.347	0.373	13.3	15.4
3	0.519	0.467	0.631	17.2	18.2
	0.636	0.374	0.448	14.4	15.8
	0.739	0.592	0.579	20.2	21.7
4	0.457	0.403	0.488	9.6	10.5
5	0.665	0.469	0.516	14.8	17.7
	0.293	0.356	0.291	5.3	6.1
	0.468	0.312	0.329	9.5	10.8
6	0.477	0.500	0.595	8.9	10.3
	0.413	0.288	0.414	8.5	10.0
	0.421	0.230	0.233	12.5	15.0
7	0.597	0.748	0.651	10.9	13.2
	0.310	0.285	0.276	5.1	6.1
	0.519	0.247	0.289	16.0	18.4
8	0.237	0.304	0.361	5.7	6.3
9	0.348	0.213	0.278	6.4	7.5
	0.216	0.158	0.161	4.9	5.3
10	0.571	0.337	0.217	9.5	11.9
	0.563	0.436	0.512	19.7	20.4
	0.380	0.242	0.155	13.1	15.8
11	0.487	0.440	0.279	14.0	15.4
12	0.232	0.159	0.256	4.0	4.2
13	0.231	0.147	0.132	7.7	9.9
	0.077	0.102	0.112	1.3	1.3
14	0.467	0.144	0.232	8.5	8.7
15	0.009	0.086	0.129	0.4	0.4
Average moduli:	0.435	0.340	0.364		

Comments

None.

Reference(s)

Vahey JW, Lewis JL, Vanderby R Jr (1987) Elastic moduli, yield stress, and ultimate stress of cancellous bone in the canine proximal femur. J Biomech 20:29–33 (with permission)

Compressive Property (2)

· Compressive strength ·	· Calf · Lumbar vertebra	· Trabecular bone · Locational dependence

Materials
- Holstein calves (age, 6–8 weeks)
- Lumbar vertebra
- Trabecular bone

Testing Methods and Experimental Conditions
- Cylindrical specimens
- Uniaxial compression tests

Data

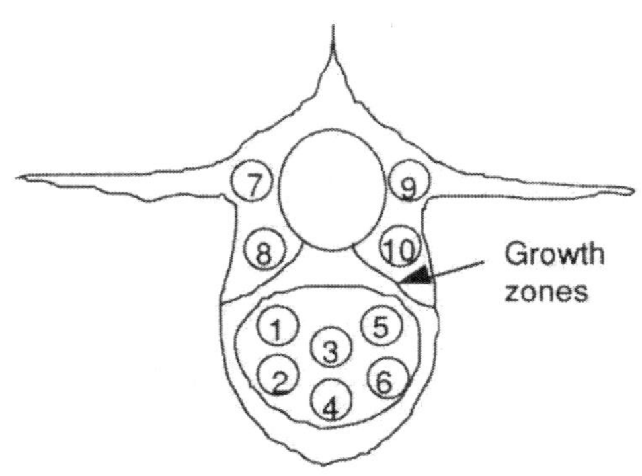

Region number of a lumbar vertebra

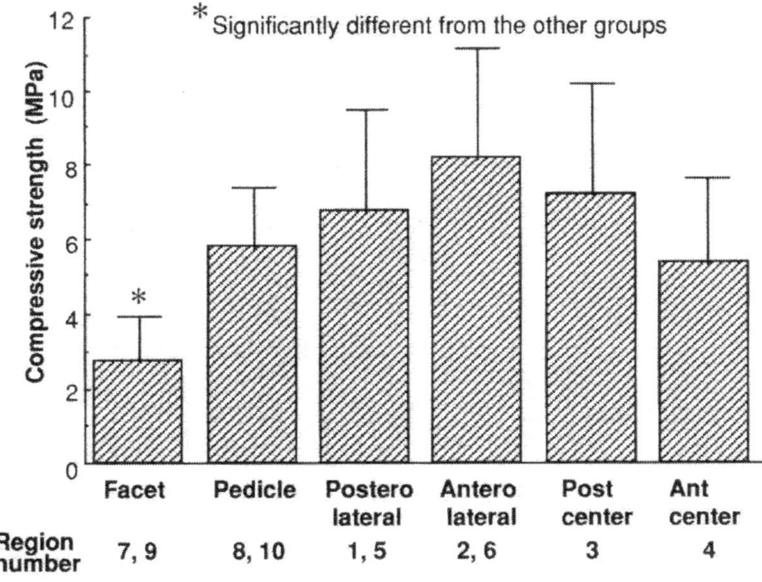

Comments
None.

Reference(s)
Swartz DE, Wittenberg RH, Shea M, White III AA, Hayes WC (1991) Physical and mechanical properties of calf lumbosacral trabecular bone. J Biomech 24:1059–1068 (with permission)

Compressive Property (3)

• Compressive strength •	• Human • Vertebra	• Trabecular bone • Mineral density

Materials
- Human (male; age, 53–80 years)
- Vertebra
- Trabecular bone

Testing Methods and Experimental Conditions
- Compression testing of cylindrical specimens
- Equivalent mineral density was determined using computed tomography
- Atomic absorption spectrophotometry and quantitative light microscopy were also used to determine the calcium weight percent and area fraction of bone, respectively

Data

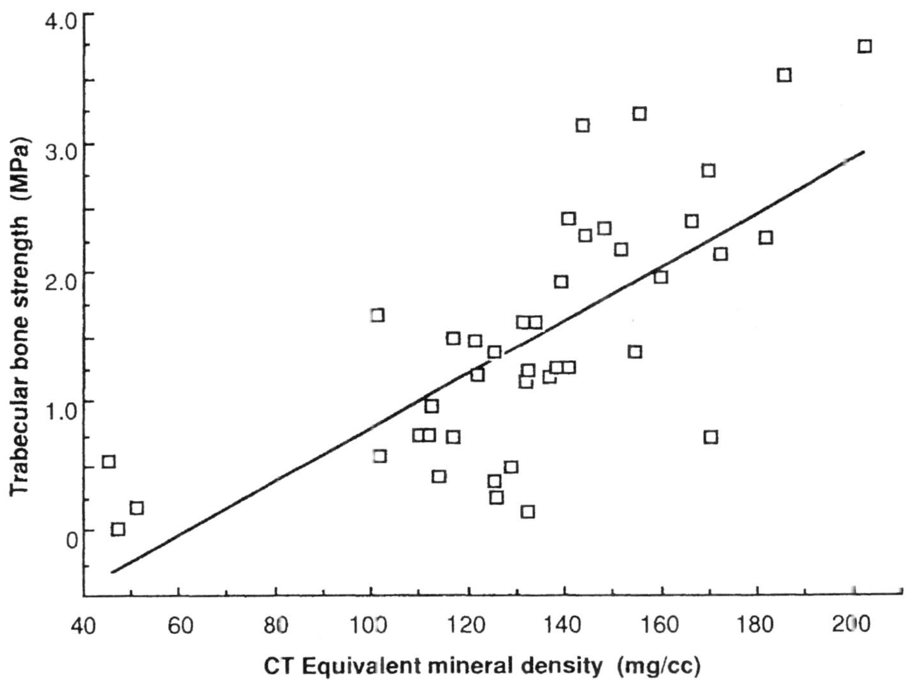

Comments
- Compressive strength (r = 0.720), elastic modulus (r = 0.574), trabecular calcium density (r = 0.780), and bone area fraction (r = 0.579) were all correlated with equivalent mineral density.

Reference(s)
Lang MS, Moyle DD, Berg EW, Detorie N, Gilpin AT, Pappas NJ Jr, Reynolds JC, Tkacik M, Waldron RL II (1988) Correlation of mechanical properties of vertebral trabecular bone with equivalent mineral density as measured by computed tomography. J Bone Joint Surg 70A:1531–1538 (with permission)

Compressive Property (4)

• Compressive strength • Elastic modulus	• Human • Tibia	• Trabecular bone • Strain rate

Materials

- Human (male; age, 65 years)
- Proximal tibia
- Trabecular bone

Testing Methods and Experimental Conditions

- Cylindrical specimen geometry (diameter, 5.5 mm; length, 8.25 mm)
- Room temperature
- Uniaxial compression
- Strain rates of 0.0001, 0.001, 0.01, 0.1, 1, and 10 per second

Data

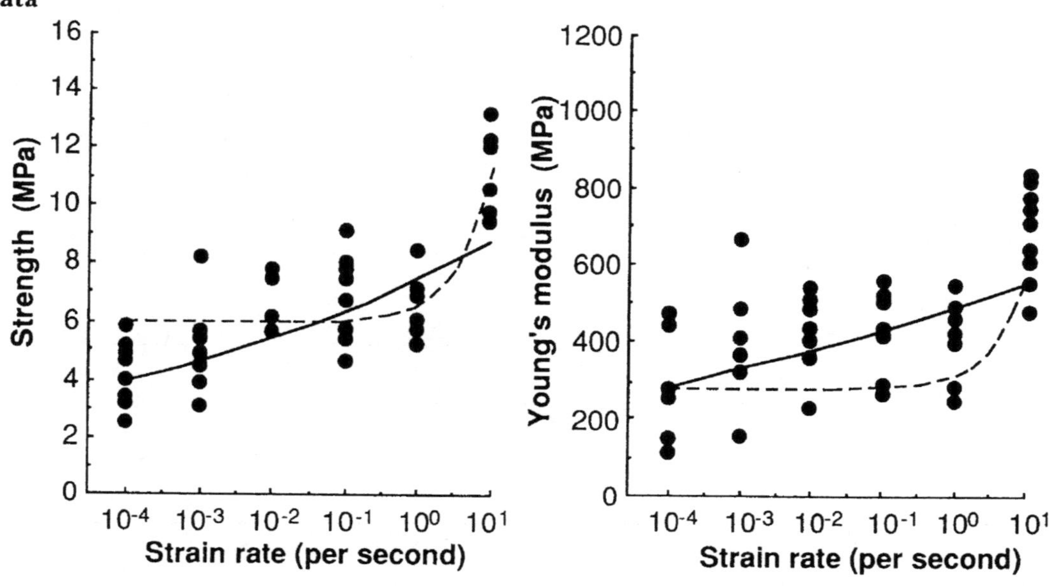

Comments

- Dashed lines and solid lines are linear and power law relationships, respectively.

Reference(s)

Linde F, Norgaard P, Hvid I, Odgaard A, Soballe K (1991) Mechanical properties of trabecular bone. Dependency on strain rate. J Biomech 24:803–809 (with permission)

Compressive Property (5)

• Compressive strength • Energy to failure	• Human • Femur	• Cancellous bone • Computed tomography

Materials
- Human (4 males and 4 females; age, 62–92 years)
- Proximal femur

Testing Methods and Experimental Conditions
- Unconstrained uniaxial compression test at a strain rate of 0.25 per second
- Computed tomography (CT) and apparent density

Data

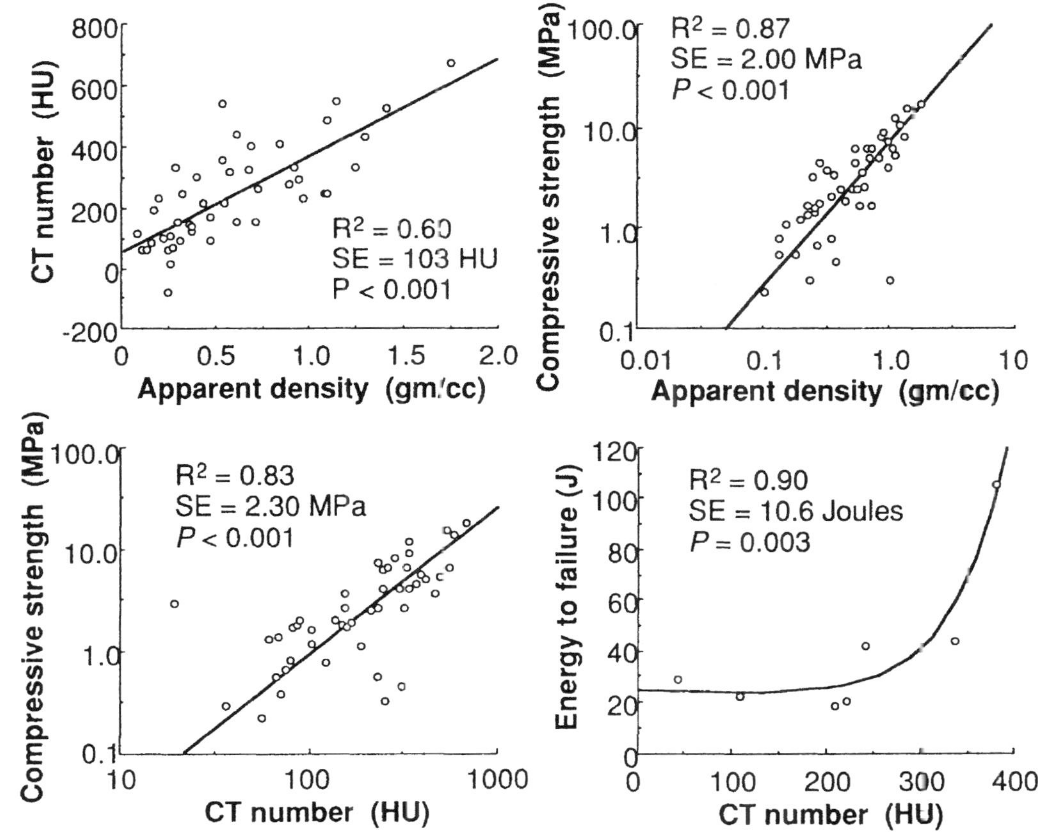

Comments
None.

Reference(s)
Esses SI, Lotz JC, Hayes WC (1989) Biomechanical properties of the proximal femur determined in vitro by single-energy quantitative computed tomography. J Bone Min Res 4:715–722 (with permission)

Compressive Property (6)

• Elastic modulus • Compressive strength	• Human • Femoral diaphysis	• Osteon •

Materials
- Human (age, 30 years; apparently free from skeletal defects)
- Femoral shaft
- Osteon

Testing Methods and Experimental Conditions
- Cross sections of diaphysis about 500 μm in thickness
- Diameter of specimen was 180-200 μm
- Three types of osteon were distinguished in terms of the orientation of collagen bundles
- Degree of calcification was determined by microradiographic technique
- Compression test at 20°C

Data

Type of osteon	Modulus of elasticity (GPa)	Ultimate compressive strength (MPa)	Shortening at breaking point (%)
a	9.3007 ± 1.6337	163.7 ± 11.7	1.88 ± 0.29
b	7.2189 ± 0.5494	98.2 ± 10.3	2.82 ± 0.57
c	7.3896 ± 1.6022	133.9 ± 9.3	2.09 ± 0.51
d	3.2939 ± 1.6722	78.3 ± 13.4	3.04 ± 0.59
e	6.3207 ± 1.8119	109.8 ± 10.1	2.46 ± 0.40
f	4.8082 ± 1.4992	87.8 ± 9.2	2.97 ± 0.36

a, fully calcified osteons of Type I ($n = 13$) ; b, osteons of Type I at the initial stage of calcification ($n = 7$) ; c, fully calcified osteons of Type II ($n = 12$) ; d, osteons of Type II at the initial stage of calcification ($n = 7$) ; e, fully calcified osteons of Type III ($n = 7$) ; f, osteons of Type III at the initial stage of calcification ($n = 7$).
Type I, osteon having marked transverse spiral course of fiber bundles in successive lamellae; Type II, osteon with fiber bundles in one lamella making an angle of nearly 90° with the fiber bundles in the next one; Type III, osteons having marked longitudinal spiral course of fiber bundles in successive lamellae.

Comments
- The ultimate compressive strength and the modulus of elasticity are greatest for osteons having transversely oriented fiber bundles.

Reference(s)
Ascenzi A, Bonucci E (1968) The compressive properties of single osteons. Anat Rec 161:377–392 (with permission)

Compressive Property (7)

• Elastic modulus • Compressive strength	• Human • Lumbar vertebra, tibia	• Spongy bone • Age effect

Materials

- Human (both sexes; age, 14–89 years)
- Lower lumbar vertebra and head of tibia
- Spongy bone

Testing Methods and Experimental Conditions

- Compression test
- Specimen ($2 \times 2 \times 4$ cm)

Data

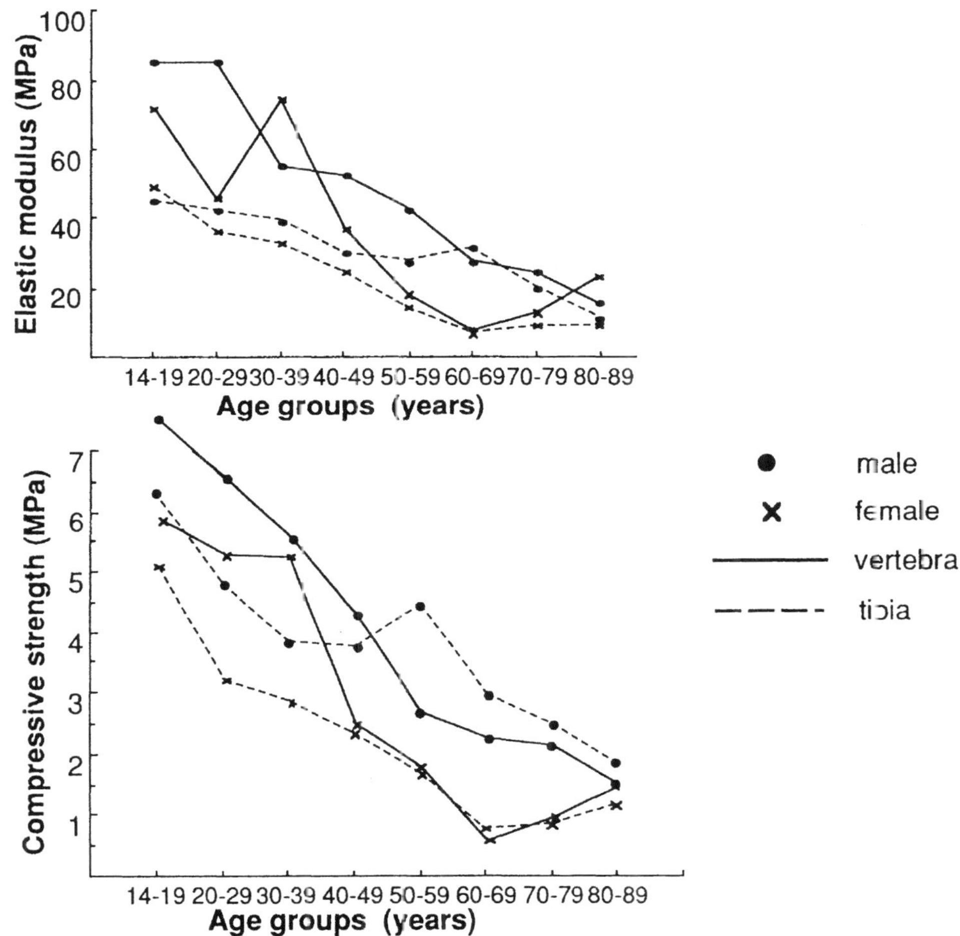

Comments

- This study was conducted to ascertain the compressive strength, strain at rupture, limit of proportionality, strain at the limit of proportionality, and the modulus of elasticity of spongy bone from vertebrae and tibiae.

Reference(s)

Lindahl O (1976) Mechanical properties of dried defatted spongy bone. Acta Orthop Scand 47:11–19 (with permission)

Compressive Property (8)

• Modulus • Anisotropy	• Human • Tibia, femur, radius	• Trabecular bone • Locational dependence

Materials

- Human (age, 55–70 years)
- Trabecular bone from major metaphyseal regions (proximal tibia, distal femur, proximal femur, distal radius, proximal humerus)

Testing Methods and Experimental Conditions

- 8 mm cube of trabecular bone under uniaxial compression at a strain rate of 0.01 per second
- Preloading (load, 40%–60% of the ultimate load; 3–6 times)
- Preyield tests in two of three orthogonal directions (AP [anterior-posterior], ML [medial-lateral], IS [inferior-superior]) and failure in the third direction
- Tested moist at room temperature
- Quantitative computed tomography (CT)

Data

See next page.

Comments

- Strength and stiffness of trabecular bone vary significantly not only within metaphyseal regions but from metaphysis to metaphysis. The power and significance of the relationships between density and modulus varied as a function of metaphyseal location. Both linear and nonlinear models were significant, suggesting that trabecular deformation occurs in response to both axial and bending loads.

Reference

Ciarelli MJ, Goldstein SA, Kuhn JL, Cody DD, Brown MB (1991) Evaluation of orthogonal mechanical properties and density of human trabecular bone from the major metaphyseal regions with materials testing and computed tomography. J Orthop Res 9:674–682 (with permission)

| | | Linear model $E = A(\text{density}) + \text{error}$ | | | | | | | | | Power function $E = A(\text{density})^p + \text{error}$ | | | | | |
| | | CT density | | | Apparent density | | | Ash density | | | CT density | | Apparent density | | Ash density | |
		n	Slope	r^2	n	Slope	r^2	n	Slope	r^2	Power	r^2	Power	r^2	Power	r^2
Proximal femur	AP	27	1.751	0.84	54	1.241	0.50	53	2.046	0.55	1.37	0.84	1.94	0.55	1.49	0.56
	ML	27	1.149	0.50	54	791	0.31	53	1.192	0.29	1.10	0.57	1.38	0.40	0.96	0.34
	IS	27	1.801	0.52	54	1.797	0.48	53	2.739	0.46	0.99	0.54	2.17	0.48	1.43	0.37
	Mean	27	1.567	0.84	54	1.276	0.54	53	1.992	0.54	1.10	0.87	1.80	0.57	1.25	0.49
Distal femur	AP	305	1.803	0.65	279	1.231	0.65	278	2.311	0.68	1.44	0.59	1.78	0.65	1.90	0.63
	ML	306	1.283	0.41	280	957	0.55	279	1.801	0.58	1.62	0.58	2.15	0.68	2.31	0.67
	IS	306	1.302	0.47	281	883	0.38	280	1.656	0.40	0.88	0.46	0.95	0.33	1.00	0.32
	Mean	305	1.456	0.77	278	1.019	0.77	277	1.921	0.80	1.21	0.79	1.45	0.80	1.57	0.78
Proximal tibia	AP	–	–	–	95	343	0.41	95	574	0.45	–	–	1.54	0.40	1.61	0.46
	ML	–	–	–	95	520	0.69	95	829	0.68	–	–	2.79	0.67	2.75	0.68
	IS	–	–	–	94	1.217	0.35	94	2.068	0.39	–	–	2.12	0.61	2.18	0.67
	Mean	–	–	–	94	693	0.52	94	1.157	0.56	–	–	2.05	0.68	2.10	0.74
Proximal humerus	AP	60	0.783	0.32	46	712	0.65	36	1.280	0.67	0.70	0.39	2.19	0.66	1.08	0.71
	ML	61	0.733	0.46	45	708	0.62	36	1.139	0.51	0.91	0.48	2.32	0.66	1.04	0.56
	IS	61	0.680	0.25	46	858	0.43	36	1.562	0.45	0.69	0.33	2.04	0.55	0.79	0.33
	Mean	60	0.738	0.46	45	761	0.72	36	1.327	0.75	0.75	0.61	2.06	0.83	0.91	0.73
Distal radius	AP	10	0.850	0.52	10	403	0.17	10	742	0.23	1.24	0.38	1.17	0.13	1.34	0.17
	ML	10	1.789	0.74	10	1.453	0.73	10	2.349	0.73	2.23	0.52	3.55	0.52	3.50	0.49
	IS	10	1.163	0.33	10	1.202	0.52	10	1.936	0.52	0.65	0.44	1.18	0.57	1.22	0.59
	Mean	10	1.267	0.92	10	1.019	0.88	10	1.676	0.91	1.14	0.90	1.80	0.89	1.86	0.91
Overall population	AP	402	1.620	0.68	484	1.162	0.72	472	1.944	0.67	1.13	0.61	2.13	0.76	1.53	0.60
	ML	404	1.098	0.43	484	865	0.61	473	1.405	0.51	1.13	0.53	2.27	0.75	1.57	0.54
	IS	404	1.399	0.58	485	1.077	0.54	473	1.819	0.52	0.96	0.58	1.64	0.63	1.23	0.55
	Mean	402	1.367	0.79	481	1.033	0.81	470	1.719	0.76	1.02	0.79	1.87	0.85	1.35	0.68

E, orthogonal modulus in MPa; CT density, Hounsfield units; apparent density, g/cm^3; ash density, g/cm^3.
The overall population represents the pooled raw data for all regions tested.

Compressive Property (9)

• Stress–strain curve • Impact and static tests	• Bovine • Femoral diaphysis	• Compact bone • Anisotropy

Materials
- Bovine
- Femur
- Compact bone

Testing Methods and Experimental Conditions
- Impact compression test using the split-Hopkinson pressure-bar technique at strain rate of about 100 per second
- Quasi-static compression test at strain rate of 10^{-3} per second
- Cylindrical specimen, 6–10 mm in diameter and 5–10 mm in length
- Specimen orientation, longitudinal direction (BA-direction), radial direction (RA-direction), and tangential direction (TA-direction)
- Room temperature

Data

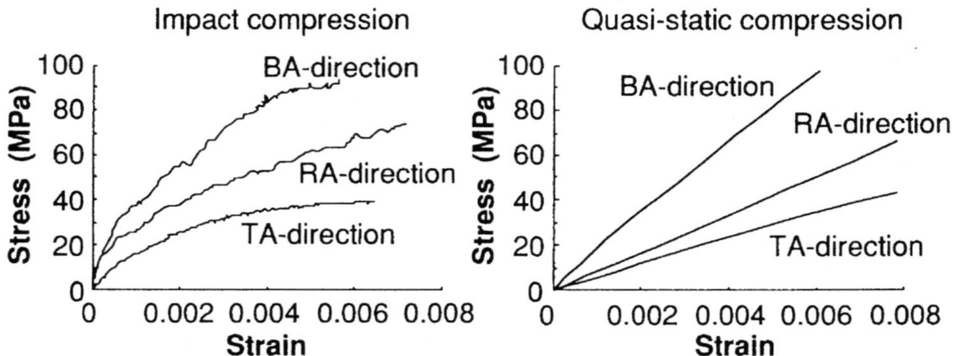

Comments
None.

Reference(s)
Kobayashi K, Tanabe Y, Hara T (1994) Anisotropic and nonlinear viscoelastic properties of bovine compact bone. Proc 1993 Ann Meet JSOB 15:23–27

Compressive Property (10)

• Stress–strain curve • Impact test	• Human • Tibial diaphysis	• Compact bone • Locational dependence

Materials
- Human (male; age, 16 years)
- Right tibia
- Compact bone

Testing Methods and Experimental Conditions
- Impact compression test using the split-Hopkinson pressure-bar technique at strain rate of about 200 per second
- Six samples were taken from the diaphysis (sample no. 1 from the most proximal location and no. 6 from the most distal)
- Cylindrical specimen, 6 mm in diameter and 6 mm in length
- Specimen orientation, longitudinal direction (BA-direction) and radial direction (RA-direction)
- Room temperature

Data

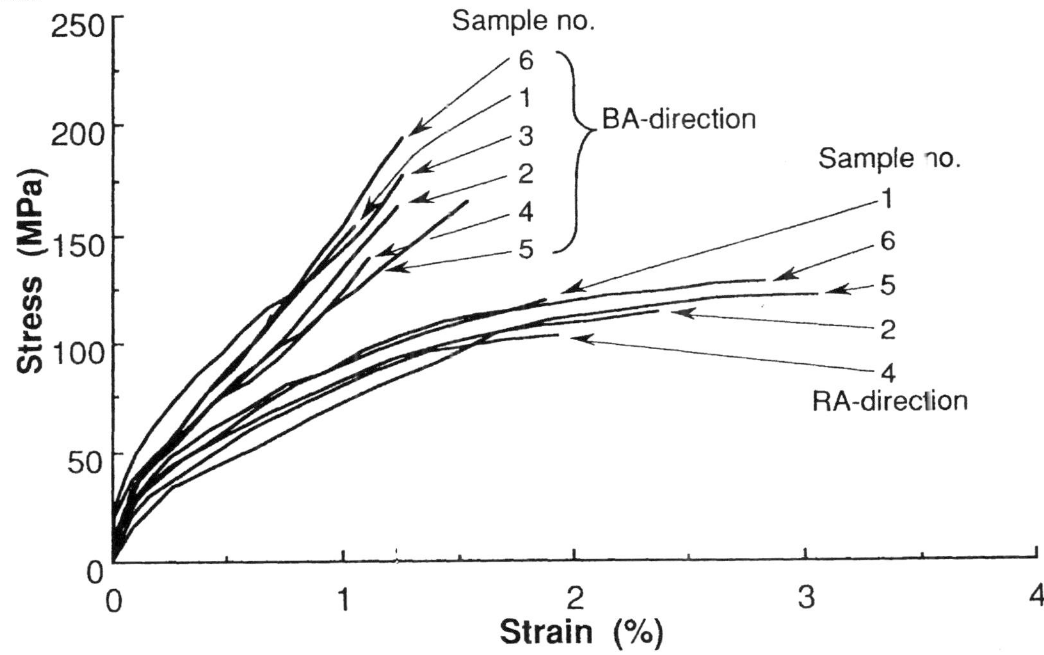

Comments
None.

Reference(s)
Kobayashi K, Tanabe Y, Koga Y, Hara T (1993) Identification of the dynamic properties of human compact bone. Theor Appl Mech 42:313–318

Compressive Property (11)

• Ultimate load •	• Human • Iliac crest	• Cancellous bone • Density

Materials
- Human (age, 48 years [mean])
- Iliac crest

Testing Methods and Experimental Conditions
- Specimens were trimmed to uniform dimensions (thickness, 7 mm; depth, 12 mm); the width of the crest was unaltered.
- Apparent densities and cross-sectional area of the cortical and cancellous bone were measured with use of a specific computed-tomographic technique
- Load to failure (ultimate load) was defined as the load at which the first obvious and abrupt reduction in load on the load-displacement curve occurred

Data

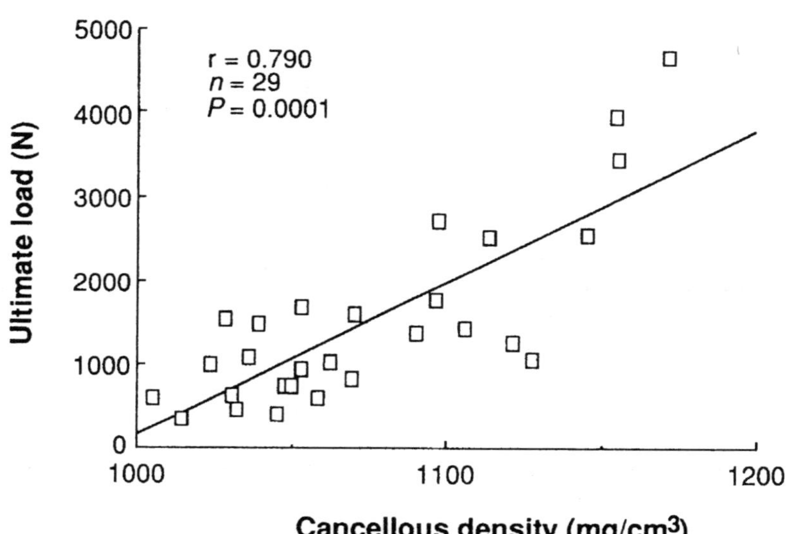

Comments
- The cancellous density was the most significant single factor in the prediction of the load to failure of the specimens (adjusted $r^2 = 0.58$; $P < 0.0001$).

Reference(s)
Smith DM, Cody DD (1993) Load-bearing capacity of corticocancellous bone grafts in the spine. J Bone Joint Surg 75A:1206–1213 (with permission)

Compressive Property (12)

• Ultimate stress • Modulus	• Human, dog • Femur	• •

Materials
- Human, and mature mongrel dogs
- Femur

Testing Methods and Experimental Conditions
- Cubic specimen ($8 \times 8 \times 8$ mm)
- Room temperature
- Uniaxial compression tests at a strain rate of 0.01% per second

Data

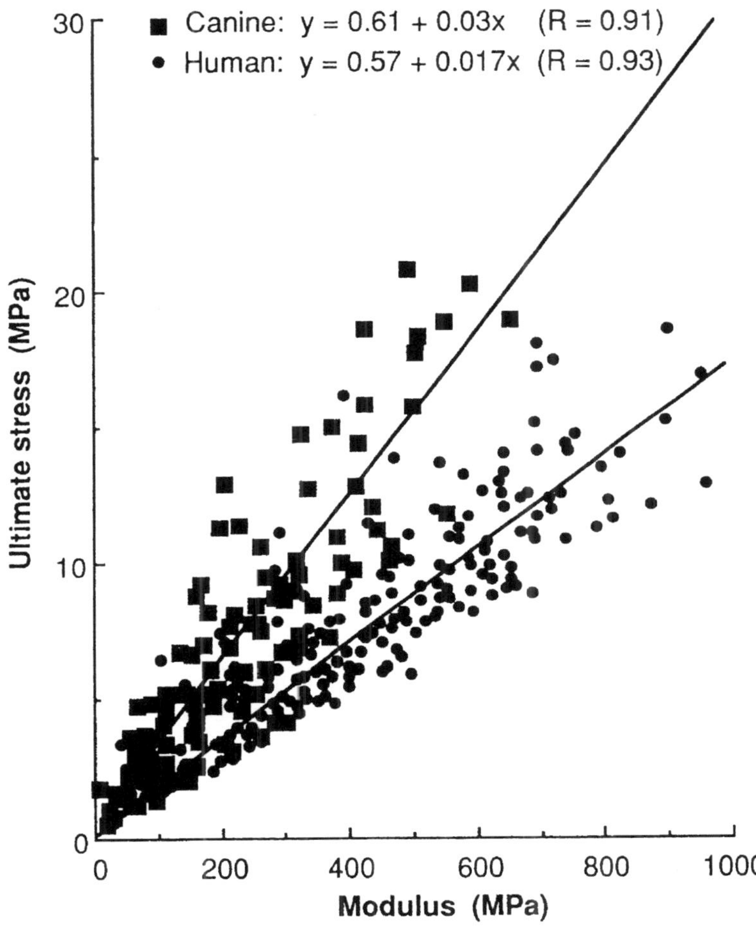

Comments
- A significant difference was seen in the canine and human relationships between ultimate stress and modulus.

Reference(s)
Kuhn JL, Goldstein SA, Ciarelli MJ, Matthews LS (1989) The limitations of canine trabecular bone as a model for human: a biomechanical study. J Biomech 22:95–107 (with permission)

Compressive Property (13)

• Yield stress • Elastic modulus	• Bovine • Femur	• Cancellous bone •

Materials

- Bovine
- Distal femur
- Cancellous bone

Testing Methods and Experimental Conditions

- Specimen cubes of cancellous bone ($10 \times 10 \times 10$ mm)
- Compression test at room temperature (25°C) at a strain rate of 0.014 ± 0.005 per second
- Specimens were preconditioned at four or five different stress levels before yield stress was reached
- Yield point was defined by a 0.0003 offset criterion

Data

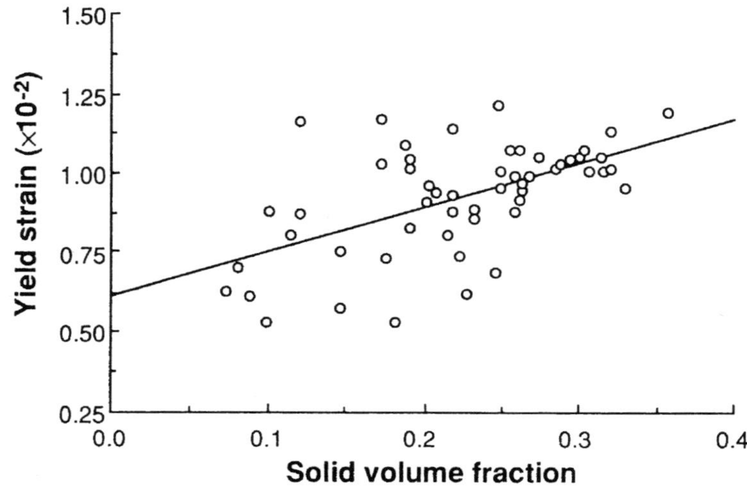

	Longitudinal direction	Overall
Structural density (g/cm^3)	0.375 ± 0.137	0.459 ± 0.133
Bone density (g/cm^3)	2.126 ± 0.094	2.072 ± 0.103
Volume fraction	0.178 ± 0.069	0.224 ± 0.068
Yield stress (MPa)	6.975 ± 4.098	8.677 ± 5.287
Young's modulus (MPa)	1036 ± 531	1026 ± 556
Normalized fabric component	0.353 ± 0.027	0.335 ± 0.034
Yield strain ($\times 10^{-2}$)	0.742 ± 0.143	0.915 ± 0.169

All data are given as mean $\pm$ SD.

Comments

- The relationship between yield stress σ_y and Young's modulus E:

$$\sigma_y = -0.375 + 0.00882E \quad (P < 0.00001, R^2 = 0.859).$$

- No significant relationship between yield strain and the degree of trabecular orientation.
- A significant positive correlation between yield strain and structural density and significant negative correlation between yield strain and bone density.

Reference(s)

Turner CH (1989) Yield behavior of bovine cancellous bone. Trans ASME J Biomech Eng 111:256–260

Compressive Property (14)

• Yield stress • Elastic modulus	• Dog • Femur, tibia	• Cortical bone • Age effect

Materials

- Canine
- Femur and tibia
- Cortical bone

Testing Methods and Experimental Conditions

- Age-related material properties
- Standard compression tests at a strain rate of approximately 0.08 per second
- Specimens: rectangular beam of $3 \times 1 \times 1$ mm oriented to the longitudinal axis of the bone
- Moistened with continuous drip of normal saline

Data

Age (weeks)	Specimens per bone	Yield strain	Yield stress (MPa)	Failure strain	Failure stress (MPa)	Ultimate strain	Ultimate stress (MPa)	Elastic modulus (GPa)	Nonelastic modulus (GPa)
Femora									
16	3	0.009	94	0.070	104	0.029	97	9.2	0.139
		(0.004)	(52)	(0.017)	(41)	(0.015)	(46)	(4.5)	(0.153)
24	2	0.016	133	0.082	99	0.041	114	7.5	-0.491
		(0.003)	(80)	(0.000)	(4)	(0.032)	(43)	(5.1)	(1.300)
24	6	0.012	185	0.087	114	0.026	161	15.5	-0.810
		(0.002)	(48)	(0.002)	(30)	(0.010)	(27)	(3.5)	(0.860)
40	3	0.010	202	0.081	178	0.023	155	16.6	-0.425
		(0.001)	(63)	(0.003)	(27)	(0.007)	(20)	(2.3)	(0.479)
48	3	0.012	205	0.075	67	0.018	186	17.6	-2.200
		(0.003)	(20)	(0.001)	(79)	(0.003)	(12)	(4.9)	(1.500)
Adult	12	0.011	229	0.056	143	0.018	217	21.0	-2.000
		(0.002)	(16)	(0.020)	(52)	(0.004)	(17)	(3.9)	(0.914)
Tibiae									
16	3	0.006	72	0.073	69	0.012	72	12.1	-0.188
		(0.000)	(7)	(0.002)	(21)	(0.003)	(10)	(1.8)	(0.468)
20	8	0.010	118	0.075	107	0.024	115	12.3	-0.173
		(0.003)	(15)	(0.002)	(11)	(0.007)	(14)	(3.8)	(0.103)
24	9	0.011	103	0.078	101	0.021	97	8.7	-0.055
		(0.001)	(20)	(0.002)	(21)	(0.007)	(23)	(2.2)	(0.036)
24	11	0.013	139	0.076	125	0.022	139	11.5	-0.247
		(0.003)	(14)	(0.002)	(11)	(0.004)	(17)	(2.0)	(0.169)
40	8	0.012	178	0.071	160	0.024	160	15.9	-0.214
		(0.003)	(29)	(0.011)	(20)	(0.009)	(19)	(6.0)	(0.761)
48	9	0.010	205	0.074	148	0.019	197	21.9	-0.914
		(0.002)	(15)	(0.005)	(21)	(0.004)	(13)	(5.1)	(0.501)
Adult	27	0.012	198	0.072	148	0.022	191	16.5	-0.827
		(0.003)	(25)	(0.006)	(34)	(0.010)	(25)	(4.3)	(0.620)
Adult	18	0.012	225	0.075	168	0.020	216	18.7	-0.883
		(0.002)	(23)	(0.002)	(25)	(0.004)	(22)	(4.0)	(0.481)

All data are given as mean (SD).

Comments

- All bones exhibited a two-phase growth cycle, an initial rapid phase (20 weeks) followed by a substantially slower growth to maturation (48 weeks).

Reference(s)

Torzilli PA, Takebe K, Burstein AH, Zika JM, Heiple KG (1982) The material properties of immature bone. Trans ASME J Biomech Eng 104:12–20

Compressive Property (15)

• Yield stress • Yield strain	• Human • Iliac crest, vertebra	• Trabecular bone •

Materials
- Human
- Iliac crest and L5 vertebra
- Trabecular bone

Testing Methods and Experimental Conditions
- Cylindrical test cores about 5mm in length and 8mm in diameter
- Obtained from the iliac crest perpendicular to the cortices
- Obtained from the body of L5 vertebra, coring being in the cranio-caudal direction
- Compressed at 1 mm/min

Data

	Iliac crest	L5 vertebra	P
Apparent density (mg/cc)	360.7 ± 118.8	275.2 ± 82.8	NS
True density (mg/cc)	2032 ± 111	1981 ± 106	NS
Yield stress (MPa)	2.61±1.76	3.39±1.69	NS
Yield energy (MPa)	108.7 ±105.5	116.6±79.6	NS
Yield strain	0.057 ± 0.02	0.051 ± 0.017	NS
$\dfrac{\text{(Yield stress)}}{\text{(App. dens)}}$ ($\times 10^{-3}$)	6.5±2.9	11.6 ±3.6	< 0.01
Number of specimens	11	11	

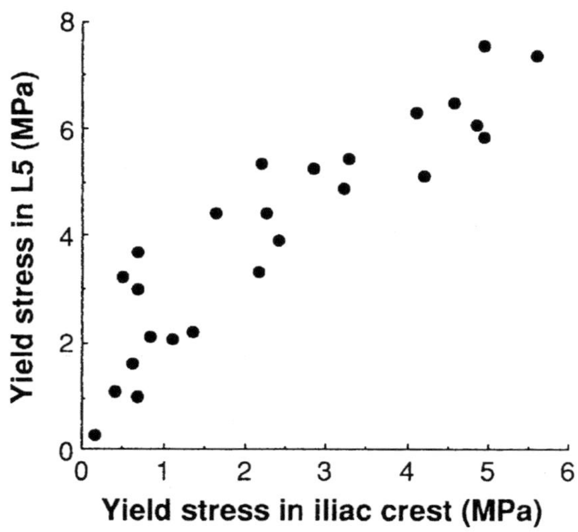

Comments
- Apparent density was the property showing the best correlation between iliac crest and L5 (r = 0.79), while yield energy was less well associated (r = 0.67).
- Mechanical properties of the iliac crest bone were very dependent upon site and did not satisfactorily predict those of the lumbar spine.

Reference(s)
Britton JM, Davie MW (1990) Mechanical properties of bone from iliac crest and relationship to L5 vertebral bone. Bone 11:21–28 (with permission)

Creep (1)

• Compliance •	• Bovine • Long bone diaphysis	• Cortical bone •

Materials

- Bovine
- Diaphysis of long bone
- Cortical bone

Testing Methods and Experimental Conditions

- Creep under torsion
- Testing was carried out using an apparatus consisting of a system of weights and pulleys and an environmental chamber
- Specimens were maintained at body temperature (37°C) in Ringer's solution
- Gauge was $20 \times 3.2 \times 3.2$ mm in size, oriented parallel to the bone long axis

Data

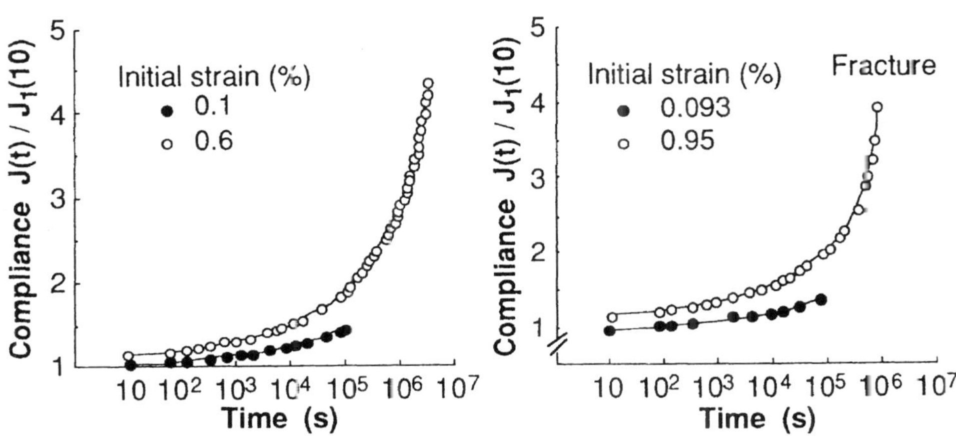

Comments

- Compliance was calculated from torque and angular displacement applying the theory of torsion for an orthotropic prism.
- The average compliance at 10 s was $J_1(10) = 2.07 \times 10^{-10}$ m^2/N.

Reference(s)

Lakes R, Saha S (1980) Long-term torsional creep in compact bone. Trans ASME J Biomech Eng 102:178–180

Creep (2)

• Creep	• Human	• Cortical bone
•	• Femur	•

Materials
- Human (male; age, 24–49 years)
- Femur
- Cortical bone

Testing Methods and Experimental Conditions
- Cylindrical specimens with a central test section, 3 mm in diameter and 8 mm in gauge length
- Temperature, 37°C
- Wrapped in water-soaked tissue paper
- Creep under cyclic load

Data

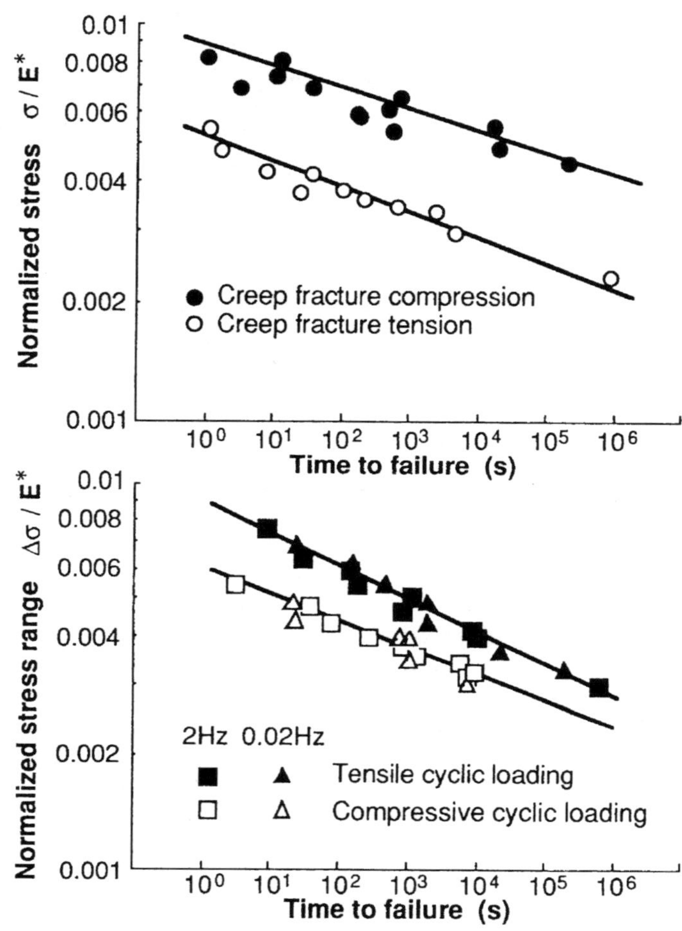

Comments
- Young's modulus E* was used to normalize the stress level for each specimen and was measured by applying a single loading cycle with a peak stress of 14 MPa at a stress rate of 28 MPa/s.

Reference(s)
Caler WE, Carter DR (1989) Bone creep-fatigue damage accumulation. J Biomech 22:625–635 (with permission)

Creep (3)

• Creep •	• Human • Femur	• Multiple loading •

Materials
- Human (male; age 21 years)
- Fresh frozen femur

Testing Methods and Experimental Conditions
- Plate specimens (thickness, 1.6 mm)
- Room temperature
- Standard constant stress creep test with multiple loading scheme (1 s ramp-up in load, 1 min hold at constant tensile load, 1 s ramp-down, 1 min hold at zero load for each cycle)

Data

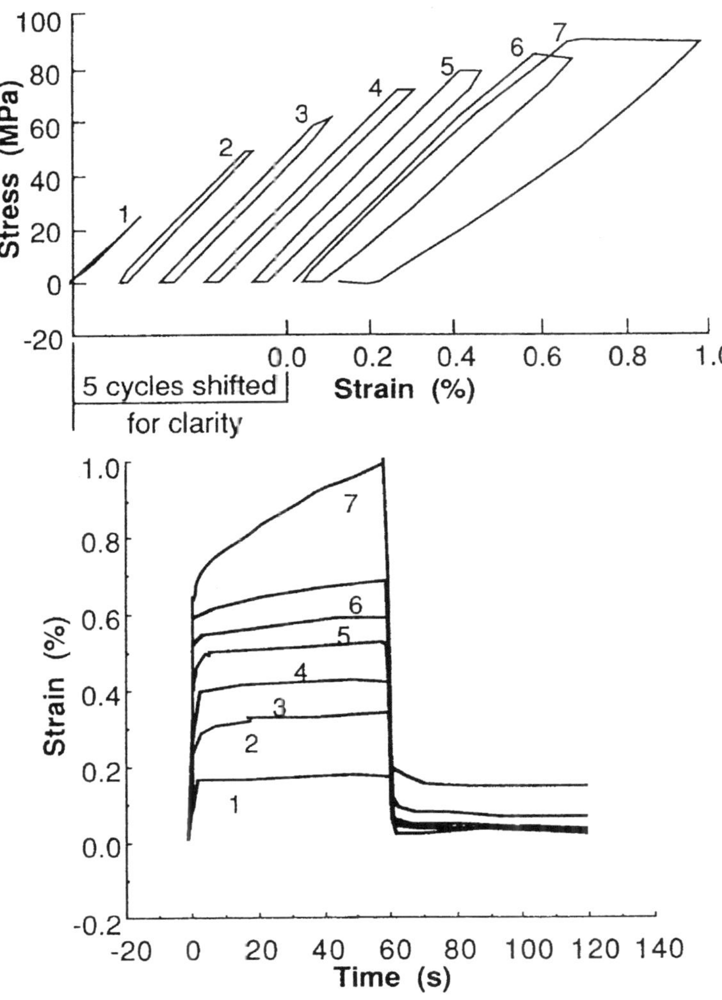

Comments
None.

Reference(s)
Fondrk M, Bahniuk E, Davy DT, Michaels C (1988) Some viscoplastic characteristics of bovine and human cortical bone. J Biomech 21:623–630 (with permission)

Dynamic Shear Modulus (1)

• Damping factor •	• Bovine • Metacarpus	• Compact bone • Locational dependence

Materials

- *Bos taurus* (mature; female)
- Metacarpus
- Compact bone

Testing Methods and Experimental Conditions

- Free torsional oscillation by pendulum
- Twenty samples were taken from the middle part from 60 mm proximal to 40 mm distal to the center of the bone (site no.1 for the most proximal and no.20 for the most distal)
- Specimens perpendicular to long axis
- Cross section, 4.0 × 2.4 mm; gauge length, 20 or 15 mm
- Fixed frequency of 1 Hz

Data

Site no.	Dynamic shear modulus ($\times 10^8$ dynes/cm^2)		Damping factor ($\times 10^{-2}$)	
	Dorsal	Ventral	Dorsal	Ventral
1	419.36 ± 3.52	375.63 ± 41.22	2.08 ± 0.48	2.30 ± 0.42
2	418.41 ± 9.77	362.18 ± 62.40	2.04 ± 0.29	2.41 ± 0.20
3	418.84 ± 27.69	359.44 ± 61.67	1.89 ± 0.33	2.54 ± 0.29
4	433.58 ± 50.86	351.14 ± 77.19	2.00 ± 0.36	2.55 ± 0.40
5	426.29 ± 38.45	357.88 ± 61.07	2.10 ± 0.33	2.61 ± 0.45
6	421.61 ± 14.63	358.74 ± 55.24	2.04 ± 0.22	2.58 ± 0.45
7	445.83 ± 33.42	358.68 ± 66.00	1.98 ± 0.41	2.56 ± 0.27
8	432.58 ± 39.33	362.10 ± 64.62	2.09 ± 0.34	2.53 ± 0.20
9	444.74 ± 48.44	359.88 ± 56.67	2.01 ± 0.43	2.46 ± 0.25
10	435.36 ± 36.10	364.91 ± 59.29	2.09 ± 0.41	2.48 ± 0.18
11	451.75 ± 37.04	363.25 ± 57.68	2.09 ± 0.35	2.41 ± 0.15
12	449.00 ± 21.94	372.99 ± 59.83	2.18 ± 0.39	2.46 ± 0.27
13	445.46 ± 24.62	366.08 ± 61.03	2.13 ± 0.28	2.34 ± 0.26
14	412.24 ± 41.57	382.03 ± 73.10	2.14 ± 0.35	2.29 ± 0.21
15	388.23 ± 32.98	396.94 ± 63.78	2.29 ± 0.38	2.33 ± 0.23
16	397.28 ± 36.26	397.19 ± 72.80	2.32 ± 0.34	2.28 ± 0.38
17	381.74 ± 29.88	413.69 ± 56.82	2.40 ± 0.32	2.13 ± 0.23
18	396.10 ± 34.18	422.98 ± 49.43	2.34 ± 0.25	2.08 ± 0.21
19	394.57 ± 42.83	437.21 ± 60.21	2.46 ± 0.19	2.09 ± 0.36
20	410.00 ± 41.75	453.97 ± 69.27	2.47 ± 0.15	1.97 ± 0.14

All data are given as mean ± SD.

Comments

- The dynamic shear modulus differs from site to site along the shaft. The dorsal aspect of the metacarpus has the highest moduli, while the ventral aspect has the highest values of the damping of dissipated energy.

Reference(s)

Ramaekers JGM (1979) The dynamic shear modulus and damping of compact bovine metacarpal bone in dependence on the topography along the bone shaft. Neth J Zool 29: 151–165 (with permission)

Dynamic Shear Modulus (2)

• Storage modulus • Damping factor	• Bovine, mallard, tortoise • Metacarpus, humerus	• Compact bone •

Materials

- Bos taurus (bovine): compact bone from right metacarpus
- Anas platyrhynchos (mallard): compact bone from right humerus
- Testudo hermanni (tortoise): bone from lateral side of hypoplastron, a bony element of the plastron
- Rana esculenta (common frog): bone from the femur

Testing Methods and Experimental Conditions

- Prismatic specimens with gauge length of 21 mm and cross section of 5×2 mm
- Torsional test by a free-oscillating torsion technique
- At temperatures lying between 110 K and room temperature
- Load frequency of 1 Hz

Data

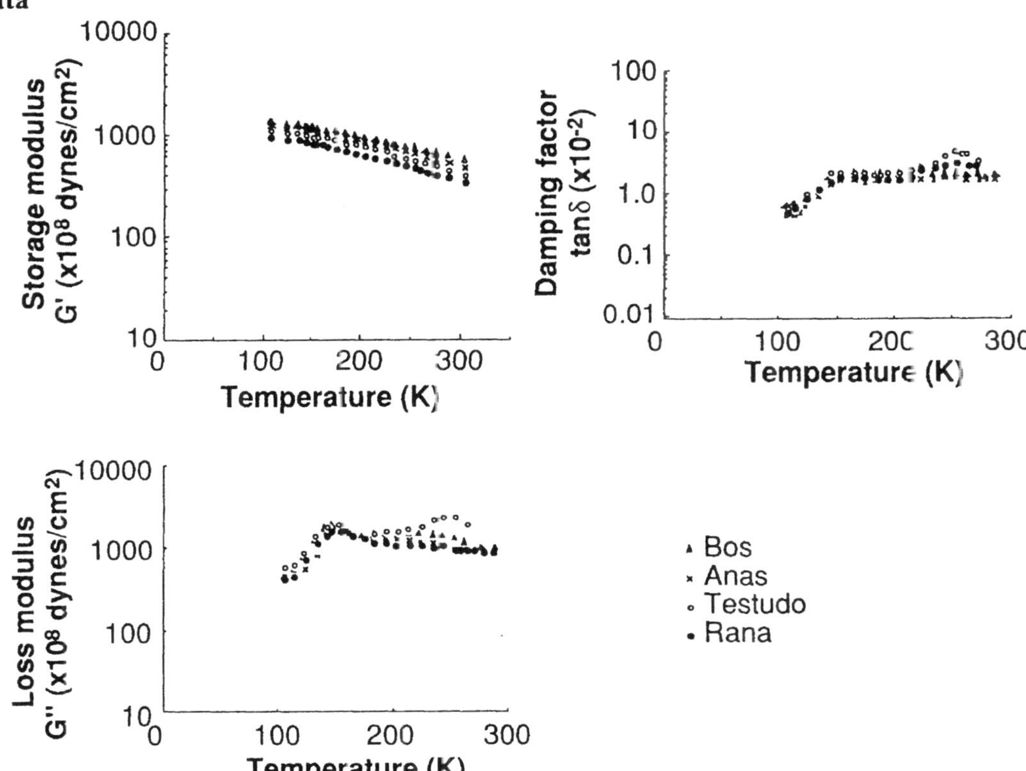

Comments

- The slope of the storage modulus (G', dynes/cm²) curves are parallel, and curves of the damping (tan δ) show a maximum at 150 K and a second around 250 K, where $G^*(=G'+iG'')$ is the dynamic shear modulus and tan $\delta = G''/G'$.

Reference(s)

Ramaekers JGM (1979) The rheological behaviour of skeletal material originating from several classes of vertebrates. Neth J Zool, 29:166–176 (with permission)

Elastic Modulus (1)

• Stiffness coefficient • Poisson's ratio	• Bovine • Femur	• Cortical bone • Locational dependence

Materials

- Bovine
- Femoral bone

Testing Methods and Experimental Conditions

- Pulse transmission ultrasound method
- Frequency of 2.25 MHz
- Twenty specimens in the form of 5-mm cubes
- Orthotropic stiffness coefficients and technical constants were determined

Data

		c_{11}	c_{22}	c_{33}	c_{44}	c_{55}	c_{66}	c_{12}	c_{13}	c_{23}	ρ
$\frac{z}{l} = 0.3$	A	12	18	22	6.7	5.5	5.9	4.3	-	17	2.0
	M	13	19	23	8.0	7.1	5.9	6.8	-	6.4	2.1
	P	18	20	23	8.0	7.0	5.6	7.4	8.7	7.4	2.1
	L	12	16	28	7.0	6.5	5.3	5.4	2.5	9.0	2.0
$\frac{z}{l} = 0.4$	A	16	22	30	6.3	6.7	5.7	10	7.2	8.5	2.1
	M	14	22	30	7.3	6.5	5.3	7.1	5.2	6.4	2.0
	P	15	16	32	7.8	7.3	5.8	8.4	3.7	4.7	2.1
	L	21	24	31	7.6	6.9	5.6	6.2	5.6	6.5	2.0
$\frac{z}{l} = 0.5$	A	14	15	25	6.9	6.4	4.7	7.3	4.9	5.0	2.0
	M	16	20	30	7.1	6.9	5.7	7.2	1.5	7.4	2.1
	P	14	17	24	7.3	6.8	5.3	7.5	4.7	6.8	2.0
	L	13	21	30	7.9	6.7	5.7	5.9	2.5	5.5	2.1
$\frac{z}{l} = 0.6$	A	13	15	21	6.3	5.8	4.8	7.7	5.7	5.9	2.0
	M	14	17	19	6.8	6.0	5.0	7.0	8.2	7.2	2.0
	P	12	17	20	7.0	6.5	5.0	1.7	4.3	5.9	2.0
	L	12	16	22	7.0	5.9	4.8	6.4	3.2	4.3	2.0
$\frac{z}{l} = 0.7$	A	13	16	18	6.2	5.7	5.2	5.2	4.5	5.8	2.0
	M	14	17	24	6.7	5.4	5.3	4.9	5.1	7.4	2.0
	P	10	23	27	6.8	4.7	4.2	-	-	-	2.0
	L	15	17	21	6.3	5.6	4.8	4.0	-	4.7	2.0
mean		14.1	18.4	25.0	7.00	6.30	5.28	6.34	4.84	6.94	
$\pm$ SD		2.4	2.7	4.3	0.67	0.67	0.46	1.78	1.93	2.67	

Stiffness coefficient c is in GPa and density ρ is in g/m^3. Parameter z is measured along the axis of the femur from the hip to the knee; l is the length of the femur. A stands for bone from the anterior quadrant, M from the medial quadrant, P from the posterior quadrant, and L from the lateral quadrant. The coordinate axes are : x_1 is the radial coordinate, x_2 is the circumferential coordinate, and x_3 is along the long axis of the bone.

E_1	E_2	E_3	G_{12}	G_{13}	G_{23}	ν_{12}	ν_{13}	ν_{23}	ν_{21}	ν_{31}	ν_{32}
11.6	14.6	21.9	5.29	6.29	6.99	0.302	0.109	0.205	0.380	0.206	0.307

Technical constants E, G are in GPa. ν is Poisson's ratio.

Comments

- The elastic relations: $\sigma_i = c_{ij}\,\varepsilon_j$ ($i, j = 1, 2, \cdots 6$).

Reference(s)

Buskirk WCV, Cowin SC, Ward RN (1981) Ultrasonic measurement of orthotropic elastic constants of bovine femoral bone. Trans ASME J Biomech Eng 103:67–72

Elastic Modulus (2)

• Stiffness coefficient • Poisson's ratio	• Human • Femur	• Cortical bone • Locational dependence

Materials
- Human
- Femur
- Cortical bone

Testing Methods and Experimental Conditions
- High frequency ultrasonic technique: 6.8 MHz for compressional waves and 4.0 MHz for shear waves
- Specimens: 3.6 mm nominal diameter and 3.6–7.2 mm lengths
- Wet condition

Data

Site			Elastic stiffness (MPa)				
Right or left	Front or back	Proportion of total length from proximal end	c_{11}	c_{12}	c_{13}	c_{33}	c_{44}
Left	Front	0.24	6516	2519	2110	7722	2179
Left	Back	0.45	6895	2386	5226	8481	2275
Right	Front	0.45	7171	3220	3944	9170	2275
	mean		6860	2700	3760	8480	2240
	± SD		330	570	1570	760	180

E_3 (MPa)	E_1 (MPa)	ν_{31}	ν_{12}	G_{31} (MPa)
5500	4990	0.39	0.20	2240

E, elastic modulus; ν, Poisson's ratio; G, shear modulus.

Comments
- The elastic relations: $\sigma_i = c_{ij}\,\varepsilon_j$ (i, j = 1, 2, $\cdots$6).
- The elastic stiffnesses for dry bovine femurs were approximately 3.5 times higher than the equivalent stiffness for wet.
- The bone was a transversely isotropic material with the axis of symmetry parallel to the longitudinal axis of the bone.

Reference(s)
Lappi VG, King MS, LeMay I (1979) Determination of elastic constants for human femurs. Trans ASME J Biomech Eng 101:193–197

Elastic Modulus (3)

• Young's modulus •	• Bovine • Femoral diaphysis	• Cortical bone • Density

Materials
- Bovine (adult)
- Mid-diaphysis of fresh femur
- Cortical bone

Testing Methods and Experimental Conditions
- A pulsed-differential ultrasonic method
- Longitudinal waves of 200 kHz frequency were utilized
- Rectangular-parallelepiped specimens longitudinally oriented
- Specimen size was 5 mm × 5 mm in cross section and ranged in length from 62 mm to 121 mm
- Tested in a water bath containing Ringer's solution at 17, 24, 30, 37, and 41°C
- Young's modulus E was calculated by substituting the wave velocity (C_l) and the density (ρ) in equation: $C_l = (E/\rho)^{1/2}$

Data

Specimen length (mm)	62	89	100	113	115	115	118	120	121
$\rho\,(\times 10^3\ \mathrm{kgm^{-3}})$	2.067	2.062	2.038	2.107	1.897	2.093	2.101	2.054	2.057
$C_l\,(\times 10^3\ \mathrm{ms^{-1}})$									
17°C	3.35	3.34	3.33	3.41	3.52	3.40	3.54	3.31	3.27
24°C	3.29	3.30	3.29	3.39	3.47	3.37	3.49	3.30	3.24
30°C	3.27	3.24	3.26	3.36	3.42	3.35	3.48	3.26	3.22
37°C	3.25	3.22	3.22	3.35	3.41	3.34	3.44	3.25	3.21
41°C	3.16	3.21	3.18	3.30	3.40	3.30	3.43	3.24	3.12
E (GPa)									
17°C	23.2	23.0	22.6	24.5	23.5	24.2	26.3	22.5	22.0
24°C	22.4	22.6	22.0	24.2	22.6	23.8	25.6	22.3	21.6
30°C	22.1	21.8	21.7	23.8	22.2	23.5	25.5	21.7	21.4
37°C	21.8	21.6	21.2	23.2	22.1	22.9	24.9	21.6	20.4
41°C	20.6	21.4	20.6	22.9	22.0	22.7	24.7	21.5	20.1

Comments
- The density of specimens were similar with one exception. Longitudinal wave velocity (C_l) appeared to be independent of specimen length.
- The average of Young's modulus (E) showed a linear decrease of approximately 0.07 GPa per °C with increasing temperature. The effect was reversible within the range of temperature tested.

Reference(s)
Bonefield W, Tully AE (1982) Ultrasonic analysis of the Young's modulus of cortical bone. J Biomed Eng 4:23–27 (with permission)

Elastic Modulus (4)

• Young's modulus • Modulus–density	• Human • Tibia, femur, humerus	• Cortical bone • Cancellous bone

Materials
- Human (45–68 [mean 60] years)
- Tibia, femur, and humerus
- Cortical bone, cancellous bone, and trabecular bone

Testing Methods and Experimental Conditions
- Ultrasonic test
- Cubic specimen for cortical bone, $5 \times 5 \times 5$ mm
- Cubic specimen for cancellous bone, $10 \times 10 \times 10$ mm
- Cylindrical specimen for trabecular bone, 5 mm diameter and 15 mm length

Data

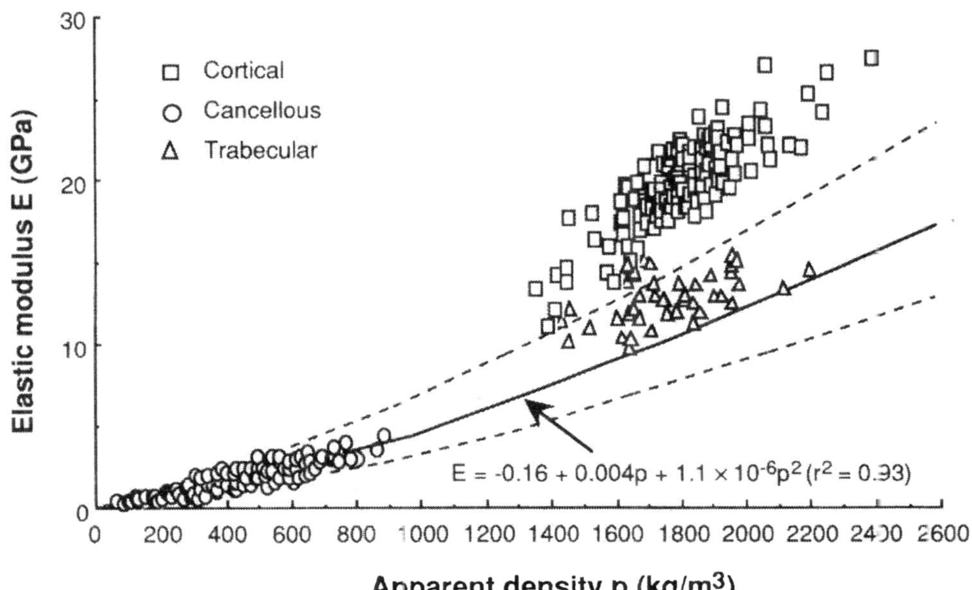

Comments
- The solid line represents a regression using a quadratic equation for cancellous bone specimens only. The dashed lines represent 95% confidence intervals of the regression.
- The majority of the values of Young's modulus of trabecular bone fall between the confidence intervals, while Young's modulus values for cortical bone fall outside the confidence intervals. In other words, the Young's modulus of trabecular bone can be extrapolated from that of cancellous bone using a quadratic relationship, while Young's modulus of cortical bone cannot be extrapolated from cancellous bone data.

Reference(s)
Rho JY, Ashman RB, Turner CH (1993) Young's modulus of trabecular and cortical bone material. Ultrasonic and microtensile measurements. J Biomech 26:111–119 (with permission)

Fatigue (1)

• Creep • Cycles to failure	• Human • Femoral diaphysis	• Cortical bone •

Materials

- Human (2 males [age, 59 and 69 years] ; 1 female [age, 25 years])
- Femoral mid-diaphysis
- Cortical bone

Testing Methods and Experimental Conditions

- Fatigue tests under tension–compression (T–C) and zero–tension (O–T) modes at 2 Hz
- Tensile creep-fracture tests at constant stress levels
- Specimens of $6 \times 6 \times 37$ mm were oriented with the long axis of the femur
- Specimens were enclosed in a high humidity environmental chamber at 37°C

Data

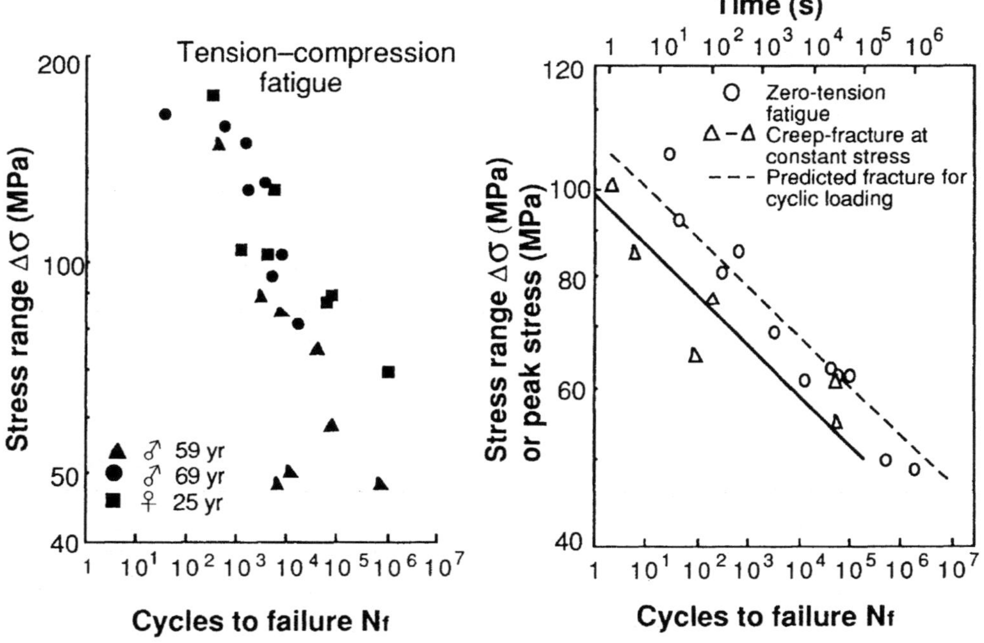

Comments

- The relationship between initial cyclic stress range and cycles to failure with the T-C specimens was consistent with that derived for low-cycle fatigue under strain control.
- Time-dependent failure model:

t_B, the time when falure occurs; $\sigma(t)$, the stress history; T_B, the time to failure in a constant stress.
- This model predicted quite well the time to failure for the O-T fatigue specimens, suggesting that creep damage plays an important role in O-T fatigue specimens.

Reference(s)

Carter DR, Caler WE (1983) Cycle-dependent and time-dependent bone fracture with repeated loading. Trans ASME J Biomech Eng 105:166–170

Fatigue (2)

• Cycles to failure •	• Human • Femoral diaphysis	• Cortical bone •

Materials
- Human (2 males [age, 82 and 84 years]; 2 females [age, 51 and 53 years])
- Femoral mid-diaphysis
- Cortical bone

Testing Methods and Experimental Conditions
- Uniaxial fatigue test in high humidity at 37°C
- Triangular strain history with strain rate of 0.01 per second
- Loading frequency was between 0.5 and 1.0 Hz

Data

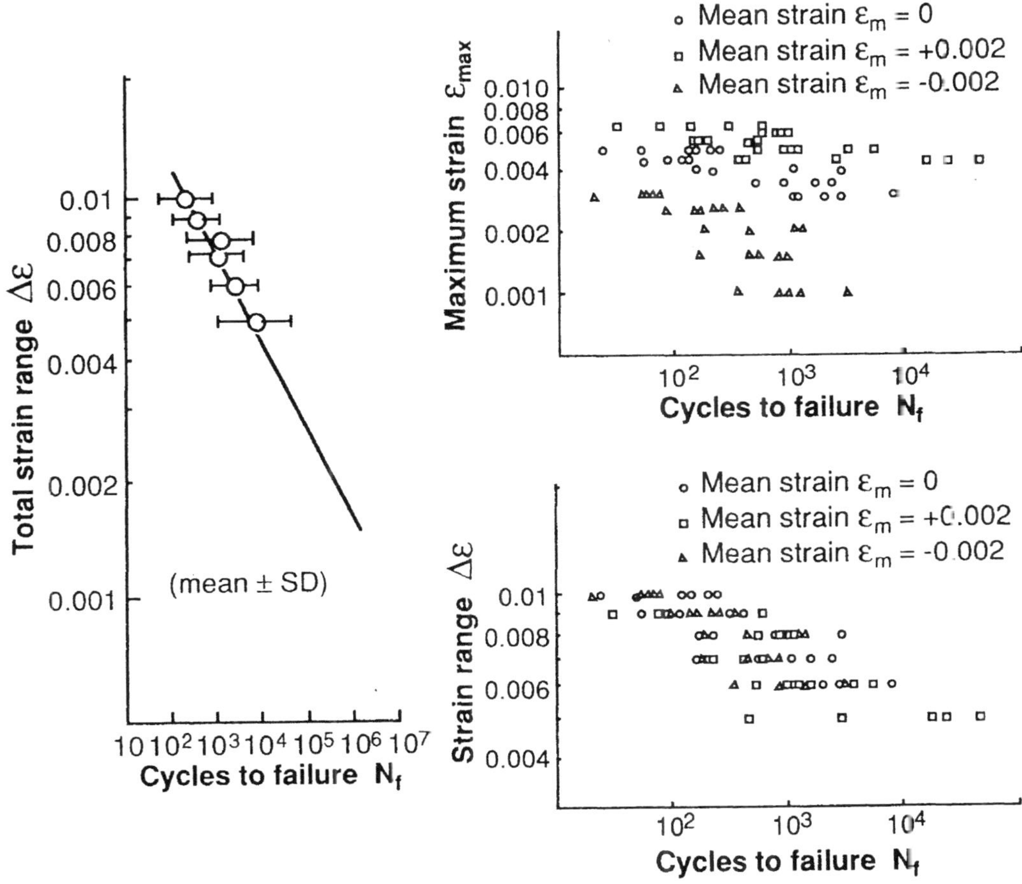

Comments
- Bone fatigue is a gradual damage process accompanied by a progressive increase in hysteresis and a loss of bone stiffness. The total number of cycles to fatigue failure was influenced only by the total strain range. Bone has extremely poor fatigue resistance.

Reference(s)
Carter DR, Caler WE, Spengler DM, Frankel VH (1981) Fatigue behavior of adult cortical bone: the influence of mean strain and strain range. Acta Orthop Scand 52:481–490 (with permission)

Fatigue (3)

• Cycles to failure • Secant modulus	• Bovine • Femoral condyle	• Trabecular bone •

Materials

- Bovine
- Distal femoral condyle
- Trabecular bone

Testing Methods and Experimental Conditions

- Specimen size, $6 \times 6 \times 7$ mm
- Room temperature, about 21°C
- Specimens were surrounded by a gauze constantly soaked in 0.9% NaCl
- Compressive fatigue tests

Data

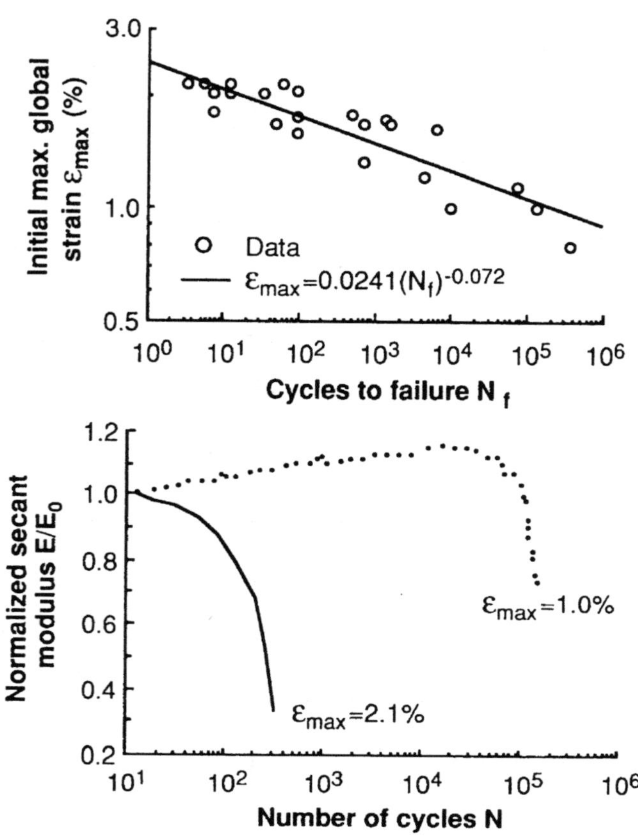

Comments

None.

Reference(s)

Michel MC, Guo XDE, Gibson LJ, McMahon TA, Hayes WC (1993) Compressive fatigue behavior of bovine trabecular bone. J Biomech 26:453–463 (with permission)

Fatigue (4)

• Fatigue life •	• Bovine •	• Cortical bone •

Materials
- Bovine
- Cortical bone

Testing Methods and Experimental Conditions
- Rotating cantilever fatigue testing
- Tests were conducted at $22° \pm 2°C$
- Physiological saline was used to keep the specimens moist
- Tests were performed at four levels of stress amplitide (60, 70, 95, and 112 MN/m^2) at each of three stress frequencies (30, 60, and 125 Hz)

Data

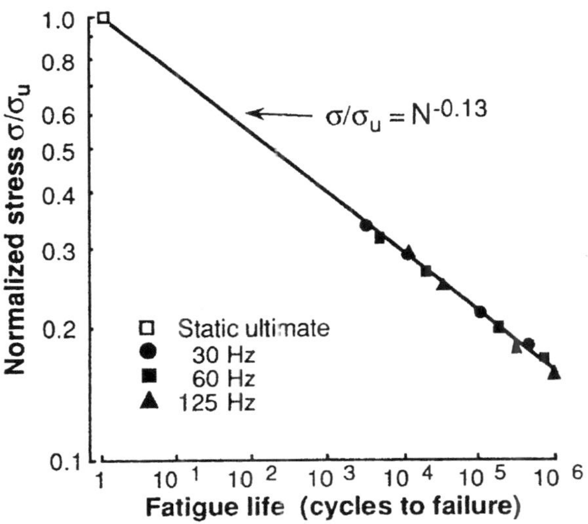

Stress frequency f (Hz)	Stress amplitude σ (MN/m²)	Fatigue life (cycles to falure) (N $\pm$ SE)	Normalized stress (σ/σ_u)
30	60	$(6.20 \pm 1.78) \times 10^5$	0.182
	70	$(1.25 \pm 0.13) \times 10^5$	0.213
	95	$(1.29 \pm 0.16) \times 10^4$	0.289
	112	$(4.33 \pm 0.64) \times 10^3$	0.341
60	60	$(9.48 \pm 1.52) \times 10^5$	0.171
	70	$(2.52 \pm 0.34) \times 10^5$	0.199
	95	$(2.54 \pm 0.43) \times 10^4$	0.270
	112	$(5.81 \pm 0.94) \times 10^3$	0.319
125	60	1.35×10^6	0.159
	70	4.33×10^5	0.185
	95	4.49×10^4	0.252
	112	1.35×10^4	0.297

Comments
- The fatigue strength increases with stress frequency.

Reference(s)
Lafferty JF, Raju PVV (1979) The influence of stress frequency on the fatigue strength of cortical bone. Trans ASME J Biomech Eng 101:112–113

Fatigue (5)

• Stress–strain curve • Fatigue life	• Bovine • Femur	• •

Materials
- Bovine
- Femur

Testing Methods and Experimental Conditions
- Specimens of 4.45 cm in length and 0.635 cm in diameter (central waisted section: 0.475 cm in diameter)
- The central axis of specimens was parallel to the longitudinal axis of femur
- Tensile fatigue testing to failure in water at 37°C

Data

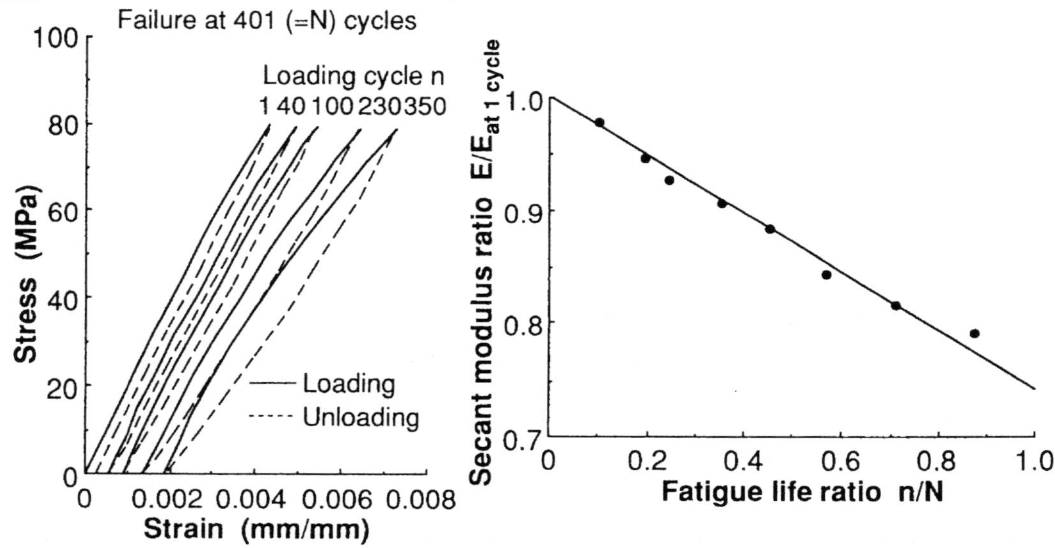

Comments
None.

Reference(s)
Carter DR, Hayes WC (1977) Compact bone fatigue damage—I. Residual strength and stiffness. J Biomech 10:325–337 (with permission)

Hardness

• Hardness •	• Human • Lumbar vertebra	• Trabecular bone • Age effect

Materials
- Human (117 males, 53 females)
- 3rd lumbar vertebra
- Trabecular bone

Testing Methods and Experimental Conditions
- Brinell hardness test (JIS-Z-2243)
- Freezing period, more than 30 days after death
- Room temperature

Data

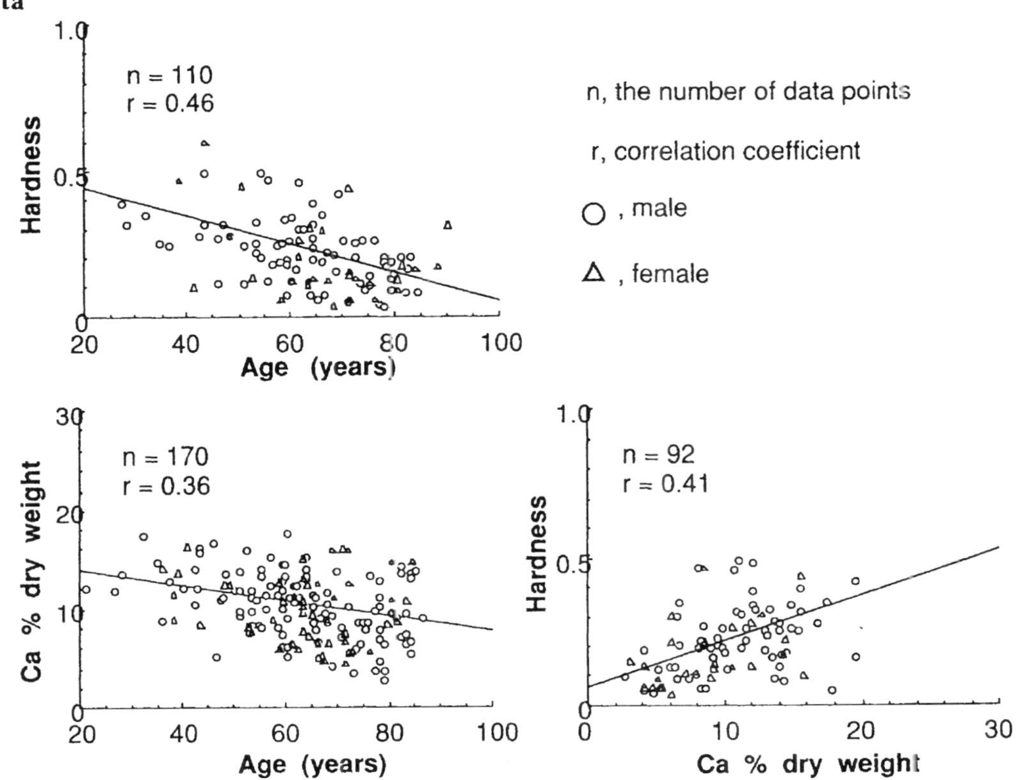

Comments
None.

Reference(s)

Wada Y, Tokuda M, Iwata J, Yagi K (1994) Evaluation of mechanical property of 3rd lumbar vertebral body. 43rd Nat Cong of Theor Appl Mech:647–650

Shear Property (1)

• Elastic modulus • Shearing strength	• Human • Femoral diaphysis	• Osteon •

Materials
- Human (age, 25 years; apparently free from skeletal defects)
- Femoral shaft
- Single osteon

Testing Methods and Experimental Conditions
- Cross sections of diaphysis about 300 μm in thickness
- Single osteon was loaded centrically with a cylindrical steel punch from the upper end of the specimen
- Osteons tested differed in the orientation of collagen bundles
- Measurements were performed at 20°C

Data

Type of osteons	Modulus of elasticity (GPa)	Ultimate shearing strength (MPa)	Movement of punch at breaking point (μm)
a	4.1513 ± 0.3963	55.9 ± 4.4	4.55 ± 0.55
b	3.4086 ± 0.3474	44.8 ± 4.3	4.46 ± 0.50
c	3.3149 ± 0.5159	46.2 ± 7.4	4.92 ± 1.10
d	3.2376^z	38.1 ± 6.1	4.10 ± 1.47

a, fully calcified osteons of Type I ($n = 8$) and II ($n = 10$); b, osteons of Type I ($n = 8$) and II ($n = 7$) at the initial stage of calcificalion; c, fully calcified osteons of Type III ($n = 8$); d, osteons of Type III at the initial stage of calcification ($n = 7$).

Type I, osteon having marked transverse spiral course of fiber bundles in successive lamellae; Type II, osteon with fiber bundles in one lamella making an angle of nearly 90° with the fiber bundles in the next one; Type III, osteons having marked longitudinal spiral course of fiber bundles in successive lamellae.

[z] This is a very approximate value because the proportional limit is reached very early.

Comments
- No significant difference was found in the ultimate shearing strength and the modulus of elasticity between osteons of Type I as opposed to those of Type II.
- Osteons having a marked longitudinal spiral course of fiber bundles in successive lamellae are least able to support shearing stress.

Reference(s)
Ascenzi A, Bonucci E (1972) The shearing properties of single osteons. Anat Rec 172:499–510 (with permission)

Shear Property (2)

• Maximum torque capacity • Maximum angle	• Dog • Femur, tibia	• •

Materials
- Dogs (male; healthy adult)
- Femur and tibia

Testing Methods and Experimental Conditions
- Torsion test at 35°C
- Specimens were thawed at 37°C after being frozen at -40°C for 48 hours

Data

Dog no.	Bone	Maximum torque capacity (Nm)		Maximum angle of torsion (degree)	
		fresh	defrosted	fresh	defrosted
1	femur	34.8	33.5	36.0	33.4
	tibia	23.2	22.6	45.8	40.1
2	femur	30.9	30.1	31.2	30.0
	tibia	22.1	21.1	38.2	37.9
3	femur	36.4	35.0	43.4	36.2
	tibia	25.2	24.4	39.8	41.8
4	femur	27.7	25.0	39.6	36.7
	tibia	19.7	18.2	35.5	36.0
5	femur	20.0	18.9	36.5	30.0
	tibia	16.3	15.8	41.3	37.9

Comments
- There were significant differences in the maximum torque capacity and maximum angle of torsion between fresh and defrosted specimens ($P < 0.05$).

Reference(s)
Stromberg L, Dalen N (1976) The influence of freezing on the maximum torque capacity of long bones, an experimental study on dogs. Acta Orthop Scand 47:254–256 (with permission)

Shear Property (3)

• Maximum torque capacity • Stiffness	• Dog • Femoral and tibial	• •

Materials

• Canine
• Diaphysis of femur and tibia

Testing Methods and Experimental Conditions

• Torsion test
• Angular velocity of 3°–12 °/s

Data

	Dog no.	Maximum torque capacity (Nm)			Stiffness (Nm/degree)		
		Inward rotation	Outward rotation	Percentage difference	Inward rotation	Outward rotation	Percentage difference
Femora	1.	31.8	30.3	4.8	1.42	1.54	-8.1
	2.	24.0	25.3	-5.3	1.40	1.33	5.1
	3.	24.0	22.0	8.7	1.30	1.40	-7.4
	4.	28.8	29.8	-1.7	1.33	1.38	-3.7
	5.	34.3	33.5	2.4	1.75	1.67	4.7
	6.	24.4	24.9	-2.0	1.67	1.69	-1.2
	7.	20.2	21.8	-7.6	1.53	1.53	0.0
Tibiae	8.	14.0	13.5	3.6	0.71	0.69	2.9
	9.	23.0	24.5	-6.3	1.00	1.08	-7.7
	10.	18.5	19.0	-2.7	0.77	0.75	2.6
	11.	18.5	20.0	-7.8	0.79	0.85	-7.3
	12.	22.0	20.0	9.5	1.00	0.96	4.1
	13.	23.0	26.0	-12.2	0.89	1.00	-11.6
	14.	19.3	21.0	-8.4	0.75	0.75	0.0
	15.	18.3	17.2	6.2	0.58	0.56	3.5
	16.	26.8	26.0	3.0	0.94	0.89	5.5
	17.	28.7	28.3	1.4	1.19	1.25	-4.9
	18.	17.6	17.9	-1.7	0.69	0.70	-1.4
	19.	39.6	38.9	1.8	1.56	1.56	0.0
	20.	30.9	28.9	6.7	1.11	1.11	0.0

Comments

• No significant effect of the direction of twist upon the torsional strength and stiffness.

Reference(s)

Netz P, Eriksson K, Stromberg L (1978) Torsional strength and geometry of diaphyseal bone: an experimental study on dogs. Acta Orthop Scand 49:430–434 (with permission)

Shear Property (4)

• Stiffness • Shear modulus	• Dog • Tibia, femur	• Age effect •

Materials
- Dogs (age, 8–44 weeks)
- Fresh tibia and femur

Testing Methods and Experimental Conditions
- Torsional test at angular velocity of 6 °/s

Data

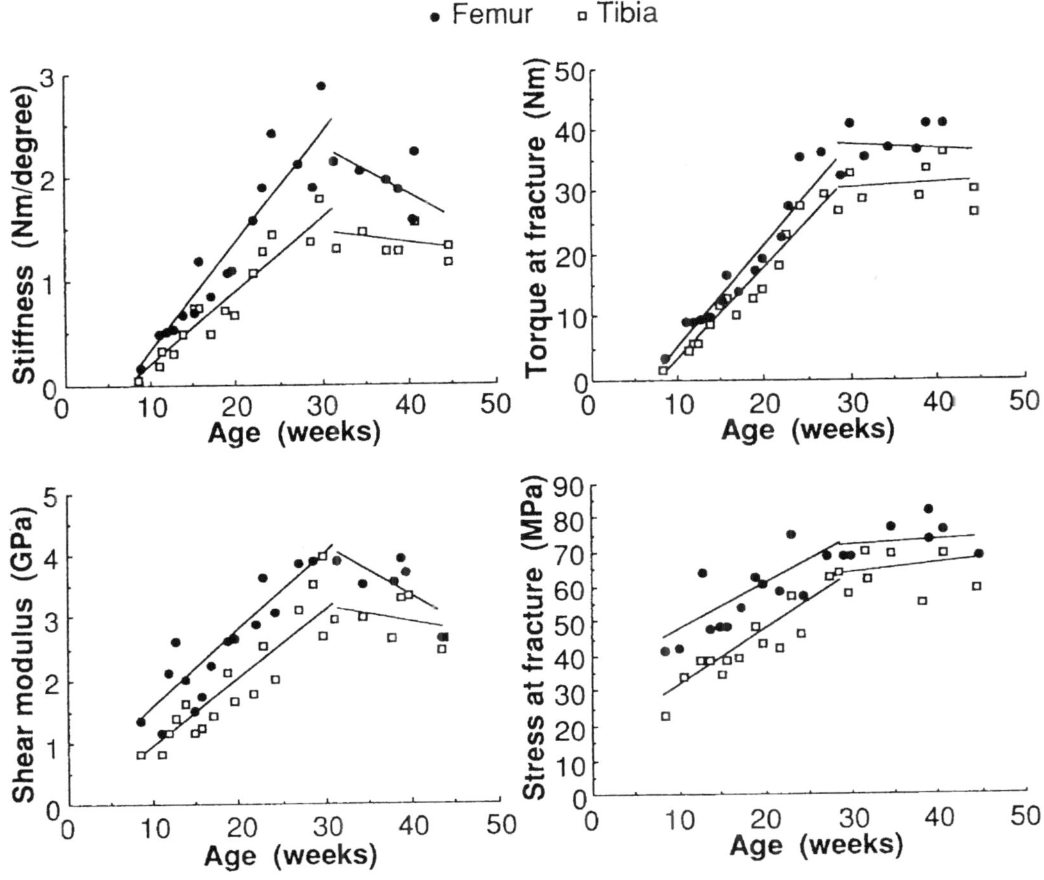

Comments
- Two regression lines were calculated for each bone type.

Reference(s)
Jonsson U, Netz P, Stromberg L (1984) Solid mechanics and strength of bone in young
dogs. Acta Orthop Scand 55:446–451 (with permission)

Tensile Property (1)

• Elastic modulus •	• Rabbit • Tibia	• Compact bone • Age effect

Materials

- New Zealand white rabbits (age, 1, 3, and 12 months)
- Tibia

Testing Methods and Experimental Conditions

- Tensile test
- Compact bone specimen from longitudinal section
- Measurement of deformation by epoxy backed foil resistance strain gauges
- Constant strain rate of $1 \times 10^{-4} - 3 \times 10^{-4}$ per second

Data

Age (month)	Porosity p (%)	Hydroxyapatite volume fraction V_h	Bone modulus E_b (GPa)	Absolute bone modulus E_{b0} (GPa)	Hydroxyapatite modulus E_h (GPa)
1	27	0.45	15.1	27.4	59.6
3	10	0.45	23.6	28.8	62.7
12	10	0.50	27.6	33.5	65.8

The average hydroxyapatite volume fraction V_h was determined by taking the density of hydroxyapatite and collagen as 3.17×10^3 kg/m^3 and 1.33×10^3 kg/m^3, respectivly. Absolute bone modulus E_{b0} and hydroxyapatite modulus E_h were calculated from

$$E_b = E_{b0}(1 - 1.9p + 0.9p^2)$$
$$E_{b0} = E_h V_h + E_c V_c$$

with measured bone modulus E_b and porosity volume p, assuming $E_c V_c \cong 0.6$ GPa.

Comments

- Stress–strain curves were linear to relatively high stress levels and showed the absence of hysteresis.
- It is concluded that the increase in modulus with age was attributed to the following factors: (a) a decrease in the total porosity content, (b) an increase in the hydroxyapatite volume fraction, and (c) an increase in the hydroxyapatite modulus.

Reference(s)

Bonfield W, Clark EA (1973) Elastic deformation of compact bone. J Mat Sci 8: 1590–1594 (with permission)

Tensile Property (2)

• Elastic modulus • Poisson's ratio	• Human • Femoral diaphysis	• Cortical bone • Locational dependence

Materials
- Human (age, 45–55 years)
- Diaphysis of femur
- Cortical bone

Testing Methods and Experimental Conditions
- Tensile tests
- Dynamic loading developed by dropping weights
- Loaded along the longitudinal axis of the femur
- Strain rate of 2×10^{-5}, 4×10^{-2}, and 1×10^2 per second (Three different loading techniques)

Data

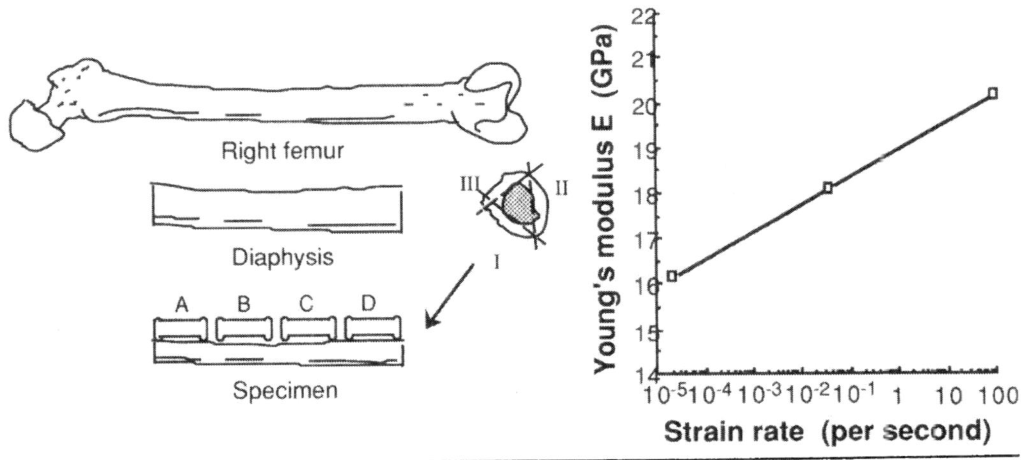

Specimen	Young's modulus E (GPa)				Poisson's ratio ν			
	A	B	C	D	A	B	C	D
I	17.21	18.32	17.09	17.21	0.355	0.380	0.370	0.350
II	18.46	18.04	17.84	17.91	0.371	0.334	0.373	0.371
III	18.46	18.04	18.29	18.47	0.363	0.367	0.371	0.355
Average value	18.04	18.13	17.74	17.86	0.363	0.360	0.371	0.359
Total average value	17.9				0.360			

All data were obtained at the strain rate of 4×10^{-2} per second

Comments
- Young's modulus values obtained for high strain are 11% and 23% higher than those obtained for intermediate and low strain rate, respectively.
- Poisson's ratio did not show any significant variation for three strain rates.

Reference(s)

Raftopoulos D, Katsamanis E, Saul F, Liu W, Saddemi S (1993) An intermediate loading rate technique for the determination of mechanical properties of human femoral cortical bone. J Biomed Eng 15:60–66 (with permission)
Katsamanis F, Raftopoulos DD (1990) Determination of mechanical properties of human femoral cortical bone by the Hopkinson bar stress technique. J Biomech 23:1173–1184 (with permission)

Tensile Property (3)

| • Maximum load | • Rat | • |
| | • Femoral bone collagen | • |

Materials
- Wistar rats (male; age, 2, 5, 15, 25 months)
- Cortical bone collagen obtained by decalcifying femoral bone

Testing Methods and Experimental Conditions
- Decalcified at 4°C for 3 weeks by suspending in 50 mM disodium-EDTA (pH7.4) that contained 1 mM p-hydroxymercuribenzoate and 10 µM phenylmethanesulfonyl fluoride
- Cut into 1 mm wide longitudinal or transverse specimens including entire cortical thickness
- Tensile test under a constant rate of deformation (10 mm/min) in a 50 mM Tris-HCl buffer (pH7.4)
- Distance between the clamps, 5.5 mm

Data

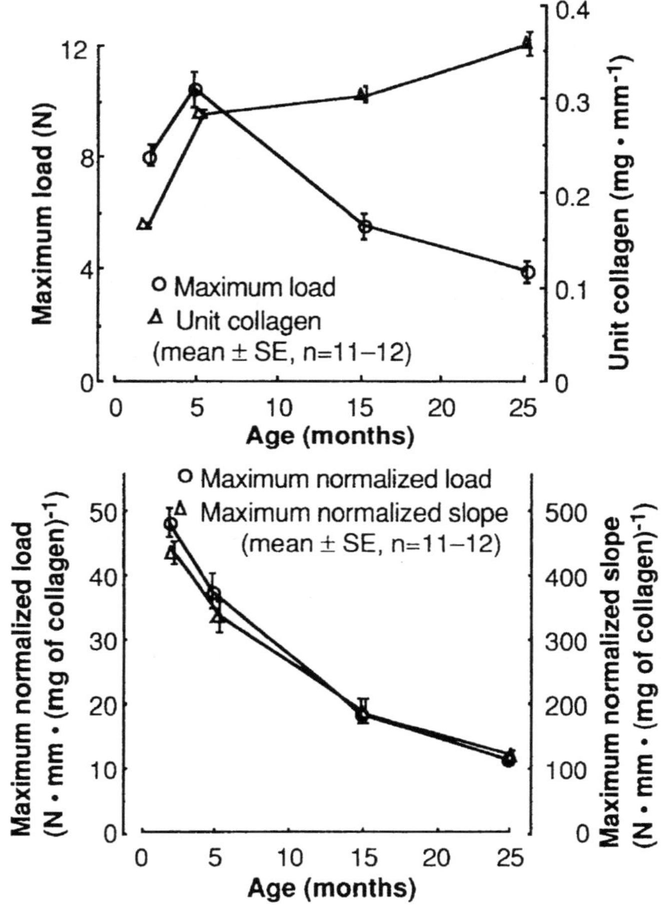

Comments
- Ultimate strength and maximum stiffness decrease with age.
- In vitro aging, produced by incubating for 0–5 months, did not change the mechanical strength of bone collagen specimens obtained from 2-month-old rats.

Reference(s)
Danielsen CC, Andreassen TT, Mosekilde L (1986) Mechanical properties of collagen from decalcified rat femur in relation to age and in vitro maturation. Calcif Tissue Int 39:69–73 (with permission)

Tensile Property (4)

• Tensile strength • Elastic modulus	• Bovine • Femur	• Strain rate • Mineral content

Materials
- Bovine
- Femur

Testing Methods and Experimental Conditions
- Tensile tests at strain rates from 0.00013 to 0.16 per second
- Rectangular prismatic test specimens, 19 mm in length and 1.8 mm^2 area in section
- Specimens were kept wet with salt solution at all times

Data

Variable	Reconstruction	Regression ss	P Anovar	P H	P ash	$P\dot{\varepsilon}$	H 1-0	Ash 3-7	$\varepsilon \times 3$
$\sigma_{ult} =$	$107.4 - 42.7H + 18.8\dot{\varepsilon}$	0.80	< 0.001	< 0.001	–	< 0.001	66	–	53
	or $38.9 + 9.5A + 18.4\dot{\varepsilon}$	0.81	< 0.001	–	< 0.001	< 0.001	–	56	52
$\sigma_y =$	$101.9 - 40.3H + 19.5\dot{\varepsilon}$	0.81	< 0.001	< 0.001	–	< 0.001	65	–	57
	or $36.2 + 9.2A + 19.7\dot{\varepsilon}$	0.85	< 0.001	–	< 0.001	< 0.001	–	58	59
$\varepsilon_y =$	$45.7 - 4.37H + 6.64\dot{\varepsilon}$	0.74	< 0.001	< 0.1	–	< 0.001	11	–	44
	or $38.7 + 0.97A + 6.67\dot{\varepsilon}$	0.74	< 0.001	–	< 0.05	< 0.001	–	9	44
$E =$	$22.8 - 5.47H + 1.01\dot{\varepsilon}$	0.45	< 0.001	< 0.001	–	< 0.1	31	–	13
	or $13.5 + 1.37A + 0.75\dot{\varepsilon}$	0.51	< 0.001	–	< 0.001	NS	–	31	10

σ_{ult}, ultimate tensile stress (MPa); σ_y, stress at yield (MPa); ε_y, strain at yield (strain $\times$ 10^4) E, modulus of elasticity (GPa); H, proportion of Haversian tissue using 0–1 classification; $\dot{\varepsilon}$, log strain rate (0.0001 per second = 0, 0.001 per second = 1, etc.); A, ash weight (%); Regression ss, fraction of total sums of squares accounted for by the regression; P Anovar, probability of the regression analysis of variance; P H, P ash, $P\dot{\varepsilon}$, probability of the t-test for the contribution of amount of Haversian tissue, ash, and strain rate to regression; H 1–0, percentage change in value of the variable produced by decreasing reconstruction from 1 to 0 at minimum strain rate; Ash 3–7, percentage change in value of the variable produced by increasing ash from 63 to 67%; $\dot{\varepsilon} \times 3$, percentage change in value of the variable produced by increasing strain rate a thousandfold when there is no reconstruction, or ash is 63%.

Comments
None.

Reference(s)
Currey JD (1975) The effects of strain rate, reconstruction and mineral content on some mechanical properties of bovine bone. J Biomech 8:81–86 (with permission)

Tensile Property (5)

• Tensile strength • Elastic modulus	• Human, bovine • Femur	• Osteon •

Materials
- Human (male; age, 30 and 80 years) and ox (age, 2 years)
- Femoral shaft
- Osteon

Testing Methods and Experimental Conditions
- Samples were taken from longitudinal sections (thickness, 20–50 μm)
- Isolated middle portion of each sample was subjected to direct tensile force parallel to the longitudinal axis
- Some samples were tested after decalcification
- All measurements were performed at a temperature of 20°C

Data

Subjects and age	Types of osteons	State of samples	No.	Ultimate tensile strength (MPa)	Modulus of elasticity (GPa)	Elongation at breaking point (%)
Man; Age, 30	OL and FC	dry	7	193.4 ± 32.0	23.38 ± 6.98	2.15 ± 0.55
	OL and LDC	dry	6	189.4 ± 28.2	19.98 ± 7.49	4.42 ± 1.41
	OL and FC	wet	12	120.1 ± 14.8	11.66 ± 5.78	6.84 ± 2.86
	OL and LDC	wet	12	104.4 ± 17.0	5.99 ± 2.94	9.40 ± 2.72
	OCD and FC	wet	9	101.8 ± 16.8	5.48 ± 2.56	10.29 ± 4.03
	OCD and LDC	wet (d)	6	88.8 ± 14.5	4.38 ± 1.93	9.80 ± 0.92
	OL and FC	wet (d)	5	83.6 ± 13.7	1.03 ± 0.34	21.90 ± 7.15
	OL and LDC	wet	4	86.0 ± 11.7	1.45 ± 0.61	20.62 ± 5.10
Man; Age, 80	OL and FC	wet	8	106.9 ± 7.9	10.65 ± 3.83	7.68 ± 2.39
	OL and LDC	wet	5	92.8 ± 12.4	4.83 ± 3.26	10.50 ± 2.54
	OCD and FC	wet	14	86.4 ± 8.3	6.09 ± 3.51	8.23 ± 3.45
	OCD and LDC	wet	8	84.3 ± 6.8	3.88 ± 0.72	14.28 ± 3.28
Ox; Age, 2	OL and FC	dry	5	204.8 ± 11.3	17.71 ± 1.90	3.45 ± 0.38
	OL and LDC	dry	5	190.1 ± 19.0	13.07 ± 2.39	9.32 ± 1.18
	OL and FC	wet	5	118.8 ± 7.3	14.66 ± 2.81	8.30 ± 1.10
	OL and LDC	wet	6	110.5 ± 4.7	5.35 ± 1.21	11.14 ± 1.41

No., number of measured osteons; OL, osteons whose fiber bundles had a longitudinal course with the angles of fibers in successive lamellae practically the same; OCD, osteons whose fiber bundles in successive lamellae formed an angle of 90°; LDC, osteons at lowest degree of calcification; FC, fully calcified osteons; d, decalcified.

Comments
- Tensile strength and modulus of elasticity increase when osteon specimens are dried.
- Modulus of elasticity in tension of the organic matrix corresponds to that of collagen.
- The tensile properties of osteons seem independent of the age of subjects.
- Human and ox osteons reveal the same tensile behavior.

Reference(s)
Ascenzi A, Bonucci E (1967) The tensile properties of single osteons. Anat Rec 158:375–386 (with permission)

Tensile Property (6)

• Tensile strength • Ultimate strain	• Human • Femoral diaphysis	• Cortical bone • Age effect

Materials
- Human (47 healthy individuals including 25 males and 22 females; age, 20–102 years)
- Mid diaphysis of right femur
- Cortical bone

Testing Methods and Experimental Conditions
- Tensile tests were performed at 19° to 21°C at a strain rate of 0.03 per second keeping specimens moist
- Elastic modulus, plastic modulus, ultimate stress, and ultimate strain were determined in a standard manner
- Fracture energy, plastic and elastic energy were calculated from stress-strain curves
- Mineral content, porosity, and other histological parameters were determined using sections adjacent to the fracture site after mechanical tests

Data

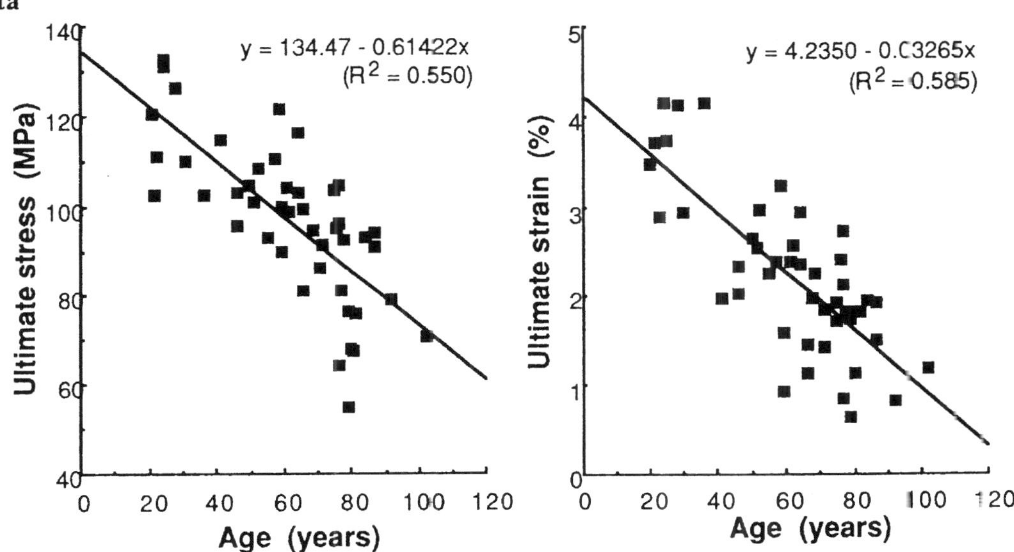

Comments
- The mechanical properties deteriorated markedly with age. Ultimate stress, ultimate strain, and energy absorption decreased by 5%, 9%, and 12% per decade, respectively. The porosity of bone increased significantly with age, while the mineral content was not affected. Amount of Haversian bone increased with age. Changes in porosity accounted for 76% of reduction in strength.

Reference(s)

McCalden RW, McGeough JA, Barker MB, Court-Brown CM (1993) Age-related changes in the tensile properties of cortical bone. J Bone Joint Surg 75A:1193–1205 (with permission)

Tensile Property (7)

• Tension and compression •	• Human • Femur	• Compact bone •

Materials
- Human (13 males and 14 females; age, 19–87 years)
- Femur
- Compact bone

Testing Methods and Experimental Conditions
- Tensile test specimen: $35 \times 3 - 8 \times 3.1$ mm
- Compressive test specimens: 5.1 mm in diameter and 7.5 mm in length
- Moist conditions at room temperature
- Strain rate of approximately 0.001 per second

Data

Relations between ultimate stress σ (MPa) and mineral density M (mg/mm^3)	No. of samples n	Correlation coefficient r	Index of determination r^2
$\sigma_T = 5.14e^{1.761M}$	105	0.87	0.75
$\sigma_c = 5.36e^{2.017M}$	91	0.92	0.85

Relations between mineral density M (mg/mm^3), ultimate tensile stress σ_T (MPa), and Age A (years)	No. of samples n	Correlation coefficient r	Index of determination r^2
$M = 1.37 - 0.0028A$	25	-0.41	0.17
$\sigma_T = 63.4 - 0.312A$	24	-0.59	0.35

Subscripts T and C denote tension and compression, respectively.

Comments
- 75% of variance in ultimate tensile stress, and 85% of variance in ultimate compressive stress are accounted for by variation in mineral density.

Reference(s)
Smith CB, Smith DA (1976) Relation between age, mineral density and mechanical properties of human femoral compacta. Acta Orthop Scand 47:496–502 (with permission)

Tensile Property (8)

• Tension and compression • Elastic modulus	• Human • Femur	• Cancellous bone • Density

Materials
- Human (male; age, 84 years)
- Femur
- Frozen cancellous bone

Testing Methods and Experimental Conditions
- Tensile and compressive testing
- Specimen, 10 mm in diameter and 10 mm in height
- Strain rate of 0.01 per second
- Moistened with saline solution

Data

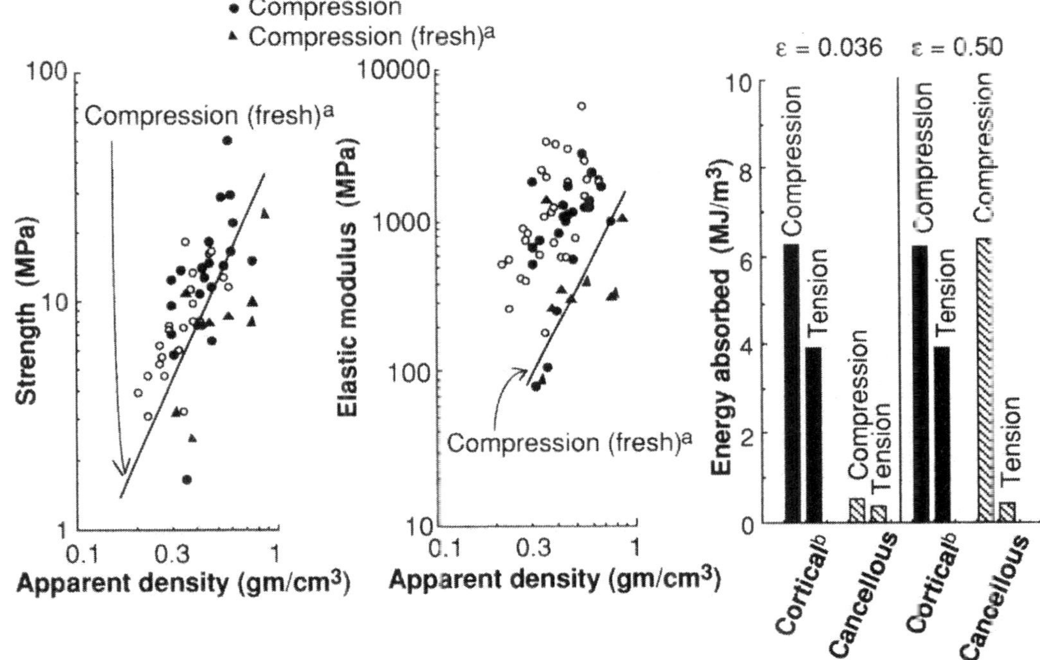

Comments
- The energy absorption capacity of cortical and cancellous bone is determined at low strains ($\varepsilon = 0.036$) and at high strains ($\varepsilon = 0.50$).
- The primary difference in mechanical properties of cancellous bone tested in tension and compression is the energy absorption capacity.

Reference(s)

Carter DR, Schwab GH, Spengler DM (1980) Tensile fracture of cancellous bone. Acta Orthop Scand 51:733–741 (with permission)

[a]Carter DR, Hayes WC (1977) The compressive behavior of bone as a two-phase porous structure. J Bone Joint Surg 59A:954–962 (with permission)

[b]Reilly DT, Burstein AH (1975) The elastic and ultimate properties of compact bone tissue. J Biomech 8:393–405 (with permission)

Tensile Property (9)

• Ultimate stress • Impact energy	• Human • Femur	• Compact bone •

Materials
- Human (adult)
- Femur
- Compact bone

Testing Methods and Experimental Conditions
- Tensile impact tests
- Dimensions of specimens followed ASTM specifications for the tensile impact testing of plastics (D-1822-61T, Type L)
- Specimens were submerged in Ringer's solution at room temperature

Data

Condition		Fresh	Embalmed
No. of samples		49	42
Ultimate stress (MPa)		126.3 ± 33.1	98.6 ± 25.5
Impact energy (J/m^2)		18790 ± 7355	14308 ± 5517
Maximum strain (%)		1.15 ± 0.30	0.982 ± 0.225
Modulus of elasticity (GPa)	Tangent	14.5 ± 3.4	
	Secant	12.4 ± 2.9	12.1 ± 1.6

All data are given as mean ± SD.

Comments
None.

Reference(s)
Saha S, Hayes WC (1976) Tensile impact properties of human compact bone. J Biomech 9:243–251 (with permission)

Yield Surface

• Axial and torsional load • Yield stress	• Human • Femur	• •

Materials

- Human
- Femur

Testing Methods and Experimental Conditions

- Specimens, 6 mm in length and 2.5 mm in diameter
- Specimen's central axis was parallel tc the longitudinal axis of femur
- Surfaces of specimens were kept wet with saline solution
- Tests of combined axial and torsional loads at controlled loading or displacement rates

Data

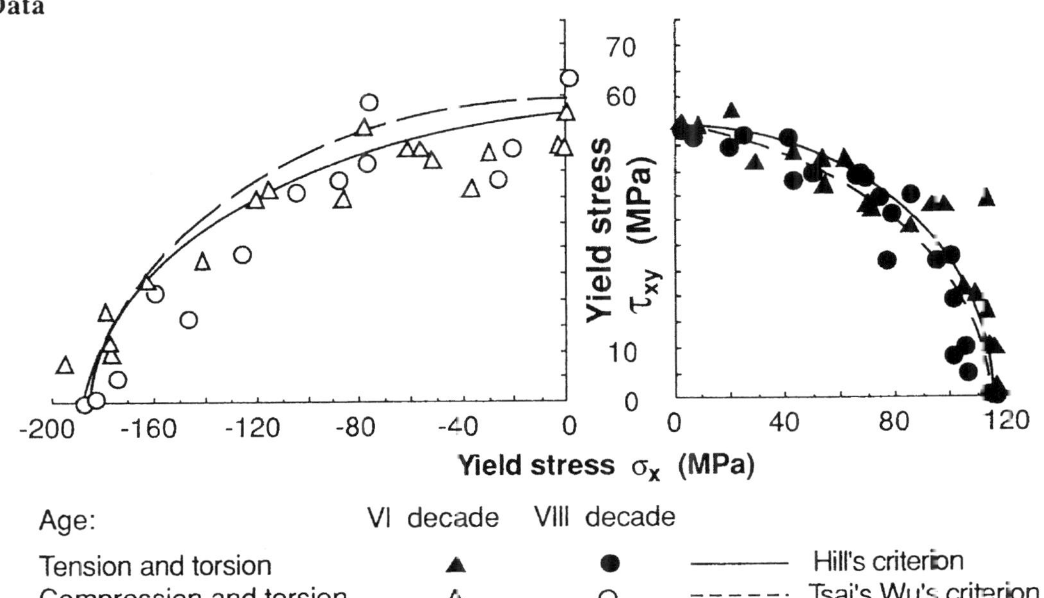

Comments

None.

Reference(s)

Cezayirliogle H, Bahniuk E, Davy DT, Heiple KG (1985) Anisotropic yield behavior of bone under combined axial force and torque. J Biomech 187:61–69 (with permission)

Compressive Property (1)

• Compressive strength • Stress–strain curve	• Human • Enamel, dentin	• •

Materials
- Human
- Enamel and dentin

Testing Methods and Experimental Conditions
- No description

Data

Material	Compressive strength (MPa)
Enamel	384
Dentin	297

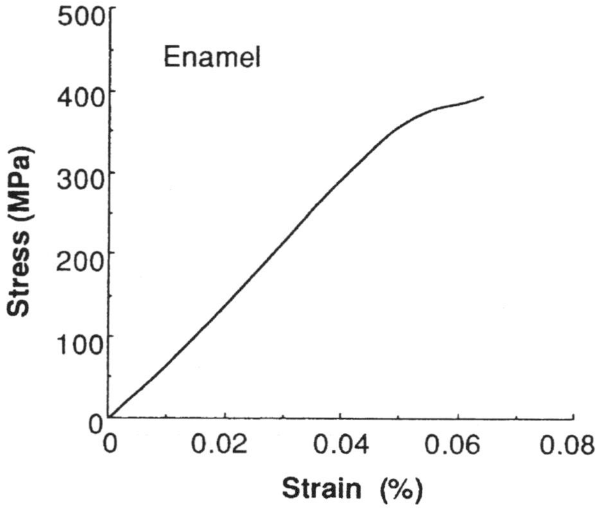

Comments
None.

Reference(s)
Craig RG (1993) Restorative dental materials, 9th edn. Mosby, St Louis, p 69 (with permission)

Compressive Property (2)

• Elastic modulus •	• Human • Dentin	• •

Materials
- Human
- Dentin

Testing Methods and Experimental Conditions
- Optical strain gauge

Data

Elastic modulus ($\times 10^4$ MPa)
1.47[a]
1.78[b]
1.52[c]
1.52[c]
1.48
1.77
1.43[a, b]
1.39
1.77[d]
Avg. 1.57 ± 0.14

[a]The long axis of the specimen was perpendicular to the occlusal surface: in the remaining specimens the long axis was parallel to the occlusal surface.

[b]The rate of loading was 234 MPa/min; the rate of loading for the remaining specimens was 23.4 MPa/min.

[c]These values represent separate determinations of the elastic modulus on identical specimens.

[d]Specimen was dried for 30 hours at 105°C.

Comments
None.

Reference(s)
Craig RG, Peyton FA (1958) Elastic and mechanical properties of human dentin. J Dent Res 37:710–718 (with permission)

Compressive Property (3)

| • Elastic modulus
• Compressive strength | • Human
• Enamel, dentin | •
• |

Materials

- Human
- Enamel and dentin

Testing Methods and Experimental Conditions

- No description

Data

Tooth	Structure	Modulus of elasticity $(\times 10^4$ MPa)	Proportional limit (MPa)	Modulus of resilience $(\times 10^4$ J/m^3)	Strength (MPa)
Molar	Dentin	1.2	148	9.4	305
	Enamel (cusp)	4.6	224	5.5	261
Bicuspid	Dentin	1.4	146	7.7	248
	Enamel	—	—	—	—
Cuspid	Dentin	1.4	140	7.1	276
	Enamel (cusp)	4.8	194	4.1	288
Incisor	Dentin	1.3	125	6.0	232

Comments

None.

Reference(s)

Stanford JW, Weigel KV, Paffenbarger GC, Sweeney WT (1960) Compressive properties of hard tooth tissues and some restorative materials. J Am Dent Assoc 60:746–756 (with permission)

Compressive Property (4)

• Elastic modulus • Poisson's ratio	• Human • Enamel, dentin	• •

Materials
• Human
• Enamel, dentin, and connective tissue membrane

Testing Methods and Experimental Conditions
• No description

Data

Material	Young's modulus (MPa)	Poisson's ratio	Density ($\times 10^{-6}$ kg/mm^2)
Hydroxyapatite	34300	0.28	3.1
Enamel	130340	0.30	2.9
Dentin	14700	0.30	4.0
Cortical bone	14700	0.30	1.3
Cancellous bone	490	0.30	1.3
Connective tissue membrane	98	0.45	–

Comments
• Data used for finite element analysis.

Reference(s)
Moroi HH, Okimoto K, Moroi R, Terada Y (1993) Numeric approach to the biomechanical analysis of thermal effects in coated implants. Int J Prosthodont 6:564–572 (with permission)

Compressive Property (5)

• Stress–strain curve •	• Human • Enamel, dentin	• •

Materials
- Human
- Enamel and dentin

Testing Methods and Experimental Conditions
- Cylindrical specimens
- Deformation rate of 0.025–0.051 mm/min

Data

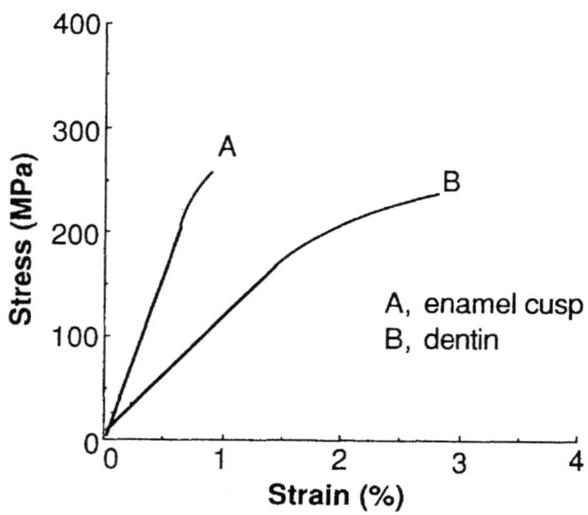

Comments
None.

Reference(s)
Stanford JW, Weigel KV, Paffenbarger GC, Sweeney WT (1960) Compressive
properties of hard tooth tissues and some restorative materials. J Am Dent Assoc 60:746–
756 (with permission)

Elastic Modulus

• Shear modulus • Poisson's ratio	• Human • Enamel, dentin	• •

Materials

- Human
- Enamel and dentin

Testing Methods and Experimental Conditions

- Mechanical and ultrasonic techniques

Data

Material	Young's modulus ($\times 10^4$ MPa)	Bulk modulus ($\times 10^4$ MPa)	Shear modulus ($\times 10^4$ MPa)	Poisson's ratio
Dentin (mechanical)	1.8	. . .	. . .	. . .
Enamel (mechanical)	8.1[a]	. . .	. . .	. . .
Enamel (ultrasonic)	7.7	6.5	2.9	0.30

[a]Average of cusp and side.

Comments

- No description on mechanical technique.

Reference(s)

Hall DR, Nakayama WT, Grenoble DE, Katz JL (1973) Elastic constants of three representative dental cements. J Dent Res 52:390 (with permission)

Fracture Toughness

• Fracture toughness •	• Human • Enamel	• •

Materials

- Human
- Enamel
- Max; Maxillary
- Man; Mandibular

Testing Methods and Experimental Conditions

- Conducted with a Vickers' diamond indenter in a Micromet hardness tester (Adolph I. Buchler, Evanston, IN, USA)
- Crack lengths and hardness impression diagonals were measured

Data

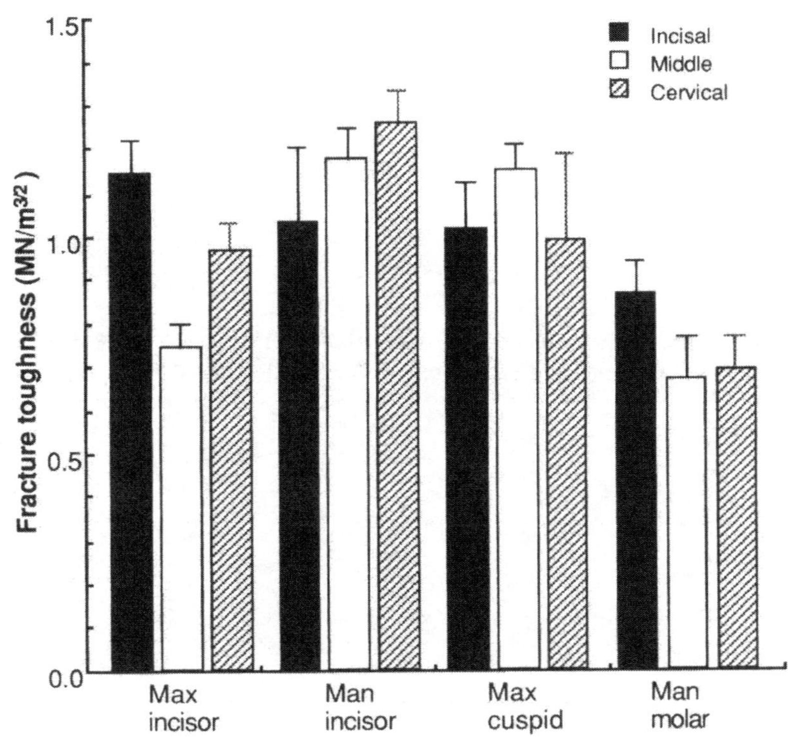

Comments

None.

Reference(s)

Hassan R, Caputo AA, Bunshah RF (1981) Fracture toughness of human enamel. J Dent Res 60:820–827 (with permission)

Hardness

• Hardness	• Human	•
•	• Enamel, dentin	•

Materials
- Human
- Enamel and dentin

Testing Methods and Experimental Conditions
- Microhardness tester was used with a Knoop diamond indenter

Data

Tooth type	Position of section[a] below cusp tip (mm)	Location of indentations	Knoop hardness numbers (KHN) in kg./mm^2			
			Enamel		Dentin	
			mean ± SD	No. of indentations	mean ± SD	No. of indentations
Left maxilla 2nd molar	2.02	2.35 mm from buccal	371 ± 35	52	77 ± 13	5
	4.71	1.69 mm from mesial	390 ± 32	21	78 ± 3	32
	5.40	1.78 mm from lingual	368 ± 32	23	67 ± 4	51
	8.04	1.71 mm from distal	---	---	68 ± 4	29
Left mandible 1st premolar	0.51	1.44 mm from buccal	364 ± 43	28	66 ± 11	6
	2.77	2.96 mm from buccal	292 ± 28	17	62 ± 6	17
	5.23	2.57 mm from buccal	309 ± 45	9	71 ± 5	19
	5.92	1.43 mm from buccal	308 ± 47	6	65 ± 4	15
Right mandible 1st premolar	0.78	2.14 mm from buccal	344 ± 49	22	60 ± 6	8
	2.92	2.14 mm from buccal	327 ± 20	15	72 ± 6	26
	4.98	1.14 mm from buccal	345 ± 21	9	61 ± 5	25
	5.67	0.91 mm from buccal	335 ± 25	6	64 ± 5	18

[a]Sectioned parallel to occlusal.

Comments
None.

Reference(s)
Craig RG, Peyton FA (1958) The microhardness of enamel and dentin. J Dent Res 37:661–668 (with permission)

Mechanical Mobility (1)

• Damping parameter •	• Human • Periodontal membrane	• •

Materials
- Human
- Periodontal membrane (maxillary central incisor)

Testing Methods and Experimental Conditions
- Portable measuring device

Data
A new mobility triangle figure (MT figure)

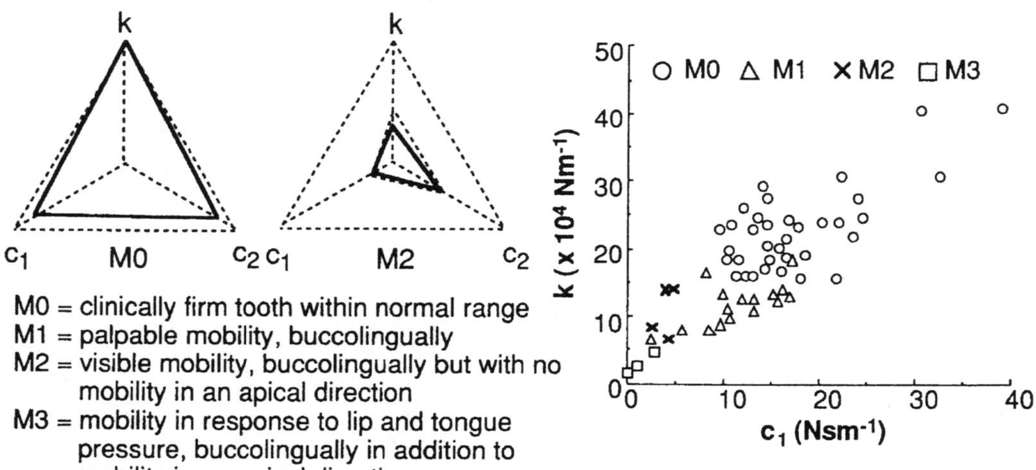

M0 = clinically firm tooth within normal range
M1 = palpable mobility, buccolingually
M2 = visible mobility, buccolingually but with no
    mobility in an apical direction
M3 = mobility in response to lip and tongue
    pressure, buccolingually in addition to
    mobility in an apical direction

Comments

- Mechanical mobility λ: $1/\lambda = 1/\{1/(c_1 + k/j\omega) + 1/c_2 + 1/j\omega m_2\} + j\omega m_1$

  k, spring constant; m, mass; c, damping parameter; ω, angular frequency.

Reference(s)
Oka H, Yamamoto T, Saratani K, Kawazoe T (1989) Application of mechanical mobility of periodontal tissues to tooth mobility examination. Med Biol Eng Comput 27:75–81 (with permission)

Mechanical Mobility (2)

• Mobility •	• Human • Periodontal membrane	• •

Materials
- Human (seven males; age, 24–26 years; good periodontal health)
- Periodontal membrane

Testing Methods and Experimental Conditions
- Transducer (Wilcoxon Research, Bethesda, MD, USA) consisted of a piezoelectric force gauge and a piezoelectric accelerometer
- Maxillary central incisor (#9) and a mandibular incisor (#25)

Data

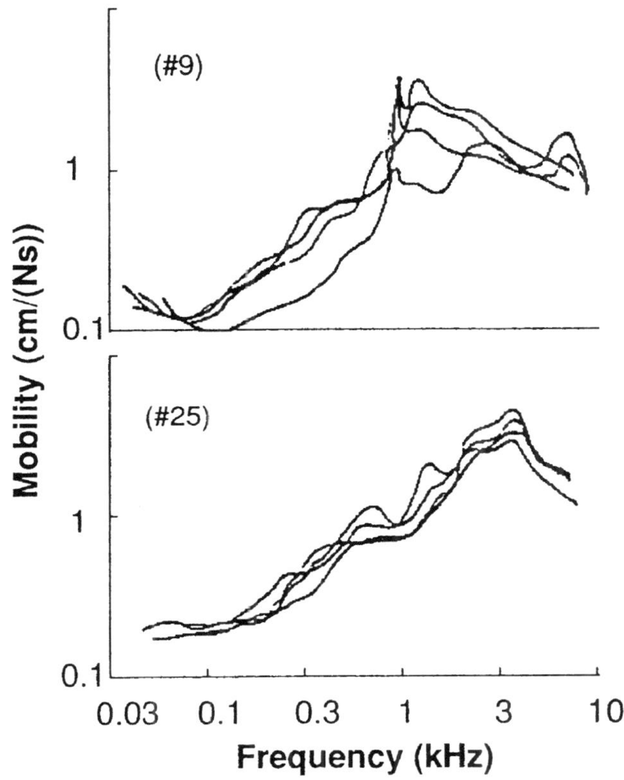

Comments
- Mechanical model which approximates mobility data obtained from teeth measured in vivo was proposed.

Reference(s)
Noyes DH, Solt CW (1973) Measurement of mechanical mobility of human incisors with sinusoidal forces. J Biomech 6:439–442 (with permission)

Shear Property

<table>
<tr><td>• Shear strength
•</td><td>• Human
• Enamel, dentin</td><td>•
•</td></tr>
</table>

Materials

• Human
• Enamel and dentin

Testing Methods and Experimental Conditions

• Single plate punch sampling apparatus

Data

Material	Shear strength (MPa)
Enamel	93.1 ± 1.5
Dentin	132.3 ± 2.1
Amelodentinal junction	138.2 ± 2.0

Comments

None.

Reference(s)

Smith DC, Cooper WEG (1971) The determination of shear strength — a method using a micro-punch apparatus. Brit Dent J 130:333–337 (with permission)

Tensile Property

• Tensile strength •	• Human, bovine • Enamel, dentin	• •

Materials
- Human and bovine
- Enamel and dentin

Testing Methods and Experimental Conditions
- Specimens were kept moist at $23° \pm 2°C$

Data

Material	Average tensile strength (MPa)	Number of specimens	Standard deviation (MPa)	Coefficient of variation (%)
Human enamel	10.3	9	2.6	25
Human dentin	51.7	9	10.3	20
Bovine enamel	20.7	8	5.5	28
Bovine dentin	58.6	8	11.7	20

Comments
None.

Reference(s)
Bowen RL, Rodriguez MS (1962) Tensile strength and modulus of elasticity of tooth structure and several restorative materials. J Amer Dent Assoc 64:378–387 (with permission)

Work of Fracture

<table>
<tr><td>• Mechanical work
•</td><td>• Human
• Enamel, dentin</td><td>•
•</td></tr>
</table>

Materials
• Human (age, 12–14 years)
• Enamel and dentin (noncarious premolars)

Testing Methods and Experimental Conditions
• Specimens were subjected to three point bending until fracture, using a constant strain testing machine

Data

Age (years)	Work of fracture for perpendicular specimens of dentin (10^2 J/m^2)	Age (years)	Work of fracture for parallel specimens of enamel (10^2 J/m^2)
12	4.1	12	0.1
12	1.5	12	0.103
13	6.4	12	0.091
13	1.6	12	0.098
13	1.89	12	0.061
13	1.42	13	0.266
13	1.97	13	0.188
13	3.72	13	0.12
13	0.84	13	0.151
13	4.35	14	0.204
14	3.38	14	0.058
14	2.21		
14	2.33		

Comments
• The specimen's work of fracture was defined as the work required to form a new surface of unit area.

Reference(s)
Rasmussen ST, Patchin RE, Scott DB, Heuer AH (1976) Fracture properties of human enamel and dentin. J Dent Res 55:154–164 (with permission)

4. Cells, Proteins, and Polymers

Bending Modulus

• Membrane buckling •	• Human • Red cell	• Micropipette •

Materials
- Human
- Red cell

Testing Methods and Experimental Conditions
- Micropipette aspiration test using pipettes with of inner diameter in the range of 1 to 2 μm
- The aspiration pressure and length where cell membrane buckling occurred, were measured
- Membrane bending modulus was related to the observed buckling pressures *vs* pipette radii

Data

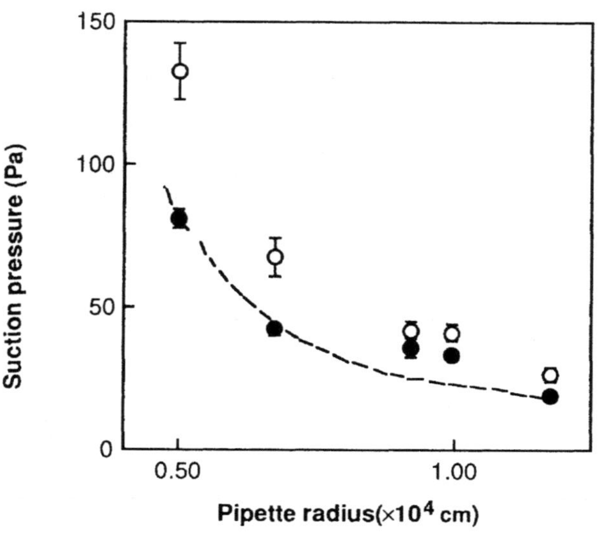

Comments
- Data for aspiration pressure where the cells became unstable, folded and moved up the pipette ○, and where cell membrane appeared to buckle initially ●, plotted against pipette inner radius; dotted line, the correlation with the theoretical prediction for the onset of buckling.
- The membrane bending modulus is estimated to be 1.8×10^{-12} dyn-cm.

Reference(s)
Evans EA (1983) Bending elastic modulus of red blood cell membrane derived from buckling instability in micropipet aspiration tests. Biophys J 43:27–30 (with permission)

Deformability(1)

• Overall strain • Shear stress	• Human • Red cell	• Modulus of elasticity •

Materials
- Human
- Red cell

Testing Methods and Experimental Conditions
- The fresh human red cells were settled on a clean glass coverslip
- The cells attached at a single point to the surface (point attachment) or two or more points (line attachment)
- The coverslip was inverted and placed in the flow channel which had been connected to the syringe pump and filled with the test fluid (phosphate-buffered Ringer's solution or Eagle's solution)
- The overall strain in the direction of flow of the cells was measured
- Shear stress at the surface between 0.1 and 1 Pa
- Temperature: $25° \pm 1°C$

Data

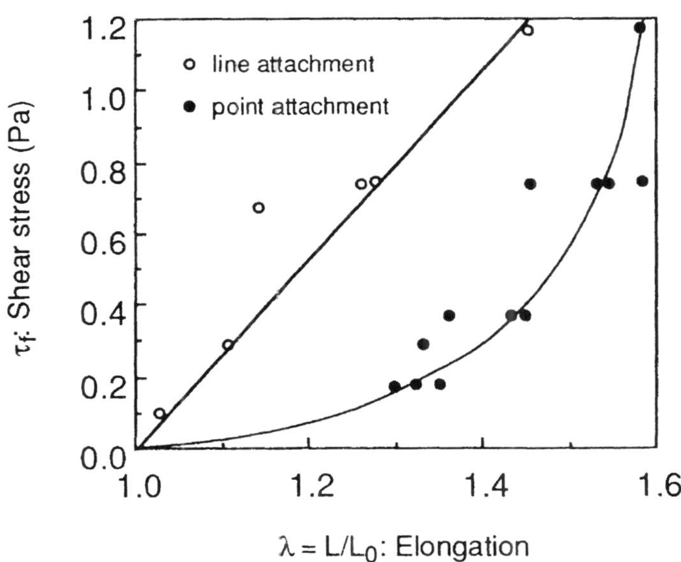

Comments
- The elongation of a red cell with a line attachment is less than that of a cell with a point attachment.
- The cell with a line attachment is pulled into a rectangular shape with a rounded trailing edge while the cell with a point attachment is shaped more like a tear drop.
- An analysis of cell deformation as a function of shear stress gives a modulus of elasticity for cell membrane of approximately 10^3 Pa if the membrane thickness of 0.01 µm is assumed.

Reference(s)
Hochmuth RM, Mohandas N (1972) Uniaxial loading of the red-cell membrane. J Biomech 5:501–509 (with permission)

Deformability(2)

• Shear stress • Shape change	• Human • Red cell	• •

Materials

• Human
• Red cell

Testing Methods and Experimental Conditions

• Fresh human blood, anticoagulated with EDTA (1.5 mg/ml)
• Cone-plate devices
• Their physiological responses and active and passive shape changes were observed
• Temperature: 23°C

Data

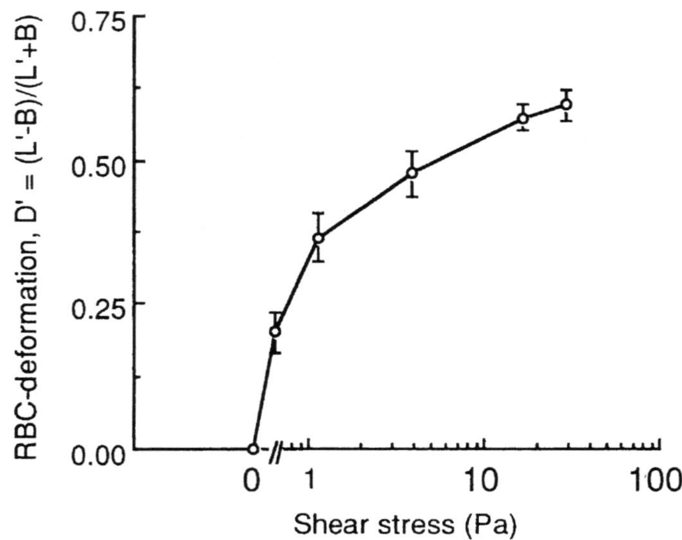

Comments

• Deformation
  $D' = (L' - B)/(L' + B)$,
  D': apparent deformation; L', apparent length of the ellipsoid; B, width of the ellipsoid.
• The cell shape changed to a thin ellipsoid at high shear stress.

Reference(s)

Schmid-Schönbein H, Rieger H, Zander R (1976) Microrheology of erythrocytes and
platelets: physiological basis and consequences for the design and the operation of
extracorporeal circulatory devices. In: Davids SG, Engell HC (eds) Physiol and Clin
Aspects of Oxygenator Design, 163–176 (with permission)

Distensibility(1)

• Compressibility modulus • Isotropic tension	• Human • Red cell	• Micropipette • Area expansion

Materials
- Human (healthy)
- Red cell

Testing Methods and Experimental Conditions
- Osmotically preswollen red cells were used
- Cells suspended in a phosphate-buffered solution (pH 7.4) contained 0.5% serum albumin
- Micropipette measurements were performed with a pipette 1.8-2.4 μm in diameter
- Temperature range: 2°–60°C
- The increase in surface area is calculated from the increase in the projection length and the assumption of constant cell volume

Data

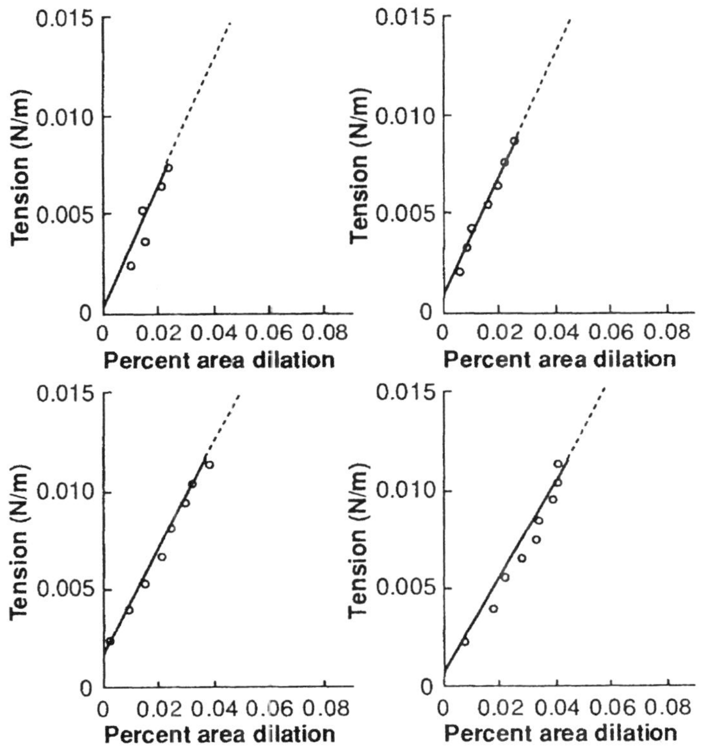

Comments
- Typical cell tension *vs* area expansion curves at 25°C.
- The elastic, area compressibility moduli at 25°C from the slope of the tension-area expansion curve was 0.288 ± 0.050 (SD) N/m.
- This value is about 4×10^4 times the elastic coefficient for shear rigidity; therefore, the membrane behaves in a two-dimensional, incompressible manner in shear and elongation.
- The maximum isotropic tension at 25°C is 0.010–0.012 N/m and at 50°C is 0.003–0.004 N/m.

Reference(s)
Evans EA, Waugh R, Melnik L (1976) Elastic area compressibility modulus of red cell membrane. Biophys J 16:585–595 (with permission)

Distensibility(2)

• Compressibility modulus • Isotonic tension	• Human • Red cell	• Micropipette • Mass transport equation

Materials
- Human (healthy)
- Red cell

Testing Methods and Experimental Conditions
- Osmotically preswollen red cells were used
- Cells were suspended in a phosphate-buffered solution (pH 7.4) containing 0.5% serum albumin
- Micropipette aspiration experiments were performed
- The area change was calculated from pipette diameter, aspirated length, initial diameter of the spherical portion of the cell outside the pipette
- The volume change was calculated from the model of osmotic equilibrium using mass transport equations
- Temperature: 25°C

Data

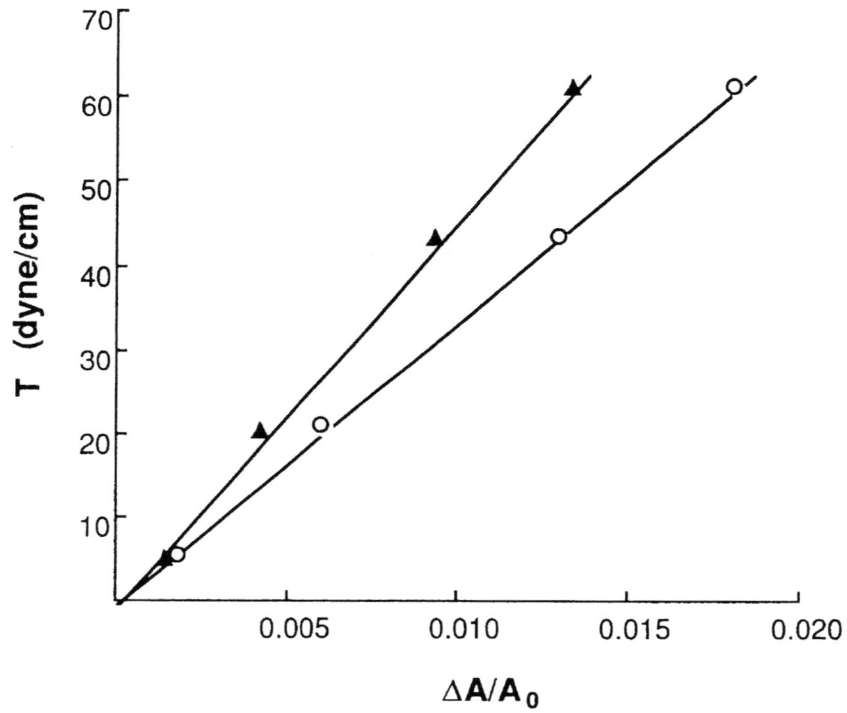

Comments
- The isotropic tension is plotted against the fractional change in area: T, isotropic tension; ΔA, change in area; A_0, total area.
- Lower curve: ΔV(change in volume) $\equiv 0$; Upper curve: with osmotic volume correction.
- From the slope of upper curve, the elastic area compressibility modulus is 450 dyne/cm.

Reference(s)
Evans EA, Waugh R (1977) Osmotic correction to elastic area compressibility measurements on red cell membrane. Biophys J 20:307–313 (with permission)

Distensibility(3)

• Elastic constant •	• Human • Red cell	• Micropipette •

Materials
- Human
- Red cell

Testing Methods and Experimental Conditions
- Comparison of the results of fluid shear-deformed, point-attached red cells and micropipette aspiration of red cell disks using a new material concept
- The data of shear-deformed cells was from Hochmuch and Mohandas (1972)
- The data of micropipette aspiration was from Rand and Burton (1964)
- A two-dimensional elastomer material concept of the red cell membrane was applied

Data

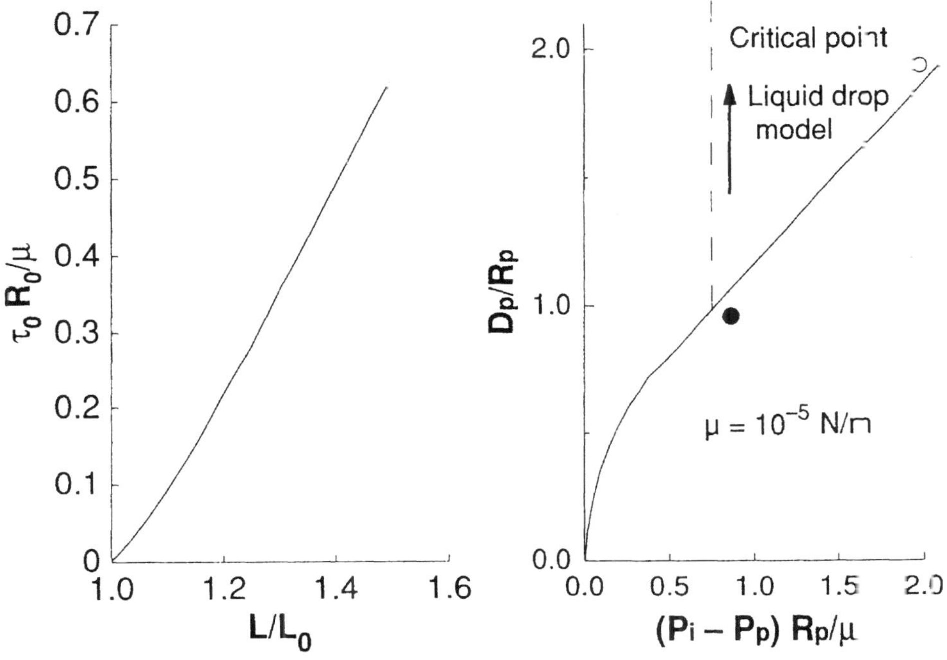

Comments
- Left figure shows the dimensionless ratio $\tau_0 R_0/\mu$ as a function of the total length to original length ratio.

  τ_0, shear stress; R_0, radius of the undeformed disk; μ, elastic constant.

- In right figure, the dimensionless ratio $(P_i - P_p) R_p/\mu$ representing the pressure difference required to aspirate the cell a distance D_p into pipette, is shown as a function of the dimensionless distance D_p/R_p; R_p, radius of the micropipette.
- The elastic constant is of the order 10^{-5} N/m for both cases.

Reference(s)

1. Evans EA (1976) New membrane concept applied to the analysis of fluid shear and micropipette-deformed red blood cells. Biophys J 13:941-954 (with permission)
2. Hochmuth RM, Mohandas N (1972) Uniaxial loading of the red-cell membrane. J Biomech 5:501–509
3. Rand RP, Burton AC (1964) Mechanical properties of the red cell membrane I. Membrane stiffness and intracellular pressure. Biophys J 4:115-135

Shear Modulus

• Shear Modulus •	• Bovine • Endothelial cell	• Micropipette •

Materials
- Bovine
- Cultured aortic endothelial cell (7th to 9th passage)
- The passage time ranged between 2 and 85 days

Testing Methods and Experimental Conditions
- Cultured bovine aortic endothelial cells were suspended in modified Dulbecco medium containing 25 m mole/l Hepes buffer and 20% fetal bovine serum
- Trypsin-detached cells from coverslip and mechanically-detached cells were used
- Micropipette aspiration tests were performed with a pipette of inner diameter ranging from 2.0 to 3.4 μm
- Temperature: $24 \pm 0.5°C$

Data

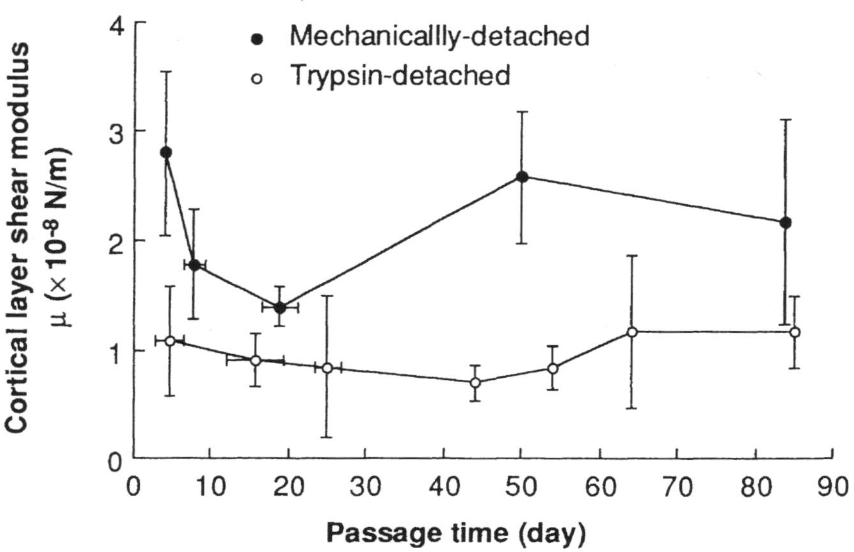

Comments
- The cortical layer elastic shear modulus, μ was calculated from the following equation:

$$\Delta P = (\mu/R)[(2L/R\text{-}1) + \ln(2L/R)].$$

  ΔP, pressure difference; R, pipette radius; L, aspirated length.

Reference(s)
Sato M, Levesque MJ, Nerem RM (1987) An application of the micropipette technique to the measurement of the mechanical properties of cultured bovine aortic endothelial cells. ASME J Biomech Eng 109:27–34

Stiffness(1)

• Resistance to deformation •	• Human • Red cell	• Micropipette •

Materials
- Human
- Red cell

Testing Methods and Experimental Conditions
- Cells were suspended in one of three solutions (0.6%, 0.9%, 1.2% NaCl)
- Cells suspended in 0.9% NaCl were checked for biconcave shape and determinations were made on the rim and biconcavity of the cell
- Cells suspended in 0.6% NaCl were swollen into a spheroidal
- Cells suspended in 1.2% NaCl were shrunken
- The cell membrane was pulled into the micropipette until the cell membrane deformation became stable

Data

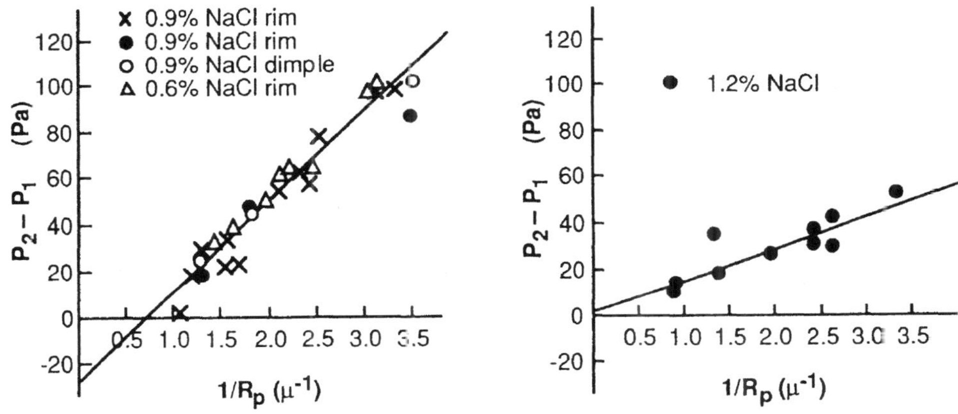

Comments
- Left figure: Results for normal biconcave cells and cells swollen to ellipsoids; P_1, pressure inside the pipette; P_2, pressure of the liquid; R_p, pipette radius.
- Right figure: Results for shrunken and crenated cells.
- The regression lines indicate a membrane resistance to deformation.
- Stiffness of hypertonically crenated cells was less than that of biconcave discs or hypotonically swollen cells.
- Crenated cells showed zero pressure gradient and a stiffness, probably due to pure bending, equivalent to 0.007 ± 0.001 (mean $\pm$ SE) dyne/cm.
- Normal and swollen cells showed a pressure gradient of 2.3 ± 0.8 (mean $\pm$ SE) mmH$_2$O and a stiffness, due to bending and tension in the membrane, equivalent to 0.019 ± 0.002 (mean $\pm$ SE) dyne/cm.

Reference(s)
Rand RP, Burton AC (1964) Mechanical properties of the red cell membrane I. Membrane stiffness and intracellular pressure. Biophys J 4:115-135 (with permission)

Stiffness(2)

• Stiffness parameter • Shear stress	• Bovine • Endothelial cell	• Micropipette • Culture

Materials
- Bovine
- Cultured aortic endothelial cell (7th to 9th passage)

Testing Methods and Experimental Conditions
- Cells exposed to a fluid-imposed shear stress of 1.0, 3.0, and 8.5 Pa with parallel flow chamber
- Exposure time: 0.5, 1, 2, 4, 12, or 24 hours
- Micropipette aspiration tests were performed with a pipette of inner diameter ranging from 2.0 to 3.4 μm
- Medium: aerated, modified Dulbecco medium containing 25mmol/l Hepes buffer and 20% fetal bovine serum and antibiotics
- Temperature: $37 \pm 0.5°C$

Data

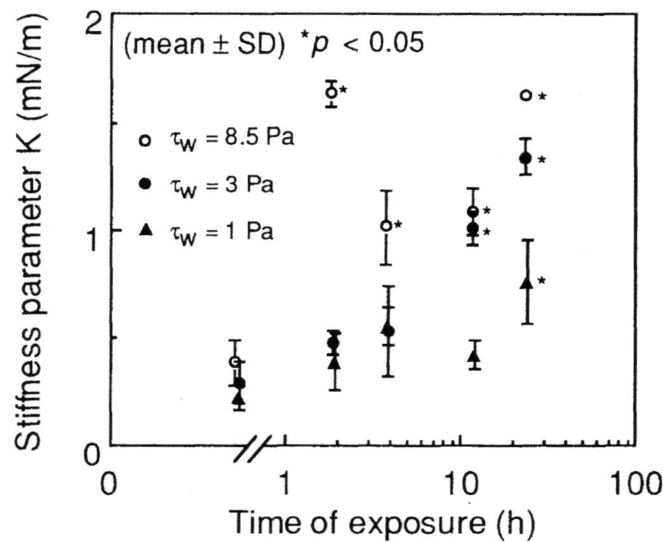

Comments
- Effect of exposure time on the stiffness parameter K for the three shear stresses:

$$K = R \times \Delta P/(L/R),$$

  ΔP, the pressure difference; R, the pipette radius; L, aspirated length.
- Cells exposed to shear stress demonstrated a mechanical stiffness significantly greater than that of control cells, which was dependent on both the level of shear stress and the duration of exposure.

Reference(s)
Sato M, Levesque MJ, Nerem RM (1987) Micropipette aspiration of cultured bovine aortic endothelial cells exposed to shear stress. Arteriosclerosis 7:276–286 (with permission)

Thermoelasticity

• Expansivity • Shear modulus	• Human • Red cell	• Micropipette • Compressibility

Materials
- Human
- Red cell

Testing Methods and Experimental Conditions
- Flaccid and osmotically swollen red cells were used
- Cells suspended in phosphate-buffered saline containing bovine serum albumin (0.5 g/100 ml)
- Thermal area expansivity: The cell was held with the micropipette at a constant suction pressure; the temperature change produced a reversible change in length of the aspirated cell projection
- Elastic shear modulus at different temperatures: the modulus was measured by aspirating the cell in the dimple region into the micropipette
- From an initial pressure of 20.0–30.0 Pa, the pressure was increased until the cell began to fold or buckle
- Temperature range, 2°–50 °C

Data

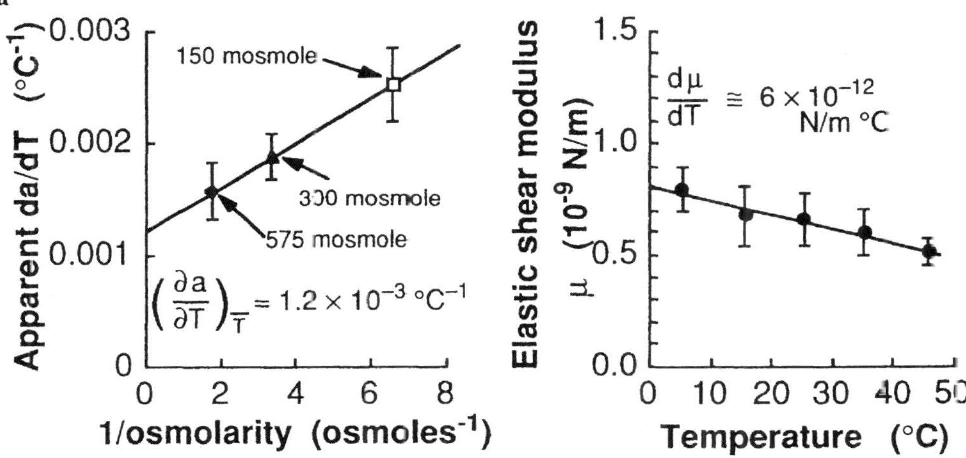

Comments
- Left figure: slopes of the apparent area change with respect to temperature *vs* inverse osmotic strength; a, membrane area; T, temperature.
- The thermal area expansivity: $1.2 \times 10^{-3}\ °C^{-1}$.
- The change in membrane area ΔA: $\Delta A \approx 2\pi R_p\, \Delta L\, (1 - R_p/R)$.
  R_p, pipette radius; ΔL, aspirated length; R, radius of the outer spherical portion of the cell.
- Right figure: the elastic shear modulus as a function of temperature.
- The change in the elastic shear modulus with temperature, 6×10^{-12} N/m°C.
- Predicted relation between the pipette suction pressure and the aspirated length:
  $\Delta P = (\mu/R_p) \{(2L/R_p) - 1) + \ln(2L/R_p)\}$.
  ΔP, pipette suction pressure; μ, elastic shear modulus; L, aspirated length.
- The area compressibility modulus at different temperatures is also shown.

Reference(s)
Waugh R, Evans EA (1979) Thermoelasticity of red cell membrane. Biophys J 26:115–132 (with permission)

Viscoelasticity(1)

| • Recovery time | • Human | • Micropipette |
| • | • Leukocytes | • |

Materials
- Human (age, 28–36 years)
- Leukocytes

Testing Methods and Experimental Conditions
- Leukocytes were suspended in a saline solution with 12 ml Tris and bovine serum albumin (0.25 g/100 ml)
- Individual cells were aspirated into a pipette with inner radius of 5–7 μm until they completely entered
- After the holding period, the cell was gently pushed out of the pipette
- The recovery times, major and minor diameter of the cells were measured
- Temperature: 21–23°C

Data

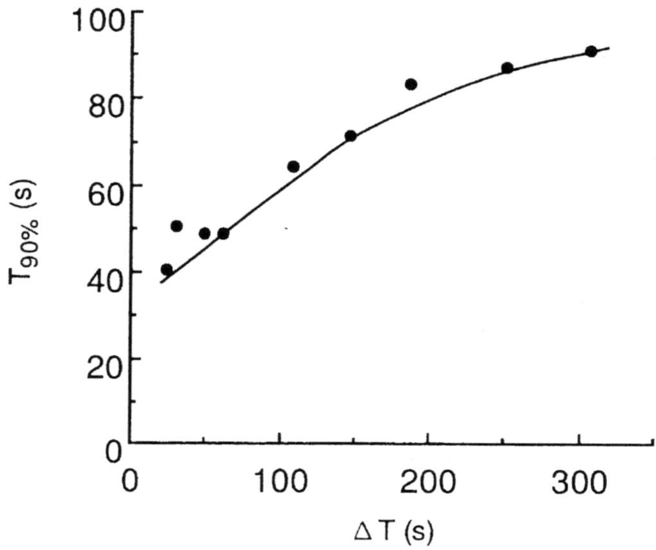

Comments

- The rate of recovery was dependent on the holding period ΔT: $T_{90\%}$, time of 90% recovery.

Reference(s)
Sung K-LP, Dong C, Schmid-Schönbein GW, Chien S, Skalak R (1988) Leukocyte relaxation properties. Biophys J 54:331–336 (with permission)

Viscoelasticity(2)

• Recovery time •	• Human • Red cell	• Micropipette • Temperature dependence

Materials
- Human (healthy, adult)
- Red cell

Testing Methods and Experimental Conditions
- Cells suspended in phosphate-buffered saline plus 0.1% human serum albumin
- Attached cell elongation was produced by aspirating the rim of the cell into a micropipette and then withdrawing the pipette
- Cell deformation and subsequent recovery process were obserbed
- Temperature: 6°C to 37°C

Data

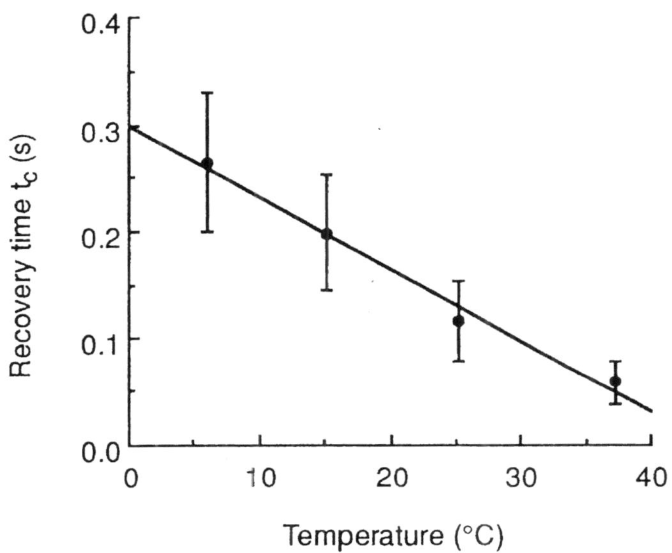

Comments
- The recovery time constant at 6, 15, 25, 37°C: The error bars represents a linear best-fit to the experimental data points.

Reference(s)
Hochmuth RM, Buxbaum KL, Evans EA (1980) Temperature dependence of the viscoelastic recovery of red cell membrane. Biophys J 29:177–182 (with permission)

Viscoelasticity(3)

• Residual deformation • Force relaxation	• Human • Red cell	• Micropipette •

Materials
- Human
- Red cell

Testing Methods and Experimental Conditions
- Cells suspended in a phosphate-buffered saline containing human albumin
- Albumin concentrations:0.0% to1.0%
- Force relaxation and permanent deformation processes were investigated with two techniques: micropipette aspiration test and whole cell extension test
- Whole cell deformation was produced by aspirating diametrically opposite points on the erythrocyte rim into micropipettes and displacing the pipettes so as to produce a symmetrically extended cell

Data

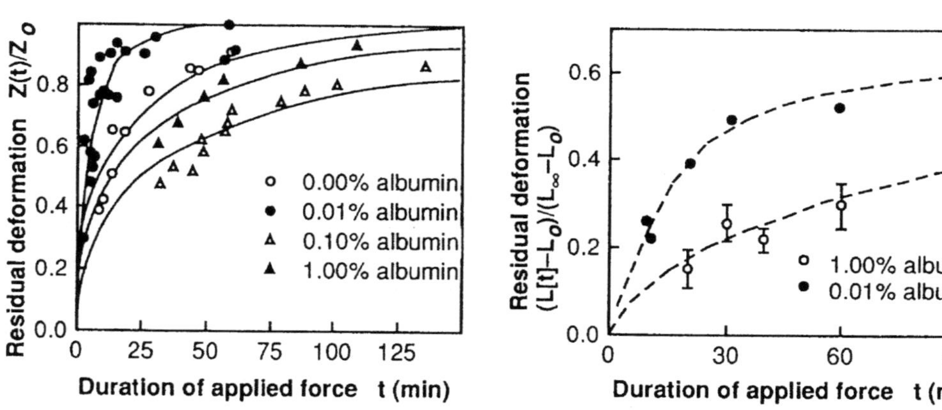

Comments
- Left figure: The residual deformation in the micropipette aspiration test is defined as the bump height produced after t minutes of aspiration $Z(t)$ scaled by the length of the cells projection into the pipette Z_0.
- Right figure : The residual deformation in the whole cell extension test is defined by the dividing the change in the overall cell length at time t, $L(t)-L_0$, by the maximum possible length change, $L_\infty-L_0$.
- Data from these experiments determine a characteristic time constant force relaxation τ which is the ratio of a surface viscosity to the elastic shear modulus.
- Values for τ of 10.1, 40.0, 62.8, and 120.7 min are measured at albumin concentrations of 0.0, 0.01, 0.1, and 1%, respectively, in the micropipette aspiration test.
- τ increases with increasing concentration of albumin in the whole cell extension test.

Reference(s)
Markle DR, Evans EA, Houchmuth RM (1983) Force relaxation and permanent deformation of erythrocyte membrane. Biophys J 43:27–30 (with permission)

Viscoelasticity(4)

• Shear modulus • Viscosity	• Human • Red cell	• Micropipette • Constitutive equation

Materials
- Human
- Red cell

Testing Methods and Experimental Conditions
- Cells suspended in phosphate-buffered saline plus 0.1% human serum albumin and 0.1% glucose
- Attached cell elongation was produced by aspirating the rim of the cell into a micropipette with an internal diameter of 0.5 µm and then withdrawing the pipette
- Cell deformation and the subsequent recovery process were observed
- Analysis with a constitutive equation which describes the finite deformation at constant area of two-dimensional viscoelastic solids

Data

$\mu \times 10^6$ (N/m)	t_c (s)	$\eta \times 10^7$ (N·s/m)
4.93	0.111	5.47
5.81	0.135	7.84
3.15	0.140	4.41
4.60	0.114	5.24
3.17	0.090	2.85
5.72	0.105	6.01
4.59	0.100	4.59
5.58	0.121	6.75
3.94	0.142	5.59
4.46	0.152	6.78
3.24	0.145	4.70
3.28	0.125	4.20
2.96	0.150	4.44
4.34	0.113	4.90
4.44	0.104	3.52
4.28 ± 0.97	0.123 ± 0.020	5.15 ± 1.31

Average values ± SE are given for all 15 cells.

μ, shear modulus of surface elasticity; t_c, recovery time constant; $\eta = t_c \cdot \mu$, surface viscosity.

Comments
None

Reference(s)
Hochmuth RM, Hampel, III WL (1979) Surface elasticity and viscosity of red cell membrane. J Rheology 23:669–680 (with permission)

Viscoelasticity(5)

• Shear modulus • Viscosity	• Vertebrate animal • Red cell	• Micropipette •

Materials
- Amphiuma, frog, turkey, toadfish, snake, turtle, iguana, opossum, tree shrew, and human
- Red cell

Testing Methods and Experimental Conditions
- Cells suspended in modified Ringer's solution
- Aspirated length into the micropipette and applied negative pressure were measured
- The membrane surface shear moduli and surface viscosity coefficient were obtained
- Temperature : 23°–25°C

Data

Class	Genus/species (common name)	Shear modulus (dyne/cm)		Number of cells
		Average	SD	
Amphibia	*Amphiuma means* (congo snake)	0.1365	0.028	15
	Rana cateslueiana (bullfrog)	0.0732	0.013	17
Aves	*Meleagris gallopavo* (wild turkey)	0.0812	0.0073	22
Mammalia	*Didelphis virginiana* (opossum)	0.0153	0.0038	18
	Homo sapiens (man)	0.0161	0.0044	28
Osteichthyes	*Opsanus tau* (toadfish)	0.0671	0.023	22
Reptilia	*Pseudemys scripta* (pond turtle)	0.240	0.082	27
	Chrysemys picta (western painted turtle)	0.194	0.018	5
	Lampropeltis getulus (king snake)	0.208	0.037	10

Animal	Relaxation half time (s)		Viscosity (dyne-s/cm)		Number of cells
	Average	SD	Average	SD	
Amphiuma	0.23	4.5×10^{-2}	1.3×10^{-2}	3.8×10^{-3}	5
Frog	0.28	9.1×10^{-2}	6.7×10^{-3}	2.2×10^{-3}	17
Turkey	0.21	7.6×10^{-2}	5.8×10^{-3}	2.3×10^{-3}	19
Opossum	0.063	4.3×10^{-2}	3.4×10^{-4}	2.7×10^{-4}	8
Toadfish	0.29	4.4×10^{-2}	8.7×10^{-3}	2.8×10^{-3}	4
Turtle	0.46	1.7×10^{-2}	3.7×10^{-2}	2.2×10^{-2}	22
Snake	0.20	8.4×10^{-2}	1.2×10^{-2}	5.8×10^{-3}	4

Comments
- Upper table: elastic shear moduli.
- Lower table: surface viscosity data.
- Three types of nucleated cells (amphibian, avian, and bony fish) exhibited similar hyperelasticity with shear moduli of $7–8 \times 10^{-2}$ dyne/cm; the surface viscosity coefficients were $6–9 \times 10^{-3}$ dyne-s/cm.
- The properties of nucleated cell membranes for reptiles were higher; the shear moduli were in range $2–3 \times 10^{-1}$ dyne/cm, and the surface viscosity was $4–10 \times 10^{-2}$ dyne-s/cm.
- In contrast, membranes of non nucleated mammalian cells all exhibited lower shear moduli, 1.5×10^{-2} dyne/cm, and surface viscosity coefficients, $<10^{-3}$ dyne-s/cm.

Reference(s)
Waugh R, Evans EA (1976) Viscoelastic properties of erythrocyte membranes of different vertebrate animals. Microvas Res 12:291–304 (with permission)

Viscoelasticity(6)

• Viscoelastic model •	• Human • Leukocyte	• Micropipette •

Materials
- Human (age, 22–29)
- Leukocyte

Testing Methods and Experimental Conditions
- Cells suspended in a saline solution containing 0.1 g/100 ml EDTA, 0.25 g/100 ml serum albumin
- Micropipette aspirating experiments performed with a pipette of internal radius between 1.1 and 1.8 μm at 22°C
- Time-dependent deformation analysis using a viscoelastic model

Data

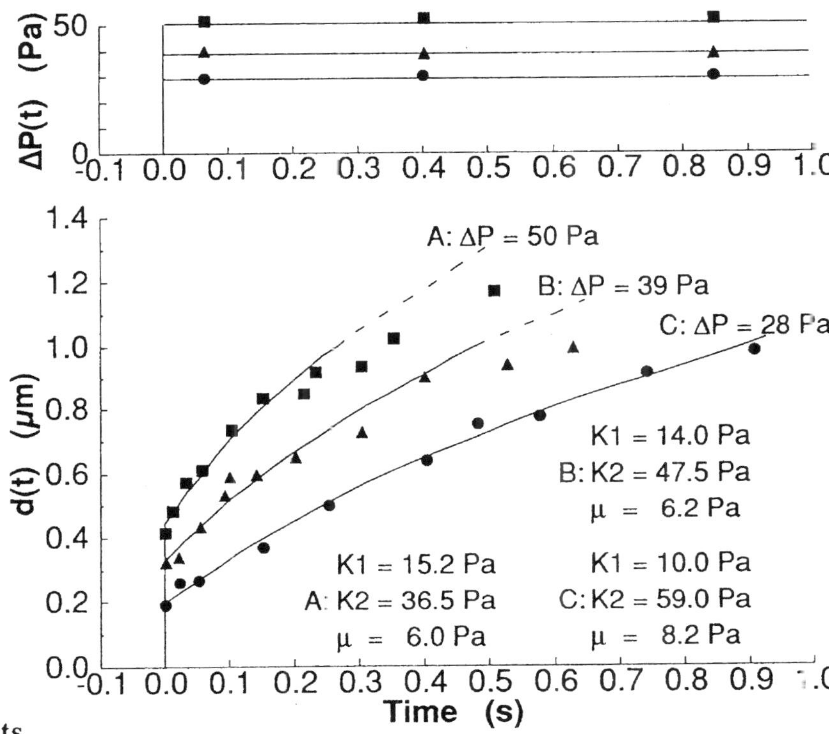

Comments
- In the figure: d(t), aspirated length; ΔP(t), aspiration pressure.
- The viscoelastic model is followed

$$\sigma'_{ij} + (\mu/K_2)\,(d\sigma'_{ij}/dt) = e'_{ij}K_1 + (e'_{ij}/dt)\,\mu\,(1 + K_1/K_2).$$

  σ'_{ij}, stress deviater; e'_{ij}, strain deviater; $(d\sigma'_{ij}/dt)$ and (e'_{ij}/dt), time derivatives of $d\sigma'_{ij}$ and e'_{ij}; K_1 and K_2, elastic coefficients; μ, viscous coefficient.
- $K_1 = 27.5 \pm 11.9$ Pa, $K_2 = 73.7 \pm 34.6$ Pa, $\mu = 13.0 \pm 5.4$ Pa.s (mean $\pm$ SD) for 75 neutrophils.
- The values for monocytes and eosinophils fall into the same range.

Reference(s)
Schmid-Schönbein GW, Sung K-LP, Tözeren H, Skalak R, Chien S (1981) Passive mechanical properties of human leukocytes. Biophys J 36:243–256 (with permission)

Viscoelasticity(7)

• Viscoelastic model • Shear modulus	• Human • Red cell	• Micropipette •

Materials
- Human
- Red cell

Testing Methods and Experimental Conditions
- Red cells suspended in phosphate-buffered saline plus 0.1% bovine serum albumin and 0.1% glucose
- Cells were attached to cover slip and cell deformation was produced by aspirating a small portion of the rim of the red cell into the micropipette and withdrawing the micropipette, then the cell was released
- The cell shape and recovering time were measured
- Correlation with a first order viscoelastic constitutive relation

Data

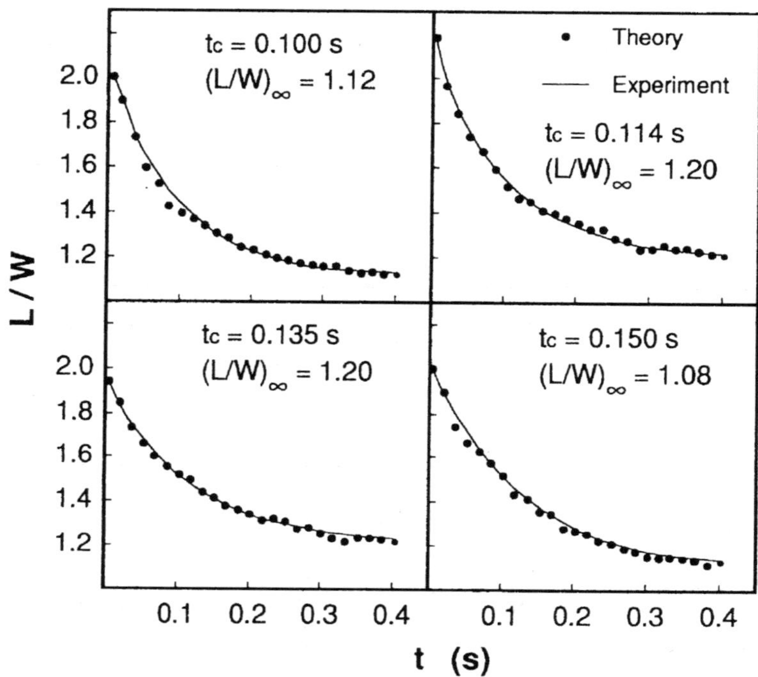

Comments
- The ratio of length (L) and width (W) is subscripted by ( )$_\infty$ for long times.
- Time constant, t_c; $t_c (\eta/\mu) \approx 0.10\text{-}0.13$ s.

  η, coefficient of viscosity for surface shear; μ, surface elastic shear modulus.
- Measurements for the shear modulus μ of 6×10^{-6} N/m (Waugh, 1977) give a value for the surface viscosity of red cell membrane as a viscoelastic solid material of $\eta = (6\text{-}8) \times 10^{-4}$ poise•cm.

Reference(s)

1. Hochmuth RM, Worthy PR, Evans EA (1979) Red cell extensional recovery and the determination of membrane viscosity. Biophys J 26:104-114 (with permission)
2. Waugh RE (1977) Temperature dependence of the elastic properties of red blood cell membrane. Ph.D thesis. Duke University (with permission)

Viscoelasticity(8)

• Viscoelastic model • Young's modulus	• Human • Red cell	• Micropipette • Viscosity

Materials
- Human
- Red cell

Testing Methods and Experimental Conditions
- Cells were suspended in a 0.65% buffered (pH = 6.2) saline solution
- The pressure and time to suck a cell into the micropipette were measured
- A viscoelastic model was used

Data

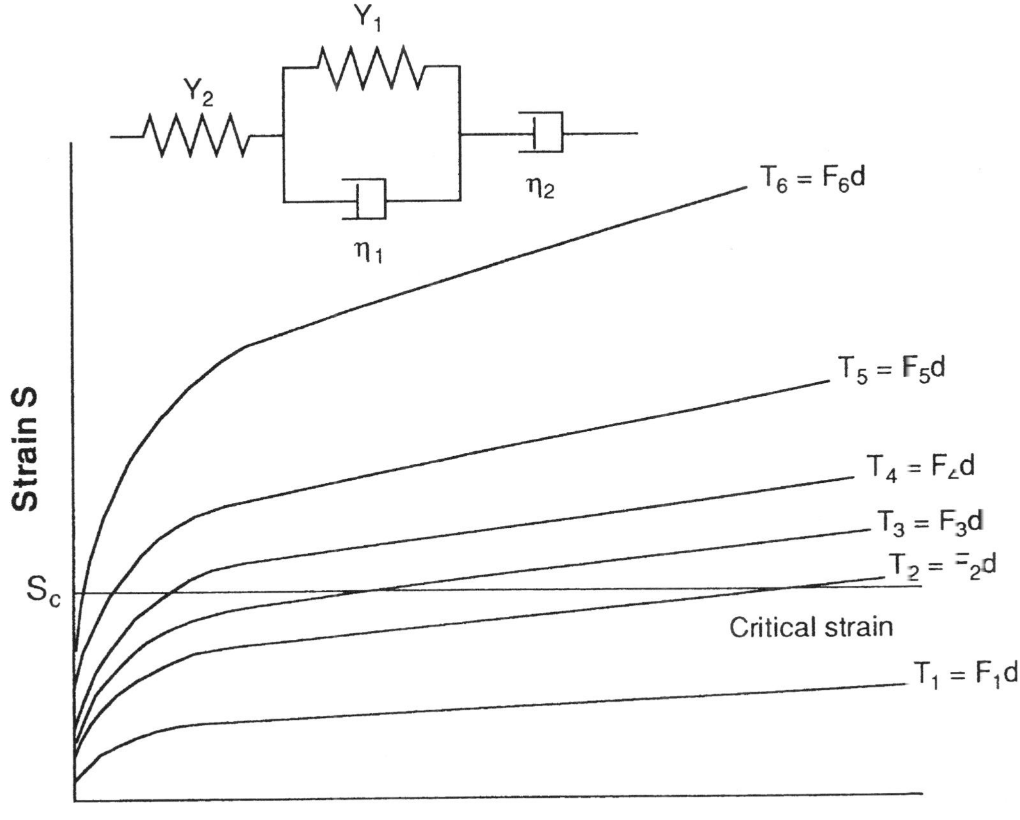

Comments
- The mechanical model consists of a purely elastic element, a purely viscous component and parallel arrangement of each (Y, Young's modulus; η, viscosity).
- A plot of strain S *vs* time. T, stress; F, stretching force; d, membrane thickness
- Critical strain S_c, 0.08–0.42.
- Young's modulus Y, approximately 0.1–10 MPa.
- Viscosity η, approximately 10^7–10^{10} poise.

Reference(s)
Rand RP (1964) Mechanical properties of the red cell membrane II Viscoelastic breakdown of the membrane. Biophys J 4:303–316 (with permission)

Young's Modulus

• Shear stress •	• Bovine • Endothelial cell	• Micropipette •

Materials

- Bovine
- Cultured aortic endothelial cell

Testing Methods and Experimental Conditions

- Comparison with two mathematical models suggested by micropipette technique: Force model and punch model
- Data from Sato M et al (1987)

Data

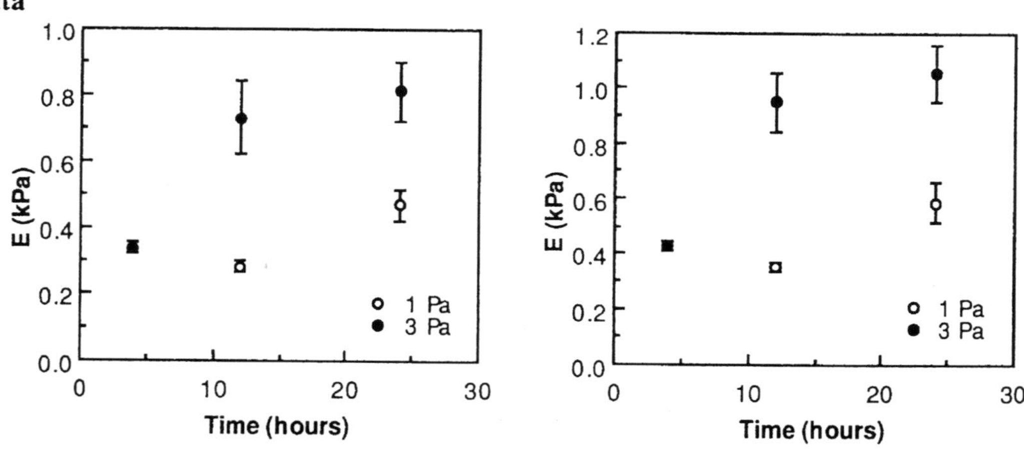

Comments

- Young's modulus E was calculated from the following equation:

$$E = ([3a\Delta P] / [2\pi L]) \times \Phi.$$

  a, inner pipette radius; ΔP, aspiration pressure; L, aspirated length; Φ, wall function.
- Force model (left figure): the stress imposed by the micropipette on the cell surface is assumed to be constant over the annular region.
- Punch model (right figure): the constrain of a constant stress in the annular region is removed.
- The punch model yielding slightly higher value for Young's modulus.
- The Young's modulus is increased due to the exposure to shear stress.
- This effect appears to depend upon both the level and time of exposure.

Reference(s)

1. Theret DP, Levesque MJ, Sato M, Nerem RM, Wheeler LT (1988) The application of a homogeneous half-space model in the analysis of endothelial cell micropipette measurements. ASME J Biomech Eng 110: 190–199
2. Sato M, Levesque MJ, Nerem RM (1987) Micropipette aspiration of cultured bovine aortic endothelial cells exposed to shear stress. Arteriosclerosis 7:276–286

Elastic Modulus (1)

• Elongation dependence •	• Bovine elastin • Polypentapeptide	• Cross-linked •

Materials
- Bovine ligamentum nuchae elastin
- X^{30}-polypentapeptide, which is obtained by 30 Mrad irradiation cross-linking of poly(Val-Pro-Gly-Val-Gly), a high polymer of the pentapeptide repeat of human, bovine, porcine, and chick elastins

Testing Methods and Experimental Conditions
- Ligamentum nuchae elastin and X^{30}-polypentapeptide elastomer were allowed to stand at 0% extension and then to successively high extensions immersed in water at 25°C

Data

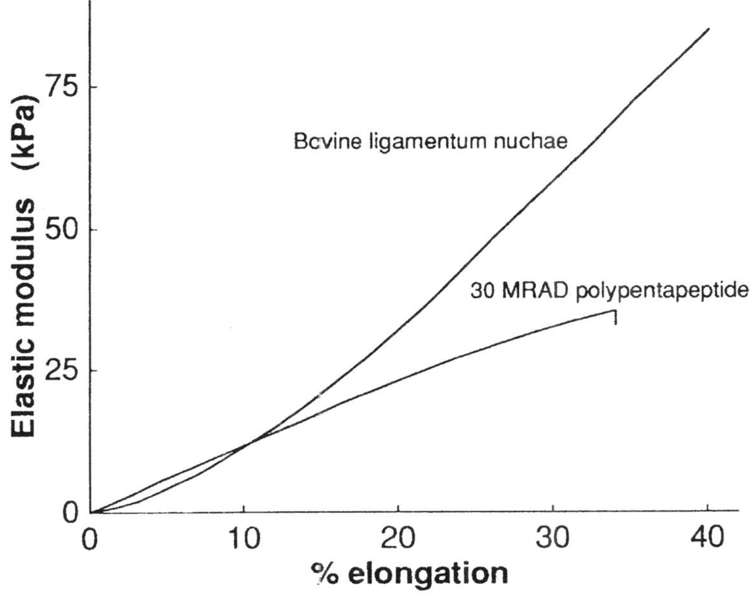

Comments
- X^{30}-polypentapeptide: elastic modulus is a little over 100 kPa.
- Bovine ligamentum nuchae elastin: about 20% extension, elastic modulus is 260 kPa.

Reference(s)
Urry DW (1984) Protein elasticity based on conformations of sequential polypeptides: the biological elastic fiber. J Protein Chem 3:403–436 (with permission)

Elastic Modulus (2)

• Relaxation time •	• Alginate gel •	• Electrolyte solution •

Materials

- Alginate (Alg) gels, which are obtained by gelation of sodium alginate (a polysaccharide found in intercellular substances of brown algae) in $CaCl_2$, $SrCl_2$, $BaCl_2$ and $Pb(NO_3)_2$ solutions at a concentration of 0.3 M

Testing Methods and Experimental Conditions

- Alg gel rods fixed by clamps were suspended in electrolyte solution in which the gels were equilibrated
- Strains were imposed on the gels in steps and the tensile stresses were measured at 25°C by strain gauges connected to a constant direct source and a microvoltmeter
- By curve fitting methods, equilibrium elastic moduli Ge, spontaneous elastic moduli Ge + Gr and relaxation times τ were observed

Data

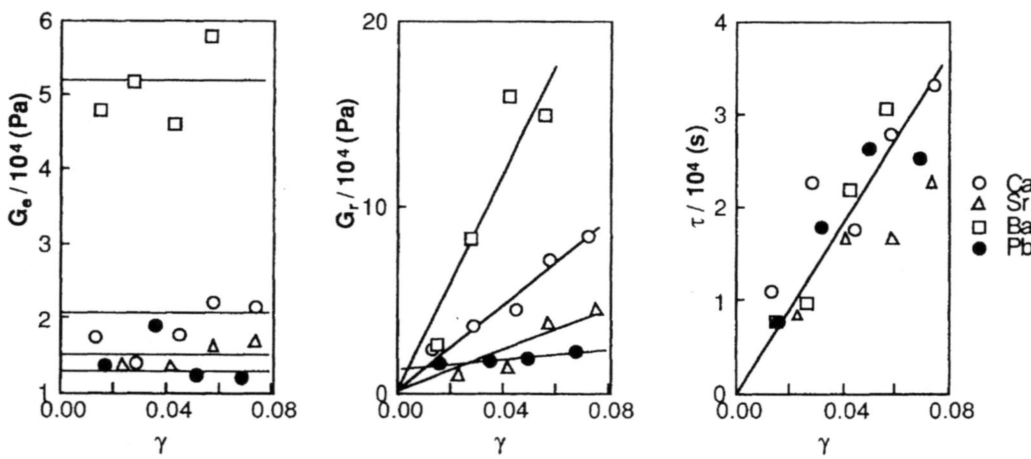

Comments

- Ge was almost constant for low strain.
- Gr and τ increased with increasing γ.

Reference(s)

Yonese M, Baba K, Kishimoto H (1988) Stress relaxation of alginal gels crosslinked by various divalent metal ions. Bull Chem Soc Jpn 61:1857–1863 (with permission)

Force-Elongation Relation (1)

• Elastic modulus •	• Bovine ligamentum • Elastin	• Effect of chlorine •

Materials
- Bovine ligamentum nuchae elastin

Testing Methods and Experimental Conditions
- Ligamentum nuchae elastin in water at 37°C was allowed to stand at given extensions (0% and 40%)
- The water was exchanged with a solution of 20% Clorox (Clorox Company, Oakland, CA, USA), in which Clorox contains 9.5% chlorine by weight
- After 1 h of treatment, stress–strain runs were performed on each sample with determination of the zero force length

Data

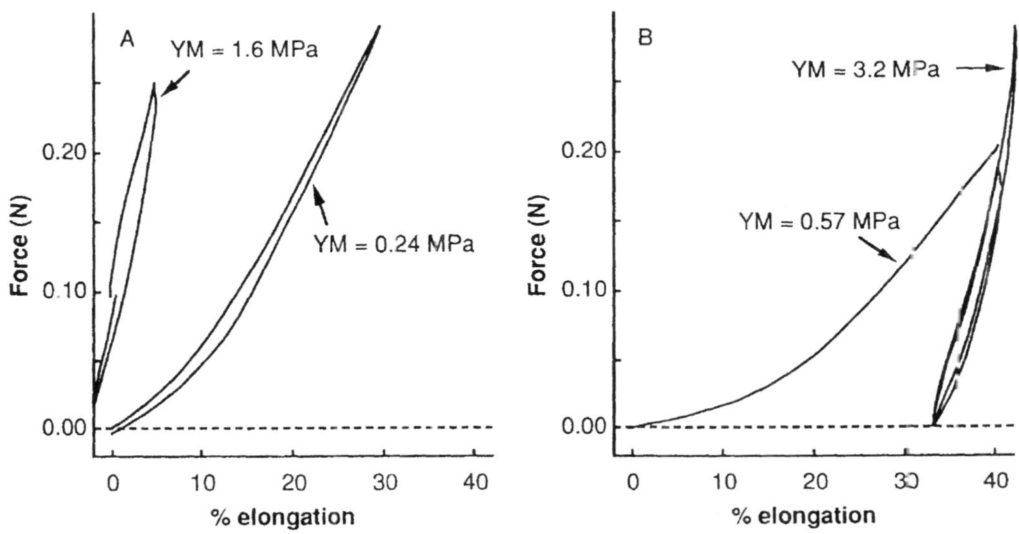

Comments
- Stress–strain curves of bovine ligamentum nuchae elastin before and after 1-h treatment with 20% Clorox solution. The treatments were at the extensions of 0% (A) and 40% (B).
- Fixation and increase in elastic modulus were taken to be due to the addition of covalent cross-links between polypeptide chains.

Reference(s)
Urry DW, Haynes B, Thomas D, Harris RD (1988) A method for fixation of elastin demonstrated by stress/strain characterization. Biochem Biophys Res Commun 151:686–692 (with permission)

Force–Elongation Relation (2)

• Elastic modulus •	• Elastin • Polypentapeptide	• Heating •

Materials

- X^{20}-polypentapeptide elastomer, which is obtained by 20 Mrad irradiation cross-linking of poly (Val-Pro-Gly-Val-Gly), a high polymer of pentapeptide repeat of human, bovine, porcine, and chick elastins

Testing Methods and Experimental Conditions

- Series of stress–strain curves were determined at 40°C to an extension of 60%. Between subsequent curves, the sample was returned to zero extension and heated at 80°C for 24 h in the stress–strain apparatus
- Natural logarithm (ln) of elastic modulus was plotted against the time of heating at 80°C to determine a half-life for the loss of elastic modulus due to the thermal denaturation

Data

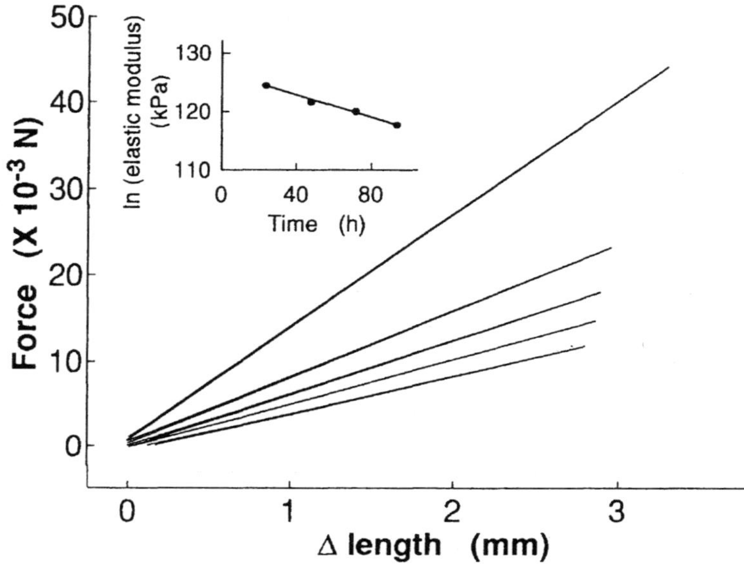

Comments

- Heating at 80°C resulted in a loss of elastic modulus due to the thermal denaturation of X^{20}-polypentapeptide.
- A plot of the ln (elastic modulus) versus time at 80°C, shown as the insert, gave a half-life of about 70 h.

Reference(s)

Urry DW (1988) Entropic elastic processes in protein mechanisms. I. Elastic structure due to an inverse temperature transition and elasticity due to internal chain dynamics. J Protein Chem 7:1–34 (with permission)

Force–Elongation Relation (3)

• Elastic modulus • Length	• Bovine ligamentum • Elastin	• Heating •

Materials
- Bovine ligamentum nuchae elastin

Testing Methods and Experimental Conditions
- Series of stress–strain curves were determined at 40°C to an extension of 60%. Between subsequent curves, the sample was returned to zero extension and heated at 80°C for 24 h in the stress–strain apparatus
- Natural logarithm (ln) of elastic modulus was plotted against the time of heating at 80°C to determine a half-life for the loss of elastic modulus due to the thermal denaturation

Data

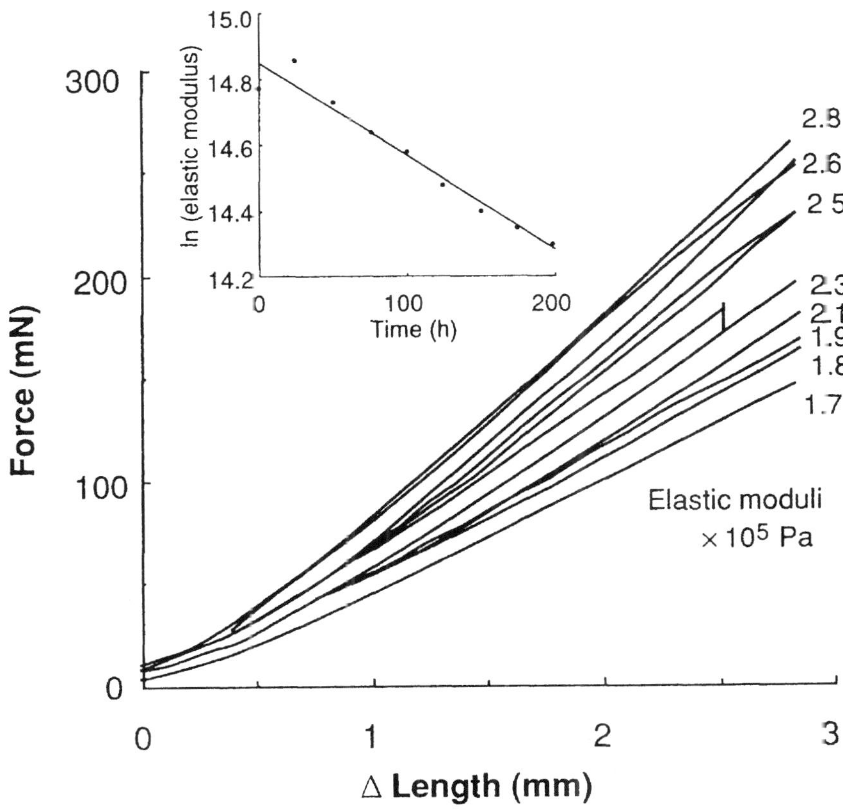

Comments
- Stress–strain data on ligamentum nuchae elastin as a function of time at 80°C and 0% extension.
- Heating at 80°C resulted in a progressive loss of elastic modulus due to a slow thermal denaturation of ligamentum nuchae elastin.
- A plot of ln (elastic modulus) versus time at 80°C shown as the insert gave a half-life of about 10 days.

Reference(s)
Urry DW (1988) Entropic elastic processes in protein mechanisms. II. Simple (passive) and coupled (active) development of elastic forces. J Protein Chem 7:81–114 (with permission)

Force–Elongation Relation (4)

• Hysteresis • Stress–strain relation	• Bovine ligamentum • Elastin	• Effect of chlorine •

Materials
• Bovine ligamentum nuchae elastin

Testing Methods and Experimental Conditions
• Ligamentum nuchae elastin in water at 37°C was allowed to stand at given extensions 0% and 40%
• The water was exchanged with a solution of 20% Clorox (Clorox Company, Oakland, CA, USA), in which Clorox contains 9.5% chlorine by weight
• Stress–strain curves were carried out to successively higher extensions before and after 1 h of Clorox treatment

Data

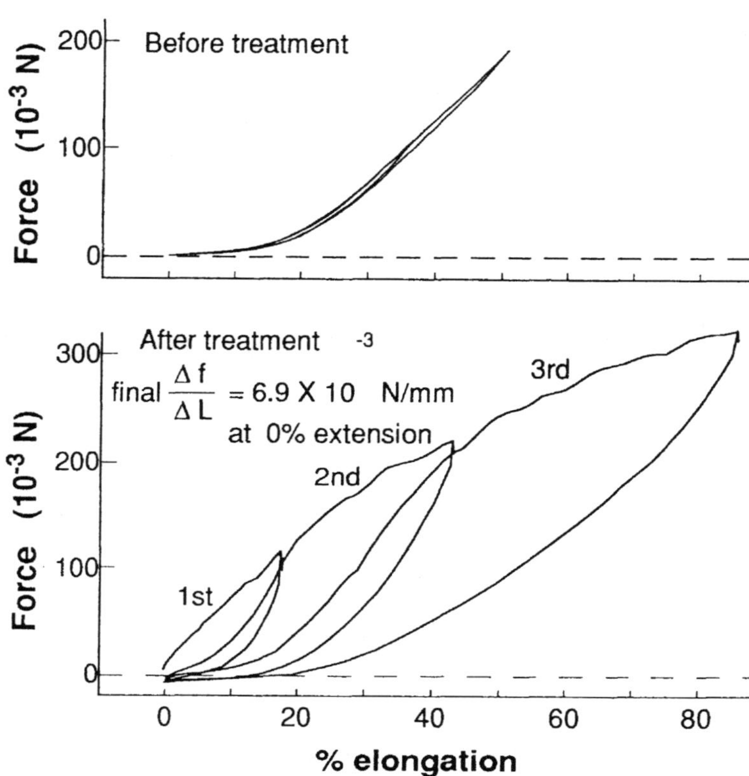

Comments
• In case before treatment, there was a minimal hysteresis and the curves were characteristically convex toward the abscissa.
• In case after treatment, there was a dramatic, largely irreversible hysteresis and the curves became concave with respect to the abscissa.

Reference(s)
Urry DW, Haynes B, Thomas D, Harris RD (1988) A method for fixation of elastin demonstrated by stress/strain characterization. Biochem Biophys Res Commun 151: 686–692 (with permission)

Force–Elongation Relation (5)

• Young's modulus •	• Bovine ligamentum • Elastin	• Superoxide •

Materials

- Bovine ligamentum nuchae elastin exposed to superoxide (O_2^-)

Testing Methods and Experimental Conditions

- Ligamentum nuchae elastin was monitored for 12 h at 37°C, with a xanthine oxidase superoxide generating system being replenished every 2 h
- Natural logarithm (ln) of elastic modulus was plotted against time exposed to superoxide to determine a half-life for the loss of elastic modulus due to oxidative destructuring

Data

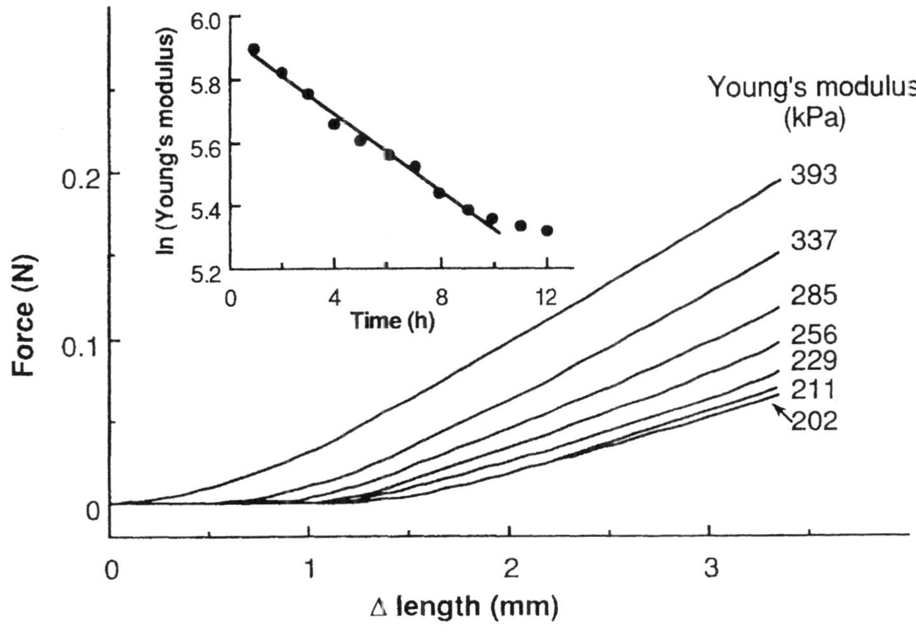

Comments

- Ligamentum nuchae elastin exposed to superoxide exhibited a systematic loss of elastic modulus due to oxidation of elastin.
- A plot of the ln (elastic modulus) versus time shown as the insert gave a half-life of 12 h.

Reference(s)

Urry DW (1988) Entropic elastic processes in protein mechanisms. II. Simple (passive) and coupled (active) development of elastic forces. J Protein Chem 7:81–114 (with permission)

Hydrodynamic Radius

· ·	· Bovine ligamentum · Elastin	· Light scattering ·

Materials

- Bovine neck ligamental elastin
- α-Elastin: Soluble fragmentation product of elastin

Testing Methods and Experimental Conditions

- Quasi-elastic light scattering photometry
- Light source: Argon ion laser (Coherent, Innova 304, Palo Alto, CA, USA), 514.5 nm
- Scattering angle: 30°–120°
- 60-channel single clipped correlator (Malvern, K7027, Malvern, Worcs, UK)
- α-Elastin concentration: 0.08-0.8 mg/ml-water
- Temperature: 10°–40°C

Data

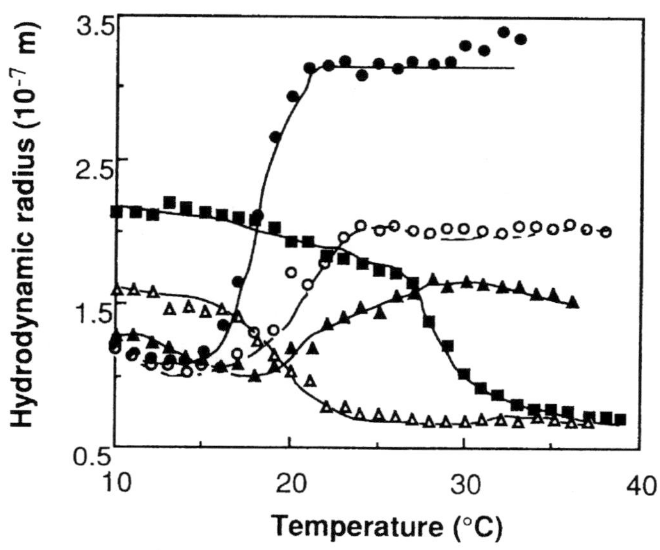

Comments

- Hydrodynamic radius as a function of temperature for various α-elastin concentrations (mg/ml): ○ (0.08), ● (0.11), ▲ (0.2), △ (0.28), ■ (0.8).
- Near the critical point of the coacervation process (α-elastin concentration: 0.11 mg/ml, temperature: 21.5°C), the hydrodynamic radius showed show a remarkable increase.
- Far from the critical point, the hydrodynamic radius showed a gradual decrease on raising the temperature.

Reference(s)

Miyakawa K, Ito Y, Kaibara K (1993) Dynamic light scattering study of coacervation of α-elastin. J Phys Soc Jpn 62:2511–2515

Length–pH Relation

| • Shortening | • Poly-L-glutamic acid | • pH dependency |
| • | • | • |

Materials

- Water-insoluble films of poly-L-glutamic acid (PGA), which is obtained by cross-linking with glycerol at pH 4.8

Testing Methods and Experimental Conditions

- A dried film strip (dimension: $0.03 \times 2.0 \times 25$ mm) was fixed with two small plastic bars at both ends, one end attached to a fixed position and the other to a small glass bead (load: 0.20g in water), and then the film was immersed in water (film length L_0: 58.0 mm for salt free and 53.0 mm for 0.2 M NaCl) and their relative lengths measured as a function of pH

Data

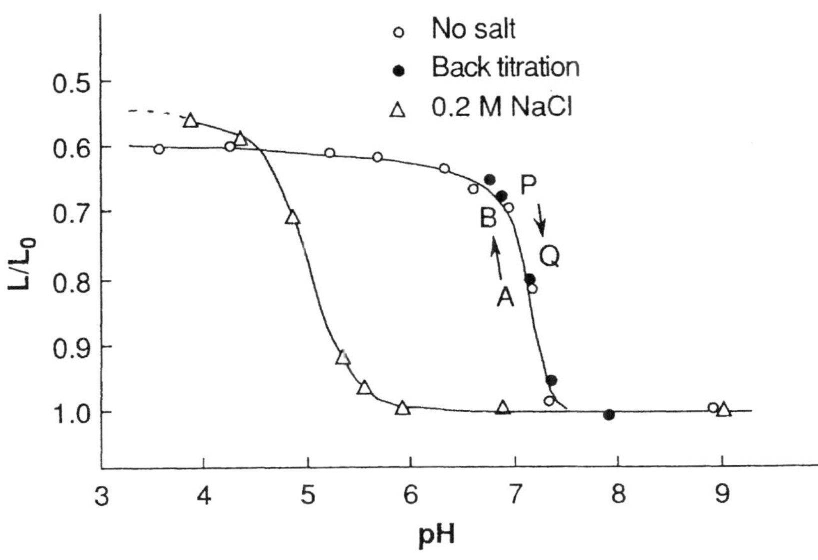

Comments

- The film in salt-free solution elongated in alkaline pH and contracted in acidic pH, demonstrating a completely reversible process.
- The titration curve of the film shifted toward a lower pH with added salt (0.2 M NaCl).

Reference(s)

Noguchi H, Yang JT (1964) Helix-coil transition of poly-L-glutamic acid film. Biopolymers 2:175–183 (with permission)

Mechanochemical Property (1)

• Force •	• Elastin • Glu-polypentapeptide	• pH effect •

Materials

- X^{20}-4%-Glu-polypentapeptide, which is obtained by 20 Mrad irradiation cross-linking of poly [4 (Val-Pro-Gly-Val-Gly), (Val-Pro-Gly-Glu-Gly)] where the Val residue in the pentapeptide repeat of human, bovine, porcine, and chick elastins is replaced in 1 out of 5 pentamers by a Glu residue

Testing Methods and Experimental Conditions

- At 37°C, at constant length and at the completion of a pH 2.1 phosphate-buffered saline (PBS) thermoelasticity experiment involving X^{20}-4%-Glu-polypentapeptide, the changes in force due to changing the pH to 7.4 and then back to 2.1 were monitored with time

Data

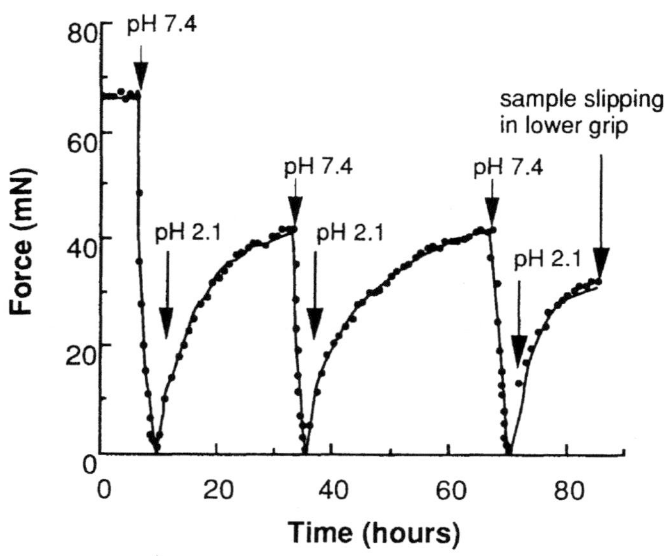

Comments

- Mechanochemical coupling exhibited by X^{20}-4%-Glu-polypentapeptide due to changes in pH at 37°C in PBS.
- At constant length X^{20}-4%-Glu-polypentapeptide developed force at pH 2.1 and relaxed at pH 7.4.

Reference(s)

Urry DW, Haynes B, Zhang H, Harris RD, Prasad KU (1988) Mechanochemical coupling in synthetic polypeptides by modulation of an inverse temperature transition. Proc Natl Acad Sci USA 85:3407–3411 (with permission)

Mechanochemical Property (2)

• Force •	• Elastin • Polypentapeptide	• Cross-linked • Ionic strength

Materials

- X^{20}-polypentapeptide, which is obtained by 20 Mrad irradiation cross-linking of poly(Val-Pro-Gly-Val-Gly), a high polymer of the pentapeptide repeat of human, bovine, porcine, and chick elastins

Testing Methods and Experimental Conditions

- X^{20}-polypentapeptide elastomer was equilibrated in solution at 25°C and stretched to 28% extension
- The force was monitored at the constant extension (28% extension at 25°C in phosphate buffered saline [PBS])
- Sample dimensions at 25°C in PBS at zero load were $0.7 \times 3.56 \times 5.72$ mm

Data

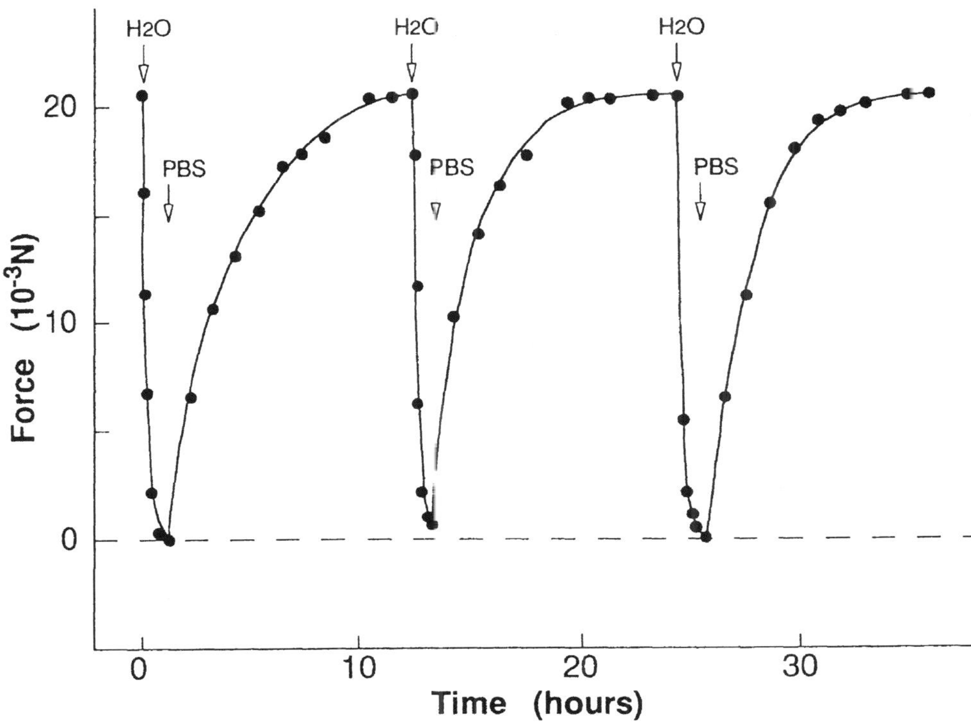

Comments

- Force changes at constant extension due to changing the medium between PBS (pH 7) and water at 25°C.
- Mechanochemical coupling in response to changes in ionic strength of the medium was exhibited.
- Complete reversibility of relaxation and contraction was observed.

Reference(s)

Urry DW, Harris RD, Prasad KU (1988) Chemical potential driven contraction and relaxation by ionic strength modulation of an inverse temperature transition. J Am Chem Soc 110:3303–3305 (with permission)

Mechanochemical Property (3)

• Length change •	• Elastin • Glu-polypentapeptide	• Cross-linked • pH effect

Materials

- X^{20}-4%-Glu-polypentapeptide, which is obtained by 20Mrad irradiation cross-linking of poly[4(Val-Prp-Gly-Val-Gly), (Val-Pro-Gly-Glu-Gly)] where the Val residue in the pentapeptide repeat of human, bovine, porcine, and chick elastins is replaced in 1 out of 5 pentamers by a Glu residue

Testing Methods and Experimental Conditions

- At 37°C, at constant force and at the end of a pH 2.1 thermoelasticity experiment to 35°C, the X^{20}-4%-Glu-polypentapeptide elastomer was placed under constant load of 1.5 g and the pH was varied between 3.3 and 4.3 while monitoring the resulting length change

Data

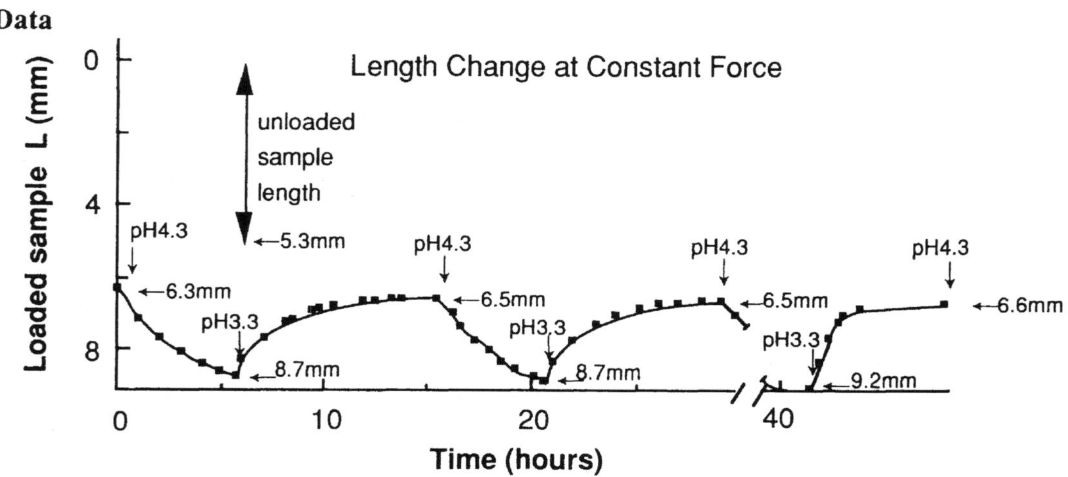

Mechanochemical coupling exhibited by x^{20}-4%-Glu-polypentapeptide due to changes in pH at 37°C in phosphate-buffered saline

Comments

- At constant force X^{20}-4%-Glu-polypentapeptide contracted, lifting weight (1.5 g) at pH 3.3, and relaxed, lowering the weight at pH 4.3

Reference(s)

Urry DW, Haynes B, Zhang H, Harris RD, Prasad KU (1988) Mechanochemical coupling in synthetic polypeptides by modulation of an inverse temperature transition. Proc Natl Acad Sci USA 85:3407–3411 (with permission)

Mechanochemical Property (4)

• Length change •	• Elastin • Polypentapeptide	• Cross-linked • Ionic strength

Materials

- X^{20}-polypentapeptide, which is obtained by 20 Mrad irradiation cross-linking of poly(Val-Pro-Gly-Val-Gly), a high polymer of the pentapeptide repeat of human, bovine, porcine, and chick elastins

Testing Methods and Experimental Conditions

- X^{20}-polypentapeptide elastomer was equilibrated in solution at 25°C and stretched to 20% extension and the length was monitored at constant force (1.6 g obtained at 20% extension in phosphate-buffered saline (PBS) at 25°C)
- Sample dimensions at 25°C in PBS at zero load were, 0.7 x 3.76 × 6.06 mm

Data

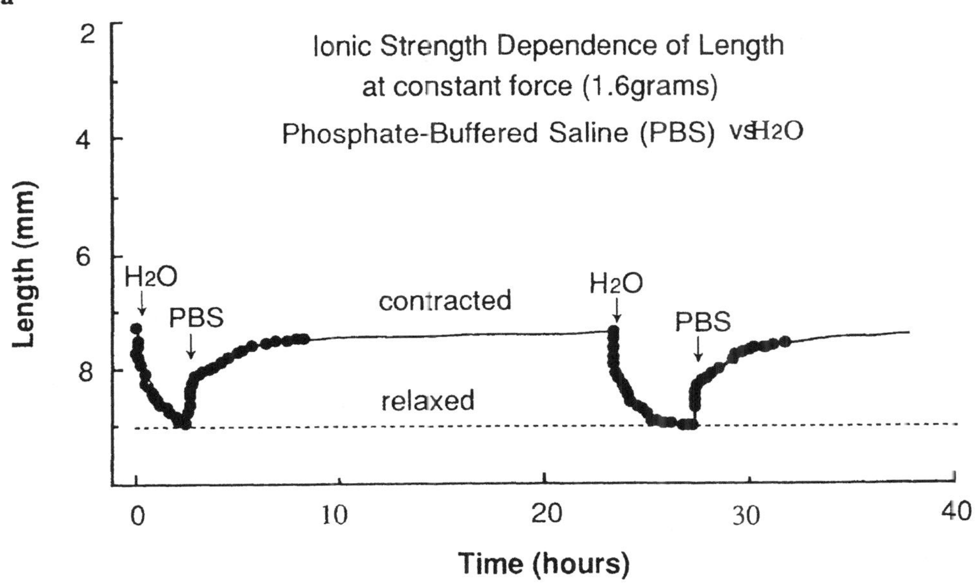

Length change at constant force due to changing the medium between PBS (pH 7) and water at 25°C

Comments

- Mechanochemical coupling in response to changes in ionic strength of the medium was exhibited.
- Complete reversibility of relaxation and contraction was observed.

Reference(s)

Urry DW, Harris RD, Prasad KU (1988) Chemical potential driven contraction and relaxation by ionic strength modulation of an inverse temperature transition. J Am Chem Soc 110:3303–3305 (with permission)

Mechanochemical Property (5)

• Length change •	• Poly-L-glutamic acid •	• Crosslinked • pH effect

Materials

- Water-insoluble films of poly-L-glutamic acid (PGA), which is obtained by cross-linking with glycerol at pH 4.8

Testing Methods and Experimental Conditions

- A dried film strip (dimension: 0.03 x 2.3 x 36 mm) was fixed with two small plastic bars at both ends, one end attached to a fixed position and the other to a small glass bead (load: 0.38 g in water), and then the film was immersed in water (film length L_o: 57.4 mm) and alternately suspended in 50 ml HCl (pH 2) and NaOH (pH 12) for 5 min

Data

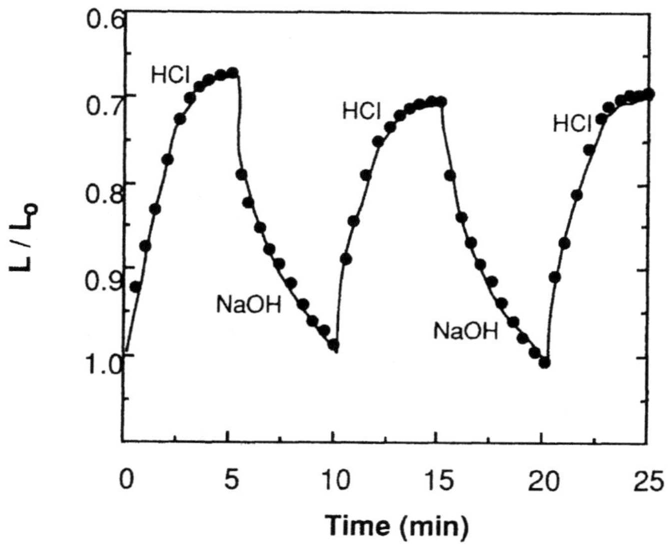

Comments

- Reversible contraction of glycerol-cross-linked PGA film in alternating acid (pH 2) and alkaline (pH 12) media.
- Films contracted in HCl (pH 2) and elongated in NaOH (pH 12).
- The process was completely reversible.

Reference(s)

Noguchi H, Yang JT (1964) Helix-coil transition of poly-L-glutamic acid film. Biopolymers 2:175–183 (with permission)

Specific Volume (1)

• Compressibility •	• Calf skin • Gelatin	• •

Materials
- Calf skin
- Gelatin

Testing Methods and Experimental Conditions
- The protein dissolved in distilled water (protein concentration 0.5%–1.0%) is dialyzed against the solvent at 4°C
- The properties were measured at the temperature 25°±0.01°C

Data

Molecular Properties of the Proteins Used		Values
Molecular weight M	$(\times 10^3)$	100
Specific volume v_0	(ml/g)	0.689
Adiabatic compressibility β_S	$(\times 10^{18}\ cm^2/N)$	−2.50
Isothermal compressibility β_T	$(\times 10^{18}\ cm^2/N)$	1.92

Comments
None.

Reference(s)
Gekko K, Hasegawa Y (1986) Compressibility-structure relationship of globular proteins. Biochemistry 25: 6563–6571 (with permission)

Specific Volume (2)

• Compressibility •	• Egg white • Lysozyme	• •

Materials

- Egg white
- Lysozyme

Testing Methods and Experimental Conditions

- The water-soluble protein is deionized by dialysis against distilled water at 4°C and lyophilized after passage through a mix-bed ion-exchanger column (Amberlite IR-120 and IRA-400) to obtain the isoionic protein
- The properties were measured at 25°±0.01°C

Data

Molecular properties of the proteins used	Values
Molecular weight M $(\times 10^3)$	14.3
Specific volume v_0 (ml/g)	0.712
Adiabatic compressibility β_S $(\times 10^{18}\ cm^2/N)$	0.467
Isothermal compressibility β_T $(\times 10^{18}\ cm^2/N)$	0.773

Comments

None.

Reference(s)

Gekko K, Noguchi H (1979) Compressibility of globular proteins in water at 25°C. J Phys Chem 83:2706–2714 (with permission)

Specific Volume (3)

• Compressibility •	• Whale muscle • Myoglobin	• •

Materials
- Whale skeletal muscle
- Myoglobin

Testing Methods and Experimental Conditions
- The water-soluble protein is deionized by dialysis against distilled water at the temperature 4° C and lyophilized after passage through a mix-bed ion-exchanger column (Amberlite IR-120 and IRA-400, Rohm and Haas Co, Philadelphia, PA, USA) to obtain the isoionic protein
- The properties were measured at $25° \pm 0.01°C$

Data

Molecular properties of the proteins used			Values
Molecular weight M	$(\times 10^3)$		17.0
Specific volume v_0	(ml/g)		0.747
Adiabatic compressibility β_S	$(\times 10^{18}$	$cm^2/N)$	8.98
Isothermal compressibility β_T	$(\times 10^{18}$	$cm^2/N)$	13.1

Comments
None.

Reference(s)
Gekko K, Noguchi H (1979) Compressibility of globular proteins in water at 25°C. J Phys Chem 83:2706–2714 (with permission)

Stress Relaxation

• Repeated strain	• Calcium alginate gel	•
•	•	•

Materials

- Calcium alginate (CaAlg) gel, which is obtained by gelation of sodium alginate (a polysaccharide found in intercellular substances of brown algae) in 0.3 M $CaCl_2$ solution

Testing Methods and Experimental Conditions

- CaAlg gel rods fixed by clamps were suspended in electrolyte solutions in which the gels were equilibrated
- Strains were imposed on the gels in steps and the tensile stresses were measured at 25°C by strain gauges connected to a constant direct source and a microvoltmeter

Data

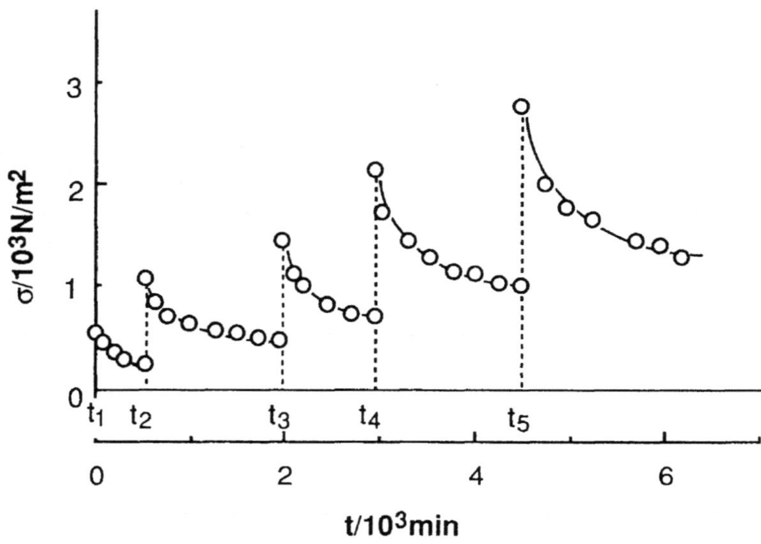

Stress relaxation curve for CaAlg gel. The Value t_k shows time when kth strain is imposed.

Comments

- The stress under each strain showed a long relaxation and attained an equilibrium state after 20 h.

Reference(s)

Yonese M, Baba K, Kishimoto H (1988) Stress relaxation of alginate gels crosslinked by various divalent metal ions. Bull Chem Soc Jpn 61:1857–1863 (with permission)

Stress–Strain Relation

• Young's moduli • Anisotropy	• Elastin • Polypentapeptide	• Cross-linked •

Materials

- Cross-linked polypentapeptide elastomer, which is obtained by chemically cross-linking reaction between poly[4(Val-Pro-Gly-Val-Gly), (Val-Pro-Gly-Lys-Gly)] and poly[4(Val-Pro-Gly-Val-Gly), (Val-Pro-Gly-Glu-Gly)] where the Lys or the Glu is introduced into a high polymer of the pentapeptide repeat of human, bovine, porcine, and chick elastins
- Porcine aortic elastic fiber

Testing Methods and Experimental Conditions

- Force–strain curves were recorded by placing one end of the specimen in the load cell clamp that was detached from the load cell. The other end of the specimen was clamped to the platform. The load cell clamp was then attached to the load cell with an initial length of 1 mm and no initial tensile or compressive force. The drive was turned on and the specimen stretched at a rate of 0.5 mm/s
- Elastic fibers were hydrated in a chamber of 100% humidity

Data

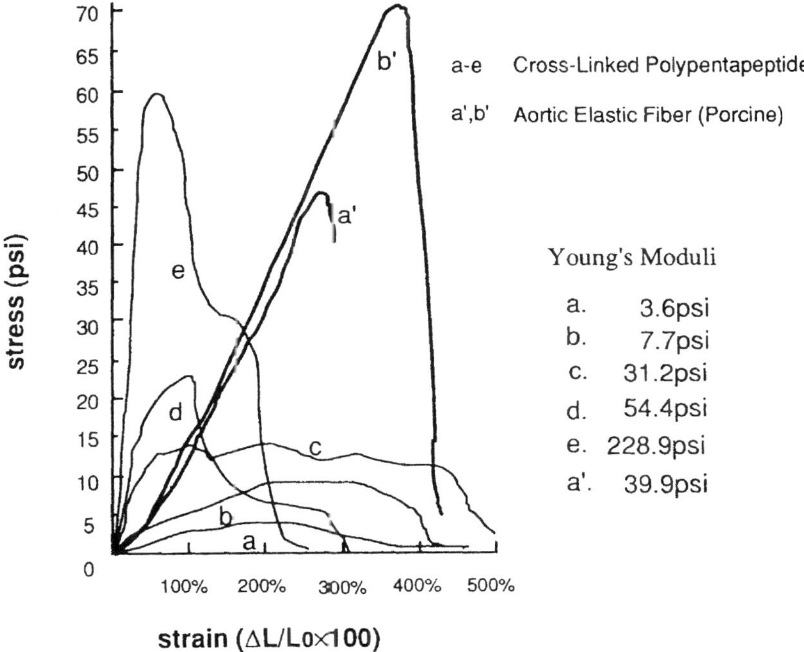

- The water content of cross-linked polypentapeptide in curve a was 70% and sequentially lowered in curves b through e
- Experiments were performed at room temperature

Comments

- Cross-linked polypentapeptide was an anisotropic, fibrillar elastomer.

Reference(s)

Urry DW, Okamoto K, Harris RD, Hendrix CF, Long MM (1976) Synthetic, cross-linked polypentapeptide of tropoelastin. Biochemistry 15:4083–4089 (with permission)

Tensile Strength (1)

• Elastic modulus •	• Sheep skin • Collagen	• Noncross-linked • Cross-linked

Materials
- Noncrosslinked dermal sheep collagen (N-DSC)
- Glutaraldehyde-cross-linked dermal sheep collagen (G-DSC), which is obtained by cross-linking of N-DSC with glutaraldehyde
- Hexamethylene diisocyanate cross-linked dermal sheep collagen (H-DSC), which is obtained by cross-linking of N-DSC with hexamethylene diisocyanate

Testing Methods and Experimental Conditions
- Each sample of N-DSC, G-DSC and H-DSC was incubated with collagenase in Tris-HCl buffer and the degradation was discontinued at the desired time interval by addition of EDTA
- Stress–strain curves of partially degradated samples and control samples were determined by uniaxial measurements using an Instron (High Wycombe, UK) mechanical tester after hydrating the sample for at least 30 min in phosphate-buffered saline containing EDTA at room temperature

Data

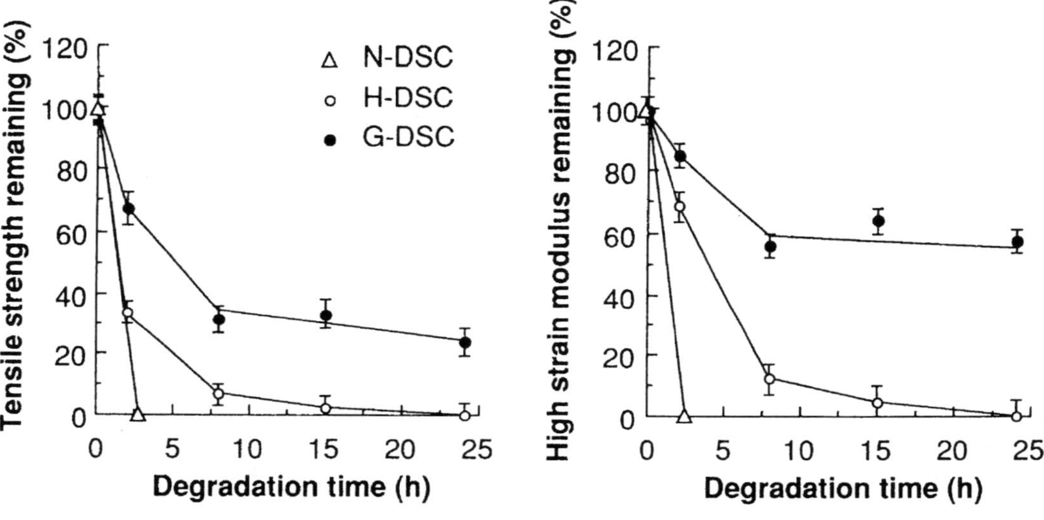

Comments
- Changes in mechanical properties were expressed as a percentage of the initial value of either tensile strength or high strain modulus, respectively.

Reference(s)
OldeDamink LHH, Dijkstra PJ, VanLuyn MJA, Vanwachem PB, Nieuwenhuis P, Feijen J (1995) Changes in the mechanical properties of dermal sheep collagen during in vitro degradation. J Biomed Mater Res 29:139–147 (with permission)

Tensile Strength (2)

• Ultimate elongation • Elastic modulus	• Sheep skin • Collagen	• Noncross-linked • Cross-linked

Materials
- Noncrosslinked dermal sheep collagen (N-DSC)
- Glutaraldehyde-cross-linked dermal sheep collagen (G-DSC), which is obtained by cross-linking of N-DSC with glutaraldehyde
- Hexamethylene diisocyanate cross-linked dermal sheep collagen (H-DSC), which is obtained by cross-linking of N-DSC with hexamethylene diisocyanate

Testing Methods and Experimental Conditions
- Stress-strain curves of N-DSC, G-DSC, and H-DSC samples were determined by uniaxial measurements using an Instron (High Wycombe, UK) mechanical tester
- Tensile test samples (30.0 x 6.0 x 0.8 mm) were hydrated for at least 30 min in phosphate buffered saline at room temperature
- The tensile strength, the elongation at alignment, the elongation at break, the low strain modulus, and the high strain modulus of the sample were calculated from five independent measurements

Data

Sample	Tensile strength (MPa)	Elongation at alignment (%)	Elongation at break (%)	Low strain modulus (MPa)	High strain modulus (MPa)
N-DSC	19 ± 1	68 ± 6	141 ± 7	2.0 ± 0.3	25 ± 3
G-DSC	18 ± 1	61 ± 3	160 ± 5	4.6 ± 0.7	15 ± 1
H-DSC	17 ± 2	63 ± 7	163 ± 5	3.5 ± 0.4	15 ± 1

All mechanical properties were measured in fivefold and are given as mean ± SD.

Comments
- No significant differences between the mechanical properties of G-DSC and H-DSC were observed.
- N-DSC had the same tensile strength and elongation at alignment as G-DSC and H-DSC, but had a lower elongation at break and a lower low strain modulus.
- The high strain modulus was the highest for N-DSC.

Reference(s)
OldeDamink LHH, Dijkstra PJ, VanLuyn MJA, VanWachem PB, Nieuwenhuis P, Feijen J (1995) Changes in the mechanical properties of dermal sheep collagen during in vitro degradation. J Biomed Mater Res 29:139–147 (with permission)

Thermoelasticity (1)

• Force • Dielectric permittivity	• Elastin • Pentapeptide	• Cross-linked • Temperature

Materials

- Poly (Val-Pro-Gly-Val-Gly), which is a high polymer of the pentapeptide repeat of human, bovine, porcine, and chick elastins
- X^{20}-polypentapeptide, which is obtained by 20 Mrad irradiation cross-linking of poly (Val-Pro-Gly-Val-Gly)

Testing Methods and Experimental Conditions

- Poly (Val-Pro-Gly-Val-Gly) coacervate was injected into a coaxial line cell, and the dielectric permittivity (real part) was determined as a function of temperature over the frequency range 1 MHz to 1 GHz
- The real part of the dielectric permittivity at 3.9 MHz divided by the absolute temperature (°K) was plotted as a function of temperature
- X^{20}-polypentapeptide elastomer was stretched to a fixed extension (63%) while immersed in water and the force was monitored as a function of temperature

Data

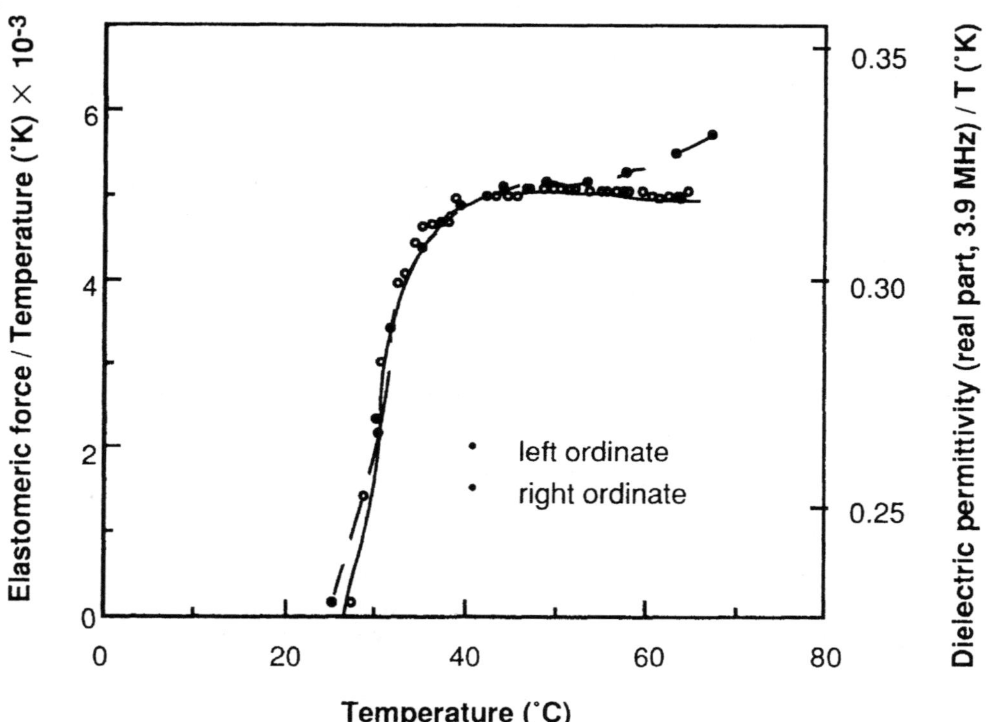

Comments

- Thermoelasticity data correlated with dielectric permittivity, real part, at 3.9 MHz.
- Development of the local intense relaxation at 3.9 MHz was correlated to the development of elastomeric force.

Reference(s)

Urry DW (1988) Entropic elastic processes in protein mechanisms. I. Elastic structure due to an inverse temperature transition and elasticity due to internal chain dynamics. J Protein Chem 7:1–34 (with permission)

Thermoelasticity (2)

• Force • Turbidity	• Bovine ligamentum • Elastin	• Temperature •

Materials

- Bovine ligamentum nuchae elastin
- α-elastin, which is an oxalic acid fragmentation product of bovine ligamentum nuchae elastin

Testing Methods and Experimental Conditions

- Turbidity of α-elastin was measured at 300 nm as a function of temperature
- Bovine ligamentum nuchae elastin was stretched to a fixed extension (60%) while immersed in water and the force was monitored as a function of temperature

Data

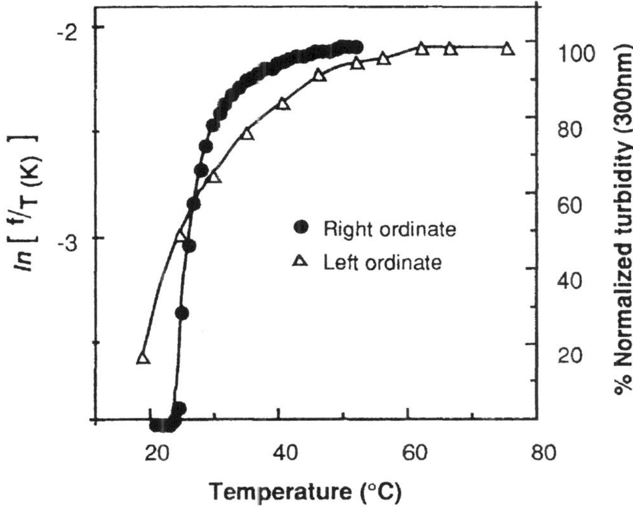

Comments

- f, elastomeric force (an internal energy component + an entropy component).
- Increase in turbidity at 300 nm was correlated to the development of elastomeric force.
- Bovine ligamentum nuchae elastin was an anisotropic, entropic elastomer.

Reference(s)

Urry DW (1987) Entropic elastomeric force in protein structure/function. Int J Quantum Biol Symp 14:261–280 (with permission)

Thermoelasticity (3)

• Force • Turbidity	• Pentapeptide • Polypentapeptide	• Cross-linked • Temperature

Materials
- Poly (Val-Pro-Gly-Val-Gly), which is a high polymer of the pentapeptide repeat of human, bovine, porcine, and chick elastins
- X^{20}-polypentapeptide, which is obtained by 20 Mrad irradiation cross-linking of poly (Val-Pro-Gly-Val-Gly)

Testing Methods and Experimental Conditions
- Turbidity of poly (Val-Pro-Gly-Val-Gly) in water was measured at 300 nm as a function of temperature and normalized with respect to maximum turbidity as 100
- X^{20}-polypentapeptide elastomer was stretched to a fixed extension (60%) while immersed in water and the force was monitored as a function of temperature

Data

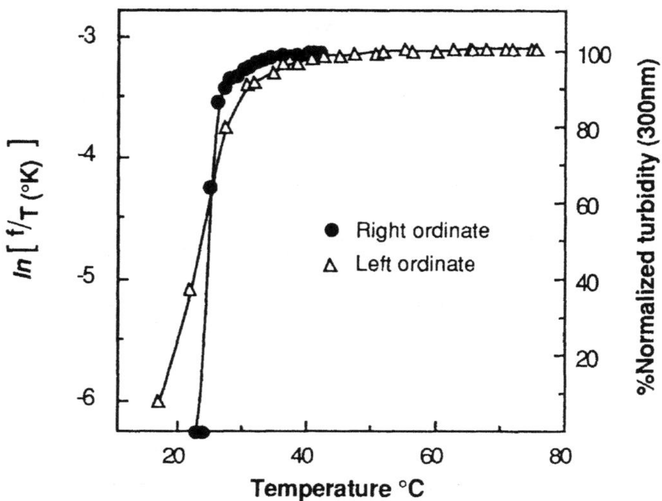

Comments
- f, elastomeric force (an internal energy component + an entropy component).
- Increase in turbidity at 300 nm was correlated to the development of elastomeric force.
- X^{20}-polypentapeptide was an anisotropic, entropic elastomer.

Reference(s)
Urry DW (1987) Entropic elastomeric force in protein structure/function. Int J Quantum Chem: Quantum Biol Symp 14:261–280 (with permission)

Thermoelasticity (4)

• Length change •	• Bovine elastin • Polypentapeptide	• Cross-linked • Temperature

Materials
- Bovine ligamentum nuchae elastin
- X^{20}-polypentapeptide (X^{20}-PPP), which is obtained by 20 Mrad irradiation cross-linking of poly (Val-Pro-Gly-Val-Gly), a high polymer of the pentapeptide repeat of human, bovine, porcine, and chick elastin
- Latex rubber (Kent Latex series 73570, Kent, OH, USA)

Testing Methods and Experimental Conditions
- Samples were pre-equilibrated overnight in water at 20°C before measurement
- Temperature was increased in 5°C to 10°C increments
- Change in elastomer length at zero load was plotted as a function of temperature
- Length at 19°C was taken as 100%

Data

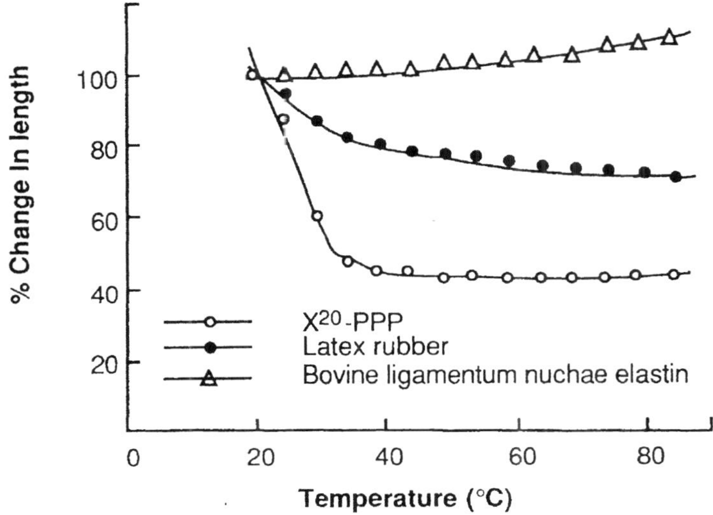

Comments
- The data for latex rubber was characteristic elastomers known as classical rubber comprised of a network of random chains.
- The data for ligamentum nuchae elastin and X^{20}-PPP of elastin were characteristic of elastomers that differ significantly from a network of random chains.

Reference(s)
Urry DW, Haynes B, Harris RD (1986) Temperature dependence of elastin and its polypentapeptide. Biochem Biophys Res Commun 141:749–755 (with permission)

Thermoelasticity (5)

• Length change •	• Bovine elastin • Polypentapeptide	• Cross-linked • Temperature

Materials

- Bovine ligamentum nuchae elastin
- X^{20}-polypentapeptide (X^{20}-ppp), which is obtained by 20 Mrad irradiation cross-linking of poly (Val-Pro-Gly-Val-Gly), a high polymer of the pentapeptide repeat of human, bovine, porcine, and chick elastins
- Latex rubber (Kent Latex series 73570, Kent, OH, USA)

Testing Methods and Experimental Conditions

- Forces obtained at 60% extension and 40°C (the 60%/40°C force) were maintained: 1.256 g for X^{20}-ppp, 15.756 g for ligamentum nuchae elastin and 22.789 g for latex rubber
- After elongation at 40°C, samples were equilibrated overnight near 20°C in water and the length was readjusted to achieve the 60%/40°C force before raising the temperature
- Temperature was increased in 5°C to 10°C increments and the length at 19°C was taken as 100%

Data

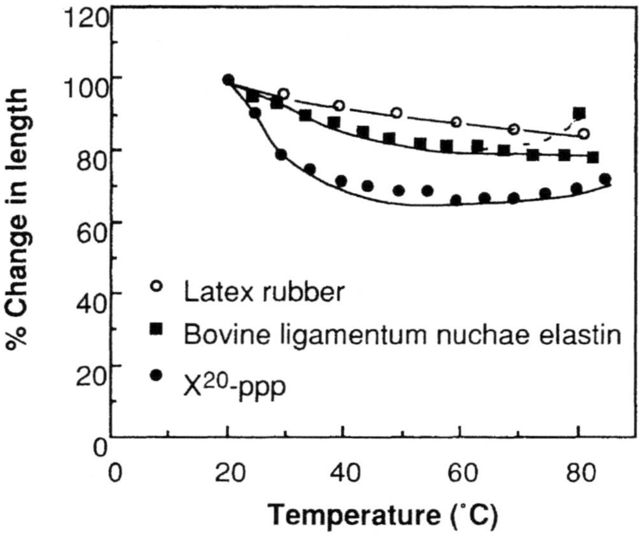

Comments

- Temperature dependence of elastomer length at constant load.
- The data for latex rubber was characteristic of elastomers known as classical rubbers comprised of a network of random chains.
- The data for ligamentum nuchae elastin and X^{20}-ppp of elastin were characteristic of elastomers that differ significantly from a network of random chains.

Reference(s)

Urry DW, Haynes B, Harris RD (1986) Temperature dependence of elastin and its polypentapeptide. Biochem Biophys Res Commun 141:749–755 (with permission)

Viscosity

• Temperature dependence •	• Bovine ligamentum • Elastin	• Concentration •

Materials
- Bovine neck ligamental elastin
- α-Elastin: Soluble fragmentation product of elastin

Testing Methods and Experimental Conditions
- Cone-plate type rotary viscometer
- Cone angle 1°34'; rotation rate 100 rpm
- α-Elastin concentration: 0.01–0.8 mg/ml in water
- Temperature: 0°–60°C

Data

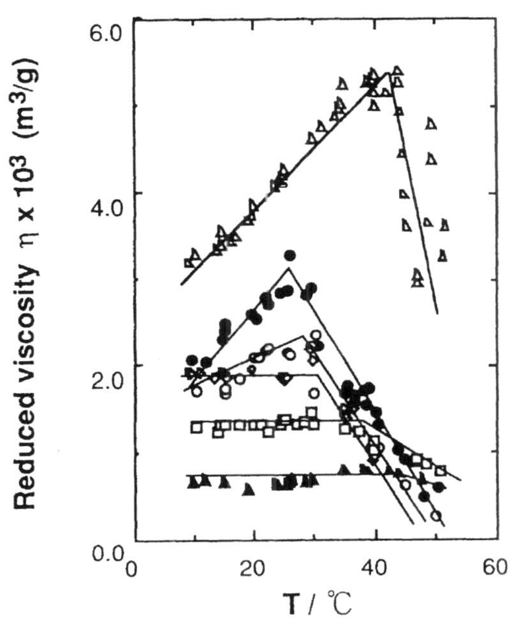

Reduced viscosity as a function of the temperature.

α -Elastin concentration (mg/ml) :

$\triangle$ (0.05), ● (0.1), ○ (0.2), ◆ (0.25), ☐ (0.4), ▲ (0.8)

Comments
- Break points refer to the temperature at which the coacervation of α-elastin -water system initiates.
- Near the critical point of coacervation process (α-elastin concentration: 0.11mg/ml, temperature: 21.5°C), the transient increases in the reduced viscosity are observed.

Reference(s)
Kaibara K (1995) Role of metal chlorides on self-assembly and function control of proteins. The salt science research foundation annual report 1993, Physiol Food Science pp 243-253

Pressure–Flow Curve

	• Human	• Hematocrit
•	• Red cell	•

Materials
- Human
- Red cells

Testing Methods and Experimental Conditions
- Pressure and flow rate relationship of human red cells suspended in standard acid-citratedextrose solution (ACD) at 25.5°C in a glass capillary of radius 472 μm

Data

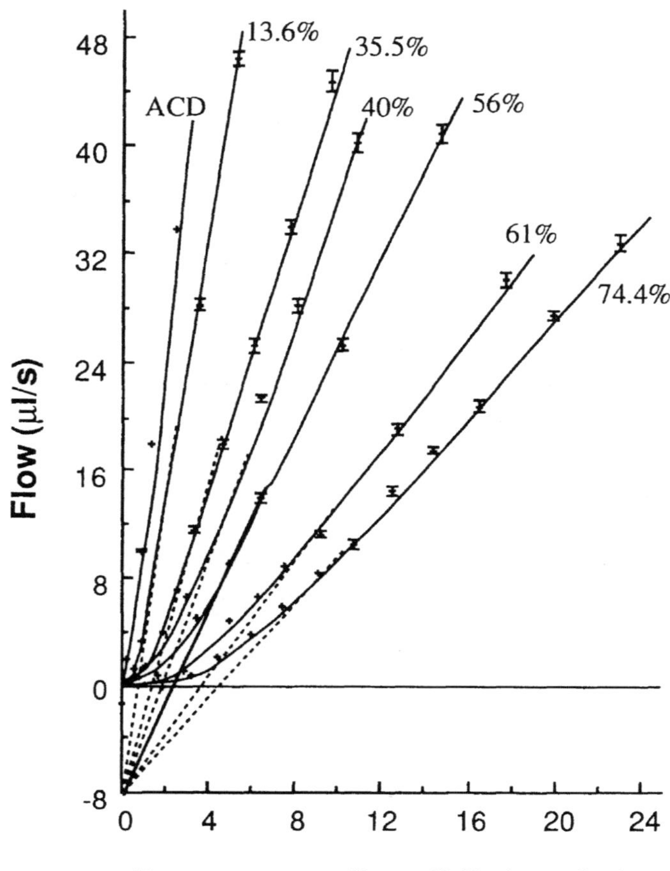

Comments
- The increase in gradient of the curve indicates the flow resistance in the glass capillary, being associated with the apparent viscosity.
- At very low pressure and flows the curves are nonlinear, but the curves gradually become linear at higher driving pressure.
- We should note that the extrapolated linear segments converge to a common point in the negative flow region.

Reference(s)
Haynes R H, Burton A C (1959) Role of the non-Newtonian behavior of blood in hemodynamics. Am J Physiol. 197:943–950 (with permission)

Shear Rate–Shear Stress Curve

	• Human • Blood	
• •		• •

Materials

- Human
- Blood

Testing Methods and Experimental Conditions

- Casson plots (square root of shear stress versus square root of shear rate) for typical normal blood, acid citrate dextrose solution (ACD) as anticoagulant, at three temperatures and four heamatocrit levels

Data

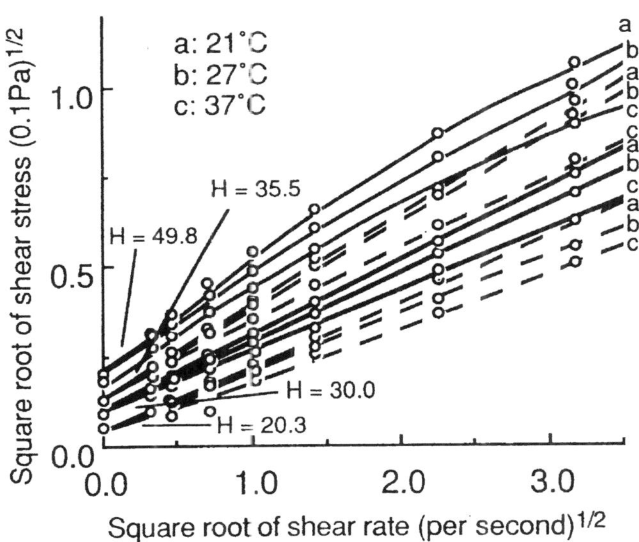

Comments

None

Reference(s)

Merril EW, Gilliland ER, Cokelet G, Shin E, Britten A, Wells RE Jr (1963) Rheology of human blood near and at zero flow. Biophysical J 3: 199–213 (with permission)

Viscoelasticity(1)

<table>
<tr><td>•
•</td><td>• Human
• Blood</td><td>•
•</td></tr>
</table>

Materials
- Human
- Blood

Testing Methods and Experimental Conditions
- The viscoelastic properties of human blood was measured by the oscillatory tube flow driven by an oscillating piston coupled to a cylinder which is terminated in tubes filled with blood
- Temperature: 25°C

Data

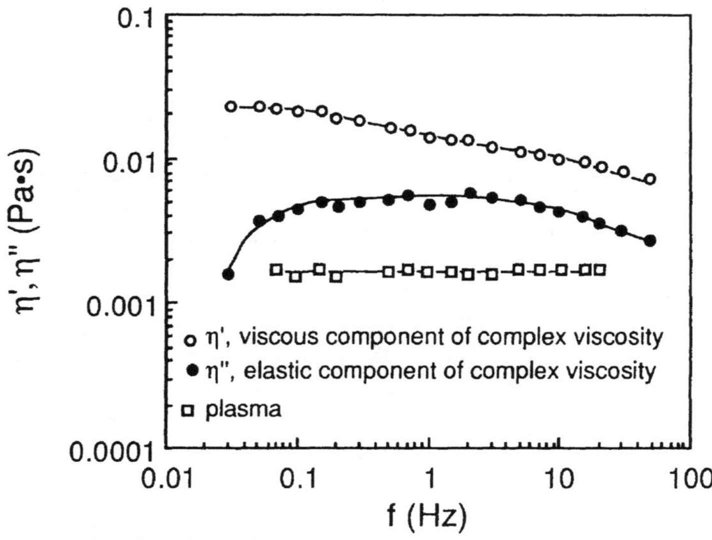

Comments
- Frequency dependence of components of the complex viscosity of blood and blood plasma.
- Plasma does not exhibit a measurable elastic component.

Reference(s)
Thurston GB (1973) Frequency and shear rate dependence of viscoelasticity of human blood. Biorheology 10:375–381 (with permission)

Viscoelasticity(2)

• Complex shear modulus •	• Human • Blood	• Oscillatory frequency •

Materials
- Human
- Blood

Testing Methods and Experimental Conditions
- Rheological properties of concentrated red cell suspensions are studied with a magneto acoustic microrheometer in which a ball is suspended in a vertically oriented cylindrical tube.
- The rheometer determines the steady state viscosity and viscoelastic properties.

Data

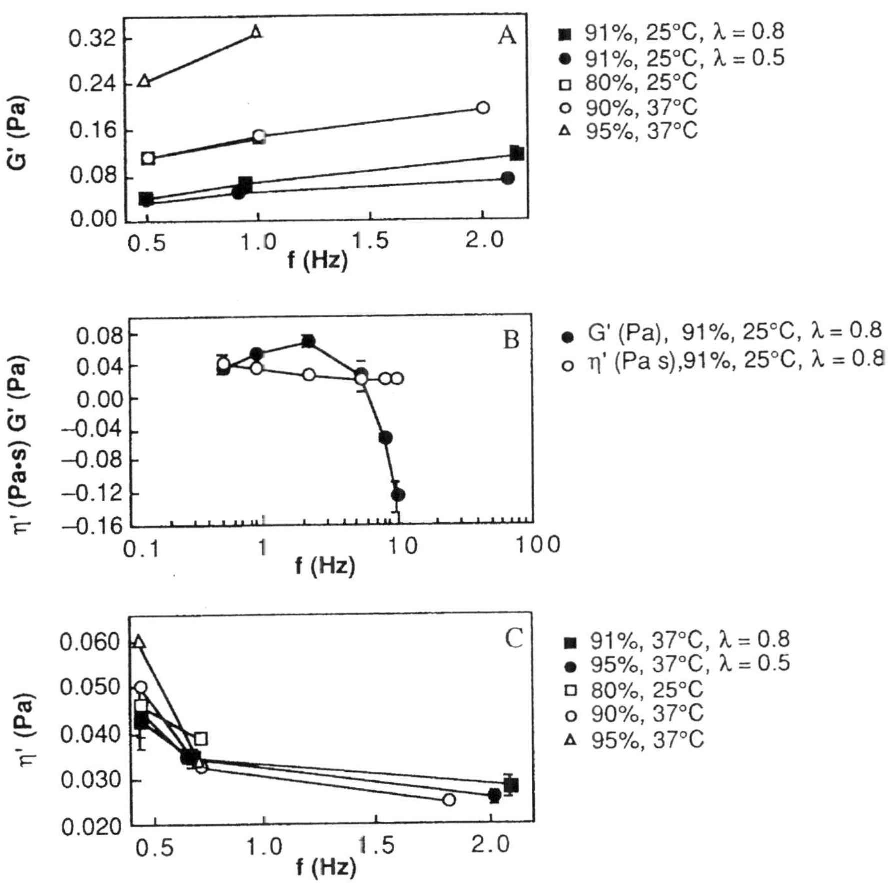

Comments
- G'; storage moduli of the complex shear modulus.
- Elastic component for packed RBCs. Variation of G' as function of oscillatory frequency.
- Inertial Effect; Variation of η' and G' as a function of oscillatory frequency.
- Viscous component for packed RBCs; Variation of η' as function of oscillatory frequency.

Reference(s)
Tran-Son-Tay R, Coffey BE, Hochmuth RM (1989) A rheological study of packed red blood cell suspensions with an oscillating ball microrheometer. Biorheology 26:143–151 (with permission)

Viscoelasticity(3)

• Elastic modulus • Loss modulus	• Human • Blood	• •

Materials
- Human
- Blood

Testing Methods and Experimental Conditions
- The elastic modulus and loss modulus during clotting of human whole blood were measured by a dynamic viscoelastic apparatus
- Data were taken for three different samples

Data

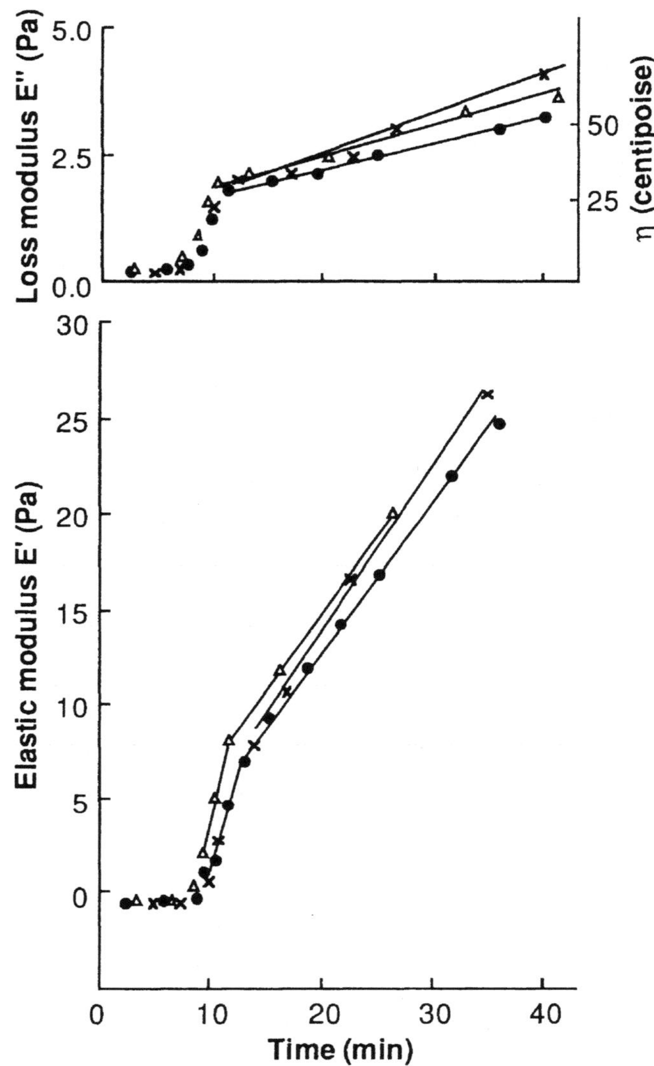

Comments
None

Reference(s)
Kaibara M and Fukuda E (1969) Non-Newtonian viscosity and dynamic elasticity of blood during clotting. Biorheology 6:73–84 (with permission)

Viscoelasticity(4)

• Red cell aggregation • Viscoelasticity	• Seal, pig, human • Blood	• •

Materials
- Seal, pig, human
- Blood

Testing Methods and Experimental Conditions
- Red cell aggregation and viscoelasticity of blood

Data

	Elephant seals ($n = 2–4$)	Ringed seals ($n = 2–3$)	Pigs ($n = 3–5$)	Humans ($n = 15–42$)
η_{pl}(mPa·s)		1.63 ± 0.08	$1.56 (n = 1)$	1.59 ± 0.08
ZSR		0.405 ± 0.009	0.460 ± 0.005	0.498 ± 0.005
MCV(μm^3)	176	122.3 ± 3.6	54	37
EM($n = 50$) ($\mu m \cdot s^{-1} \cdot v^{-1} \cdot cm$)		-1.37 ± 0.02		-1.09 ± 0.02
η' (2, shear rate [s^{-1}])	9.36 ± 0.64	8.76 ± 0.32	11.74 ± 0.64	9.23 ± 0.60
η'' (2)	2.48 ± 0.34	1.57 ± 0.06	5.74 ± 0.63	2.90 ± 0.45
η' (10)	5.89 ± 0.53	7.30 ± 0.16	9.35 ± 0.75	7.43 ± 0.56
η'' (10)	1.12 ± 0.18	0.44 ± 0.04	2.31 ± 0.55	1.26 ± 0.36
η' (50)	6.06 ± 0.41	7.15 ± 0.20	6.88 ± 0.40	5.66 ± 0.24
η'' (50)	0.28 ± 0.18	0.10 ± 0.02	0.40 ± 0.10	0.15 ± 0.05
η' (100)	5.88 ± 0.34	6.78 ± 0.20	6.06 ± 0.29	5.13 ± 0.25
η'' (100)	0.26 ± 0.21	0.02 ± 0.02	0.00 ± 0.04	0.03 ± 0.03
Total protein (g/dl)	8.2 ± 0.8	7.5 ± 0.3	7.2 ± 0.8	7.0 ± 0.9
Fibrinogen (mg/dl)	125 ± 75	172 ± 2	400 ± 100	300 ± 100
ESR (mm/hr)		0	14.4 ± 14.7	0-20
HCT (%)	34 ± 5	53 ± 8	34 ± 2	45 ± 5

Data are given as mean $\pm$ SD

η_{pl}, plasma viscosity; ZSR, zeta sedimentation ratio; MCV, mean corpuscular volume; EM, red blood cell electrophoretic mobility; η', viscous components of complex viscosity at hematocrit of 40%; η'', elastic components of complex viscosity at hematocrits of 40%; ESR, erythrocyte sedimentation rate; HCT, hematocrit for native blood

Comments
- Differences of η''/η' between Ringed seal and human blood were significant at $P < 0.01$ at all shear rates

Reference(s)
Wickham LL, Bauersachs RM, Wenby RB, Sowemimo-Coker S, Meiselman HJ, and Elsner R (1990) Red cell aggregation and viscoelaticity of blood from seals, swine and man. Biorheology 27:191–204 (with permission)

Viscoelasticity(5)

| • Storage modulus
• Loss modulus | • Horse, bovine
• Blood | • Hematocrit
• |

Materials
- Horse, bovine
- Blood

Testing Methods and Experimental Conditions
- The transient viscoelastic properties of horse and bovine blood were measured by a concentric double-cylinder viscoelastometer
- Temperature: 25°C

Data

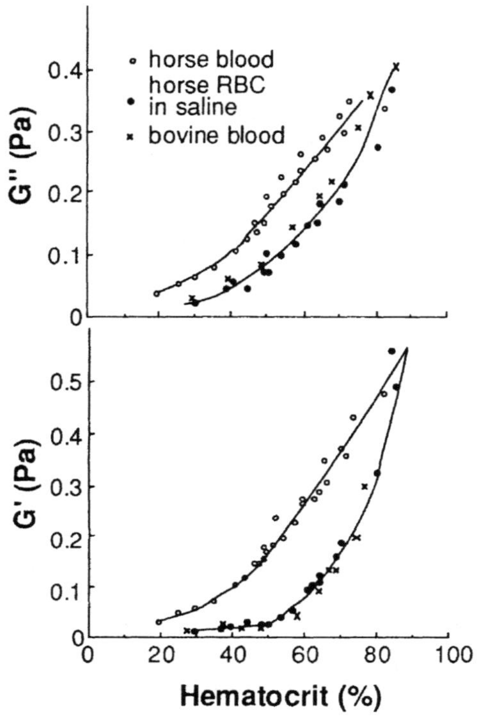

Comments
- G': the strage modulus (dynamic rigidity modulus).
  G": loss modulus.
- The values of G' and G" for horse blood, horse red blood cells suspended in saline containing 3% albumin and bovine blood plotted against the hematocrit.
- The maximum values of G' and G" were plotted for horse blood.

Reference(s)
Kaibara M, Fukada E (1982) Transient viscoelastic behavior of blood. Clin Hemorheology 2:7–11 (with permission)

Viscoelasticity(6)

• Viscosity • Shear rate	• Human • Blood	• •

Materials
- Human
- Red cells and plasma

Testing Methods and Experimental Conditions
- Viscosity of red cells suspended in saline and plasma as a function of cell concentration at several shear rates
- Data are from Brooks et al. (1970)

Data

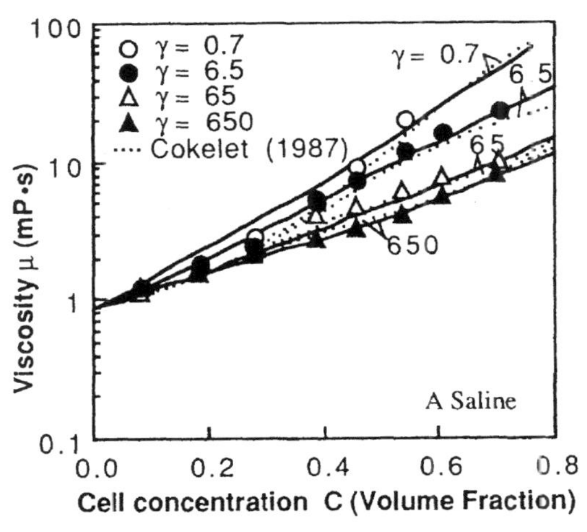

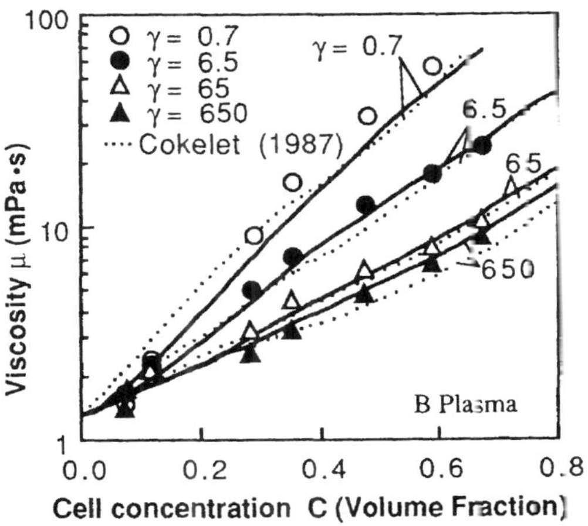

Comments
- Solid curves are evaluated from the global fit to all of the viscosity data with $\mu_f = 0.89$ and 1.24 mPa s for saline and plasma, respectively. (μ_f: viscosity of the suspending fluid)
- Dotted curves are the Quemada equation with parameter correlations.

Reference(s)

1. Zydney AL, Oliver III JD, Colton CK (1991) A constitutive equation for the viscosity of stored red cell suspensions: effect of hematocrit, shear rate, and suspending phase. J Rheol 35(8):1639–1680 (with permission)
2. Brooks DE, Goodwin JW, Seaman GVF (1970) Interaction among erythrocytes under shear. J Appl Physiol 28:174-177

Viscoelasticity(7)

• Viscosity • Shear rate	• Human • Blood	• Hematocrit •

Materials
- Human
- Red cells

Testing Methods and Experimental Conditions
- Viscosity as a function of shear rate with varying hematocrit
- Red blood cells suspended in their own anticoagulant plasma
- Temperature: 25°C

Data

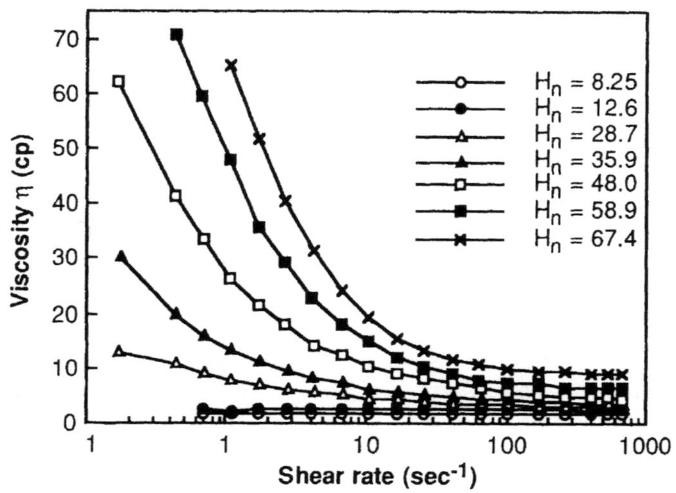

Comments
- Relationship between viscosity and shear rate for human red blood cell suspensions in homologous acid citrate dextrose (ACD)-plasma at 25°C for various volume concentrations of the erythrocytes.

Reference(s)
Brooks DE, Goodwin JW, Seaman GVF (1970) Interaction among erythrocytes under shear. J Appl Physiol 28:174–177 (with permission)

Viscosity(1)

<table>
<tr><td>•
•</td><td>• Elephant, man, dog, ...
• Blood</td><td>•
•</td></tr>
</table>

Materials
- Elephant, man, dog, sheep, goat
- Erythrocytes, serum

Testing Methods and Experimental Conditions
- Relationship between logarithm of viscosity and logarithm of shear rate for plasma, serum, and Ringer suspensions, erythrocyte (top figure)
- Relationship between logarithm of viscosity at two different shear rates (0.052 and 52 sec^{-1}) and cell percentage for elephant (bottom figure)

Data

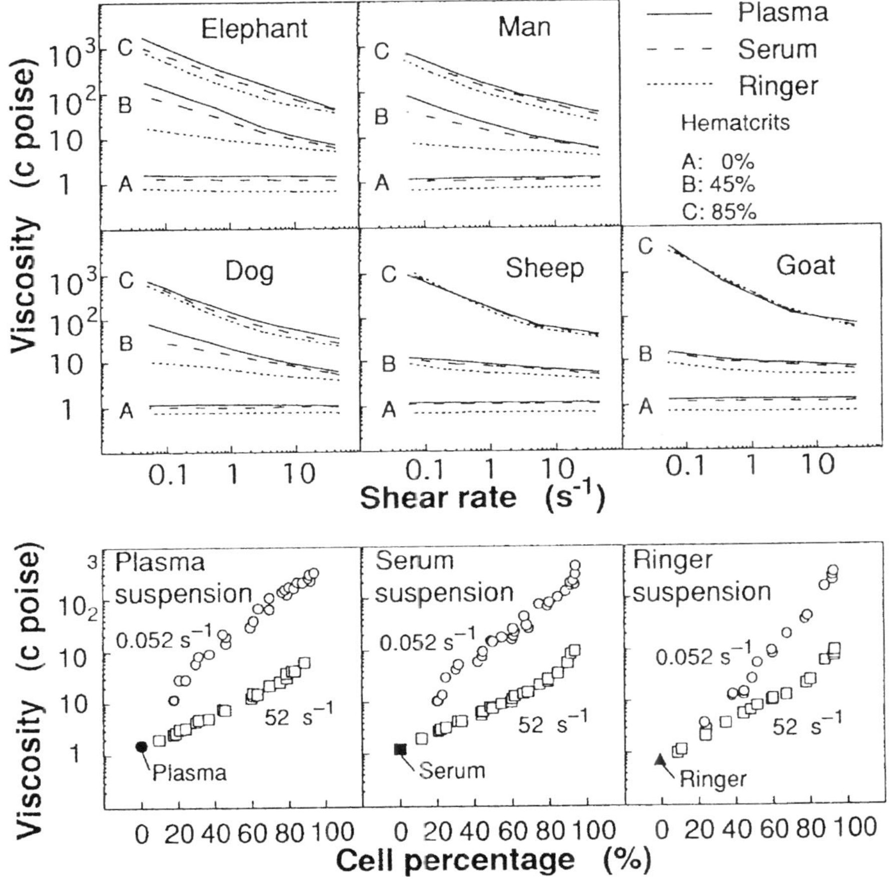

Comments
- The shear dependence in elephant, human, and dog erythrocyte suspensions is strikingly increased by the presence of serum proteins and more with the inclusion of fibrinogen.
- With sheep and goat erythrocytes, the addition of serum protein and fibrinogen does not increase the shear dependence of viscosity.

Reference(s)
Usami S, Chien S, Gregersen MI (1969) Viscometric characteristics of blood of the elephant, man, dog, sheep, and goat. Am J Physiol 217(3):884–890 (with permission)

Viscosity(2)

<table>
<tr><td>•
•</td><td>• Human
• Blood</td><td>•
•</td></tr>
</table>

Materials
- Human
- Blood

Testing Methods and Experimental Conditions
- Ox, Fahraeus-Lindqvist effect for defibrinated ox blood at 38°C calculated from the data of Kümin

Data

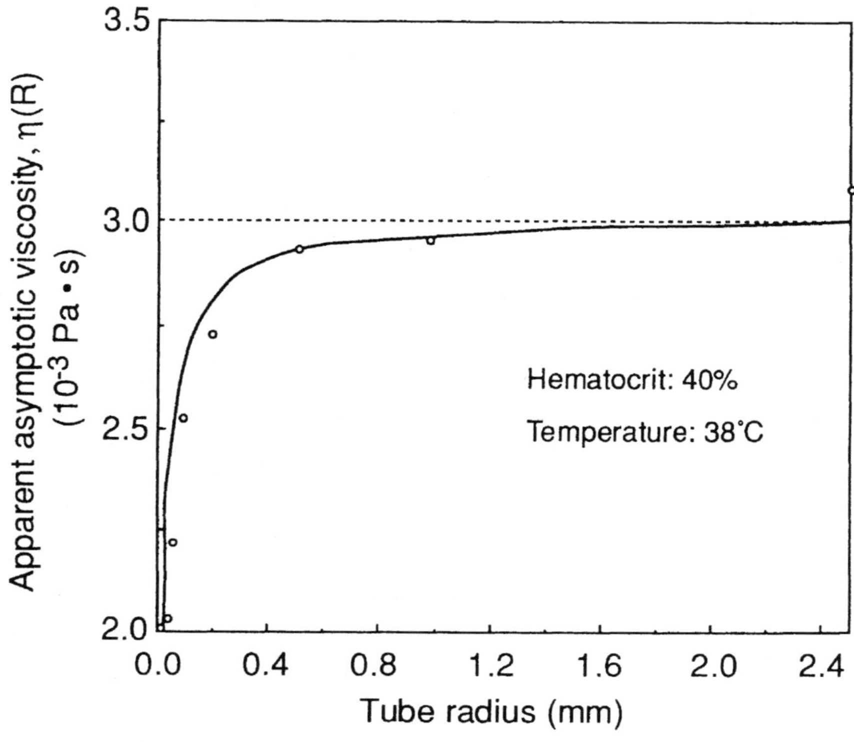

Comments
- The Smooth curve is a least squares fit and the broken line is the value for a tube of infinite radius.

Reference(s)
Haynes RH (1960) Physical basis of the dependence of blood viscosity on tube radius. Am J Physiol 198:1193–1200 (with permission)

Viscosity(3)

	• Human • Blood	• Hematocrit • Shear rate

Materials
- Human
- Blood

Testing Methods and Experimental Conditions
- Whole blood, defibrinated blood, Ringer suspension of red cells at 37°C
- Relationship between the logarithm of viscosity and the logarithm of shear rate at hematocrits (H) of 90%, 45%, and 0%
- The points were calculated from the regression equation

Data

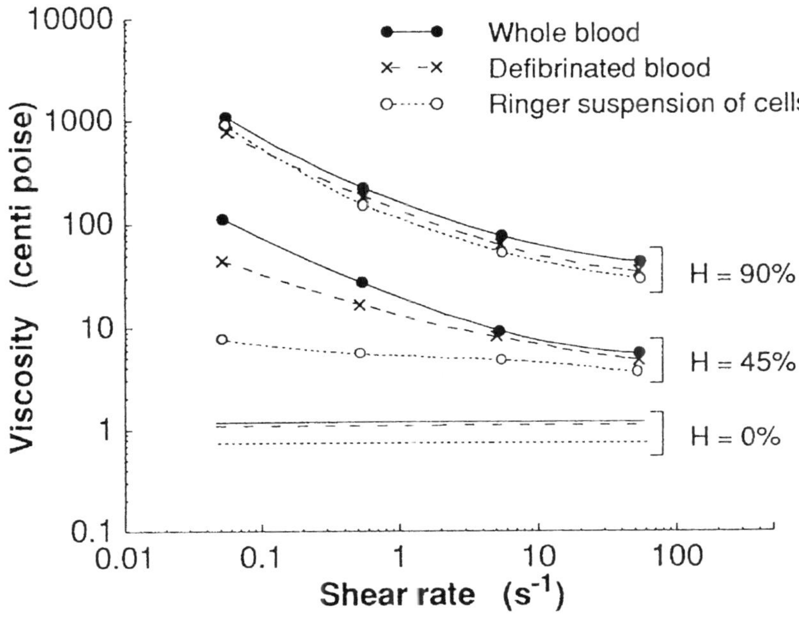

Comments
- The lines marked H = 0 represent the data on plasma (solid line), serum (broken line), and Ringer solution (dotted line).
- Note the Newtonian behavior of the suspending media and the relative extents of departure from Newtonian behavior in the three blood systems at H = 45% and at H = 90%

Reference(s)
Chien S, Usami S, Taylor HM, Lundberg JL, Gregersen MI (1966) Effects of hematocrit and plasma proteins on human blood rheology at low shear rates. J Appl Physiol 21(1):81–87 (with permission)

Viscosity(4)

| •
 • | • Human
 • Blood | • Shear rate
 • |

Materials
  • Human
  • Blood

Testing Methods and Experimental Conditions
  • Apparent viscosity of human blood as a function of shear rate

Data

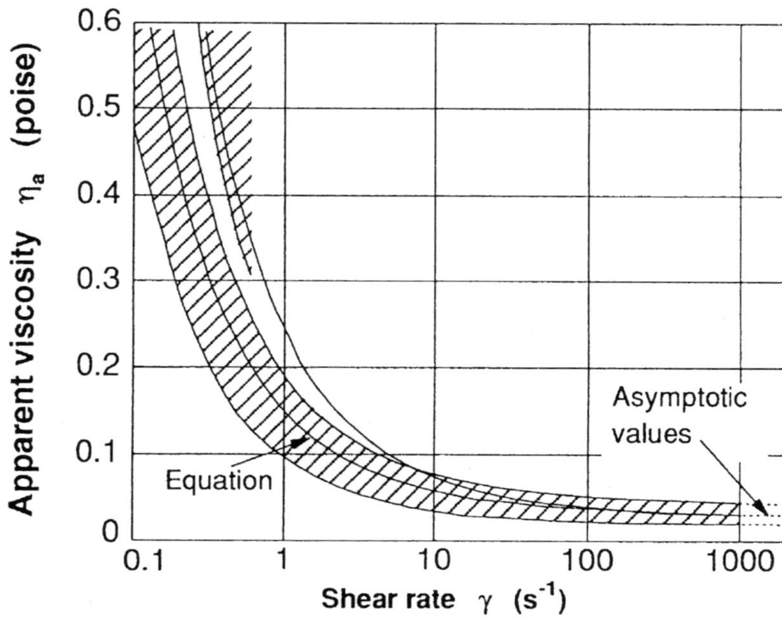

Comments

  • Equation $(\tau/\eta_0)^{1/2} = 1.53\,\gamma^{1/2} + 2.0$.

  τ, shear stress; η_0, plasma viscosity; γ, shear rate.
  • Overall variation of apparent viscosity of human blood with shear rate or the
    assumption that the viscosity of plasma is 0.012 poise.

Reference(s)
  Whitemore RL (1968) Rheology of the circulation. Pergamon Press, Oxford, p 67 (with
  permission)

Viscosity(5)

• Relative viscosity •	• Human • Blood	• Hematocrit •

Materials
- Human
- Blood

Testing Methods and Experimental Conditions
- Influence of hematocrit on the relative viscosity of normal cells suspended in saline with the shear rate over the range 42 to 212 per second

Data

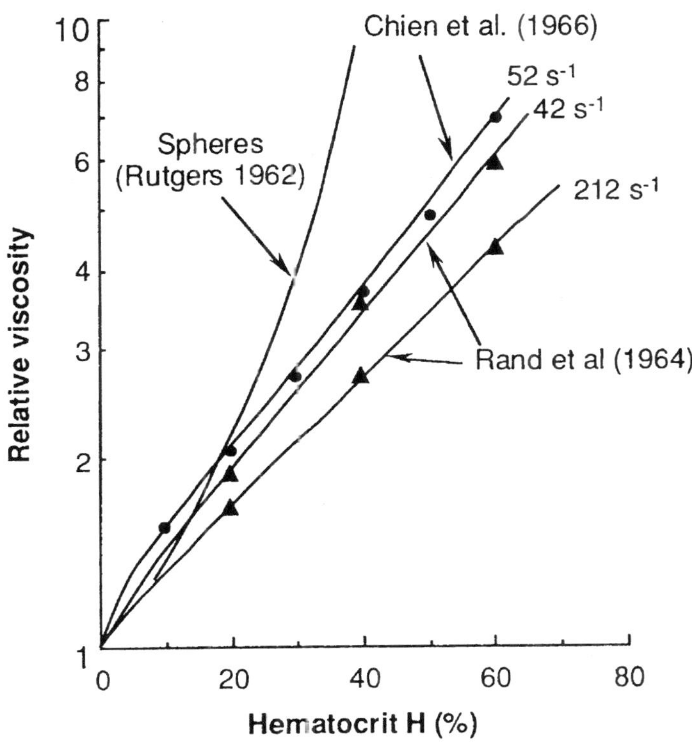

Comments
- The relative viscosity defined as the ratio of viscosity to that of plasma is plotted against the hematocrit.
- The viscosity of suspensions of rigid, smooth, dispersed, equisized spheres is higher than that of human blood at higher hematocrit.

Reference(s)

1. Whitemore RL (1968) Rheology of the circulation. Pergamon Press, Oxford, p 71 (with permission)
2. Rutgers R (1962) Rheol Acta 2: 202
3. Rand PW, Lacombe E, Hunt HE, Austin WH (1964) Viscosity of normal human blood under normothermic and hyperthermic condition. J Appl Physiol 19: 112–122
4. Chien S, Usami S, Taylor HM, Lundberg JL, Gregersen M (1966) Effects of hematocrit and plasma proteins on human blood rheology. J Appl Physiol 21: 81–87

Yield Shear Stress

•	• Human	• Hematocrit
•	• Blood	•

Materials
- Human
- Blood

Testing Methods and Experimental Conditions
- Yield stress of blood with viscous hematocrit

Data

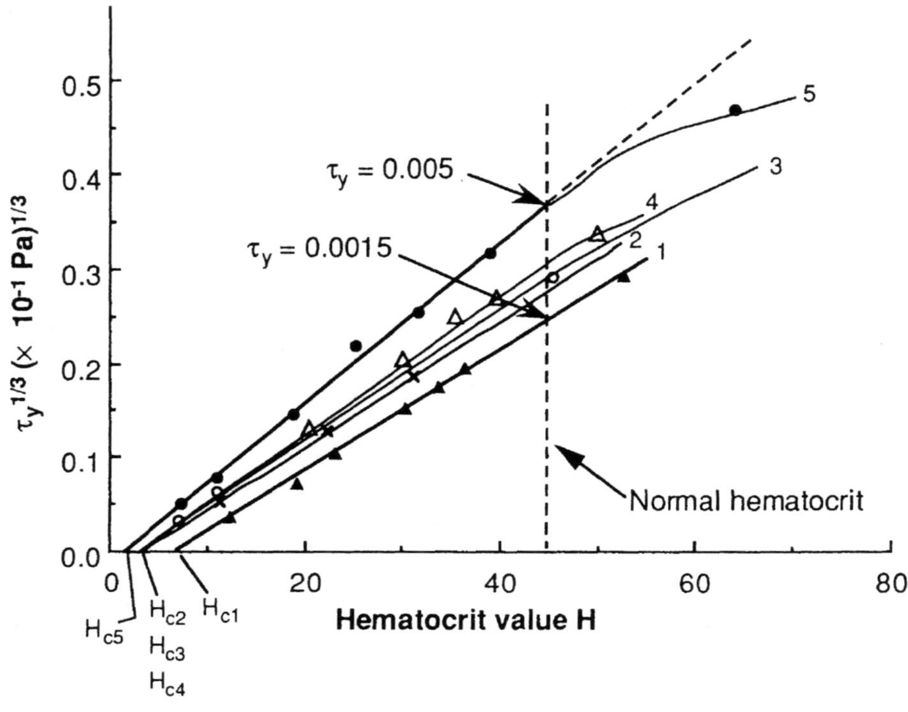

Comments

- The cube root of the yield shear stress (τ_y) is plotted against the hematocrit for 5 different blood samples.
- There are critical hematocrit values H_c below which there is no yield stress, and above H_c the cube root of yield stress is linear with the hematocrit.

Reference(s)

Merril EW, Gilliland ER, Cokelet G, Shin E, Britten A, Wells RE Jr (1963) Rheology of human blood near and at zero flow. Biophysical J 3:199–213 (with permission)

Constituents

<table>
<tr><td>•
•</td><td>• Pig
• Synovial fluid</td><td>•
•</td></tr>
</table>

Materials
- Pig
- Synovial fluid in knee joint

Testing Methods and Experimental Conditions
- Hyaluronic acid constituents analysis (concentration: Morgan-Elson method, molecular weight: Lorent equation)
- Protein constituents analysis by the Lorry method and cellulose acetate electrophoresis
- Analysis of phospholipids by thin-layer chromatography

Data

	Percentage	
Protein	Pig[2], $n = 30$	Human[3]
Albumin	57.2 ± 0.92	55–70
α1-globulin	4.53 ± 0.59	6–8
α2-globulin	13.1 ± 0.68	5–7
β-globulin	6.91 ± 0.53	8–10
γ-globulin	7.26 ± 0.62	10–14

Pig data are given as mean $\pm$ SE.

	Percentage	
Phospholipid	Pig[2], $n = 6$	Canine[4], $n = 8$
Phosphatidyl cholines	54.7 ± 2.23	44.7 ± 3.90
Sphingomyelins	28.2 ± 1.17	12.2 ± 2.23
Lysophosphatidyl cholines	3.23 ± 0.98	10.9 ± 2.83
Phosphatidyl ethanolamines	1.05 ± 0.62	15.1 ± 3.25
Phosphatidyl inositols	$>1.80 \pm 0.62$	7.3 ± 2.08
Phosphatidyl serines		7.4 ± 2.14

Comments
- Constituents of synovial fluid:

  Hyaluronic acid[1], 119 mg/dl (Molecular weight Mw $= 4.77 \times 10^6$);

  Total proteins[2], 3.4 g/dl;

  Total phospholipids[3], 19 mg/dl.
- Protein constituents in pig synovial fluid are similar to those in human synovial fluid.
- Main phospholipid in pig synovial fluid is phosphatidyl choline.

Reference(s)

1. Higaki H, Murakami T (1995) Role of constituents in synovial fluid and surface layer of articular cartilage in joint lubrication (Part 2): Boundary lubrication by proteins. J Jpn Soc Tribologists 40:7
2. Higaki H, Murakami T, Nakanishi Y (1995) Boundary lubricating ability of protein and phospholipid in natural synovial joints. Trans Jpn Soc Mech Engrs C 61:238–243
3. Sasada T, Tsukamoto Y, and Mabuchi K (1988) Biotribology. Sangyo-Tosho, Tokyo, p 41
4. Hills BA, Butler BD (1984) Surfactants identified in synovial fluid and their ability to act as boundary lubricants. Ann Rheum Dis 43:641 (with permission)

Viscoelasticity(1)

• Complex modulus • Complex dynamic viscosity	• Human • Synovial fluid	• Normal • Diseased

Materials
- Human
- Synovial fluid

Testing Methods and Experimental Conditions
- The viscoelastic parameters of synovial fluids (SF) for normal and diseased conditions were obtained using oscillatory rheometry

Data

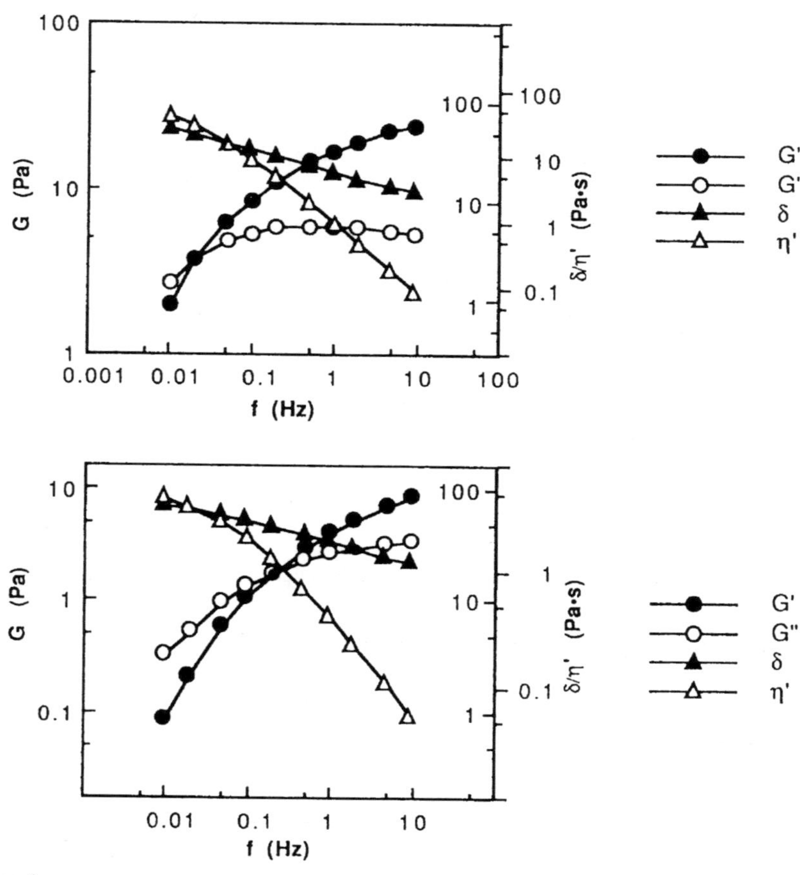

Comments
- The dynamic moduli and viscosity of "near-normal" SF *vs* frequency.
- The dynamic moduli and viscosity of SF from a patient with ligament defects *vs* frequency.
- G': elastic component of complex modulus
  G": viscous component of complex modulus

  η': viscous component of complex dynamic viscosity

  δ: Tan^{-1} G"/G'.

Reference(s)

Safari M, Bjelle A, Gudmundsson M, Hogfors C, Granhed H (1990) Clinical assessment of rheumatic diseases using viscoelastic parameters for synovial fluid. Biorheology 27:659–674 (with permission)

Viscoelasticity(2)

• Shear viscosity • Shear modulus	• Human • Synovial fluid	• Normal • Diseased

Materials
- Human
- Synovial fluid

Testing Methods and Experimental Conditions
- The viscoelastic parameters of synovial fluids for normal and diseased conditions were obtained using oscillatory rheometry
- η': viscous components of the complex viscosity
- η'': elastic components of the complex dynamic viscosity

Data

	Strain ranges	Equipment capacity %	η' (mPas)	η'' (mPas)	Relaxation time (s)	Zero shear visco. h_o (Pas)	Loss tangent
Near normal	0.0056-0.1237	1-22	366-513	789-1406	0.65-5.82	9.56-33.55 1)	0.34-0.5
Ligament def.	0.0281-0.1462	5-26	137-262	179-458	0.59-0.89	2.43-6.06 1)	0.53-0.68
Treated R.A.	0.0168-0.2531	3-45	41-312	42-669	0.61-2.75	1.59-12.88 1)	0.43-0.64
Meniscus def.	0.0562-0.2756	10-49	14-94	4-77	0.016-0.09	$7.53\text{-}458 \times 10^{-3}$ 2)	1.022-1.6
R.A. seroneg	0.0844-0.5625	15-100	20-33	11-25	0.022-0.03	$1.27\text{-}111 \times 10^{-3}$ 3)	1.34-1.97
R.A. seropos	0.1875-1.250	15-100	5-15	0.05-9	< 0.022	$8.93\text{-}44 \times 10^{-3}$ 3)	1.79-4.68

1) evaluated at 0.01 Hz, 2) evaluated at 0.05 Hz, 3) evaluated at 0.1 Hz, T = 27°C

Ligament def.; Ligament defects, Treated R.A; Treated rheumatoid arthritis, Meniscus def.; Meniscus defects, R.A. seroneg; rheumatoid arthritis seronegative, R.A. seropos; rheumatoid arthritis seropositive

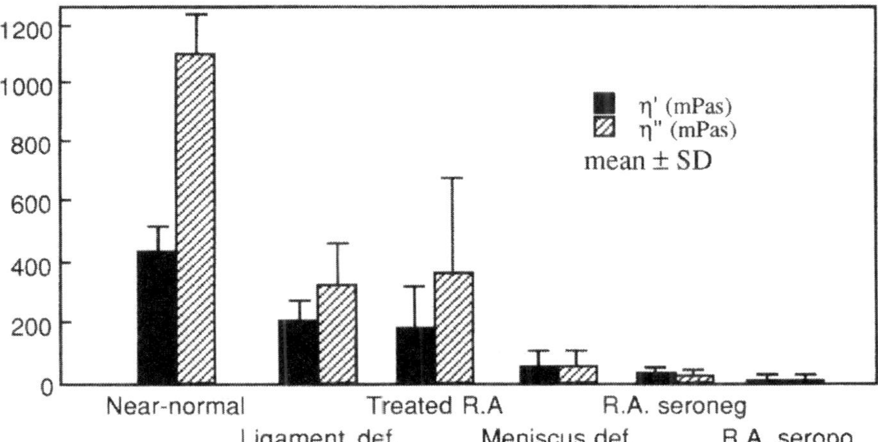

Comments
- Details of the measurement in oscillatory rheometry for various SF.
- η', η'' and loss tangent were obtained at a frequency of 2 Hz.
- The relaxation time and h_0 were obtained from frequency sweep.
- Mean values and standard deviations for η' and η'' of SF from various knee joints.

Reference(s)
Safari M, Bjelle A, Gudmundsson M, Hogfors C, Granhed H (1990) Clinical assessment of rheumatic diseases using viscoelastic parameters for synovial fluid. Biorheology 27:659–674 (with permission)

Viscoelasticity(3)

• Shear viscosity • Shear modulus	• Human • Synovial fluid	• Normal • Diseased

Materials

- Human
- Synovial fluid

Testing Methods and Experimental Conditions

- The viscoelasticities of synovial fluid in normal and diseased states were measured by a low-shear rotational viscometer and a mechanical spectrometer

Data

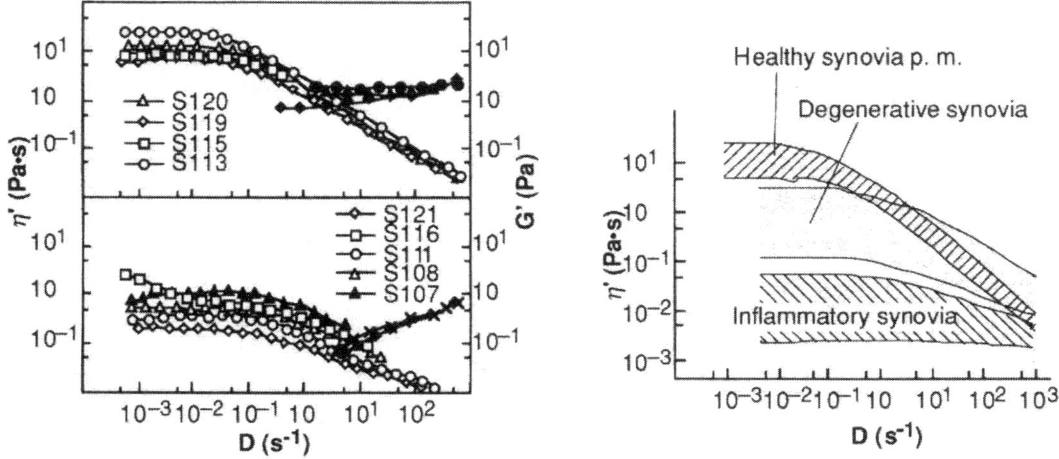

Comments

- Shear viscosity η' and shear modulus G' of normal (top left) and pathological (bottom left) synovial fluids as a function of shear rate D.
- Shear viscosity η' ranges of healthy, degenerative and inflammatory synovial fluids, and all rheological parameters deteriorate in the disease state.

Reference(s)

Schurz J, Ribitsch V (1987) Rheology of synovial fluid. Biorheology 24:385–399 (with permission)

Viscosity

• •	• Pig • Synovial fluid	• Cone-plate viscometer •

Materials
- Pig
- Synovial fluid in knee joint

Testing Methods and Experimental Conditions
- Viscosity measurement by cone/plate viscometer at 15°, 25° and 37°C
- Mixture of synovial fluids in several pig knee joints

Data

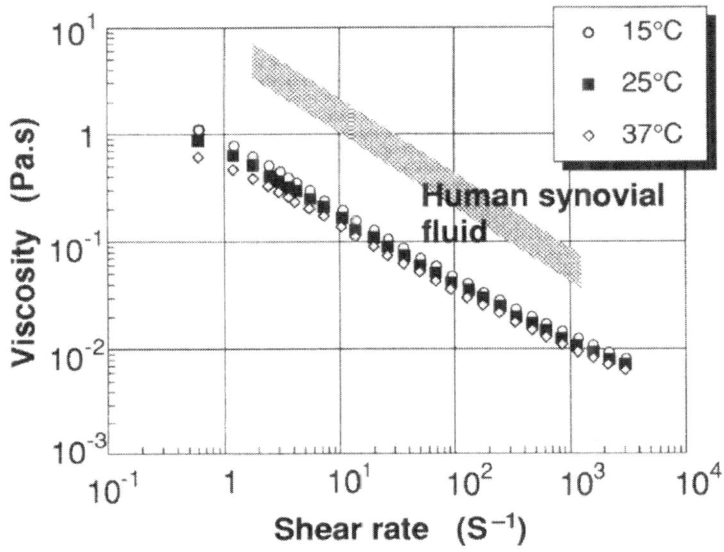

Comments
- Synovial fluid in pig knee joint shows marked non-Newtonian behavior (shear rate; 0.6–3000 s⁻¹).
- The viscosity is lower than that of human joints from reference 2.

Reference(s)

1. Murakami T, Ando H, Higaki H (1995) Viscosity of synovial fluid and related solutions. Preprint 4th Bioengineering Symposium. Jpn. Soc. Mech. Engrs:63–64
2. Sasada T, Tsukamoto Y, Mabichi K (1988) Biotribology, Sangyo-Tosho, Tokyo, p 44

Author Index

Made in the USA
Monee, IL
07 July 2026

56553856R00249